Oc

To John —

a small thankyou for your work on, cover for Kingfisher's Fire,

With best wishes

from

Barbara + Peter

John Kirk Townsend

Collector of Audubon's

Western Birds and Mammals

Barbara and Richard Mearns

Barbara C Mearns
Richard H. Mearns

including
J.K.Townsend's
"A Narrative of a Journey Across the Rocky Mountains to the
Columbia River, and a Visit to the Sandwich Islands, Chili, &c." (1839)

with commentary and zoological notations
by Barbara and Richard Mearns

Barbara and Richard Mearns, Dumfries

First published 2007

Copyright © 2007 Barbara and Richard Mearns

Published by Barbara and Richard Mearns, Connansknowe, Kirkton, Dumfries, DG1 1SX
www.mearnsbooks.com

Illustrations © copyright as credited, except those provided by Barbara and Richard Mearns

British Library Cataloguing in Publication Data
A catalogue record of this book is available from the British Library

ISBN 978-0-9556739-0-0

Design by Julian Gillespie

Printed in Scotland by Print 2000 Ltd., Glasgow

John Kirk Townsend (1809-1851)

"In the year 1834, accompanied by my friend Mr. Thomas Nuttall, I made the overland journey across the continent of North America to the Pacific … from St Louis to Fort Vancouver, gun in hand … Our views were directed solely to natural history; Mr. Nuttall's to botany, mine to zoology."

John Kirk Townsend

"I fear that expedition will prove a failure"   John James Audubon

Photographic Credits

All wildlife, habitat and location photography by Barbara and Richard Mearns except for the following: Jack Jeffrey Photography: p. 161 (Red-tailed Tropicbird), p. 167 (I'iwi), and p. 242 ('Apapane); Ginny Vroblesky: p. 317 (J.K. Townsend's gravestone); and Donald T. Little: p. 317 (Charlotte Townsend's gravestone).

Townsend's bird and mammal specimens: AMNH: Craig Chesek © American Museum of Natural History. **ANSP**: Will Brown Photography © Academy of Natural Sciences of Philadelphia. **LM**: Nathan Pendlebury © National Museums Liverpool. **OUM**: Barbara and Richard Mearns © Oxford University Museum. **USNM**: Brian K. Schmidt © Smithsonian Institution (Division of Birds).

Additional images are from the collection of B. and R. Mearns or were supplied by, or reproduced by, kind permission of the following: **American Museum of Natural History, New York**: p. 293 (Titian Ramsay Peale, self portrait. Neg 323811). **Ewell Sale Stewart Library, The Academy of Natural Sciences of Philadelphia**: pp. iii, 281 (Coll. 457. J.K. Townsend colour miniature); p. 19 (The Academy 1826-40, from Nolan 1909: 11); p. 50 (Coll. 247. Description of a new species of Bunting); p. 176 (Coll. 404. No. 9. Townsend letter, 2nd May 1835); p. 191 (QL696 A2T7 California Condor, plate 1 of Townsend's *Ornithology of the United States* (1839)); p. 268 (*Journal of the Academy of Natural Sciences of Philadelphia*); p. 283 (W.S.Vaux); p. 285 (S.G. Morton); p. 286 (QL696 A2T7, title page and Crested Caracara, plate IV from Townsend's *Ornithology of the United States* (1839)); p. 290 (Coll. 404. No. 1. McEuen's list, 25th October 1839); p. 292 (Coll. 457. Charles Pickering); **Bishop Museum Archives, Honolulu**: p. 160 (CP 29158. Palace of Kamahamah, artist unknown, May 1826); p.162 (G 96731. King Kamehameha III, photographer unknown, undated); p. 163 (CP 37459. Vallée du Paree by Fisquet, 1845-1852); p. 166 (SP 201503. Sugar plantation, Koloa, Kauai, by Peter Hurd, 1841); p. 245 (SP 201502. Feather cape in Bishop Museum, photo by W.T. Brigham, ca. 1899). **Hawaiian Mission Children's Society Library, Honolulu**: p. 247 (Rev C. Forbes). **Library of Congress, Washington, D.C**: p. 16 (Westtown School); p. 282 (Ezra Michener), p. 305 (William Cooper). **Missouri Historical Society, St Louis, Missouri**: p. 45 (Neg. OutStL 0825. Independence Courthouse, from Meyer's *Universum*, 1851-54). **The Trustees of the National Library of Scotland, Edinburgh**: p. 225 (Clatsap Skull, plate 46 of S.G. Morton's *Crania Americana* (1839)). © **National Museums Liverpool** (*World Museum Liverpool*): pp. 328, 330 (Derby list). **Collection of The New-York Historical Society, New York**: p. 21 (NYHS 1863.18.013. Townsend's Bunting); p. 64 (NYHS 1863.17.390. Lark Bunting etc); p. 69 (NYHS 1863.17.394. Lazuli Bunting etc); p. 80 (NYHS 1863.17.350 Mountain Plover); p. 141 (NYHS Z3333 [S-30]. Townsend's Big-eared Bat etc); p. 181 (NYHS 1863.17.353. Black-capped Chickadee etc); p. 183 (NYHS 1863.17.393. Townsend's Warbler etc); p. 213 (NYHS Z3327. California Ground Squirrel). **Oregon Historical Society, Portland, Oregon**: p. 36 (OrHi 3632. Nathaniel Wyeth); p. 101 (OrHi 83420. Jason Lee); p.128 (OrHi 89932 Fort Walla Walla in 1847 by J.M. Stanley, from *Pacific Railroad Report*, 1853-56); p. 131 (OrHi 98723. The Dalles by J.M. Stanley, from *Pacific Railroad Report*, 1853-56); p. 133 (OrHi 5981 Passage of the Dalles by C.E. Watkins, 1867); p. 139 (OrHi 76375. Fort Vancouver, from H.J. Warre sketchbook, undated); p. 140 (OrHi 098538. John McLoughlin); p. 153 (OrHi 21681. Astoria, from G. Franchère 1831); p. 190 (OrHi 102217. Willamette Falls, Drayton sketch, ca. 1841); p. 200 (OrHi 20035. Rev Samuel Parker, 1835); p. 212 (OrHi 707. Peter Skene Ogden); p. 222 (OrHi 1645. Narcissa Whitman, by O.W. Dixon, undated). **Museum of American Art of Pennsylvania Academy of the Fine Arts, Philadelphia**: p. 18 (The Artist in His Museum by Charles Willson Peale, 1822). **Princeton University Library, Princeton, New Jersey**: p. 106 (Record Id 8564. "Four-striped Ground Squirrel" by J.J. Audubon, 1841). **Smithsonian Institution, Washington, D.C**: p. 295 (Townsend's preserving powder). **Townsend Family Papers**: p. 10 (John Kirk Townsend); p. 15 (seven family portraits); p. 309 (John Kirk and Charlotte Townsend); p. 313 (John Kirk and Priscilla Townsend); p. 314 (Elizabeth Townsend); **R.P. Truman**: p. 275 (William MacGillivray). **Walters Art Museum, Baltimore, Maryland**: p. 88 (37.1940.159. Trappers' Rendezvous by Alfred Jacob Miller, 1837).

The authors have made every attempt to contact the copyright holders of all the images. In the few instances where they have been unsuccessful, they invite the copyright holders to contact them.

Contents

Maps

Preface and acknowledgements

Our book has been a while in the making, originally inspired by our research into Townsend, Nuttall, Audubon and other naturalists when compiling *Audubon to Xantus: the lives of those commemorated in North American bird names* (1992). After completing *The Bird Collectors* (1998) more time gradually became available to devote to the Townsend project.

One of the most exciting aspects has been following in the footsteps of Townsend, tracing his westward route over the Rocky Mountains, much of which later developed into the famous Oregon Trail. In 1993 we journeyed along the Platte River westwards to the Snake River – with memorable side trips to the Craters of the Moon National Monument, Rocky Mountain National Park, Grand Teton National Park and the wonderful Pawnee National Grasslands. In 2002 we picked up the trail where we had left it in the Sawtooth Mountains and continued westwards to Fort Walla Walla and down the Columbia River to the Pacific, continuing yet further westwards with a pelagic trip out of Westport, Washington. Much has changed on the lower Columbia over the past 170 years, and it was necessary, we thought, to hike on Mount Adams and Mount Rainier to properly appreciate the land as it used to be. As we travelled, we became more familiar with western birds and mammals, many of them known from our earlier forays into Alaska, California, Arizona and New Mexico. When we went to the Hawaiian Islands in 2005 the endemic birds were all quite new to us.

Although we took photographs of the habitats and some of the wildlife to help illustrate this book, on Townsend's journey there was not even an artist to record such things and so his *Narrative of a Journey across the Rocky Mountains* (1839) appeared without any etchings of the scenery or wildlife. And yet it so happened that almost all the new birds and mammals discovered and collected by Townsend were superbly painted in watercolours. Not only that, they were painted by the best wildlife artist of the day – none other than John James Audubon – appearing in *The Birds of America* and *The Viviparous Quadrupeds*, the two publications that illustrate the majority of Townsend's discoveries. We are a little surprised that until now, no one has combined Audubon's artwork with Townsend's expedition narrative to form a fully illustrated zoological expedition journal.

It is a matter of some regret that there appears to be little surviving correspondence between the principal characters in the Townsend story. No letters from Townsend to Audubon, and only one from Audubon to Townsend. No letters from Townsend to Thomas Nuttall and only one from Nuttall to Townsend. Also, along with generations of other biohistorians, we have been frustrated by the loss of most of Audubon's journals in a warehouse fire in New York in 1845, by the destruction of many of the Rev John Bachman's papers in Charleston during the Civil War, the accidental burning of William MacGillivray's papers in a fire in Australia, and the comparatively recent destruction of Nuttall's journals by unappreciative relatives in the 1950s. Nevertheless, we are grateful that much has survived and our thanks are extended to an array of kindly staff from the libraries and/or museums of the following institutions who helped in so many ways but are too numerous to name individually: Alabama Department of Archives and History, Montgomery, Alabama; Alexander Library, Edward Grey Institute, Oxford; American Museum of Natural History, New York; American Philosophical Society, Philadelphia; Bishop Museum Archives, Honolulu, Hawaii; Carnegie Museum, Pittsburgh; Field Museum, Chicago; Hawaiian Mission Children's Society Library, Honolulu, Hawaii; Historical Society of Pennsylvania, Philadelphia; Huntingdon Library, San Marino, California; Inverness Museum and Art Gallery, Scotland; Kelvingrove Museum, Glasgow; Museum of Comparative Zoology, Ernst Mayr Library and Houghton Library, Harvard University, Massachusetts; Museum für Naturkunde, Humboldt-Universität zu Berlin, Berlin; Musselman Library,

Gettysburg College, Pennsylvania; National Natuurhistorisch Museum, Leiden; The Natural History Museum, London (and the Sub-department of Ornithology at Tring); New-York Historical Society, New York; Oregon Historical Society, Portland, Oregon; Oxford University Museum of Natural History, Oxford, UK; Pratt Museum of Natural History, Amherst College, Massachusetts; Princeton University Library, Princeton, New Jersey; Royal Botanic Gardens, Kew; Royal Museums of Scotland, Edinburgh; Smithsonian Institution Archives, Washington, D.C.; United States National Museum of Natural History, Washington, D.C.; University Museum of Zoology, Cambridge, UK; Vassar College, Poughkeepsie, New York; West Chester University of Pennsylvania and Westtown School, Pennsylvania.

Special thanks must be extended to past and present staff of the Academy of Natural Sciences of Philadelphia where the bulk of Townsend's surviving bird and mammal specimens are housed and where there is much archival material concerning him. David Agro, Ned Gilmore, Leo Joseph, Nathan H. Rice, Earle E. Spamer, Carol Spawn and colleagues were unfailingly helpful, never forgetting Robert McCracken Peck who was a consistent source of advice and encouragement.

Much the same applies to the past and present staff of the Smithsonian Institution's Division of Birds where Storrs Olson, James Dean, Craig Ludwig, Gary R. Graves, Brian K. Schmidt and colleagues were consistently diligent and cooperative in responding to our requests. William Cox, Tracy Robinson, Keith Gorman and fellow archivists at the Smithsonian were also most obliging. For his long-term support Storrs Olson must be particularly thanked, having shown us Townsend specimens in 1995, and provided helpful comments concerning Hawaiian birds and other wildlife.

In Britain, we must thank past and present staff of the following museums that house Townsend specimens: Cambridge University Museum of Zoology, Cambridge (R.J. Symonds); Liverpool Museum (Clemency Fisher and Malcolm Largen); Oxford University Museum, Oxford (Jane Pickering, Sammy De Grave and Mrs M. Nowak-Kemp); and BMNH Sub-department of Ornithology, Tring (Michael Walters).

We are deeply indebted to fellow biohistorians for their specialist knowledge and remarkable enthusiasm in responding to our enquiries. Those not already mentioned by name include: the late Joe Ewan for early advice, Errol Fuller for encouragement (and example) to publish this book ourselves, C. Stuart Houston, Saskatchewan; Christian Jouanin, Paris; Mary LeCroy at the American Museum of Natural History, New York; Bob McGowan at the Royal Museums of Scotland, Edinburgh, Roberta J. M. Olson at the New-York Historical Society, Harald Pieper at Zoologisches Museum, Kiel; Robert Ralph, Aberdeen.

On a personal level, we want to thank the many friends who provided invaluable assistance, especially the following: Ginny Vroblesky, along with Charles and Sue Staines, for making research trips to the Academy of Natural Sciences of Philadelphia on our behalf, and Ginny for locating and photographing Townsend's grave in Washington, D.C. In Idaho, Bill and Joan Mattox, and their friends at the World Center for Birds of Prey at Boise, gave us hospitality on our journey westwards (and discussed Swainson's Hawks). Jim Denny and Jack Jeffrey helped us get to grips with the endemic birds of Kaua'i and Hawai'i. In Scotland we must thank Rosie Rutherford for her historical knowledge of Quakers and her comments on selected chapters, Volkmar Nix for German translation, Joan Curzio and Colin Kinloch for once again being our American bankers, and Julian Gillespie for all the graphic design – his perseverance and professionalism making this book everything we wanted it to be.

Finally, it was a privilege and pleasure to work with two of John Kirk Townsend's relatives: Donald Townsend Little and Dana Dunbar King. We whole-heartedly thank them for their generous and unstinting supply of material and photographs from the family archives that were fortuitously being transcribed and researched during the time that this book was being compiled. We thank them too for comments on early drafts of parts 1 and 2, for undertaking various research requests on our behalf, and for being so supportive. We hope to meet them some day.

Our other commitments have turned this into a very protracted project. If the passage of time has resulted in the unintentional omission of any key helpers we can only apologise and trust that you will forgive us.

Abbreviations

ADAH	Alabama Department of Archives and History, Montgomery
AMNH	American Museum of Natural History, New York
ANSP	Academy of Natural Sciences of Philadelphia
AOU Check-list	American Ornithologists' Union 1998. *Check-list of North American Birds*, 7th edn. and supplements up the 47th Supplement (2006)
APS	American Philosophical Society, Philadelphia
BMNH	The Natural History Museum, London and Tring
Cambridge	University Museum of Zoology, Cambridge, UK
FMNH	Field Museum, Chicago
HSP	Historical Society of Pennsylvania, Philadelphia
JKT	John Kirk Townsend
Journal ANSP	*Journal of the Academy of Natural Sciences of Philadelphia*
Kew	Royal Botanical Gardens, Kew
Library Co	Library Company of Philadelphia
LM	Liverpool Museum, National Museums and Galleries on Merseyside
MCZ	Museum of Comparative Zoology, Harvard University, Cambridge, Massachusetts
MS Journal	J.K. Townsend's journal, 1834-37
MVZ	Museum of Vertebrate Zoology, University of California, Berkeley
Narrative	J.K. Townsend's *Narrative of a Journey across the Rocky Mountains &c.* (1839)
NYHS	New-York Historical Society, New York
OUM	Oxford University Museum of Natural History, Oxford, UK
Proceedings ANSP	*Proceedings of the Academy of Natural Sciences of Philadelphia*
PUL	Princeton University Library, Princeton, New Jersey
RMS	Royal Museums of Scotland, Edinburgh
SI Archives	Smithsonian Institution Archives, Washington, D.C.
USNM	United States National Museum of Natural History, Washington, D.C.

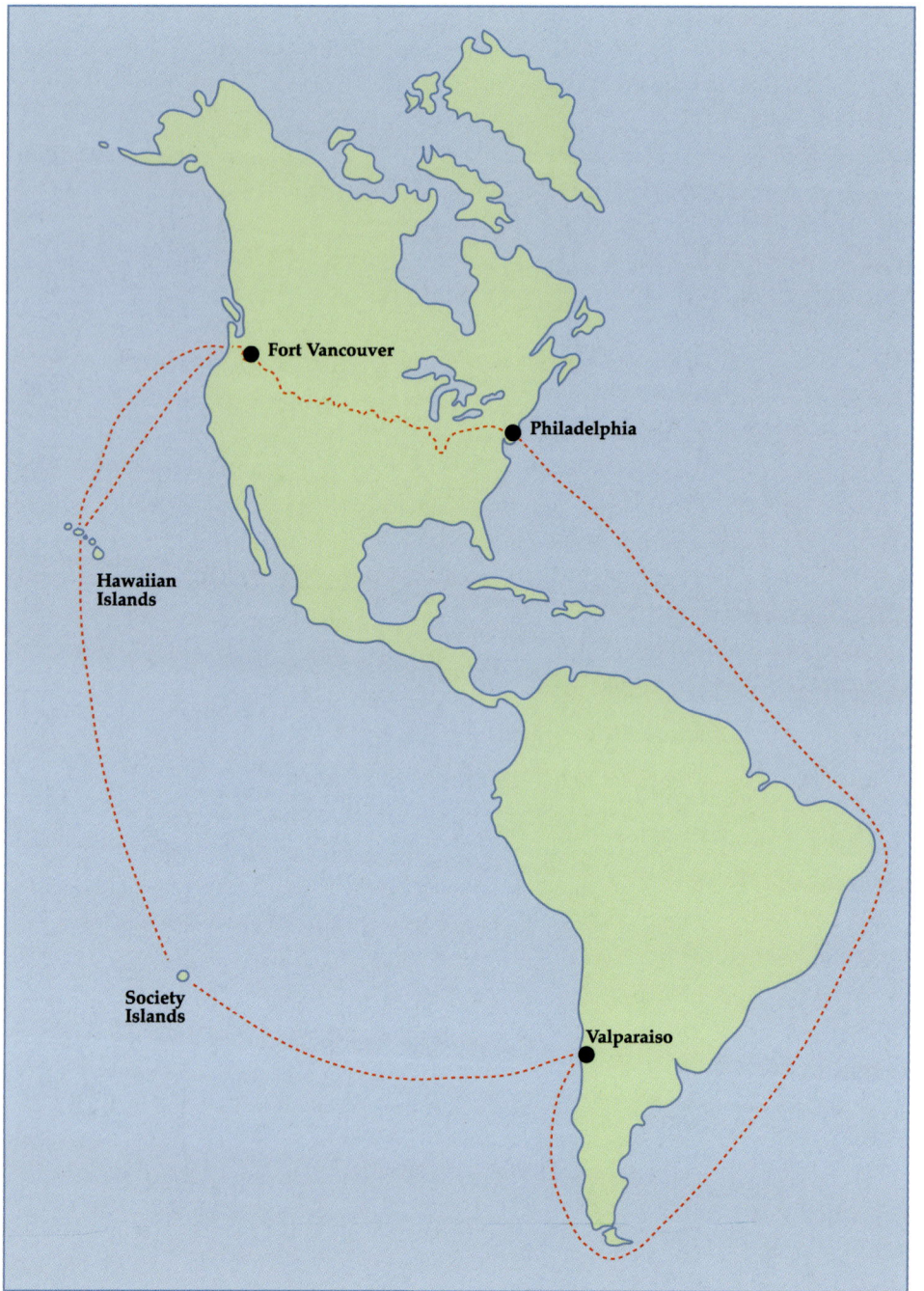

Map 1. Townsend's travels, 1834-37.

John Kirk Townsend

Chronology

Date	John Kirk Townsend	Other Events
1805		Lewis and Clark overland expedition arrives at the Pacific in November and spends the winter at Fort Clatsop, near the mouth of the Columbia River.
1808		First volume of Alexander Wilson's *American Ornithology* is published.
1809 10th August	John Kirk Townsend is born in Philadelphia.	
1812-23	Grows up in Philadelphia and spends time at West Chester, Pennsylvania. Makes collection of local birds, often with cousin William P. Townsend. Attends Westtown Boarding School, Chester County (27th March to 27th September 1819; and 28th December 1822 to 29th September 1823).	Academy of Natural Sciences of Philadelphia (ANSP) founded 25th January 1812. Alexander Wilson dies 23rd August 1813 in Philadelphia. Final volume of his *American Ornithology* completed by George Ord in 1814. Thomas Nuttall, in between expeditions, bases himself at Philadelphia 1808-1823.
1824		On 5th April 1824 Audubon makes first unsuccessful attempt in Philadelphia to win support for a project illustrating all the birds of America.
1825		Charles Bonaparte's edition of Wilson's *American Ornithology* (1825-33) begins to appear. David Douglas first arrives on the Columbia River to collect plants.
1826		Audubon arrives in Britain in July to search for an engraver for *The Birds of America*. ANSP holds first meeting at new building at 12th and Sansom Street on 9th May.
1827		Audubon's plates for *The Birds of America* (1827-38) begin to appear.
1829	Townsend gets to know Nuttall, and in the summer he probably meets Audubon for the first time (or in 1830 or 1831 during Audubon's other visits to Philadelphia).	Nuttall at the ANSP, January-March and December. Audubon returns from Britain at the end of April after an absence of three years and visits Philadelphia seeking subscribers for *The Birds of America*.
1830		Early in the year, Audubon is in Philadelphia but once again the Academy refuses to subscribe to his work. Audubon returns to Britain in April and engages William MacGillivray as his scientific editor for the *Ornithological Biography*. In May, David Douglas arrives for second time on the Columbia River.

Date	John Kirk Townsend	Other Events
1831		First volume of Audubon's *Ornithological Biography* published in Edinburgh.
		Audubon is in Philadelphia in September where the Academy and American Philosophical Society at last subscribe to *The Birds of America*. In October he is in Charleston, South Carolina, where he meets Rev John Bachman.
		Nuttall's first volume, on 'Land Birds', for his *Manual of Ornithology* (1832) is on sale early, in December 1831.
1832		Bonaparte and Jardine's edition of Wilson's *American Ornithology* (1832) appears.
		Nathaniel Wyeth undertakes first overland expedition to the Columbia River (1832-33).
1833		
11th May (or 12th June)	Collects Townsend's Bunting - New Garden, Chester Co.; the skin goes to Ezra Michener.	Audubon leads an expedition to Labrador, June-August.
24th September	Elected a member of ANSP.	
27th September	Writes description of Townsend's Bunting (not published) and "renews his acquaintance" with Audubon when the artist passes through Philadelphia.	
1834	Townsend invited by Nuttall to join overland expedition to the Columbia River.	Nuttall in Philadelphia January-March. His second volume, on 'Water Birds', for his *Manual of Ornithology* is published.
13th March	Townsend sets off from Philadelphia with Nuttall to join the Wyeth Expedition at Independence, Missouri.	
24th March	Arrives at St Louis.	
14th April	Arrives at Independence. Townsend and Nuttall observe a new bunting for which Townsend writes a formal description (undated and never published). This bird, Harris's Sparrow, was eventually described by Nuttall in 1840. Townsend's first consignment of bird specimens is sent back to Philadelphia, around 21st April.	Audubon sails to Britain in April to work on the *Ornithological Biography* and check on the production of plates for *The Birds of America*.
28th April	Townsend and Nuttall depart from Independence with the second Wyeth Expedition to the Columbia River.	
1st June	Arrives at Laramie's Fork.	
22nd June to 2nd July	The expedition halts at the Trapper's Rendezvous, near the Green River. Townsend writes home saying he has collected 35 species of birds.	
15th July to 6th August	The expedition halts for three weeks to build Fort Hall.	David Douglas killed in Hawaiian Islands, 12th July.
3rd September	Arrives at Fort Walla Walla.	
16th September	Arrives at Fort Vancouver.	

Date	John Kirk Townsend	Other Events
11th December	Townsend and Nuttall set sail for the Hawaiian Islands.	
1835		
5th January	Arrives at O'ahu.	
10th February	Sails to Kaua'i with Nuttall and spends five weeks there.	
26th March	Townsend and Nuttall leave Hawaiian Islands and return to Columbia River.	
16th April	Arrives back on Columbia River.	
20th May	Townsend makes a short trip to the Falls of the Willamette with Wyeth; collects a California Condor.	
end September	Townsend's first main collection shipped to Philadelphia via Hawaiian Islands. Takes on surgeon's duties at Fort Vancouver, employed by the Hudson's Bay Company.	Nuttall, Dr Gairdner and Daniel Lee leave for Hawaiian Islands where Nuttall spends four months.
October	Meets Rev Samuel Parker who arrives at Fort Vancouver a few days before 16th October.	
1836		
January	Visit to Fort William, Sauvie Island – takes Indian body from tree.	
16th March	Hands over surgeon's duties to Dr Tolmie who arrives at Fort Vancouver from Fort Langley, Fraser River.	Nuttall on California coast, March to 8th May, then sails round Cape Horn on the *Alert*.
11th April	Writes "Description of the Birds of the Columbia River Region" for Rev Parker.	
25th June	Goes upriver to Fort Walla Walla with Peter Ogden and the summer brigade.	
6th July	The brigade arrives at Fort Walla Walla. Townsend received well by Mr Pambrun, others travel onwards.	First main consignment of Townsend's bird skins reported, on 12th July, to have arrived by ship at Philadelphia.
25th July	Townsend sets off for the Blue Mountains.	
30th July	Returns to Fort Walla Walla and collects in the vicinity throughout August.	
1st September	Meets Marcus and Narcissa Whitman and Rev Spalding who have travelled overland to Fort Walla Walla. They give him letters from Philadelphia.	
3rd September	Townsend travels downriver, arrives at Fort Vancouver five days later.	
13th September		Audubon has returned from Britain and travels from New York to Philadelphia in a vain attempt to acquire some of Townsend's birds.
20th September		Nuttall arrives in Boston on the *Alert* and a few days later gives Audubon six new birds.
26th September	Townsend bases himself at Fort George (Astoria).	
6th - 7th October	Makes collecting trip to Cape Disappointment.	

Date	John Kirk Townsend	Other Events
14th October	Goes to Young's Bay to see the remains of Fort Clatsop built by the Lewis and Clark party.	In mid October, Audubon and Nuttall at last examine Townsend's birds. Edward Harris buys 93 of them for Audubon.
17th October	Three-day excursion northwards from Chinook into Willapa Bay.	
21st October	Returns to Fort George.	On 23rd October Audubon writes a triumphant letter to Bachman to tell him the news about the skins.
5th November	Returns to Fort Vancouver where he packs up his collections in preparation for leaving the west coast.	
15th November		"Description of Twelve New Species of Birds, chiefly from the vicinity of the Columbia river" is prepared under Townsend's name and read in Townsend's absence at the Academy. Published in *Journal ANSP* (1837).
17th November		Audubon arrives Charleston, South Carolina, to paint Townsend and Nuttall skins.
21st November	Townsend sails downriver from Fort Vancouver on the barque *Columbia*, crossing the sand bar at the mouth of the Columbia River on 30th November and heads for the Hawaiian Islands.	
23rd December	Arrives at O'ahu.	
1837		William Cooper publishes description of *Plecotus townsendii* (Townsend's Big-eared Bat) in the *Annals of the Lyceum of Natural History of New York* (1837). Conrad's "Description of new Marine Shells" collected by Nuttall and Townsend is read at the Academy in January and February. Published in *Journal ANSP* (1837).
9th February	Sails to Maui, arrives 14th February.	
17th February	Arrives Hawai'i Island, next day visits Capt James Cook memorial.	
25th February	Sails back to O'ahu, arriving on 27th February.	
18th March	Sails for Tahiti on the *Europa*.	
9th April	Arrives at Tahiti, departs on 5th May.	
13th June	Arrives at Valparaiso, Chile. Collects in vicinity but his departure delayed by a severe illness.	
5th August		Audubon arrives in Britain for his last visit.
22nd August	Departs from Valparaiso on brig *B. Mezick* on a non-stop passage to Philadelphia.	
17th November	Arrives home at Philadelphia.	
1838	Works on his *Narrative of a Journey across the Rocky Mountains etc.* He later starts on his *Ornithology of the United States of North America.*	
June		Lord Derby makes selection of Townsend skins sent to Audubon to sell in Europe. Last of the folio engravings for *The Birds of America* is completed.

Date	John Kirk Townsend	Other Events
7th August		Bachman's "Description of several New Species of American Quadrupeds" containing descriptions of Townsend's mammals is read at ANSP. Published in *Journal ANSP* (1839).
September	Stays at Ezra Michener's home, New Garden, West Chester, writing his *Narrative*.	
1839	Townsend's *Narrative* is published, soon sells out but is not reprinted in the United States.	Randall's "Catalogue of the Crustacea brought by Thomas Nuttall and J.K. Townsend, from the West Coast of North America and the Sandwich Islands" published in *Journal ANSP* (1839). Human skulls collected by Townsend on the Columbia River form an important section in S.G. Morton's *Crania Americana* (1839).
5th March	Townsend's "Description of a New Species of CYPCELUS [Vaux's Swift], from the Columbia River" is read at ANSP. Published in *Journal ANSP* (1839).	
2nd April	Townsend's "Description of a New Species of SYLVIA [MacGillivray's Warbler]" is read at ANSP. Published in *Journal ANSP* (1839).	
May		Fifth and final volume of Audubon's *Ornithological Biography* is published.
July	Collecting birds on the New Jersey coast.	
10th September	Townsend's "List of the Birds Inhabiting the Region of the Rocky Mountains, the Territory of the Oregon, and the North West Coast of America" and "Note on SYLVIA *Tolmoei*" are read at ANSP. Published in *Journal ANSP* (1839).	Audubon leaves Britain in September and returns to the United States to settle in New York.
25th October	Townsend and Thomas McEuen make a list of the bird specimens that were lent to Audubon.	
December	Becomes a curator at the ANSP.	
1840	London edition of *Narrative* published as *Sporting Excursions in the Rocky Mountains, including a journey to the Columbia River, and a visit to the Sandwich Islands, Chili, etc.* Only one part of his *Ornithology of the United States of North America* is ever issued (apparently in 1840, even though title page is dated 1839).	Audubon begins work on *The Viviparous Quadrupeds of North America*. By January he has secured the assistance of the Rev John Bachman who assumes responsibility for the text. The first of seven volumes of Audubon's octavo edition of *The Birds of America* appears. Nuttall's second edition of *Manual of the Ornithology of the United States* appears with revised and updated 'Land Birds'.
18th February		First meeting of ANSP is held in its new building at Broad and Sansom Streets.
March	Townsend and a friend make a trip to Arkansas that lasts about eight weeks.	
December	Resigns as a curator at the Academy and moves to Washington to take up a post at the National Institute.	

Date	John Kirk Townsend	Other Events
1841		
13th September	Donates 121 bird specimens to the National Institute, mainly from Washington, Pennsylvania and New Jersey.	
12th October	Townsend and Charlotte Holmes of Cape May Courthouse are married by the mayor of Philadelphia.	
December		Nuttall leaves the United States to settle in England.
1842		John Cassin joins ANSP as Honorary Curator of Birds.
February	Townsend is still working as a taxidermist at the National Institute buying up birds and donating a few more birds and quadrupeds.	
April		The Audubon family moves into Minnie's Land, New York.
12th September	Donates 237 birds as well as plants, insects, shells and fishes from New Jersey to the National Institute.	
1843	Regularly attends National Institute meetings.	Audubon, Edward Harris, J.G. Bell and companions make an expedition to the Upper Missouri as far as Fort Union, April-August. They arrive at Fort Union on 12th June.
April	Townsend's only child, Francis (Frank) Holmes Townsend, is born about 8th April.	
1844		Seventh and last volume of Audubon's octavo edition of *The Birds of America* is completed.
April		Congress fails to support the National Institute and it goes into decline.
7th June	Townsend still at the National Institute, very busy with hummingbirds and insects.	
1845		Audubon's first volume of *The Viviparous Quadrupeds* of North America is published.
18th March	Still Recording Secretary at the National Institute meetings, but few members attend.	
8th September	Acts as Recording Secretary at the National Institute for last time. Thinks of returning to Philadelphia.	
December	Becomes a curator at ANSP again and studies to become a dentist.	
1846	Continues as a curator at the Academy until December.	Second folio volume of *The Viviparous Quadrupeds* is published. Smithsonian Institution established at Washington.
1847	Publishes "Sketches of a voyage, and residence in the South Sea Islands" in *Literary Record and Journal of the Linnaean Association of Pennsylvania College*.	
November	Attends Pennsylvania College 1847-48 session, November to March.	Nuttall returns to the United States again for a short visit and stays off and on in Philadelphia for a few weeks from December to March 1848.
2nd December		William Baird marries Harriet Holmes (Townsend's sister-in-law).

Date	John Kirk Townsend	Other Events
1848	Receives medical degree from Pennsylvania College, Philadelphia.	Third and last folio volume of *The Viviparous Quadrupeds* is published.
October	Publishes "Popular monograph on the Accipitrine Birds of N.A." in *Literary Record and Journal of the Linnaean Association of Pennsylvania College*.	Edward Harris presents 119 bird specimens to ANSP, including many Townsend types.
1849	In January, plans to sell his mammal skins in Europe; asks Spencer Baird for a certificate regarding quality of the specimens.	
1850	Publishes a note "On the Giant Wolf of North America" in *Proceedings ANSP* (1850).	
January	Donates all his mammal skins to the ANSP. Elected a Life Member of ANSP.	
5th July		S.F. Baird appointed Assistant Secretary of the Smithsonian Institution.
1851	Townsend listed as M.D. Dentist in an 1851 Philadelphia city directory.	
27th January		Audubon dies at Minnie's Land, New York, aged 66.
6th February	Townsend dies at Washington, aged 41. Afterwards buried at Congressional Cemetery, Washington.	
15th May		Samuel Morton dies aged 52.
1852		Bachman completes the last volume of text for *The Viviparous Quadrupeds*.
1859		
10th September		Nuttall dies at Nutgrove, England, aged 73.
1872		
19th October		William Baird dies aged 55.
1876		
23rd October	Townsend's wife, Charlotte Holmes, dies aged 58.	
1887		
19th August		S.F. Baird dies aged 64.
1911 12th February	Townsend's son, Frank, dies in Philadelphia and is buried at Cape May Courthouse.	
1937		Townsend and Audubon bird skins in the former Harris collection are saved from destruction (Street 1948).

Townsend and Audubon
– two very different characters

"John K. Townsend was evidently a genius whom force of circumstances prevented from reaching his proper place in ornithological annals."

Witmer Stone (1916) *Condor* 18: 10.

T he memory of John Kirk Townsend is still cherished by the Townsend and Kirk families, who have preserved many of his papers and letters as well as a few artefacts brought back from his travels, either by passing them down from generation to generation or by presenting them to the Academy of Natural Sciences of Philadelphia. They count Townsend's *Narrative of a Journey across the Rocky Mountains* (1839) as a wonderful resource that allows them an intimate insight into his character and ability. Of the family members who actually knew him, none have been as publicly elegant in their praise of him as his brother-in-law Mahlon Kirk (in Stone 1903):

"Since his death over half a century has passed, but his lovable qualities and scientific attainments are as fresh and green now as I recall them in the first development of my manhood. His personality was most attractive. His courtesy, kindness of heart, and his brilliant conversational powers, fortified with a vivacious intellect and a fund of knowledge covering almost all subjects, made him a delightful companion and endeared him to every one who came within his influence."

Since then Townsend has been somewhat forgotten by the wider American public, interest in him being revived only from time to time as different editions of his *Narrative* have appeared. Yet Townsend is worthy of so much more. He is owed a lengthy tribute because of his singular place in the early annals of North American ornithology, because of his close association with Thomas Nuttall, John James Audubon and other famous naturalists, and not least because his exploits never received the recognition or analysis that they deserved during his lifetime.

As we shall see, Townsend was a sensitive and devoted family man, an amiable raconteur and an intelligent and trustworthy zoologist. Most significantly, for this book and in terms of his lasting contribution to North American zoology, he was a first-rate collector, not only because he was a fine shot but because he prepared his bird and mammal specimens to the highest of standards. Fellow naturalists commented on his speed and agility in preparing skins, and his innovative preparatory techniques. Scott B. Wilson, who examined some of Townsend's Hawaiian birds in 1886, declared that "Most of those from that excellent collector are in a capital state of preservation, though now over 50 years old" (Wilson and Evans 1890-99, part II, p. 3.). Indeed his specimens survive to this day, scattered through museums in the United States and Europe.

By the time Townsend returned from his great journey across the Rocky Mountains he knew the birds of North America better than anyone of his generation (both Thomas Nuttall and John James Audubon being much older than Townsend) and was also familiar with a wide range of mammals. He wrote up his new discoveries in a careful, scientific and lucid style and has been justly praised by succeeding American zoologists, particularly those that knew him, or worked closely with his specimens

John Kirk Townsend
(Family papers).

and his descriptions. In the spring of 1833, when Townsend was only twenty-three, Audubon noted that "His zeal for the study of ornithology was unrelented. I saw this in his fine eye whilst he with enthusiastic glee spoke to me of a new bird lately procured by himself" (in Deane 1909). Spencer Fullerton Baird, who became the most influential American naturalist of the latter half of the 1800s, insisted that Townsend was one of "the cleverest people" known to him (S.F. Baird to W. Baird, 4th August 1846). Witmer Stone, Curator of Birds at the Academy of Natural Sciences of Philadelphia, felt that as an ornithologist Townsend "was equal to any this country has produced, a painstaking and reliable observer and a fluent, scholarly writer" (Stone 1903). Later, after speaking to people who had known him, Stone (1916: 10-11) had this to add:

"John K. Townsend was evidently a genius whom force of circumstances prevented from reaching his proper place in ornithological annals. Had he had the financial backing that Edward Harris was ever ready to provide for Audubon, or had there been salaried scientific positions in those early days by which an ornithologist could make a living, the name of Townsend would have been among the leaders in American ornithology. I have talked with his cousin who remembers him dressed in the furs and skins that he brought from the far west, and with his brother-in-law who knew him in the intimacy of family relationship; and I have read the opinions of [John] Cassin in his confidential letters to [S.F.] Baird and all testify to his high character and ability."

That Townsend never achieved true greatness as a zoologist is due to a certain lack of ambition and scientific vision, and the absence of good fortune. Theodore Palmer (1928: 299) described Townsend as "A brilliant young ornithologist who lived in advance of his time" but he might equally have said that Townsend flourished at the wrong period or in the wrong country where he was simply eclipsed by the presence and stature of Audubon who dominated the period in which Townsend strove to make a name for himself. That Townsend lacked vision was demonstrated by the missed opportunity in the Hawaiian Islands where he was casual about recording the birds, seeing the avifauna there as unrelated to North America and therefore less deserving of his time and effort. In the light of subsequent species extinctions in the islands, a more detailed study would have been of enormous historical and scientific value.

William Baird offers perhaps the best appraisal of Townsend. Baird knew him as both ornithologist and family member, in good times and bad, so in a private letter to his brother he cannot but have given his honest opinion when he wrote:

"[Those at the Academy of Natural Sciences of Philadelphia] may think that Townsend knows so much about birds that his word would be enough. He certainly knows more than any of them but his knowledge is certainly not as great as it ought to be, he does not appear to have that critical nicety of observation which a first rate ornithologist ought to have. This however is entirely sub rosa [private], as he is one of the best fellows I ever met with, and has been a good friend to me, and will take any trouble on my account" (W. Baird to S.F. Baird, 16th July 1843. SI Archives).

In other words Townsend was an unusually good ornithologist but not a great one, an assessment that we would agree with. He just did not have that urgent and compelling need to keep publishing his scientific discoveries, partly perhaps because of early discouragements. He may also have imbued that resistance to intellectual inquiry and distrust of higher education that was common to the early Philadelphia Quakers, as well as their dislike and contempt for self-promotion.

Today, the name of Townsend is best known to wildlife enthusiasts because of Townsend's Warbler, Townsend's Solitaire, Townsend's Ground Squirrel, Townsend's

Chipmunk, Townsend's Mole, Townsend's Vole and the White-tailed Jackrabbit *Lepus townsendi*. Caution is required with other such eponyms: Townsend's Shearwater, the plant genus *Townsendia* and Port Townsend, are each named after different Townsends[1]. The genus *Townsendia*, created by W.J. Hooker in 1834, is particularly likely to mislead because 1834 was the year that Townsend crossed the Rocky Mountains with Thomas Nuttall, it was Hooker who later received some of the plants collected by Townsend, and the genus now contains a plant that rejoices in the name of Nuttall's Townsend Daisy *Townsendia nuttallii*!

If Townsend is mentioned now in a biohistorical context it is usually because of his link with Audubon, and such commentaries are most frequently by art enthusiasts who naturally concentrate on Audubon as an artist and the creator of *The Birds of America* (1827-38) and *The Viviparous Quadrupeds* (1845-48). They do not emphasise the fact that Townsend provided nearly a tenth of the bird specimens and a similar proportion of the mammals that Audubon and his son portrayed, as well as much of the information about them. More often than not they uncritically repeat Audubon's oft-quoted statements that Townsend was lazy and careless, that his observations were sparse, that he saw species that he did not collect, and that he became unhelpful, without ever asking themselves if Audubon was being truthful or was perhaps even to blame for Townsend's shortcomings.

The majority of commentators examine Audubon through rose-tinted glasses, considering him to be a sort of patron saint of bird art who could do no wrong (e.g. Ford 1965, a rather pro-Audubon biography). Audubon was a great showman-cum-salesman and his propaganda is evidently still working today. Why else would it be said (Tyler 2000: 137) that he "discovered and published literally dozens of new birds and animals [sic]", a statement that seems to follow widely held beliefs but does not stand up to critical examination. The "great backwoodsman" discovered comparatively few birds for himself, and as far as we can tell, no mammals whatsoever. Of the ninety or so birds Audubon published as new, fully three-quarters are not now credited to him because they had already been described by others, or turned out to be young or immature examples of well known species. Audubon is credited as the describer of twenty-two bird species that occur in North America north of Mexico (*AOU Check-list* 1983) but he only discovered (or worked out) thirteen of them for himself. Moreover, to overstretch the point, these thirteen include those that were actually found by other expedition members, so the number discovered personally by Audubon diminishes to half a dozen or so. This is a somewhat startling statistic – and the data for the mammals is even more sobering.

The Reverend John Bachman described about 14 new species of North American mammals in scientific papers published in Philadelphia, Boston and London. When Bachman and Audubon collaborated on *The Viviparous Quadrupeds* "Audubon and Bachman" are credited as the describers of a further seven new species – but all the scientific analysis and descriptions were by Bachman, not by Audubon at all. Indeed Bachman, when discussing Audubon with Edward Harris, cautioned that "In his Quadrupeds – Tell it not in Gath – He [Audubon] never collected or sent me one skin from New York to Louisiana along the whole of the Atlantic States" (Bachman to E. Harris, 24th December 1845, in Herrick 1968, 2: 269). So, not only did he not help Bachman with obtaining common mammals, there were no new mammals discovered during Audubon's expeditions to Labrador, Texas or up the Missouri (though he did, of course, obtain information for the text on the habits and distribution of many species). Townsend's record of discovery is rather more commendable though difficult to summarise because some of his birds were described and named in his absence: roughly speaking he knowingly discovered about eleven bird species and he collected another eight birds and ten new mammals that were worked up by other naturalists (see appendix 2). Some of his observations on habitat and behaviour are indeed sparse but this was often because of the rapid pace of travel to which he was obliged to adhere.

John James Audubon (from Herrick's *Audubon the Naturalist* (1938)).

"AUDUBON, a man of whom America may well be proud, and who has given to the world a work of magnificence which will forever remain a monument of his unsurpassing genius, perseverance and research" J. K. Townsend (1839g: iii).

Very few commentators ever acknowledge that Audubon was proud, vain and self-conceited (Buchanan 1868, in Herrick 1938 1: 20), devious (Harris 1941: 29), arrogant, proud, egocentric, highly critical of other ornithologists, and reluctant to share the glory or financial reward in ornithological matters (Peck 2000: 73, 74). One of Audubon's most unattractive habits was his constant cruel criticism, directed at his contemporaries, often for failings that were so obviously his own. For instance, he hated those who added to the ever-increasing synonymy of American birds! He also had a general disregard for the value of scientific specimens once he had finished with them and it is a wonder that so many of the Townsend specimens that he painted have survived to this day.

There is no doubt in our minds that Audubon was an artistic genius – a man of enormous talent, charm and energy, who achieved his great ambition partly because of these virtues, but also because of those character traits that are often possessed by successful entrepreneurs: a ruthless determination to succeed, with scant regard for the needs, hopes and ambitions of those that are used along the way. But this book is not about Audubon. It is about Townsend: his epic journey across North America to the Columbia River, his time in the Hawaiian Islands, his voyage home, his contributions to North American zoology, the fate of his collections and his subsequent career. His somewhat reluctant association with the great artist is only part of the story.

1. Townsend's Shearwater *Puffinus auricularis* was named for Charles Haskins Townsend (1859-1944) (see Mearns and Mearns 1992: 441-446). Port Townsend was named by Captain George Vancouver in 1792 for a British General, the Marquis of Townshend (Meany 1923: 220). *Townsendia* was created by W.J. Hooker in his *Flora Boreali-Americana* (1832-1840); in volume ii, p. 16 (dated to 1834 by BMNH) Hooker says: "I have named the Genus in compliment to *David Townsend, Esq.* of West Chester, Pennsylvania, who having imbibed the most ardent love of Botany from his friend and instructor Dr. Darlington of the same city, has devoted his leisure hours to the science with eminent success. The plant now under consideration is perfectly worthy of bearing his name, because he has studied and ably discriminated the numerous Pennsylvanian species of the allied Genus *Aster*." John Kirk Townsend is known to have had cousins living in West Chester but whether or not there was a relationship with David Townsend is not known.

Townsend's early life in Philadelphia

"He seems to have an innate propensity [to go gunning for birds] that cannot be conquered by advice."

Townsend's mother, circa 1826-30.

Philadelphia in 1800 was the largest city in the United States, with a population approaching 70,000. It was then at its peak as a maritime trading port and was known the world over for sailing ships that exported flour, timber and furs and returned with tea from China, molasses from the West Indies, and porcelain and cutlery from Britain. Although Philadelphia had recently lost its capital status to Washington, much of that city was still on the drawing board and its famous Mall was still partly an uninviting swamp. Many congressmen longed for the capital to return to Philadelphia, then considered the most beautiful city in North America. In time, Philadelphia would be overtaken by New York and other cities in both size and prosperity, but when Townsend knew it, during the first part of the 19th century, it was still enjoying a golden age of trade, business, industrial development and scientific advancement. All this was achieved despite the trauma and upheaval of the recent War of Independence, regular outbreaks of cholera and occasional financial depression.

Townsend was fortunate that throughout his lifetime, from 1809 to 1851, Philadelphia remained the great cultural heart of the country, the centre for both arts and sciences. In terms of natural history, Philadelphia could boast of Peale's Museum, the best and most scientifically arranged of the early emporia, as well as the Academy of Natural Sciences that by 1850 possessed a collection of birds that was larger than any other in North America – and most of those in Europe. During this time, the city saw the publication, in quick-fire succession, of seminal works on geology, botany, entomology and ornithology that included Alexander Wilson's *American Ornithology* (1808-1813) and

Philadelphia, view of Second Street north from Market Street, circa 1800, from Weiss and Zeigler's *Thomas Say* (1931).

Philadelphia, view of the Court House at second and Market Streets, circa 1829, from Weiss and Zeigler's *Thomas Say* (1931).

Thomas Nuttall's *Genera of North American Plants* (1818). It is also relevant to our story that Philadelphia had a motley assortment of medical schools and the most active and successful printing, engraving and publishing industry.

Townsend's ancestors[1] emigrated from England, possibly from Cirencester in Gloucestershire[2]. All that is known with certainty is that John Townsend (1646-1727) after a spell in Kent County, Delaware, purchased a plot of ground in Philadelphia in 1715 and built himself a log house in the woods that then existed there (the site in time becoming the corner of Sixth and Walnut Streets, in the grounds of Independence Hall). The Kirks originated from Lurgan in Ireland. Alphonsus Kirk took passage from Belfast to the New World in 1688, arriving in Jamestown on 12th January 1689.

John Kirk Townsend was born in Philadelphia on 10th August[3] 1809, the son of Charles Townsend, a clock and watchmaker, and Priscilla (Kirk) Townsend. He was almost certainly delivered at 105 Chestnut Street, the family home from 1803 to 1822, that was a short distance from the busy wharves on the Delaware River and adjacent to Franklin Court where Benjamin Franklin, his daughter and son-in-law and their children all lived during Benjamin Franklin's older years. The two families may have known each other, at least on business terms, because the Townsend shop was visited regularly by a host of prominent citizens who were eager to have their timepieces set or repaired. Old Philadelphians such as Stephen Girard, Robert Vaux, Nicholas Waln and Samuel Fisher were well known to the family and young John would have grown up knowing persons of means and accomplishment, in a general atmosphere of Quaker piety and good works. In 1823, when they moved a little westwards to 138 South 10th Street they were still only a few blocks away from the Academy of Natural Sciences, a much more significant building than Independence Hall in terms of John Kirk Townsend's destiny.

John was one of twelve children, five of whom died in infancy. In effect, he had three older brothers and three younger sisters. Elisha, Edward and Charles all became dentists at some point in their careers. Elisha began in his father's clock and watch making business where he learned the skills and dexterity that would serve him so well in his dental practice. He yearned to be an actor but after a season in New Orleans returned to Philadelphia because the society that enveloped him was not to his taste. At the age of twenty-eight he turned to dentistry, perhaps because mass production was already beginning to affect the clock trade, and he became particularly distinguished as a co-founder and Dean of the Philadelphia College of Dental Surgery, and then

The Townsend Family

Priscilla Kirk Townsend 1785 - 1862
Charles Townsend 1777 - 1859

Elisha 1804 - 1858

Edward 1806 - 1896

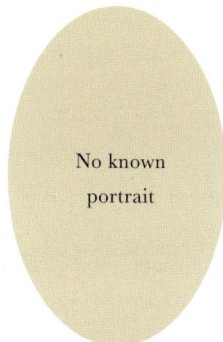

No known portrait

Charles 1807 - 1866

John Kirk 1809 - 1851

Hannah 1812 - 1851

Mary 1814 - 1851

Elizabeth 1825 - 1869

co-founder and first Dean of the Pennsylvania College of Dental Surgery. Edward was first a pharmacist and then a dentist but his real vocation lay elsewhere and he became a noted campaigner for prison reform. After years as a prison visitor and trustee, he served in later life as Warden of the Eastern Penitentiary of Philadelphia, retiring at the age of seventy-five after eleven years of service. For more than forty years he also had a great concern for the blind: he was one of the managers of the Wills Eye Hospital and the Pennsylvania Institution of the Blind, being president of the latter at the time of his death. Charles evidently earned his living as a watchmaker for thirty years, becoming a dentist only at the age of fifty-two but was soon known as a particularly fine operator, unequalled in the city for artificial work; he took over Elisha's practice when his brother died.

Both Hannah and Mary (Molly) had "female problems" and died within ten days of one another, before they reached forty. Mary had a uterine tumour; she became blind and was bedridden most of her adult life. Hannah also became bedridden, intermittently if not permanently, from at least the age of twenty-two according to the letters that she wrote to John while he was travelling. Elizabeth (Lizzie), thirteen years younger than Hannah and eleven years younger than Mary, devoted her early years to looking after her beloved sisters. She did not marry until three years after her sisters had died, marrying a distant cousin, Mahlon Kirk, at the age of twenty-nine – a late age for that era.

The Townsends were not wealthy but were well educated and intellectual with the high moral and ethical standards of their Quakerism. It was a faith that John Kirk Townsend enjoyed throughout his life, once writing home from the Columbia River, "I am a quaker yet. I glory in it & I mean always to continue to be a quaker..." (JKT to C. Townsend, 11th April 1836. Family papers). It must surely have been a severe disappointment to the Townsends when he married out. In fact it was scandalous: they would have expected him to have married another Quaker, and not only that, they would have expected him to have married a woman within their own meeting, with the same faith and rules of practice. Even so, there was never any question of a rift between John and his parents; they remained a close and affectionate family.

Most of the family seem to have been prolific writers. They kept journals, wrote poetry and were frequent letter writers. John's father kept a diary of his travels to Quaker meetings in New York State, Ohio and eastern Pennsylvania. His mother kept a diary in her apron and would add to it in her private moments, keeping track of her children's activities, but its primary purpose was as a tool for self-improvement, for which she reverently looked to God to provide strength. Mary produced *The Lives of Insects; or Conversations upon Insects, between an Aunt and her Nieces* (1844)[4] and Hannah wrote a *History of England in Verse* published posthumously in 1852. The family letters show them to have been a close-knit and caring family, caring not just for each other and an array of relatives in and around the city, but for their friends and neighbours too. Both Hannah and Elizabeth were involved in the anti-slavery movement, and the rest of the family no doubt supported its aims. Indeed, the Townsends belonged to the Green Street Meeting that was attended by Lucretia Mott (1793-1880), a famous abolitionist as well as an outspoken early supporter of women's rights. The two girls participated in abolitionist activities with her, and John sends his love to James and Lucretia Mott in a family letter from Washington dated 6th June 1845.

John went to Westtown Boarding School in Chester County, thirty miles west of the city, a highly regarded Quaker establishment that gave proper attention to the natural sciences. Notable former pupils included Thomas Say (a distinguished zoologist whose name is commemorated by Say's Phoebe), Samuel G. Morton (a lifelong friend of John's and a prominent member of the Philadelphia Academy of Natural Sciences), and John Cassin (a famous taxonomist at the Academy, commemorated by birds such as Cassin's Vireo and Cassin's Kingbird). Both John's parents had attended the school and

Westtown School.

so too did several of his siblings. John was there at the age of nine (from 27th March to 27th September 1819) and again at the age of thirteen (from 28th December 1822 to 29th September 1823). At the grand age of sixteen and a half, when a younger sister was at Westtown, he described it as "a place in which I have spent some of the happiest days of my life" (JKT to H. Townsend, 7th April 1826. Family papers). When not at that school, John probably completed his education at home and at less expensive schools, perhaps including the Friends' schoolhouse that was only 400 yards from 105 Chestnut Street. He went on to study as a pharmacist and physician, though whether or not he gained any formal qualifications at this stage is not clear[5], he may simply have gained experience through an apprenticeship; his older brother Edward had done this, becoming apprenticed to a druggist on Front Street after finishing at Westtown.

John (hereafter referred to simply as Townsend or abbreviated to JKT) was devoted to birds at a young age, for even as a school boy he was collecting birds and learning the skills that would make him one of the best taxidermists of the day. His early field excursions were often spent with his cousin William P. Townsend of West Chester and together they built up a collection of all the local birds they could muster. In Chester County he also fell in with Ezra Michener, a character whose natural history collection grew to contain 500 specimens of birds, mammals and reptiles, and of whom Townsend once confided that, "there is not in the whole world a friend I esteem more highly" (JKT to H. Townsend, 27th June 1834. Family papers).

Nothing seemed to curb his passion for birds, not even the reproach of his mother who loathed the constant killing and worried for his future. In a family letter written around this time while he was away from home, she made no attempt to suppress her disapproval: "[John's brother Charles] says John is well, goes gunning every morning and had had success at which I do not rejoice, but he seems to have an innate propensity to it that cannot be conquered by advice – I hope a few more years will quell it and a weight of business which must soon devolve on him (if he expects a livelihood) take its place and leave this amusement …" (P. Townsend to M. Townsend, undated but circa 1826-30. Family papers). Even when he was away in the far West she would write to him: "thousands of times have I wish'd thou would never again hold a gun, but thou knows well all thy mother's fears" (P. Townsend to JKT, 3rd March 1836. ANSP).

Besides the regular field trips, he would also have been a frequent visitor to Peale's Museum, founded by artist and showman Charles Willson Peale in the 1780s. When Townsend was a lad the museum was but a short walk away in Philosophical Hall[6], in the yard beside Independence Hall. In 1811, the museum consisted of four sections: a Marine Room (with shells, corals, fishes, lizards and turtles); a Quadruped Room (with over 200 mammals, including a Bighorn Sheep and Pronghorn brought back by the Lewis and Clark expedition); a Long Room (with over 1000 North American and foreign birds mounted in glass cases, and more than 100 of Peale's portraits of prominent Americans); and a Back Room (full of Native American artefacts). For a time, Peale also kept a menagerie in the yard outside, with monkeys and parrots and most famously, two unruly Grizzly Bears brought back from the west by Lieutenant Pike. We can only speculate about the effect that the museum had upon young Townsend. How often did he wander around the bird room under the watchful eye of Peale's portraits of Lewis and Clark? What was he thinking as he saw the very creatures and tribal gear collected by the two explorers on their epic overland journey to the Pacific (on a route that he would later partially follow himself)? Was he inspired to go westwards after seeing the birds and mammals brought back by Major Long's expedition to the edge of the Rocky Mountains (1819-20)? Did he pick up tips on taxidermy from old man Peale, the best and most experienced taxidermist in the country?

Townsend had a life-long association with one[7] of Peale's sons, Titian Ramsay Peale, who was ten years his senior, but there is no indication that they became particularly close friends. They knew each other in Philadelphia, and Townsend would have heard

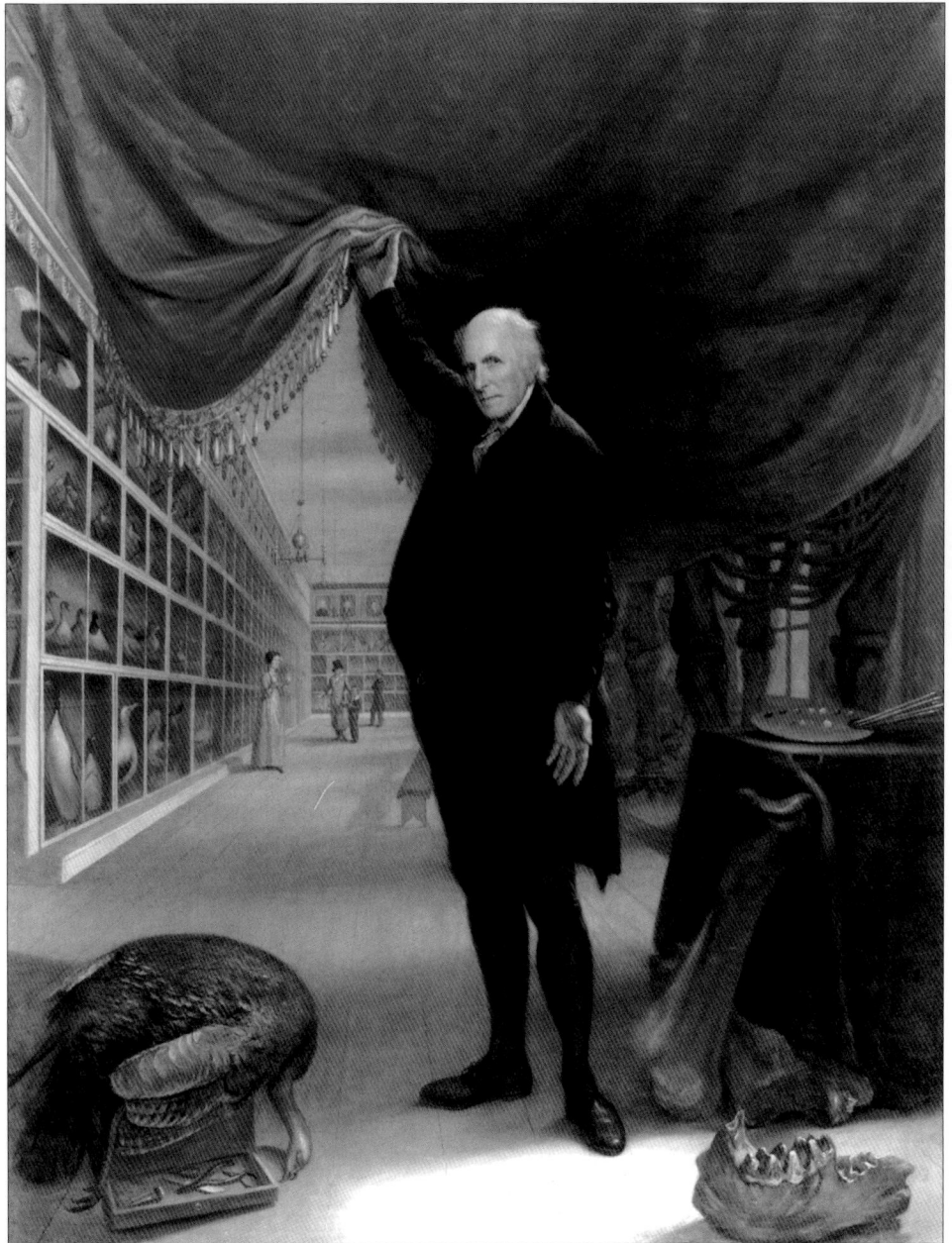

Charles Willson Peale in his museum, 1822.

first hand of his exploits when young Peale returned from Long's expedition on which he had sketched and served as assistant naturalist. Much later, the two of them worked together in Washington, following Peale's return from Wilkes's US Exploring Expedition to the southern oceans, Hawaiian Islands and Pacific Northwest.

At some stage in his youth Townsend must have begun to attend the nearby meetings of the Academy of Natural Sciences. The Academy had begun rather humbly in 1812 with just a handful of members, of which Say is now the best known. In 1826, after a series of temporary premises, a building was purchased at the corner of 12th and Sansom streets, a peculiar structure that had been a place of worship by a society of Swedenborgians but was adapted to house the ever-increasing collections and reference library. The Academy remained here until 1840 so this was the building to which Townsend's first great consignment of western birds and mammals was delivered and the building to which he returned in triumph after sailing around Cape Horn. It was here too that he met the distinguished men of science of Philadelphia, many of whom would become lifelong friends, and one or two, such as George Ord, who were more difficult to tolerate.

Unfortunately, he was not old enough to have met Alexander Wilson (1766-1813) but it was Wilson's *American Ornithology* that was then the most up-to-date reference book on the market, covering the majority of eastern species. It completely overshadowed the earlier works of Mark Catesby, John Abbot and Louis Vieillot and was not superseded for several decades, not even when Audubon's vastly expensive *The Birds of America* was issued as a cheaper octavo edition (1840-44). This was because new editions of Wilson's work kept appearing, being updated, for example, by Charles Bonaparte in 1832, and by S.F. Baird in 1878. Even if Townsend could not then afford his own copy of Wilson's original edition he would have often looked through the nine-volume set at the Academy or at the homes of wealthier friends.

The Academy of Natural Sciences of Philadelphia, 1826 - 1840.

There appears to be no record of when Townsend first met Audubon. It is unlikely to have been in 1824, when Townsend was only fifteen years old, the year that Audubon made a failed attempt to win support in Philadelphia for his outrageously ambitious plan to make life-size illustrations of all the birds of North America. In 1826 Audubon went instead to Edinburgh, and subsequently to London, to set up production of the massive work and only returned to the United States in the summer of 1829, after an absence of almost three years. Early in the following year he visited the Academy in Philadelphia but once more its officials refused to subscribe to *The Birds of America* and it was not until the next year that he met with any success there. So the first meeting between Townsend and Audubon was probably in the summer of 1829, or in either of the two following years.

It was around this time that Townsend also became acquainted with Thomas Nuttall, the most experienced field botanist in the United States who was also by dint of his extensive travels the most knowledgeable ornithologist of the day, not excluding Audubon. In 1829 Nuttall spent January to March at the Academy and was there again in December. In December 1831 the first volume of Nuttall's *Manual of the Ornithology of the United States and Canada* appeared (dated 1832), dealing exclusively with the land birds, and one can imagine Townsend reading through every page with wonder and delight, waiting eagerly for the second volume on the water birds (1834), and yearning to explore beyond the bounds of the east coast where he had probably never strayed any further than Cape May, New Jersey, a bird migration hot spot on the Atlantic flyway.

Whenever Townsend could escape from the city he headed for the marshes to hunt for ducks and shorebirds, or into the woods and fields to get acquainted with the passerines. It was not long before he made a significant mark. In May 1833, while Townsend was collecting around Ezra Michener's home near New Garden, he shot a bird the like of which has never been seen again, which was subsequently named Townsend's Bunting. Later that year, on 24th September, Townsend was elevated to full membership of the Academy, his proposers being Charles Pickering, Thomas McEuen and P.B. Goddard[8] who recognized that they had an exceptional talent amongst them. A few days later Townsend wrote his description of Townsend's Bunting and though his version was never published it served to "renew his acquaintance" with Audubon.

In 1833 Nathaniel Wyeth invited Nuttall to join him on his second overland expedition to the Columbia River, and by the end of January 1834 Nuttall had asked Townsend to come along too. Nuttall was fully aware of the Herculean task ahead of him, just with the botanical discoveries that were sure to be made, and reckoned that a young energetic marksman was needed to make the most of the zoological side of things. Townsend's family were rather apprehensive about his safety, but it was an opportunity that could not be missed, and he quickly set about preparing for the journey. He would need a good gun and a spare gun, plenty of powder and shot, arsenic to preserve the bird and mammal skins, a large container of alcohol in which to preserve reptiles, a telescope and some ornithological reference material (see 27th May 1836 in *Narrative*), and he would certainly need some money. He approached the Academy and

TOWNSEND'S BUNTING

One of the most intriguing birds ever collected by Townsend was his single specimen of a bird that has become known as Townsend's Bunting. It was featured in one of Audubon's finest watercolours of a passerine, the same specimen being drawn from three different angles, yet its current status is still undecided.

The circumstances of the bird's discovery are spelt out by Townsend in his manuscript description that was passed to Audubon with the heading "EMBERIZA AUDUBONI. Audubon's Bunting". On the back of the description Audubon wrote a short introductory paragraph, evidently intending that the whole of it should be published but it never was. This is perhaps because it became easier and perhaps quicker for Audubon to include a description of the bird in the second volume of his *Ornithological Biography* that appeared in the following year. Audubon was unable to call it after himself (as Townsend had intended) so he named it after its discoverer. Many years later, Ruthven Deane was able to examine an Audubon manuscript received from an Audubon relative and printed the contents in the *Auk* (1909: 269-272):

"On my reaching Philadelphia bent on going to the Floridas once more, I had the pleasure of renewing my acquaintance with John K. Townsend Esq. of that city. His zeal for the study of ornithology was unrelented. I saw this in his fine eye whilst he with enthusiastic glee spoke to me of a new bird lately procured by himself. I saw this bird and accepted it to make the drawing now before you, and as its habits are yet unknown, I merely can give a copy of Mr. Townsend's letter to me on the subject.

EMBERIZA AUDUBONI.

Audubon's Bunting.

I obtained this bird, (which I have honored with the name of our distinguished countryman) in New Garden, Penn. on the 12th of June 1833. It was first observed sitting listlessly upon a fence rail, but upon being approached flew to the top of an adjacent tree from which it emitted a succession of lively notes somewhat resembling the song of the Indigo Bird (Fringilla cyanea) [Indigo Bunting] but louder and more varied. Its flight was performed by short quick jerks of the wings and undulations of the body. It was with extreme difficulty that I approached sufficiently near to shoot, it being very shy and watchful and passing rapidly from tree to tree. Anxiety to procure it prevented my observing its habits more particularly. I have since visited the spot repeatedly but have never seen another individual.

[description of bird]

I was at first inclined to consider this species as identical with the Black-throated Bunting (Fringilla americana) [Dickcissel] setting aside the very considerable dissimilarity which I observed in its habits, voice &c. More particular observations however, and a careful comparison of the individual with descriptions and specimens has convinced me beyond the shadow of a doubt that my bird is new, and in this belief I am sustained by Mr. Audubon than whom there cannot be better authority.

▶

Townsend's Bunting. Detail from plate
CCCC of *The Birds of America*.

There is a species described by Vieillot under the name of *Fringilla grisea [Passer griseus* Grey-headed Sparrow, an African species] and said to inhabit the U. States which somewhat resembles the present in its markings, but upon comparison they will be found specifically distinct.

John K. Townsend, Philad.
Sept. 27th, 1833."

A few more details about the bird's discovery were included in Ezra Michener's 'Insectivorous Birds of Chester County, Pennsylvania' (1863: 303):

> 85. Genus *Euspiza*, (Bon., 1838.)
>   1. *E. americana*, (Gmel.,) Bon., black-throated sparrow. Summer resident, common; granivorous. "Their food consists of seeds, eggs of insects, and in the early part of summer of caterpillars and small coleopterous insects." (Nutt.)
>   2. *E. townsendii*, (Aud.,) Bd., Townsend's sparrow. Summer resident, unique; granivorous. Its food inferred from analogy to be, as in other sparrows, seeds and insects.
>   *Observation.*—The following extract from my diary may be properly introduced here:
>   "NEW GARDEN, 11*th of 5th month*, 1833.—This morning my friend John K. Townsend, in company with John Richards, while in quest of birds for my cabinet, shot a bunting in William Brown's cedar grove, near New Garden meeting-house, which is believed to be a nondescript. We have given it the provisional name of *Euspiza albigula*, or white-throated bunting. The following brief description was drawn from the recent bird before it was skinned:
>   Male. Upper mandible, black; middle edge, white; lower, light blue, with a longitudinal stripe extending half way from the point to the base; head, dark plumbeous; cheeks and breast, lighter plumbeous; line over the eye, white; back, varied with black and brown; the first and second primaries, equal and longest; the two lesser coverts edged with paler; throat, white, margined with black, extending down upon the breast, beneath which is a small spot of ocherous; sides, plumbeous; belly and vent, brownish white; length, 5¾ inches; extent 9 inches." (The sex was determined by subsequent dissection.)

There is a discrepancy in the dates given by Michener and Townsend. Michener is more likely to have given the correct date because it was taken from his diary, whilst Townsend may have relied on his memory as his letter was written later in the year.

The mounted specimen was in Michener's collection for twenty-four years. In 1857 it was deposited at the Smithsonian Institution – just as well, because in 1869 Michener's natural history collection was presented to Swarthmore College that later burnt down. The bunting skin still exists in the Smithsonian Institution (USNM 10282) and there are different theories about its identity. It could be a Dickcissel that lacked the normal yellow carotenoid pigments in its feathers, having only the browns, blacks and grays produced by melanin pigments (Parkes 1985). But if it is just an abnormally plumaged Dickcissel why didn't Townsend recognise the song? Perhaps it is simply a hybrid, one of its parents being a Dickcissel. In the *AOU Check-list of North American Birds* it has long been placed in Appendix C (for forms of doubtful status or of hybrid origin that have been given a formal scientific name); along with a comment that it is generally placed in the genus *Spiza* although its peculiarities cannot be accounted for by hybridism or apparently by individual variation.

the American Philosophical Society and signed an agreement (that later came back to haunt him) by which each institution would grant him $125 in exchange for a share of the discoveries. To raise more money he may have sold his own natural history collection. This is suggested by Hannah who wrote, "Perhaps thou remembers at the time of leaving home, there were 2 birds belonging to T Walmsley yet remaining of thy numerous collection" (H. Townsend to JKT, 29th November 1835. ANSP). This rather implies that the greater part had been sold, though he could just have donated it to the Academy or presented it to someone, such as his cousin William, for safe keeping.

Although Townsend confessed at the very start of his *Narrative* that "the temptation to visit a country hitherto unexplored by naturalists was irresistible" not all the family were so keen on the idea. Edward Garrigues, an uncle, announced that "not all the beaver skins, nor all the birds either, would ever tempt me to go west of the Mountains." And Aunt Lydia just could not understand the attraction of exploring new territory and thought it "great nonsense that thee must go so far to look for birds, when ... there's birds enough round here and in Chester County" (H. Townsend to JKT, 29th November 1835. ANSP).

1. Much of the information about the Townsend family, their occupations and addresses, has been researched and kindly supplied by Donald Townsend Little and Dana Dunbar King. They also provided transcribed family correspondence and diaries. These sources are referred to under the umbrella title of 'Family papers' - see Bibliography.

2. According to Edward Townsend (an older brother of John Kirk Townsend) their grandsire (paternal great-great-grandfather) John Townsend (1646-1727) arrived from England in 1682 with William Penn, the founder of Philadelphia but there is no record to support this claim. A Richard Townsend of Cirencester, Gloucestershire, was recorded as being imprisoned in 1660, 1662 and 1675 for upholding his Quaker beliefs by refusing to support the parish priest and bear arms. Richard sailed with Penn and may have been the brother of the above named John Townsend. Ann Townsend, wife of Edward Townsend (brother of JKT) had a great-great-grandfather named David Ogden who was with Penn when he first arrived in Philadelphia (Donald Townsend Little *pers. comm.*).

3. There has been some confusion concerning the exact month of his birth. This seems to be due to Stone (1903: 2) who gives 10th October, even though he elsewhere (1936: 617) changes to 10th August. Palmer (1928: 299) gives the birth date as 10th October but points out that Townsend's grave marker says 10th August. Mearns and Mearns (1992: 447) give 10th October, based on Stone (1903: 2) and until recently, the Townsend family papers gave 10th October, presumably based on Stone's error. There is no doubt that **August** is the correct month: While at Fort Walla Walla in 1836 Townsend wrote in his journal "Aug. 10th. I am this day 27 years of age ..." Further support is given by Townsend's grave marker in the Congressional Cemetery in Washington, D.C. that says simply "John K. Townsend / Born Aug. 10. 1809. / Died Feb. 6. 1851."

4. In the preface Mary Townsend says that she had been interested in insects since childhood. It was originally intended for the benefit of her nieces and the children of a few friends but she was persuaded to give it a wider circulation. One of her aims was to reduce cruelty to insects by children but it also helped to pass her bed ridden days. Because of her blindness the book was written with great difficulty with the aid of an indented card.

5. We have been unable to trace any details of Townsend's pharmaceutical or medical training other than his 1847-48 session at the medical school at Pennsylvania College, Philadelphia. He undoubtedly possessed considerable knowledge and experience before setting off in 1834 to the Columbia River, otherwise he would not have been employed as a surgeon by the Hudson's Bay Company at Fort Vancouver.

6. Peale's Museum was later in Independence Hall itself, upstairs in the very room where the Declaration of Independence was signed. When Townsend was eighteen, the museum was transferred to the Philadelphia Arcade on Chestnut and 6th/7th Street. In 1838 it moved again, this time to 9th and Sansom Streets. Here, a mastodon, excavated and reconstructed by Peale, stood as a stunning centrepiece in the enormous hall, 233 feet long by 64 feet wide.

7. Another son, Rembrandt Peale, painted two portraits of Townsend's eldest brother Elisha. The first, painted in 1844, is at the Philadelphia College of Dental Surgery, Leon Levy Library, School of Dental Medicine, University of Pennsylvania. The second, painted in 1855, is in New York in a private collection of the estate of George T. Arden.

8. In January 1840 Dr Paul Beck Goddard made the first successful attempt at indoor photography, improving upon Louis Daguerre's process by introducing bromine to improve the sensitiveness of the plates. It is conceivable that it was Goddard who took the daguerreotype of Townsend that is the earliest known portrait of the naturalist.

Townsend and Nuttall
set off westwards

"Nuttal [sic] & Young Townsend had left for the Pacific under charge of Mr. Wirth of Boston – I fear that expedition will prove a failure."

J.J. Audubon to Maria Martin, 6th April 1834 (Corning 1930: 22).

Imagine how exciting it must have been to travel across North America in 1834. Once beyond Independence, Missouri, Townsend would have had little idea of the habitats, birds and mammals waiting to be discovered, nor how the wild and exotic native tribes would react to their presence. It was a time before unfamiliar diseases and wars had depleted and changed the people and their way of life, when game was abundant and when the prairies, forests and woodland were in pristine condition, and had scarcely ever been sketched or painted.

Artists had only just begun to depict the West, capturing and romanticising what they saw, and they were to have a profound effect on the opening up of the wilderness. On the way to the Rocky Mountains in 1819, Major Long took with him Samuel Seymour and Titian Ramsay Peale to sketch the topography, tribal way of life and natural history. The Prussian explorer Prince Maximilian Wied who went up the Missouri in 1833 had with him the Swiss artist Karl Bodmer. Sir William Drummond Stewart (whom Townsend travelled with for part of his journey in 1834) would later hire Alfred Jacob Miller to make a record of a hunting trip to the Rockies. But there were no such artists on Townsend's epic journey westwards – it was a commercial venture, and it is a great wonder that there were even the two naturalists among the party. On this journey there was only Townsend's scratchy quill pen to depict in words the native encampments, the snowy peaks of the Wind River Mountains, the lava beds of the Snake River Plain and the mighty gorge of the Columbia River, and to describe their daily progress ever westwards.

And so, early in 1834, twenty-four year old Townsend set off with Nuttall from Philadelphia, heading by coach to Pittsburgh, then sternwheeler to St Louis, where they headed overland for a while towards Independence, the usual launch pad for westward expeditions. He took frequent opportunity to send letters back to his family, partly because he had never before been away from them for such a long period and partly because he knew that the further west they travelled the more difficult it would be to keep in touch. In a letter to his sister Hannah, he made it pretty clear that he did not like one of their stopping places: "Pittsburg [sic] is truly a vile, vile place. I have been in it but an hour & am wholly disgusted with it & wish to leave it as quick as I can … I don't like Pittsburg" (JKT to H. Townsend, 17th March 1834).

His journal sets the scene for the beginning of the journey, a portion that was excluded from his published *Narrative* because his book does not begin until after their arrival at St Louis:

On the 13th. day of March [1834], Mr. Nuttal [sic] & myself took our seats in the coach for Pittsburg [sic], where we arrived on the 17th. without any thing occurring worthy of notice. The winter had just broken up & the roads were somewhat rough & sloppy which occasioned considerable delay. We were moreover very tired, having ridden constantly day & night, & felt glad of the opportunity of having an abiding place for a season. At Pittsburg, I found several friends whom I had known in Philadelphia & made some pleasant visits at their houses. The city I did not admire – it appeared to have been well laid out however, somewhat in the manner of Philadelphia, but every thing about it, even to the inhabitant, had a begrimed & filthy appearance which was any thing but pleasing to me. This is accounted for by the circumstances of the factories, for which this town is remarkable, burning large quantities of bituminous coal & thereby loading the atmosphere with black & sooty particles which are constantly falling & soiling & smirching every object with which they come in contact.

We remained but one day in Pittsburg – which however was quite enough for us. It afforded us time to pay all the visits which we wished, & to write letters to our friends. On the 18th. we took passage in the steamboat Boston for St. Louis, Missouri & at 9 in the morning embarked & went puffing along the Ohio river towards our destination … On the 19th. we stopped for two hours in Marietta, Ohio, a pretty town situated at the junction of the Ohio & Muskingum rivers … On the 22nd. our boat touched at Louisville, Kentucky. I was very much pleased with this town. It is laid out very much in the same manner as Pittsburg, but is as clean & neat as the other is disgusting & filthy … On the 24th. at about daybreak we passed the mouth of the Ohio & entered the Mississippi. When I took my first view of the river we had passed the great bend, & a more turbulent, muddy & encumbered stream I never beheld. The water is absolutely thick with mud; & upon allowing some to stand in a pitcher for an hour, the bottom of the vessel is covered with the deposit to the depth of at least an inch, while the decanted liquor is still more turbid than I ever saw the waters of the Schuylkill at any time. The river is at this time covered as far as the eye can see, with driftwood, & many trees of gigantic size are constantly passing us. These added to the snags & sawyers with which the river abounds, renders the navigation exceedingly difficult & often dangerous … Owing to the continual washing of the river banks & the consequent encroachment of the stream, it would be folly for any one to establish themselves, with a view to permanent residence, within a considerable distance of it, & therefore the country in the vicinity must ever remain as it now is, wild & untrammelled, "as when it came forth from the hand of its great Creator" (MS Journal, 13th March to 21st May 1834. ANSP).

Everything was fresh and exciting to Townsend and for the most part he took great delight in his new surroundings. Nuttall was an old hand, having travelled thousands of miles in the eastern half of North America, and had passed this way before. And yet Nuttall has gained an undeserved reputation for incompetence in the wilderness just because he admitted to getting lost a few times. This can happen to anyone, though real mountain men would never admit to being lost, only to not knowing where they had been for a few weeks! It may be true that Nuttall was incapable of living off the land because he seldom burdened himself with a gun, and when he did take one along he is said to have used the barrel to dig up plants or as a store for seeds. But considering the length and extreme difficulty of the many journeys that he had already undertaken, the trip to the Pacific Northwest, accompanied by so many others, presented no particular difficulties for him even though he was now approaching fifty – and did not like horses! Near Boonville, only a few days after Nuttall and Townsend set off westwards together, Townsend related the following amusing story to his family:

"We were overtaken yesterday by a man on horseback, leading another ready saddled, who was kind enough to offer us the use of his spare horse to ride alternately to the place of our destination 8 miles – the man happened to be bound for the same place & our riding was rather an accommodation to him. I urged our frd. N. [friend Nuttall] to mount the horse as I did not feel tired, but with all my persuasion I c^d. not induce him to do so until I had ridden 5 miles. I then insisted upon it & he led the beast to a large log & bestrode it. I knew he disliked riding, w^d. always rather walk 20 miles than ride 5 - & I expected he make a strange figure. I held the horse while he mounted, & tried to make it stand perfectly still, poor creature it was gentle as Miriam. The stirrups were too short for him, but he did not know that, so it made no difference of course, he thrust his feet through them to the heel & his knees approached the pommel of the saddle. He seized the bridle with both his hands. I gave Rosinante a smart cut with a stick & away went the naturalist jolting & bouncing like a Merry Andrew. The moment he was out of hearing I gave vent to my imprisoned feelings & laughed until I was so weak that I could scarcely walk. In a short time I overtook him & found him riding slowly beside his companion, but his countenance expressed the misery he endured. He besought me to hold the horse while he dismounted, but I told him I was not in the least tired, & would quite as leave walk - & then darted ahead to be out of the way. Upon looking back shortly afterwards I observed him making the most tremendous exertions to pull his horse's head from the ground, who was walking quietly along & cropping the short grass by the road side. Finding this of no avail he suddenly released both his feet from the stirrups & commenced digging them into the poor animal's sides. The horse bolted suddenly, & fearful that he w^d. unseat my good friend & do him an injury I stepped up & released him. Poor N. he looked as he trudged along the road again like a reprieved criminal & how he will ever make his way across the Rocky Mountains on horseback I cannot conceive. I shall be particular to select for him the most gentle animal I can find, but still I believe he will walk the greater part of the way" (JKT to C. Townsend, 9th April 1834. ANSP).

Nuttall and Townsend seem to have got on very well together. Nuttall was rather shy and reserved in new company but could become eloquent and expansive on his favourite subjects when at ease with people he knew well. Townsend was half his age, enthusiastic, quick to learn, and though he evidently teased the older man a little, he was good company for Nuttall who would otherwise have been alone in his scientific pursuits.

Thomas Nuttall, 1824.

THOMAS NUTTALL (1786-1859)

Of the two men, Nuttall was the senior naturalist. He was senior in years, being nearly twenty-three years older than Townsend, and senior in ornithological experience, knowing far more about the habits, distribution and abundance of a greater range of birds than Townsend because of his frequent far-reaching earlier travels. Nuttall was also more knowledgeable about plants – his main sphere of interest.

Nuttall was born in Yorkshire, England, on 5th January 1786. He trained as a printer, but went to the United States at the age of twenty-two to pursue his passion for botany. Not long after arriving in Philadelphia, he made friends with Benjamin Barton and William Bartram, the former sponsoring some of his early collecting forays. Over the next twenty-five years, with only occasional visits back to England, he explored the botany of much of the eastern part of the United States and mid-west, travelling down the Ohio, through Kentucky, Tennessee, North and South Carolina, Georgia, northern Florida and Alabama; across the Great Lakes and down the Mississippi to New Orleans; and up the Missouri as far as Fort Union near the Yellowstone River. By 1834 he had travelled further through the wilds of North America than any other naturalist of his generation. He travelled many, many times further overland than Audubon, whose main collecting expeditions were boat-based trips of comparatively short duration.

Besides Nuttall's significant books and papers concerning plants, he wrote *A Manual of the Ornithology of the United States and Canada*, published in two parts, the first on land birds (1832), the second on water birds (1834). To help make ends meet, he had taken to lecturing on botany at Harvard University, resenting the restrictions on his freedom, but nevertheless sticking to it for ten years. When Boston businessman Nathaniel Wyeth returned from his first venture to the Columbia River he brought back plant specimens for his friend Nuttall, (who described them in a 60-page article in the *Journal of the Academy of Natural Sciences of Philadelphia* (1834)). These plants so whetted Nuttall's appetite for the far west that when Wyeth offered to let him join a second trip to the Columbia he readily abandoned the security of his Harvard position.

Nuttall and Townsend travelled together beyond the Columbia River, on to the Hawaiian Islands. Then, after a spring and summer back on the Columbia, Nuttall and Townsend parted company: Townsend remaining in the vicinity of Fort Vancouver, Nuttall revisiting the Hawaiian Islands and then returning to Boston via California where he made several notable bird discoveries (see appendix 14). After working on his botanical collections, he brought out a second edition of his *Manual* (1840), though he only updated the land birds. Into it he incorporated new information from his recent time in the west, as well as from other sources including Swainson and Richardson's volume on birds for the *Fauna Boreali-Americana* (1832) and notes from Audubon's 1837 trip along the Texas coast.

In the summer of 1841 Nuttall learned that he had inherited his uncle's estate near Liverpool. Because of his advancing years and poor financial prospects, he felt the need to abide by the peculiar terms of the will, by which he was bound to spend at least nine months of every year in England. He died at his new home on 10th September 1859 at the age of seventy-three.

Narrative of a Journey across the Rocky Mountains, to the Columbia River, and a visit to the Sandwich Islands, Chili, &c.

John Kirk Townsend

Townsend's *Narrative of a Journey Across the Rocky Mountains* (1839) and his manuscript material

"No one, we think, can read the extracts [from Townsend's journal] without a lively impression of the intelligence and zeal of the writer, who evidently adds to his qualifications the gift of imparting his observations in a most agreeable manner."

Waldie's Select Circulating Library, 1835.

Several members of the 1834 Wyeth expedition kept journals. The leader's journal entries are brief and factual, mainly referring to topographical features, distance and direction of travel, and the deployment of personnel (Wyeth 1899). At least three members of the missionary party wrote about their experiences (Edwards 1834, Lee 1916, Shepard 1986). One of the trappers included the Independence to Fort Hall section in an account of his life as a mountain man (Russell 1955). Nuttall certainly had a journal and Captain Stewart and Captain Thing may each have had one too, though none have survived. But it was Townsend who wrote most expansively and engagingly, providing by far the best account of this, or indeed any other similar journey of the period. At the start of his book he explained that "the following pages, [were] originally penned for the family-circle, and without the slightest thought of publication". He had written in such a fluent and elegant style that his texts scarcely needed any editing when it came to produce his *Narrative*.

He did not make entries every day. There are often gaps of one or two days or occasionally longer periods: there is no entry in the *Narrative* for the whole of October 1834 (for no known reason) and there is a two-month interval at Valparaiso when he was ill. Not surprisingly for a journey of this type, the dates for certain events are not always entirely consistent with those of other expedition members sometimes differing by a day or so.

Those parts of Townsend's original journal that still survive cover his journey from Pittsburgh to the Platte River (13th March to 21st May 1834) and part of his time on the Columbia River, Hawaiian Islands, and the voyage home (18th December 1835 to 15th October 1837). In addition, edited extracts from the original journal exist for the section from the Blackfoot River to Fort Vancouver (10th July 1834 to 16th September 1834). The extracts appeared in *Waldie's Select Circulating Library* 1835, pt. 2, pp. 427-432 (reproduced in Hulbert 1934: 184-226); the content differs slightly from the printed *Narrative* but the original material for this period has not been traced.

Besides his actual journal, Townsend sometimes copied out sections and sent them as letters to his family (usually addressed to his father or a sister). So there are letters from Pittsburgh (17th March 1834), Louisville (23rd March 1834); St Louis (27th March 1834); Hickory Grove (1st April 1834); Boonville (9th April 1834); Independence (21st April 1834); Kansas Agency (3rd May 1834); the Trappers' Rendezvous near the Green River (27th June 1834); O'ahu, Hawaiian Islands (6th January 1835); Columbia River (2nd May 1835); Fort Vancouver (10th September 1835 and 11th April 1836). His letters home from the early part of his trip tend to have more entries about natural history than the later ones.

Another item that Townsend kept was a catalogue of all the specimens he shot, or otherwise collected, in which he gave each new species a number (a field number), giving all specimens of the same species the same number. So, for example, all the Clark's Nutcrackers were included under No. 33. It is unlikely that the catalogue was used much when it came to write the *Narrative*, though some of the dates within it may have been useful to him. It has now been lost or destroyed but a partly reconstructed version appears in appendix 6 to this volume.

Townsend arrived back in Philadelphia in November 1837 and later gathered up his journals, retrieved his letters to the family and began to write his book. Much of it was written while staying with his friend Ezra Michener at New Garden, about thirty miles west of the city. The *Narrative* misses out the early part of the journey from Philadelphia to St Louis and also misses out a few of the journal entries written in the Hawaiian Islands and on the return voyage, probably to keep down the length of the book. Although the scientific appendix included lists of the mammals and birds seen and descriptions of the various new species discovered the main text omits much of the natural history content that appeared in his journal because he thought that it would have a limited appeal. Even at the time of writing, he was rather too apologetic about including some of his observations on birds:

"These dry & uninteresting details will probably only weary my beloved sisters & perhaps will be considered very stupid & tiresome by all who happen to read them; my only excuse is that at the time they were written I had no occupation more interesting than shooting birds & then scribbling about them" (MS Journal, 16th August 1836).

The *Narrative* was published in 1839 and quickly sold out. After a few years Townsend intended to bring out another edition that would be priced at $1.50 a copy (W. Baird to S.F. Baird, 1st February 1842. SI Archives). It never appeared, presumably because he was unable to gather sufficient subscribers to be sure that it would be a financial success.

The *Narrative* did not appear again until 1905 when it was included by R.G. Thwaites in his *Early Western Travels*, but Thwaites excluded Townsend's scientific appendix and all the material covering the trips to the Hawaiian Islands and the voyage home. Similarly reduced versions were also reprinted in 1970 (Ye Galleon Press) and in 1978 (with an introduction by Donald Jackson).

The first complete reprint, including the scientific appendix, appeared in 1999 with a seventeen-page introduction and twenty-two pages of notes by George A. Jobanek of Eugene, Oregon. Jobanek is the first editor to make any useful and serious comment on Townsend's zoological contribution, partly by referring to his own research about Townsend in the Pacific Northwest (Jobanek, 1986, 1992, 1994).

Our book is an edited version of the *Narrative*, illustrated for the first time, and with zoological commentary and additional observations drawn from Townsend's unpublished journals and letters. Whereas Townsend often chose to omit natural history observations because he felt that they would not interest the general reader, we have chosen to omit some of his important cultural observations on the native peoples of the old west. Perhaps someone will some day provide a full commentary on the *Narrative* concentrating on that aspect alone.

A guide to readers of Part 2

- The *Narrative*, and extracts from Audubon's *Ornithological Biography*, are reproduced as nearly as possible in the style that they were originally printed, with the exception of our highlighting of species' names in bold. It also differs in that almost all chapter divisions, headings and subheadings, and spaces in the text separating different days, are new, inserted by the editors to assist the reader.

- All MS Journal sources (and book and journal titles) are always reproduced in *italics*.

- Text that has been omitted is indicated by the standard three-dot elision … Omissions from the *Narrative* can vary from a few words to several pages; missing portions can be easily referred to in one of the modern reprints of the *Narrative*.

- Any additions to the *Narrative* from other sources are entered within square brackets [   ], even if the additions were written by Townsend himself in letters or scientific journals etc.

- Townsend's manuscript material shows that he usually (but not always) wrote *neighbourhood, favour, harbour* and *colour* etc. In the printed version these were edited to the current American spellings of neighborhood, favor, harbor and color. These, and other spelling inconsistencies, especially of place names, have been retained. We have refrained from the overuse of [sic] to denote errors or differences of spelling.

- Bird and mammal names are highlighted in **bold**. Current English names of bird and mammal species have been inserted in square brackets in bold (unless there has been no significant change of name in which case the original text has been highlighted). Any species only referred to in passing, for instance for a comparative size, are not in bold.

- Any scientific names included in the text by Townsend are included as they appeared in the original *Narrative*, usually without comment. Up to date scientific names appear in the index of birds and mammals. Scientific names are given in the main text only for the few species that do not appear in the appendices.

- Historical commentary concerning Hudson's Bay Company staff or native tribes has been kept to a minimum but comments can be located in the Thwaites (1905) and Jobanek (1999) versions of the *Narrative*.

Part 2 - Chapter 1

Narrative of a Journey across the Rocky Mountains, to the
Columbia River, and a visit to the Sandwich Islands, Chili, &c.
by John Kirk Townsend

First steps to Independence

*"We come from Pennsylvania; our names, Nuttall and Townsend; we are travelling to
Independence on foot, for the purpose of seeing the country to advantage, and we intend to
proceed from thence across the mountains to the Pacific. Have you any mules to sell?"*

Townsend's *Narrative*, 6th April 1834.

Arrival at St. Louis

On the evening of the 24th of March, 1834, Mr. NUTTALL and myself arrived at St.
Louis, in the steamboat Boston, from Pittsburg.

On landing, we had the satisfaction to learn that Captain WYETH was already there,
and on the afternoon of the next day [25th] we called upon him, and consulted him
in reference to the outfit which it would be necessary to purchase for the journey. He
accompanied us to a store in the town, and selected a number of articles for us, among
which were several pairs of leathern pantaloons, enormous overcoats, made of green
blankets, and white wool hats, with round crowns, fitting tightly to the head, brims
five inches wide, and almost hard enough to resist a rifle ball. [*We then walked around the
suburbs of the town & saw a number of ancient Indian mounds; they are arranged in regular order,
& a chain is formed of them extending as far as the eye can reach. We found their summits very
fertile & collected a number of rare plants on them. We then returned to the hotel where we had taken
lodging for the few days that we expect to remain. On the following morning, the 26th, Mr. N., Capt.
Wyeth & myself crossed the Mississippi to the Illinois side. The country is here entirely flat, a low
wet prairie interspersed with small ponds which are the resort of innumerable water fowl. Here I shot
several Bartram's Sandpipers (Totanus bartramius)* **[Upland Sandpiper]** *in beautiful plumage*
(MS Journal).]

The day following [27th] we saw about one hundred Indians of the Sâque [Sauk]
tribe, who had left their native forests for the purpose of treating for the sale of some
land at the Jefferson barracks. They were dressed and decorated in the true primitive
style; their heads shaved closely, and painted with alternate stripes of fiery red and deep
black, leaving only the long scalping tuft, in which was interwoven a quantity of elk
hair and eagle's feathers. Each man was furnished with a good blanket, and some had
an under dress of calico, but the greater number were entirely naked to the waist. The
faces and bodies of the men were, almost without an exception, fantastically painted,
the predominant color being deep red, with occasionally a few stripes of dull clay
white around the eyes and mouth. I observed one whose body was smeared with light
colored clay, interspersed with black streaks. They were unarmed, with the exception
of tomahawks and knives. The chief of the band, (who is said to be Black Hawk's
father-in-law,) was a large dignified looking man, of perhaps fifty-five years of age,
distinguished from the rest, by his richer habiliments, a more profuse display of trinkets
in his ears, (which were cut and gashed in a frightful manner to receive them,) and above

KANSAS

MISSOURI

ILLINOIS

Missouri River

Kansas River

● **Independence**
Arrived 14th April
Left 28th April

Blue River

**Plains Pocket Gopher
and prairie-chickens**
First seen
30th March

Sandhill Cranes
Flying over
30th March

Rocheport

Boonville **Columbia**
Fulton **Loutre Lick**
(Big Spring) **St Charles**
○ **Florissant**

Mississippi River

Illinois River

Harris's Sparrows
*"Abundant near
the town of
Independence"*
Collected 26th
April or earlier but
MS description
unpublished

**Carolina
Parakeets**
*"rather
numerous"*
9th April

Carolina Parakeets
"several flocks seen"
5th April

Missouri River

● **St Louis**
Arrived 24th March
Left 29th March

**Flocks of
Passenger Pigeons**
1000s of **American
Golden Plovers**
31st March

Steamboat *Boston*
from Pittsburgh

Greater & Lesser Prairie-Chickens
Seen displaying
1st April

OKLAHOMA

ARKANSAS

Mississippi River

TENNESSEE

14th April   9th April 7th April   29th March

1834

0   50   100 miles

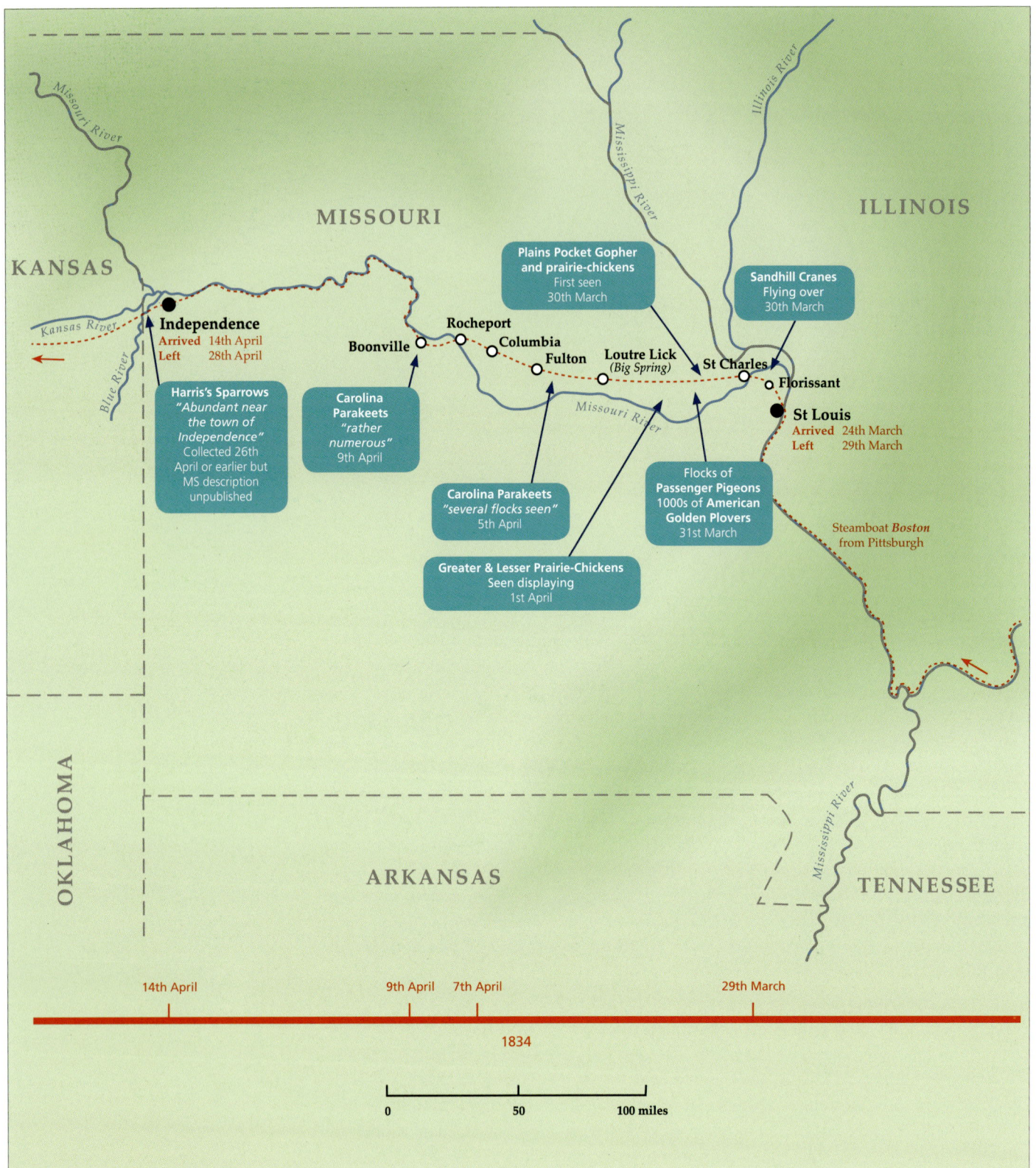

Map 2 - St Louis to Independence, 24th March - 28th April 1834.

all, by a huge necklace made of the claws of the grizzly bear. The squaws, of whom there were about twenty, were dressed very much like the men, and at a little distance could scarcely be distinguished from them. Among them was an old, superannuated crone, who soon after arrival, had been presented with a broken umbrella. The only use she made of it was to wrench the plated ends from the whalebones, string them on a piece of wire, take her knife from her belt, with which she deliberately cut a slit of an inch in length along the upper rim of her ear, and insert them in it. I saw her soon after this operation had been performed; her cheeks were covered with blood, and she was standing with a vast deal of assumed dignity among her tawny sisters, who evidently envied her the possession of the worthless baubles.

28th. – Mr. N. and myself propose starting to-morrow on foot towards the upper settlements, a distance of about three hundred miles. We intend to pursue our journey leisurely, as we have plenty of time before us, and if we become tired, we can enter the stage which will probably overtake us.

29th. – This morning our Indians returned from the barracks, where I understood they transacted their business satisfactorily. I went on board again to see them [*but I could not induce N. to accompany me as before: he has been accustomed to savages & cares not for them, or rather their appearance is no novelty to him* (MS Journal).] I feel very much interested in them, as they are the first Indians I have ever seen who appear to be in a state of uncultivated nature, and who retain the savage garb and manners of their people. They had engaged the entire covered deck for their especial use, and were lolling about in groups, wrapped in their blankets. Some were occupied in conversation, others seemed more contemplative, and appeared to be thinking deeply, probably of the business which brought them amongst us. Here and there two might be seen playing a Spanish game with cards, and some were busily employed in rendering themselves more hideous with paint. To perform this operation, the dry paint is folded in a thin muslin or gauze cloth, tied tightly and beaten against the face, and a small looking-glass is held in the other hand to direct them where to apply it. Two middle-aged squaws were frying beef, which they distributed around to the company in wooden bowls, and several half loaves of bread were circulating rapidly amongst them, by being tossed from one to another, each taking a huge bite of it. There were among the company, several younger females, but they were all so *hard favored* that I could not feel much sympathy with them, and was therefore not anxious to cultivate their acquaintance. There was another circumstance, too, that was not a very attractive one; I allude to the custom so universal amongst Indians, of seeking for vermin in each other's heads and then *eating* them. The fair damsels were engaged in this way during most of the time that I remained on board, only suspending their delectable occupation to take their bites of bread as it passed them in rotation. The effect upon my person was what an Irishman would call the attraction of repulsion, as I found myself almost unconsciously edging away until I halted at a most respectable distance from the scene of slaughter.

Onward to Independence

At noon, Mr. N. and myself started on our pedestrian tour, Captain Wyeth offering to accompany us a few miles on the way. I was glad to get clear of St. Louis, [*as I think it, without exception, the most disagreeably extortionate & uncomfortable town[1] that I have yet found. Every article that a stranger finds it necessary to purchase, the food that he eats, the little services & attention that he requires, are charged for most exorbitantly, & with it there is scarcely a grain of comfort to be found in the largest & acknowledged best hotel in the place. They seem to take advantage of the traveller's necessities & with every exertion, honest or otherwise, to fleece his pockets, & render his stay amongst them as uncomfortable as possible. I do not wish to treat them too severely, but justice obliges one to say unreservedly as a finale – that a person visiting St. Louis, Missouri, cannot depend upon an atom of comfort (at the hotels) – he cannot depend upon either his food or lodging being decent, nor upon receiving one of those many attentions which the servants & attendants of similar establishments in our state are hourly rendering him.*]

Nathaniel Wyeth.

NATHANIEL WYETH AND HIS TWO EXPEDITIONS TO THE COLUMBIA RIVER

The Wyeth family had settled in Cambridge near Boston, Massachusetts, in 1645. Nathaniel's grandfather bought an estate in 1751 that included Fresh Pond near Mount Auburn, and Nathaniel's father built a summer resort there known as Fresh Pond Hotel. The lake was the source of the family wealth: in 1824 Nathaniel became the manager of an ice-company that removed the winter ice, packed and stored it, and sold it as far afield as the West Indies. Young Wyeth invented most of the implements and techniques to make the business prosper.

First Expedition, 1832-33

Wyeth's fame now rests on the two business ventures that he made to the Pacific Northwest, the first of which started out in the spring of 1832 and ended, for Wyeth, twenty-one months later when he returned to Cambridge. He had hoped to increase his fortune by fur trading and salmon packing but instead met only with misfortune. Bit by bit all his men deserted or died, and the ship that he had sent around the Horn to the Columbia was wrecked in the Society Islands. The company that he had set up went out of business, losing not only much of Wyeth's money but also funds borrowed from his three brothers.

On the positive side, he had learned how to travel and survive in the West and he had met with many of the people then involved in that part of the country, including trappers and traders William Sublette and Thomas Fitzpatrick; the explorer Bonneville; the Scottish adventurer Captain Stewart; the Hudson's Bay Company factor at Fort Vancouver, John McLoughlin; and other characters. With this in mind, Wyeth was still convinced that his business could flourish in the Oregon, indeed on the overland return trip in August he had already made a contract with Fitzpatrick and William Sublette's brother, Milton, to transport $3000 worth of goods out to the 1834 summer rendezvous, on or before 1st July for the Rocky Mountain Fur Company.

Second Expedition, 1834-36

Wyeth lost no time in raising support for a new company, the Columbia River Fishing and Trading Company. His plan was to supply the trappers and fur traders in the mountains with merchandise in exchange for furs that he would transport further west to the Columbia River. He arranged for the 200-ton brig *May Dacre* to be loaded with trade goods and equipment for establishing a trading post and sent her from the east coast, round the Horn to the Columbia via the Hawaiian Islands. He proposed to send back the furs by sea, together with the full season's supply of salmon that his men would catch and preserve on the Columbia.

This then was the purpose of the expedition. And this was its leader: intelligent, energetic, resourceful, determined, a good superior, and still convinced of the financial rewards to be reaped. It only remained to seek out the animals and men to ease the

We proceeded over a road generally good, a low dry prairie, mostly heavily timbered, the soil underlaid with horizontal strata of limestone, abounding in organic remains, shells, coralines &c., and arrived in the evening at Florisant [Florissant], where we spent the night.

The next day [30th] Captain Wyeth left us for St. Louis, and my companion and myself proceeded on our route. We observed great numbers of the brown, or **sandhill crane**, (*Grus canadensis*,) flying over us; some flocks were so high as to be entirely beyond the reach of vision, while their harsh, grating voices were very distinctly heard. We saw trade goods westwards. Inevitably, the project attracted a few hangers on and various private adventurers among whom may be numbered Nuttall and Townsend, who had no financial interest in the trip and none of the cares of its leader. There was also a missionary party consisting of the Reverend Jason Lee and his nephew Daniel, and three other men as his support. Their intention was to take Christianity to the native tribes living in the Columbia River region. The composition of the band varied almost week by week. Some men stayed behind at Fort Hall, others left to go trapping, and, fearful of the Blackfeet, members of other tribes sometimes joined them for protection. Although numbers at times increased to about 120 only about 70 were working directly for Wyeth.

It might as well be said here rather than later, that this trip was another financial disaster. There was a competing band of traders heading west at the same time, led by William Sublette, and so the journey began at a stiff pace, since Wyeth desperately wanted to get to the rendezvous first. Unfortunately, when Wyeth arrived he found that Sublette had won the race and was bluntly told that the Rocky Mountain Fur Company had been dissolved, it was now a new company and therefore had no contract with him. It was a great setback for Wyeth (and a nuisance for Townsend and Nuttall who would have much preferred to have travelled along at a slower pace).

Wyeth had to dispose of his surplus goods and decided to build a rival trading post further westwards, on the Snake River, the aforementioned Fort Hall. The missionaries, Captain Stewart and a few others pressed on to the Columbia River. Wyeth, the two naturalists and other men followed later once the fort had been completed, arriving at Fort Vancouver on 14th September. The very next day they witnessed the *May Dacre* sailing upriver. It was not perfect timing: she had spent three months under repair in Valparaiso after being damaged by lightning, and as a result had missed the 1834 salmon season on the Columbia.

Nevertheless, his ship brought him his supplies, as well as two coopers and two smiths and he set them to work at once making fish barrels at Fort William, the new trading post that he began to build on Sauvie Island at the mouth of the Willamette River. Wyeth and some of the men spent the winter roaming the area eastwards for furs but met with little success, so he pinned his hopes on the coming salmon season. But the salmon run of 1835 was poor and only half a shipment was ever loaded. Disappointed once again, Wyeth returned overland in 1836 and the following year sold Fort Hall to the Hudson's Bay Company. He returned to the ice business, carried on with it quite successfully, but never tried any more western adventures.

For Wyeth, the second expedition meant a second round of financial setbacks and bitter memories of a run of bad luck. For Oregon, the second expedition brought out its first United States missionaries and settlers. For the Oregon Trail, Fort Hall became an important landmark and staging post. For Townsend and Nuttall, Wyeth provided a convenient means of getting across the Rocky Mountains and allowed the two naturalists to make use of the comings and goings of the *May Dacre* to travel to and fro to the Hawaiian Islands.

several flocks of the same cranes while ascending the Mississippi, several days since. At about noon, we crossed the river on a boat worked by horses, and stopped at a little town called St. Charles.

We find it necessary, both for our comfort and convenience, to travel very slowly, as our feet are already becoming tender, and that we may have an opportunity of observing the country, and collecting interesting specimens. Unfortunately for the pursuits of my companion, the plants (of which he finds a number that are rare and curious) are not yet in flower, and therefore of little use to him. [*He has however collected some & no doubt will soon be able to obtain a considerable number in the requisite state of advancement. We saw this afternoon a number of* **Pileated Woodpeckers** *(Picus pileatus). They were barking the trees in almost every direction, & constantly emitting their loud sonorous calls. I shot several of them & deposited them in my game bag in order to prepare their Skins. They are extremely tenacious of life, & as they fall wounded from their high perches, send forth the most startling & angry screams, & snap with their large & sharp beaks in a very vicious manner at the hand stretched forth to take them. When simply winged, they ran along the ground to the nearest tree & ascend the trunk with great swiftness & invariably escape unless shot a second time. 30th. This afternoon we lost sight of the timbered land – the only production of the prairie being a thick growth of long coarse grass, - the ground is undulating like the long waves of the sea. Before leaving the trees in the morning we saw several* **Red-bellied Woodpeckers** *(Picus carolinus) – The common Red headed species was frequent as also the two least ones, P. villosus & pubescens* **[Hairy and Downy Woodpeckers]**. *We saw also a large species of Squirrel, of a reddish brown color, probably the Sciurus capistratus* **[Eastern Fox Squirrel]** *- & the mounds of the Gopher or Sand rat (Mus bursarius)* **[Plains Pocket Gopher]** *were abundant. Towards evening we observed in the high grass several of the larger Pinnated Grous, - called here Prairie Hens, – (Tetrao cupido)* **[Greater Prairie-Chicken]**. *The common Partridge (Perdix virginianus)* **[Northern Bobwhite]** *appears tolerably abundant, as well as a great variety of our common birds, several species of Sparrow, Cardinals, Doves, Grackles &c &c* (MS Journal).]

Mr. N. and myself are both in high spirits. We travel slowly, and without much fatigue, and when we arrive at a house, stop and rest, take a drink of milk, and chat with those we see. We have been uniformly well treated; the living is good, and very cheap, and at any house at which we stop the inhabitants are sure to welcome us to their hospitality and good cheer. They live comfortably, and without much labor; possess a fruitful and easily tilled soil, for which they pay the trifling sum of one dollar and a quarter per acre; they raise an abundance of good Indian corn, potatoes, and other vegetables; have excellent beef and pork, and, in short, every thing necessary for good, wholesome living.

31st. – The road to-day was muddy and slippery, rendered so by a heavy rain which fell last night. This morning, we observed large flocks of wild pigeons[2] **[Passenger Pigeons]** passing over, and on the bare prairies were thousands of **golden plovers**; the ground was often literally covered with them for acres. I killed a considerable number. They were very fat, and we made an excellent meal of them in the evening. The prairie hen, or pinnated grouse **[prairie-chicken sp.]**, is also very numerous, but in these situations is shy, and difficult to be procured.

[I saw also a flock of **curious little Buntings** *of which I procured several for specimens. It is probably a new species* [identity unknown, see p.48]. *We heard to day for the first time the notes of the Pinnated Grous* **[prairie-chicken sp.]**, *a loud, mournful cooing or booming sound, resembling the syllables Kit-tum-boo, kit-tum-boo-oo, the last very strongly accented & lengthened out. They appear to be most clamorous in the mornings & evenings, large packs collecting together at this season, & their roar may be heard in every direction over the prairies. They are certainly exceedingly abundant here, but with all my skill I have not yet been able to shoot one, as they are extremely shy & vigilant & invariably spring out of distance. The* **Golden Plover** *is said to breed[3] in these prairies – makes no nest, the spot only being indicated by a slight indentation in the grass, & the eggs, - uniformly two, - are spotted* (MS Journal).]

Towards evening we were overtaken by a bluff, jolly looking man on horseback, who, as is usual, stopped, and entered into conversation with us. I saw immediately that he was superior to those we had been accustomed to meet. He did not ply us with questions so eagerly as most, and when he heard that we were naturalists, and were travelling in that capacity, he seemed to take considerable interest in us. He invited us to stop at his house, which was only a mile beyond, and as night was almost upon us, we accepted the invitation with cheerfulness. Upon arriving at his mansion [in or near Hickory Grove], our good host threw wide his hospitable doors, and then with a formal, and rather ultra-dignified politeness, making us a low bow, said, "Gentlemen,

Raccoon.

my name is P. [*Pringle* in MS Journal], and I am very happy of your company." We seated ourselves in a large, and well-furnished parlor. Mr. P. excused himself for a few minutes, and soon returned, bringing in three fine looking girls, whom he introduced as his daughters. I took a particular fancy to one of them, from a strong resemblance which she bore to one of my female friends at home … Anon, supper was brought in. It consisted of pork chops, ham, eggs, Indian bread and butter, tea, coffee, milk, potatoes, *preserved ginger*, and though last, certainly not least in value, an enormous tin dish of plovers, (the contents of my game-bag,) *fricaseed*. Here was certainly a most abundant repast, and we did ample justice to it.

I endeavoured to do the agreeable to the fair ones in the evening, and Mr. N. was monopolized by the father, who took a great interest in plants, and was evidently much gratified by the information my companion gave him on the subject.

The next morning [1st April] when we rose, it was raining, and much had evidently fallen during the night, making the roads wet and muddy, and therefore unpleasant for pedestrians. I confess I was not sorry for this, for I felt myself very comfortably situated, and had no wish to take to the road. Mr. P. urged the propriety of our stopping at least another day, and the motion being seconded by his fair daughter, (my favorite,) it was irresistible.

[*In the afternoon, as the rain had ceased, our host provided a couple of horses which we mounted & rode over the plains for the purpose of observing the Prairie Hens while in the act of emitting the singular sounds for which they are so remarkable. We left my friend N. engaged with a book which occupation he avows to be more agreeable to him than bestriding a bucephalus or carrying a gun. We had ridden about ½ a mile when we saw on a hill covered with low tangled grass about 50 of the birds of which we were in search, & whose* booming *we had heard before we left the house. The greater part of them appeared to be males as we observed almost every individual straining his throat as if in rivalry of his fellow, while the few females that appeared, - & whose note consists of a single loud* cluck, *repeated quickly & at intervals, - were walking around & among their gallants with an appearance of great unconcern, & seemed altogether to forget the dignified & august occasion upon which they had assembled. Being on horseback they allowed us to approach them sufficiently near for me to observe their motions & note their actions with some precision* [probably with a telescope, see p.58]. *The male when at rest stands with his head erected, his tail depressed & his wings rather drooping; his neck also appears to be elongated & straight from the breast, without curvature. When alarmed by the approach of danger, he squats low in the grass & only his head is seen, & he appears thus to hold himself in readiness to spring when his safety requires it. But it is when he is engaged in making his strange* booming *that he appears most interesting & at the same time most fantastical. In the operation the head & tail are first very much elevated, the body is held horizontally, & the feet are the whole time in constant & quick motion. A deep inspiration is first taken & the sides of the neck are observed to be protruded to a size larger than a man's fist, the denuded balls assuming a deep reddish orange color* [an identification feature of the **Lesser Prairie-Chicken**[4]]. *The neck is at this moment lowered so as nearly to touch the ground, & the air being suddenly exhaled causes the neck to move with a quick motion forwards & upwards. When the air is inhaled, the first syllable of the note is produced, during the instant of distension the second is heard, & the exhalation causes the third. The wing-like appendages at the back of the neck are at these times thrown straight outwards from its sides, & depressed upon the back immediately as the sound escapes. On our return I killed a very beautiful adult Marsh Hawk (*Falco cyaneus*)* [**Northern Harrier**]. *They are said to be common on these prairies & feed upon mice* (MS Journal).]

[*On the open prairies we observed the Prairie Hens or Pinnated Grous – very numerous – They are fully as large as our common heavy fowls and are most excellent eating. Ducks of several species are also exceedingly abundant in every stream we pass, so that amongst them all we are living in clover as we travel along. We load ourselves with game & when we stop have them cooked, at the same time furnishing a most pleasant meal for the family. It falls to my lot however to do all the shooting, as with all my persuasion I c*[d]. *not induce N. to carry his gun, nor even to make use of mine at any time. We are now weather bound at the Post Office – it is raining most violently, & I very much fear we*

Greater Prairie-Chicken, from Mark Catesby's *The Natural History of Carolina* (1731 - 43).

Wild Turkey by Titian Ramsay Peale (from C L Bonaparte's 1824 - 28 edition of Wilson's *American Ornithology*).

shall not be able to prosecute our journey until tomorrow *shall not be able to prosecute our journey until tomorrow* (JKT to C. Townsend, 1st April 1834. ANSP. Envelope marked "Hickory Grove Mo / Apl 2ⁿᵈ. ")]

On the following morning [2nd or 3rd] the sun was shining brightly, the air was fresh and elastic, and the roads tolerably dry, so that there was no longer any excuse for tarrying, and we prepared for our departure. Our good host, grasping our hands, said that he had been much pleased with our visit, and hoped to see us again, and when I bid good bye to the pretty Miss P., I told her that if I ever visited Missouri again, I would go many miles out of my way to see her and her sisters. Her reply was unsophisticated enough. "Do come again, and come in May or June, for then there are plenty of prairie hens, and you can shoot as many as you want, and you must stay a long while with us, and we'll have nice times; good bye; I'm so sorry you're going."

April 4th. – I rose this morning at daybreak, and left Mr. N. dreaming of weeds, in a little house at which we stopped last night, and in company with a long, lanky boy, (a son of the poor widow, our hostess,) set to moulding bullets in an old iron spoon, and preparing for deer hunting. The boy shouldered a rusty rifle, that looked almost antediluvian, and off we plodded to a thicket, two miles from the house. We soon saw about a dozen fine deer, and the boy, clapping his old fire-lock to his shoulder, brought down a beautiful doe at the distance of a full hundred yards. Away sprang the rest of the herd, and I crept round the thicket to meet them. They soon came up, and I fired my piece at a large buck, and wounded the poor creature in the leg; he went limping away, unable to overtake his companions; I felt very sorry, but consoled myself with the reflection that he would soon get well again.

We then gave up the pursuit, and turned our attention to the turkies [**Wild Turkey**], which were rather numerous in the thicket. They were shy, as usual, and, when started from their lurking places, ran away like deer, and hid themselves in the underwood. Occasionally, however, they would perch on the high limbs of the trees, and then we had some shots at them. In the course of an hour we killed four, and returned to the house, where, as I expected, Mr. N. was in a fever at my absence, and after a late, and very good breakfast, proceeded on our journey.

We find in this part of the country less timber in the same space than we have yet seen, and when a small belt appears, it is a great relief, as the monotony of a bare prairie becomes tiresome. [*I shot in the course of this day a number of birds, among which was a very beautifully plumaged Tell-tale Godwit, (Totanus melanoleucus)* [**Greater Yellowlegs**]. *I have seen several of these birds since leaving St. Louis. They frequent the small puddles & rivulets in the prairie bottoms, busily engaged in searching for worms & aquatic insects. They often swim voluntarily, & with great ease & buoyancy. The Prairie Hens are becoming scarce. We have seen very few to day, although we heard a number before we left the house in the morning* (MS Journal).]

Towards evening we arrived at Loutre Lick. Here there is a place called a *Hotel*. A Hotel, forsooth! A pig-stye would be a more appropriate name. Every thing about it was most exceedingly filthy and disagreeable, but no better lodging was to be had, for it might not be proper to apply for accommodation at a private house in the immediate vicinity of a public one. They gave us a wretched supper, not half so good as we had been accustomed to, and we were fain to spend the evening in a comfortless, unfurnished, nasty bar-room, that smelt intolerably of rum and whiskey, to listen to the profane conversation of three or four uncouth individuals, (among whom were the host and his brother,) and to hear long and disagreeably minute discussions upon horse-racing, gambling, and other vices equally unpleasant to us.

The host's brother had been to the Rocky Mountains, and soon learning our destination, gave us much unsought for advice regarding our method of journeying; painted in strong colors the many dangers and difficulties which we must encounter, and concluded by advising us to give up the expedition. [*Upon his asking who commanded the party & being told that it was Capt. Wyeth – "Why" – said he, "that's the fellow that I went with",*

& then commenced a string of vituperation against Capt. W's character which completely overthrew the last atom of my fast ebbing patience. I told him at once that Capt. Wyeth was my friend, I knew him for a worthy man, & I would not hear him abused by any one. I remarked also that we went to that house in order to seek repose, & that it was unfair to intrude conversation upon us unasked. The ruffian made some grumbling reply, and left us in quiet and undisturbed possession of our bench (MS Journal).] We had a miserable time that night. The only spare bed in the house was so intolerably filthy that we dared not undress, and we had hardly closed our eyes before we were assailed by swarms of a vile insect, (the very name of which is offensive,) whose effluvia we had plainly perceived immediately as we entered the room.

It is almost needless to say, that very early on the following morning [5th], after paying our reckoning, and refusing the landlord's polite invitation to "*liquorize*," we marched from the house, shook the dust from our feet, and went elsewhere to seek a breakfast. [*"I'd rather sup on rats & dine upon my nails", than take another meal in that detestable hovel* (MS Journal).]

Soon after leaving we came to a deep and wide creek, and strained our lungs for half an hour in vain endeavors to waken a negro boy who lived in a hut on the opposite bank, and who, we were told, would ferry us over. He came out of his den at last, half naked and rubbing his eyes to see who had disturbed his slumbers so early in the marning …

[*We observed to day several flocks of Carolina Parrots*[5], *(Psittacus Carolinensis)* [**Carolina Parakeets**] *& I killed seven of them in a few minutes. They are in beautiful plumage. Inhabit the low grounds in the neighbourhood of streams, I believe almost exclusively, & here feed upon the seeds of the Buttonwood, (Platanus [Cephalanthus] occidentalis). They go in flocks of from 10 to 25 & while flying keep a regular loud screaming, somewhat resembling the chattering note of the Red-headed Woodpecker. Immediately on alighting they become quiet, & sit very still upon the trees, seeming to wish to be as close together as possible. I found these birds, (the first I have ever seen,) quite tame, they allowed me to make three shots at them, & after each, moved from one tree to another at very short distances. I could easily have killed as many more, but I felt compelled to refrain, as I had already procured enough for examination, & I do not intend to prepare any here as I shall probably find them abundant nearer the upper Settlements. Saw to day the two Warblers Sylvia Americana* [**Northern Parula**] *& varia* [**Black-and-white Warbler**], *& shot in the afternoon one of the large Fox Squirrels of this country, (Sciurus capistratus?)* [**Eastern Fox Squirrel**]. *We were considerably annoyed to day by having to ford several deep creeks, their being no bridges across them* (MS Journal).]

Wild Turkey hunt, from an 1871 print.

6th. – Soon after we started this morning, we were overtaken by a stage which was going to Fulton, seven miles distant, and as the roads were somewhat heavy, we concluded to make use of this convenience. The only passengers were three young men from the far west, who had been to the eastward purchasing goods, and were then travelling homeward. Two of them evidently possessed a large share of what is called *mother wit*, and so we had jokes without number. Some of them were not very refined, and perhaps did not suit the day very well, (it being the Sabbath,) yet none of them were really offensive, but seemed to proceed entirely from an exuberance of animal spirits.

In about an hour and a half we arrived at Fulton, a pretty little town, and saw the villagers in their holiday clothes parading along to church. The bell at that moment sounded, and the peal gave rise to many reflections. It might be long ere I should hear the sound of the "church-going bell" again. I was on my way to a far, far country, and I did not know that I should ever be permitted to revisit my own … These reflections were soon checked, however. We took a light lunch at the tavern where we stopped. I shouldered my gun, Mr. N. his stick and bundle, and off we trudged again, westward, ho! We soon lost sight of the prairie entirely, and our way lay through a country thickly covered with heavy timber, the roads very rough and stony, and we had frequently to ford the creeks on our route, the late freshets having carried away the bridges.

Our accommodation at the farm houses has generally been good and comfortable, and the inhabitants obliging, and anxious to please. They are, however, exceedingly inquisitive, propounding question after question … The first question generally asked, is, "where do you come from, gentlemen?" We frame our answer somewhat in the style of Dr. Franklin. "We come from Pennsylvania; our names, Nuttall and Townsend; we are travelling to Independence on foot, for the purpose of seeing the country to advantage, and we intend to proceed from thence across the mountains to the Pacific. Have you any mules to sell?" The last clause generally changes the conversation, and saves us trouble … [*I shot to day a large Fox Squirrel* [**Eastern Fox Squirrel**], *several of the common Grey Squirrels (Sciurus Carolinensis) & a few Pigeons* [**Passenger Pigeon**, or possibly **Mourning Dove**]. *We found the grey Squirrels very abundant in some places, particularly in the low bottoms along water courses, but we have uniformly observed that where these abound, the large Fox Squirrel is never seen.*

7th. The Grey Squirrels continue exceedingly abundant on our route which still lies through the woods. Early in the morning & towards evening they may be seen skipping on almost every tree (MS Journal).] On last Christmas day, at a squirrel hunt in this neighborhood, about thirty persons killed the astonishing number of *twelve hundred*, between the rising and setting of the sun.

This may seem like useless barbarity, but it is justified by the consideration that all the crops of corn in the country are frequently destroyed by these animals. This extensive extermination is carried on every year, and yet it is said that their numbers do not appear to be much diminished.

About mid-day on the 7th, we passed through a small town called Columbia, and stopped in the evening at Rocheport, a little village on the Missouri river. We were anxious to find a steam-boat bound for Independence, as we feared we might linger too long upon the road to make the necessary preparations for our contemplated journey.

On the following day [8th], we crossed the Missouri, opposite Rocheport, in a small skiff. The road here, for several miles, winds along the bank of the river, amid fine groves of sycamore and Athenian poplars, then stretches off for about three miles, and does not again approach it until you arrive at Boonville. It is by far the most hilly road that we have seen, and I was frequently reminded, while travelling on it, of our Chester county. [MS Journal for 8th April includes an amusing description of Nuttall on horseback, see pp.26-27.] We entered the town of Boonville early in the afternoon, and took lodgings in a very clean, and respectably kept hotel. I was much pleased with

Boonville. It is the prettiest town I have seen in Missouri; situated on the bank of the river, on an elevated and beautiful spot, and overlooks a large extent of lovely country. The town contains two good hotels, (but no *grog shops*, properly so called,) several well-furnished stores, and five hundred inhabitants …

[*We learned, upon enquiry, that there were two steamboats some miles below, destined for Independence, & we concluded therefore to wait until they should arrive. It gave us an opportunity of writing letters to our friends, & of making several short excursions around this very pretty neighbourhood. I found the Parroquets* [**Carolina Parakeets**] *rather numerous & so very gentle that I soon procured as many as I had occasion for. I think them very stupid birds, as they seem to have no idea of escaping a threatened danger …* (MS Journal).] We saw here vast numbers of the beautiful parrot of this country, (the *Psittacus carolinensis*.) They flew around us in flocks, keeping a constant and loud screaming, as though they would chide us for invading their territory; and the splendid green and red of their plumage glancing in the sunshine, as they whirled and circled within a few feet of us, had a most magnificent appearance. They seem entirely unsuspicious of danger, and after being fired at, only huddle closer together, as if to obtain protection from each other, and as their companions are falling around them, they curve down their necks, and look at them fluttering upon the ground, as though perfectly at a loss to account for so unusual an occurrence. It is a most inglorious sort of shooting; down right, cold-blooded murder.

On the afternoon of the 9th, a steamboat arrived, on board of which we were surprised and pleased to find Captain Wyeth, and our "*plunder.*" We embarked immediately, and soon after, were puffing along the Missouri, at the rate of seven miles an hour. When we stopped in the afternoon to "wood," we were gratified by a sight of one of the enormous catfish of this river and the Mississippi, weighing full sixty pounds [*Ictalurus sp.*, perhaps Channel Catfish *Ictalurus punctatus*]. It is said, however that they are sometimes caught of at least double this weight. They are excellent eating, coarser, but quite as good as the common small catfish of our rivers. There is nothing in the scenery of the river banks to interest the traveller particularly. The country is generally level and sandy, relieved only by an occasional hill, and some small rocky acclivities …

[*11th. – This morning when we stopped to* <u>*wood*</u>, *Mr. N. called my attention to a pretty little bird which he heard singing on the shore, & which upon subsequent examination I discovered to be the* **Louisiana Water Thrush** (<u>*Turdus ludovicianus*</u> *of Audubon). Its note was very lively & agreeable, but somewhat monotonous, & emitted only while the bird was perched upon the trees: when* *he descended to the rivulet, - which he did frequently – he was altogether silent. I afterwards found this bird rather numerous in the neighbourhood of Independence. It generally affected the little forest streams & was sometimes seen on the banks of the river – when on the ground its motions were rapid & the tail frequently jetted. While singing on the trees, the tail was depressed & the head thrown very much back. It evidently breeds here, as was clearly proved by the organic development of several which I examined* (MS Journal).]

Arrival at Independence

March [= April] 20th. – On the morning of the 14th, we arrived at Independence landing, and shortly afterwards, Mr. N. and myself walked to the town, three miles distant. The country here is very hilly and rocky, thickly covered with timber, and no prairie within several miles.

The site of the town is beautiful, and very well selected, standing on a high point of land, and overlooking the surrounding country, but the town itself is very indifferent; the houses, (about fifty,) are very much scattered, composed of logs and clay, and are low and inconvenient. There are six or eight stores here, two taverns, and a few tipling houses. As we did not fancy the town, nor the society that we saw there, we concluded to take up our residence at the house on the landing until the time of starting on our journey. We were very much disappointed in not being able to purchase any mules here, all the salable ones having been bought by Santa Fee traders, several weeks since. Horses, also, are rather scarce, and are sold at higher prices than we had been taught to expect, the demand for them at this time being greater than usual. Mr. N. and myself have, however, been so fortunate as to find five excellent animals amongst the hundreds of wretched ones offered for sale, and have also engaged a man to attend to packing our loads, and perform the various duties of our camp.

The men of the party, to the number of about fifty, are encamped on the bank of the river, and their tents whiten the plain for the distance of half a mile. I have often enjoyed the view on a fine moonlight evening from the door of the house, or perched upon a high hill immediately over the spot. The beautiful white tents, with a light gleaming from each, the smouldering fires around them, the incessant hum of the men, and occasionally the lively notes of a bacchanalian song, softened and rendered sweeter by distance. I probably contemplate these and similar scenes with the more interest, as they exhibit the manner in which the next five months of my life are to be spent.

We have amongst our men, a great variety of dispositions. Some who have not been accustomed to the kind of life they are to lead in future, look forward to it with eager delight, and talk of stirring incidents and hair-breadth 'scapes. Others who are more experienced seem to be as easy and unconcerned about it as a citizen would be in contemplating a drive of a few miles into the country. Some have evidently been reared in the shade, and not accustomed to hardships, but the majority are strong, able-bodied men, and many are almost as rough as the grizzly bears, of their feats upon which they are fond of boasting.

During the day the captain keeps all his men employed in arranging and packing a vast variety of goods for carriage. In addition to the necessary clothing for the company, arms, ammunition, &c., there are thousands of trinkets of various kinds, beads, paint, bells, rings, and such trumpery, intended as presents for the Indians, as well as objects of trade with them. The bales are usually made to weigh about eighty pounds, of which a horse carries two.

I am very much pleased with the manner in which Captain W. manages his men. He appears admirably calculated to gain the good will, and ensure the obedience of such a company …

Independence, Missouri, in 1853.

Do not be misled by the wagons in this picture - none were used on the Wyeth expedition.

We were joined here by Mr. Milton Sublette, a trader and trapper of some ten or twelve years' standing. It is his intention to travel with us to the mountains, and we are very glad of his company, both on account of his intimate acquaintance with the country, and the accession to our band of about twenty trained hunters, "true as the steel of their tried blades," who have more than once followed their brave and sagacious leader over the very track which we intend to pursue. He appears to be a man of strong sense and courteous manners, and his men are enthusiastically attached to him.

Five missionaries, who intend to travel under our escort, have also just arrived. The principal of these is a Mr. Jason Lee, (a tall and powerful man, who looks as though he were well calculated to buffet difficulties in a wild country,) his nephew, Mr. Daniel Lee, and three younger men of respectable standing in society [Cyrus Shepard, C.M. Walker and A.L. Edwards], who have arrayed themselves under the missionary banner, chiefly for the gratification of seeing a new country, and participating in strange adventures.

My favourites, the birds, are very numerous in this vicinity, and I am therefore in my element. Parroquets are plentiful in the bottom lands, the two species of squirrel are abundant, and rabbits, turkies, and deer are often killed by our people. [*Yesterday I was much delighted in receiving a letter from my sisters. I have been anxiously looking for one ever since my arrival, & had almost begun to despair of seeing it. It gives a good acct. of our family & friends … We have observed within a few days several species of the Warbling Flycatchers or Vireos, (flavifrons, noveboracensis & olivaceus)* [**Yellow-throated, White-eyed** and **Red-eyed Vireo**]. *The Sylvias* [**Parulidae**] *have also arrived, & we have seen the icterocephalus, striata, varia, americana* [**Chestnut-sided, Blackpoll, Black-and-white Warblers**, and **Northern Parula**] *& others. It is somewhat singular that the Common Robin (Turdus migratorius)* [**American Robin**] *does not breed here, nor is it even seen during the summer months, although it is said to be common in the winter. The little town of Independence has within a few weeks been the scene of a brawl which at one time threatened to be attended with serious consequences, but which was happily settled without bloodshed. It had been for a considerable time the stronghold of a sect of fanatics, called Mormons, or Mormonites …*

On the afternoon of the 26th, Capt. W., Mr. N. & myself rode out to the edge of the prairie 8 miles from our lodgings where all our men have been encamped for the last two days & prepared for a start on the next day, but we discovered in the morning that several horses from our band had escaped during the night, which unavoidably detained us twenty four hours longer (MS Journal).]

1. Townsend later had a wee chuckle about his aversion to St Louis: *"How well I remember the feelings of hearty indignation under the influence of which I wrote this philipic. Now I can laugh at the whole affair as the foolish complaint of an unweaned boy. I have seen something of the busy world since then, & I cd. now visit St. Louis, & submit to these little inconveniences without murmuring"* (MS Journal, undated footnote beneath 29th March 1834).

2. In 1834 the Passenger Pigeon was still spectacularly abundant in the forests of eastern North America. Although then numbered in billions, large scale commercial shooting and netting from the 1840s through to a peak in the 1860s and 1870s reduced their numbers to a critically low level, and led to their rapid extinction. The last known individual died in Cincinnati Zoo in 1914.

3. Golden Plovers do not breed this far southwards. In Missouri they are late spring migrants and confusion has arisen because other species had started to nest at the time that they pass through. The story possibly refers to the partially completed clutches of the Upland Sandpiper which usually lays four eggs, or to the smaller clutches of two to three eggs of the Mountain Plover (which breeds further to the west) or may even refer to the two egg clutches of the Common Nighthawk that are laid on the bare ground.

4. In 1834 the two species of prairie-chicken had not yet been separated by taxonomists. Townsend's vivid description of adults in display and the observation that the air sacs were "deep reddish orange" indicates that the birds he was watching were Lesser Prairie-Chickens. This is an important (and evidently overlooked) early breeding record for central Missouri, a state at the north-eastern extremity of its former range. The Greater Prairie-Chicken has yellow-orange air sacs, and Townsend probably also saw and heard this species elsewhere in the vicinity, on wetter parts of the prairies; indeed, his description of the booming he heard on 31st March could well be that of the Greater Prairie-Chicken. Both species are now very much reduced in numbers, particularly the Lesser Prairie-Chicken which has long been absent from Missouri and is now confined to just a few parts of Colorado, Kansas, New Mexico, Oklahoma and Texas.

5. The Carolina Parakeet was the only representative of the parrot family in North America. It was slaughtered in vast numbers largely because of its destructive habits, being able to eat or destroy almost any type of bud, fruit or grain. It was also killed for its brilliant plumage, used as target practice, and captured alive for the pet trade. By the 1890s there were few parakeets left and the last known individual died in Cincinnati Zoo in 1918. The extirpation of this beautiful and engaging bird must rate as one of the saddest of all extinctions.

Townsend's first consignment of specimens

In a letter from Independence, Townsend gave the following instructions to his father:

"The box, when you receive it, you will please open & send the plants as per directions & the birds &c. to the Acad. Nat. Sciences; some of them will probably be returned to you, & these I wish presented to Dr. Michener with my best respects" (JKT to C. Townsend, 21st April 1834. ANSP).

Two days later he wrote a letter to Dr McEuen at the Academy of Natural Sciences of Philadelphia letting him know that the box of specimens was on its way (JKT to T. McEuen, 23rd April 1834. ANSP):

Independence, Mo

Ap. 23rd 1834

" Dear Dr.

My family will send you a few birds that I shot on my way to this place. I am sorry there are not more of them but my mode of travelling from St. Louis to this place, (on foot) prevented me carrying a number of rare ones that I shot. I found the Prairie Hens [prairie-chickens] immensely numerous some miles below & could easily have prepared some but I expected to find them as abundant here & concluded not to encumber myself with them; – I have been very much disappointed therefore in not being able to find one in the neighbourhood. They are said to inhabit the prairies about 8 miles above, but since our arrival here I have been so constantly engaged in preparing for the journey that I have not had time to look after them.

I have seen a number of Sandhill Cranes but always flying high – they are said to alight at night in heavy marshes in the neighbourhood of streams, but are seldom seen resting during the day. I have offered a reward for the capture of one, but none have yet been brought me.

The small Finch which I send is new to me. It is, I believe a true Emberiza, perhaps allied to the lapponica [Lapland Longspur]. Inhabits the prairies in large flocks is very shy & sings when rising like the Anthus [pipit sp.]. The specimen is a very indifferent one but I shall no doubt be able to find others. The larger Finch I am also unacquainted with, but think it possible it may be the F. leucophrys [White-crowned Sparrow] in imperfect plumage, – Should it prove to be new however, you may, if you think proper, read the inclosed description before the Academy.

The Small Woodpecker may or may not be the P. varius [Yellow-bellied Sapsucker] – it resembles it in general appearance but I think the plumage differs widely from it, the tongue of this bird is <u>rounded</u> at the end & <u>fimbriated</u>, whereas that of P. varius is, if I recollect, sagittate. The few birds that may remain after selecting what may be wanted for the Acad. I should be obliged if you would return to my father who has been directed as to the disposition of them – those that I shall send in future you will please retain for me, or dispose of them, when it can be done advantageously.

If Dr. Morton shall have returned please remember me particularly to him.

Mr. Nuttall sends his Complts.

Very truly yours

J.K. Townsend

I open my note to make a remark upon another Finch that I have just killed. I am not acquainted with it. May it not be one of the dubious species of Pennant.

It is marked No. 8.   J.K.T   ▶

The identity of the specimens

It was most likely eight species sent to the Academy rather than eight specimens. We know this from the way that Townsend operated his catalogue, where each species that he came across was given a new number. It is hard to be sure about the identity of all the species, especially as a thorough search for specimens of the three finches at the Academy was unsuccessful (Rehn 1930). Nevertheless, it is still worth considering what is known about them by analysing Townsend's letter about the specimens and other MS material:

- *"the Parrots"*

From a letter written by his sister Hannah, we learn that the box of specimens arrived safely and that it must have contained three Carolina Parakeets: "The box for the Acad. arrived safely – we unpacked it, gave the plants to Dr. Allen (who seemed very grateful for them, & desired that his love & thanks might be given when we wrote,) & the remainder sent to the Acady. It was <u>very</u> hard to part with every thing; I thought if we had only a <u>dried plant</u> it would be better than nothing – Some time after Father went to the Acad. thinking as there were 3 of the Parrots they might not be anxious to preserve all" (H. Townsend to JKT, 29th November 1835 [= 1834]. ANSP). The letter went on to say that Dr Pickering of the Academy had given one of the Carolina Parakeets to Ezra Michener, who was with the Townsend family at the time, and Michener had stuffed it and set it on the mantelpiece for them.

- *"small Finch ... allied to lapponica ... sings when rising like the Anthus"*

Rehn (1930) suggests that this was the Chestnut-collared Longspur. However, Townsend says that he collected only one specimen of this longspur and Audubon's original watercolour of it is inscribed with the date "May 28th 1834" (see Part 2 Chapter 2, note 5). Moreover, when speaking of the Chestnut-collared Longspur Townsend says that it inhabits the "prairies of the Platte" an area that he had not yet reached when the *"small Finch"* was sent back. The "small Finch" is perhaps the same as the *"curious little Buntings of which I procured several for specimens"* that he came across further down the Missouri and thought was probably a new species (MS Journal for 31st March 1834). Their identity is unknown, but they could have been Smith's Longspur *Calcarius pictus* or, less likely, McCown's Longspurs *Calcarius mccownii*, two species that do not appear later in Townsend's catalogue.

- *"larger Finch may be the F. leucophrys in imperfect plumage"*

The description of this bird that Townsend enclosed with the box of specimens is believed to be the one reproduced on page 50, that can still be found at the Academy of Natural Sciences of Philadelphia:

Unfortunately, Townsend's description (of Harris's Sparrow) was never published. Instead, the species was first officially described by Nuttall[a] who dealt with it rather more abruptly in the 1840 edition of his *Manual of Ornithology of the United States and of Canada*, p. 555 (see page 49). Nuttall picked up, as Townsend had done, on its mournful note by calling it the Mourning Finch. Neither Townsend's specimen[b] (nor Nuttall's type specimen, if there was one) can now be found (Rehn 1930; ANSP database 2005). The species acquired its common name Harris's Sparrow from Audubon's later name of Harris's Finch – a tribute to his friend and patron Edward Harris (see e.g. Mearns and Mearns 1992: 216-224).

MOURNING FINCH.

(Fringilla *querula, Nobis.)

Spec. Charact. — Face and chin black; cheeks and nape cinereous; throat spotted with dusky; belly white; above varied with black and brownish; two faint white bars on the wings.

We observed this species, which we at first took for the preceding, a few miles to the west of Independence, in Missouri, towards the close of April. It frequents thickets, uttering early in the morning, and occasionally at other times, a long, drawling, faint, monotonous and solemn note *te dē dē dē.* We heard it again on the 5th of May, not far from the banks of Little Vermilion, of the Kansa.

Harris's Sparrow, plate 484 from Audubon's octavo edition of *The Birds of America*, painted from specimens collected during Audubon's 1843 expedition up the Missouri.

The first published description of Harris's Sparrow (top left) as it appeared in Nuttall's *Manual of Ornithology* (1840).

• *"The Small Woodpecker may or may not be the P. varius"*

This must have been the Yellow-bellied Sapsucker *Sphyrapicus varius* as Townsend suggested. There is no other similar species inhabiting that area.

• *"another Finch ... May it not be one of the dubious species of [Thomas] Pennant"*

Identity unknown, perhaps one of the longspurs, or a sparrow or bunting.

Assuming the collection did indeed contain eight species, the contents may be summarised as containing specimens of Carolina Parakeet, Harris's Sparrow, Yellow-bellied Sapsucker, two "finches" that may have been longspurs, sparrows or buntings, and three unknown species. None of the Townsend specimens from this early consignment can now be traced[c].

a. By a quirk of ornithological history, Harris's Sparrow was found in the same vicinity (near the junction of the Platte and Missouri Rivers) by the explorer Maximilian, Prince of Wied, on 13th May 1834, just a few days after Townsend and Nuttall had first seen the species. Wied's description was not published until 1841, a year after Nuttall (see Harris 1919: 185-186).

b. There is no basis for Harry Harris's comment that Townsend did not mention Harris's Sparrow in the *Narrative* because it was Nuttall's discovery (see Harris 1919: 184; Jobanek 1999: 280).

c. The only known Carolina Parakeet specimen attributed to Townsend is at the Smithsonian Institution (USNM 12272). Although the specimen label records that the parakeet was collected in "Illinois" in 1834, the date seems to have been added later. Townsend's MS Journal for 1834 mentions Carolina Parakeets when he passed through Missouri but not in Illinois. All the parakeets from the first consignment of specimens went to Philadelphia, so it is more likely that USNM 12272 was collected in 1840 when Townsend made a short trip westward that included time in Illinois (see pages 287-288).

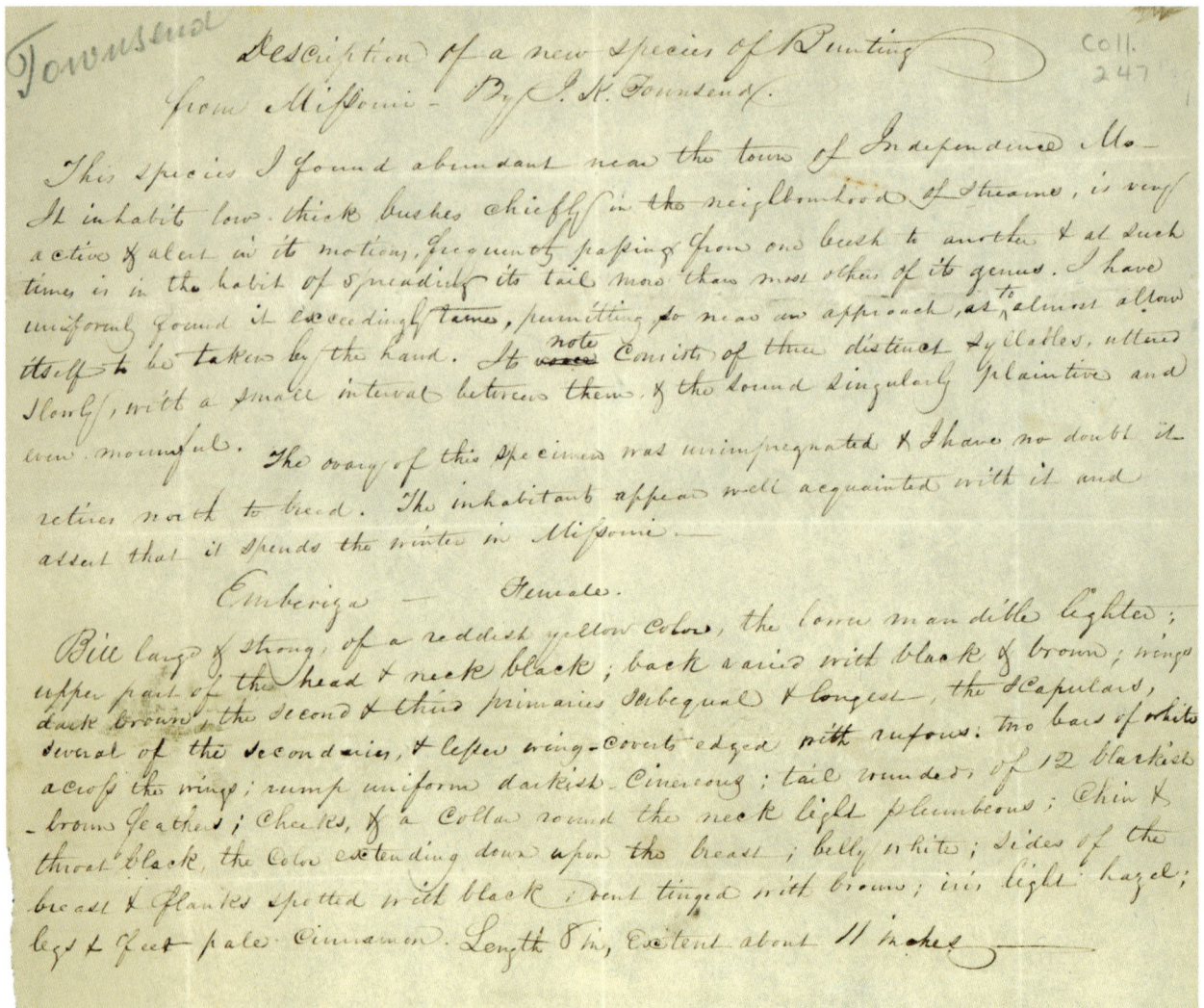

Description of a new species of Bunting

from Missouri – By J.K. Townsend.

This species I found abundant near the town of Independence Mo.
It inhabits low thick bushes chiefly in the neighbourhood of streams, is very
active & alert in its motions, frequently passing from one bush to another & at such
times is in the habit of spreading its tail more than most others of its genus. I have
uniformly found it exceedingly tame, permitting so near an approach, as to almost allow
itself to be taken by the hand. Its note consists of three distinct syllables, uttered
slowly, with a small interval between them, & the sound singularly plaintive and
even mournful.

The ovary of this specimen was unimpregnated & I have no doubt it
retires north to breed. The inhabitants appear well acquainted with it and
assert that it spends the winter in Missouri –

Emberiza - Female.

Bill large & strong, of a reddish yellow color, the lower mandible lighter;
upper part of the head & neck black; back varied with black & brown; wings
dark brown, the second & third primaries subequal & longest, the scapulars,
several of the secondaries, & lesser wing-coverts edged with rufous: two bars of white
across the wings; rump uniform darkish-cinereous; tail rounder, of 12 blackish
-brown feathers; cheeks, & a collar round the neck light plumbeous; chin &
throat black, the color extending down upon the breast; belly white; sides of the
breast & flanks spotted with black; vent tinged with brown; iris light hazel;
legs & feet pale cinnamon. Length 8 ins, Extent about 11 inches –

Independence to Laramie's Fork

"None but a naturalist can appreciate a naturalist's feelings – his delight amounting to ecstasy – when a specimen such as he has never before seen, meets his eye, and the sorrow and grief which he feels when he is compelled to tear himself from a spot abounding with all that he has anxiously and unremittingly sought for."

Townsend's *Narrative*, 31st May 1834.

On the 28th of April, at 10 o'clock in the morning, our caravan, consisting of seventy men[1], and two hundred and fifty horses, began its march; Captain Wyeth and Milton Sublette took the lead, Mr. N. and myself rode beside them; then the men in double file, each leading, with a line, two horses heavily laden, and Captain Thing (Captain W.'s assistant) brought up the rear. The band of missionaries, with their horned cattle, rode along the flanks.

I frequently sallied out from my station to look at and admire the appearance of the cavalcade, and as we rode out from the encampment, our horses prancing, and neighing, and pawing the ground, it was altogether so exciting that I could scarcely contain myself. Every man in the company seemed to feel a portion of the same kind of enthusiasm; uproarious bursts of merriment, and gay and lively songs, were constantly echoing along the line. We were certainly a most merry and happy company. What cared we for the future? We had reason to expect that ere long difficulties and dangers, in various shapes, would assail us, but no anticipation of reverses could check the happy exuberance of our spirits.

Our road lay over a vast rolling prairie, with occasional small spots of timber at the distance of several miles apart, and this will no doubt be the complexion of the track for some weeks.

In the afternoon we crossed the *Big Blue* river at a shallow ford [= Blue River, just a few miles west of Independence]. Here we saw a number of beautiful yellow-headed troopials, *(Icterus zanthrocephalus,)* **[Yellow-headed Blackbirds]** feeding upon the prairie in company with large flocks of black birds **[Common Grackles]**, and like these, they often alight upon the backs of our horses. **[Yellow-headed Blackbirds]** *were feeding on the prairie in company with the Cow Buntings* **[Brown-headed Cowbirds]**, *Grakles &c., keeping constantly near the cattle* (MS Journal).]

29th. – … Camping out to-night is not so agreeable as it might be, in consequence of the ground being very wet and muddy, and our blankets (our only bedding) thoroughly soaked; but we expect to encounter greater difficulties than these ere long, and we do not murmur.

[*30th. The Troopials still continue abundant about our cattle whenever we encamp & turn them out, & like the Cow Buntings frequently alight on their backs. The note of this species is a singularly hoarse & unmusical croak. I killed to day a female. It was plumaged very differently from the male* (MS Journal).]

A description of the formation of our camp may, perhaps, not be amiss here. The party is divided into messes of eight men, and each mess is allowed a separate tent. The captain of a mess, (who is generally an "old hand," *i.e.* an experienced forester, hunter or trapper,) receives each morning the rations of pork, flour, &c. for his people, and they choose one of their body as cook for the whole. Our camp now consists of nine messes, of which Captain W.'s forms one, although it only contains four persons besides the cook [Wyeth, Nuttall, Townsend, and presumably Captain Thing].

Yellow-headed Blackbird.

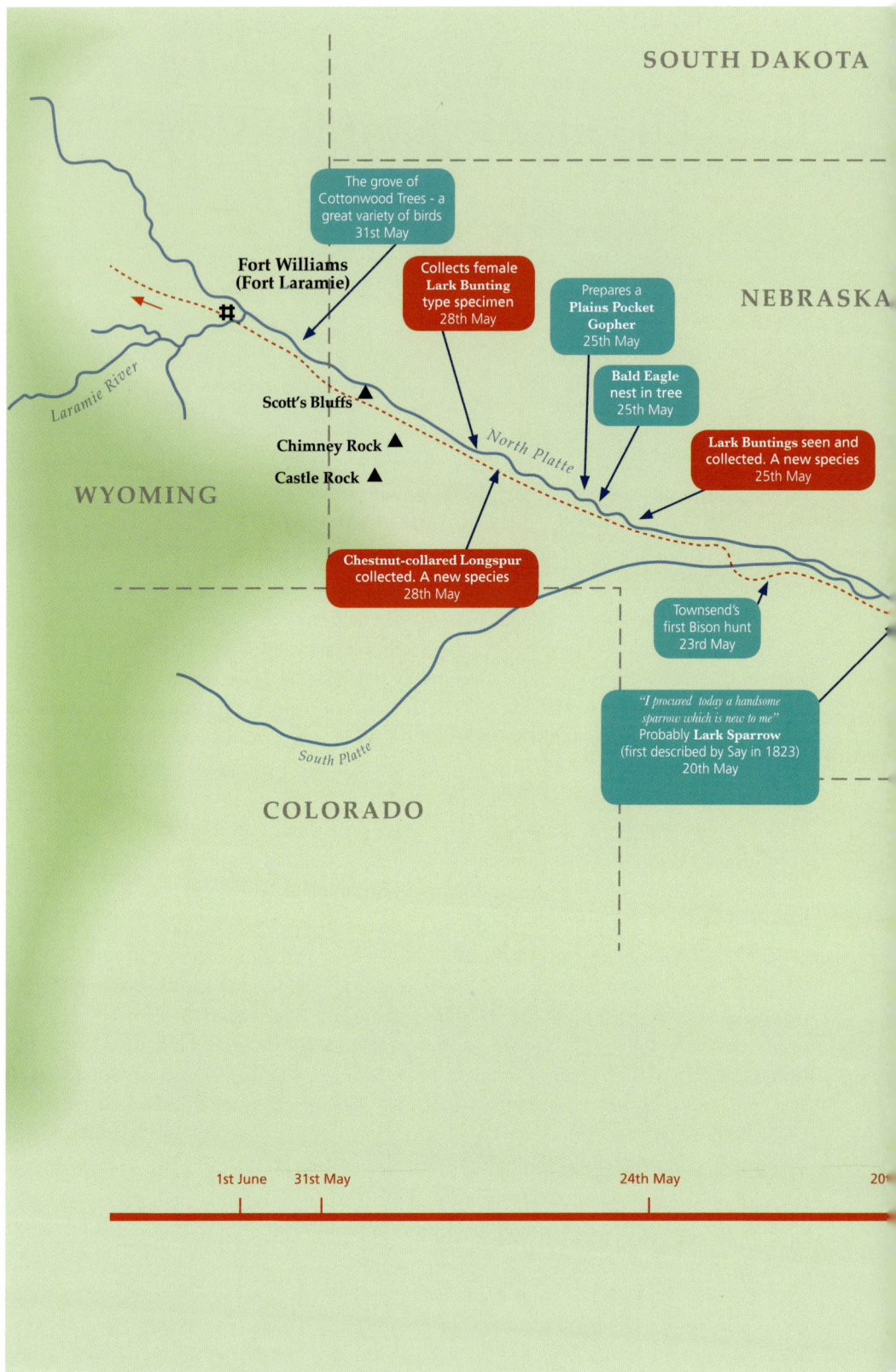

SOUTH DAKOTA

NEBRASKA

The grove of
Cottonwood Trees - a
great variety of birds
31st May

Fort Williams
(Fort Laramie)

Collects female
Lark Bunting
type specimen
28th May

Prepares a
Plains Pocket
Gopher
25th May

Bald Eagle
nest in tree
25th May

Lark Buntings seen and
collected. A new species
25th May

Scott's Bluffs

Chimney Rock

Castle Rock

North Platte

WYOMING

Chestnut-collared Longspur
collected. A new species
28th May

Laramie River

Townsend's
first Bison hunt
23rd May

"I procured today a handsome
sparrow which is new to me"
Probably Lark Sparrow
(first described by Say in 1823)
20th May

South Platte

COLORADO

1st June   31st May   24th May   20

Map 3 - Independence to Laramie's Fork, 28th April - 1st June 1834.

52

IOWA

MISSOURI

KANSAS

Sandhill Cranes
Great Blue Herons
Long-billed Curlews
seen after arrival on Platte
18th May

First sighting
of **Gray Wolves**
and Pronghorns
15th May

Harris's
Sparrow
heard again
*"not far from the bank of
Little Vermilion"*

Nuttall says that
Harris's Sparrow
was seen
*"a few miles west of
Independence"*

Yellow-headed
Blackbirds and
Brown-headed Cowbirds
abundant near Kansas
Indian villages

Swallowtail Kites
seen
1st May

● **Independence**

River Platte

Big Blue River

Little Blue River

Missouri River

Vermilion Creek

Kansas River

Blue River

May 17th May   5th May   28th April

1834

0   50   100 miles

Swallow-tailed Kite, from Nuttall's *Manual of Ornithology* (1840).

When we arrive in the evening at a suitable spot for an encampment, Captain W. rides round a space which he considers large enough to accommodate it, and directs where each mess shall pitch its tent. The men immediately unload their horses, and place their bales of goods in the direction indicated, and in such manner, as in case of need, to form a sort of fortification and defence. When all the messes are arranged in this way, the camp forms a hollow square, in the centre of which the horses are placed and staked firmly to the ground. The guard consists of from six to eight men, and is relieved three times each night, and so arranged that each gang may serve alternate nights. The captain of a guard (who is generally also the captain of the mess) collects his people at the appointed hour, and posts them around outside the camp in such situations that they may command a view of the environs, and be ready to give the alarm in case of danger ...

May 1st. – On rising this morning, and inquiring about our prospects for breakfast, we discovered that the cook of our mess (a little, low-browed, ill-conditioned Yankee) had decamped in the night, and left our service to seek for a better. He probably thought the duties too hard for him, but as he was a miserable cook, we should not have much regretted his departure, had he not thought proper to take with him an excellent rifle, powder-horn, shot-pouch, and other matters that did not belong to him ...

[*I saw this morning several Swallow-tailed Hawks, Falco furcatus)* **Swallow-tailed Kites**[2]] *sailing over the prairie. It is a very beautiful & graceful species, & I felt a strong inclination to obtain one of them, but was unable to kill them as they kept quite out of reach of my shot* (MS Journal).]

We encamped this evening on a small branch of the Kanzas river. As we approached our stopping place, we were joined by a band of Kanzas Indians, (commonly called *Kaw* Indians.) ... The whole of the following day we remained in camp, trading buffalo robes, *apishemeaus* [rush mats], &c., of the Indians ... I have observed, that among the Indians now with us, none but the chief uses the pipe. He smokes the article called *kanikanik*, – a mixture of tobacco and the dried leaves of the Poke plant (*Phytolacca decandra*) [Pokeweed *P. decandra*]. I was amused last evening by the old chief asking me in his impressive manner, (first by pointing with his finger towards the sunset, and then raising his hands high over his head,) if I was going to the mountains. On answering him in the affirmative, he depressed his hands and passed them around his head in both directions, then turned quickly away from me, with a very solemn and significant *ugh*! He meant, doubtless, that my brain was turned; in plain language, that I was a fool ... [*One of our men brought me to day a Marmot (Arctomys)* **[ground squirrel** or **prairie dog**[3]] *which I think a new species* (MS Journal).]

[Nuttall noted that "On the 2nd of May, in our western tour across the continent, around the Kansa Indian Agency, we now saw abundance of the Yellow-headed Troopial **[Yellow-headed Blackbird]**, associated with the Cow-Bird **[Brown-headed Cowbird]**. They kept wholly on the ground in companies, the males, at this time, by themselves. In loose soil they dig into the earth with their bills in quest of insects and larvae, are very active, straddle about with a quaint gait, and now and then, in the manner of the Cow Bird, whistle out with great effort, a chuckling note sounding like *ko-kukkle-'ait*, often varying into a straining squeak, as if using their utmost endeavor to make some kind of noise in token of sociability. Their music, is however, even inferior to the harsh note of the Cow Bird" (Nuttall 1840:187-188).]

We were on the move early next morning [3rd May], and at noon arrived at the Kanzas river, a branch of the Missouri. This is a broad and not very deep stream, with the water dark and turbid, like that of the former. As we approached it, we saw a number of Indian lodges, made of saplings driven into the ground, bent over and tied at top, and covered with bark and buffalo skins. These lodges, or wigwams, are numerous on both sides of the river. As we passed them, the inhabitants, men, women, and children, flocked out to see us, and almost prevented our progress by their eager

54

Major Long's Expedition to the Rocky Mountains, 1819-20

Some of the habitats that Townsend and Nuttall were now traversing, and others that they were about to encounter, had been studied by the naturalists attached to Stephen Long's expedition (and Nuttall himself had been much further up the Missouri, as far as Fort Union in 1811). Long's expedition was an elaborate but under-funded affair, equipped with small troublesome steamboats, some military support and a team of scientists and artists. The journey down the Ohio River to St Louis, then up the Missouri River via St Charles and Boonville was tedious and frustrating and they fell far short of their target, the Yellowstone. The winter was spent on the Missouri at Engineer Cantonment, located above the mouth of the Platte and about five miles below Council Bluffs. In the spring of 1820, as the twenty-two man party set off westwards with new targets, Thomas Say was serving as zoologist, Titian Peale as assistant naturalist, Samuel Seymour as artist, and Dr Edwin James as surgeon, botanist and geologist. They followed the Platte River through present day Nebraska, taking the south fork down into Colorado to the edge of the Rocky Mountains. After briefly exploring the Front Range they headed back, in two teams, one down the Arkansas River and the other down the Canadian River, both finishing up at Fort Smith, Arkansas (where Nuttall had been in April 1819). In the last few days of the journey, the zoological results were compromised to an unknown extent when three deserters went off with items that included Say's notebooks.

The botanical and vertebrate discoveries and observations were published in James's *Account of an Expedition from Pittsburgh to the Rocky Mountains* (1823), and in a variety of other publications by other authors. Say included most of the insects in his three-volume *American Entomology* (1824-28). The quantity and quality of the scientific research was disappointing, largely as a result of the difficulties inherent in any such undertaking, but compounded by their rapid pace and the fact that it was such a severe drought year. Even so, it is worth remarking that in Townsend's ten week dash through what is now Missouri, Kansas and Nebraska, he managed to find several birds new to science, in an area in which Say and Peale had both spent an entire year.

The bird and mammal specimens from the Long Expedition were deposited at Peale's Museum in Philadelphia where Townsend would have seen them, so he had a fair idea of what to expect as he travelled westwards from Independence and he too collected most of the species listed below.

New birds and mammals described from specimens from the Long Expedition:

Dusky Grouse *Dendragapus obscurus* (Say)

Long-billed Dowitcher *Limnodromus scolopaceus* (Say)

Band-tailed Pigeon *Patagioenas fasciata* (Say)

Say's Phoebe *Sayornis saya* (Bonaparte)

Western Kingbird *Tyrannus verticalis* Say

Rock Wren *Salpinctes obsoletus* (Say)

Orange-crowned Warbler *Vermivora celata* (Say)

Lazuli Bunting *Passerina amoena* (Say)

Lark Sparrow *Chondestes grammacus* (Say)

Yellow-headed Blackbird *Xanthocephalus xanthocephalus* (Bonaparte)

Lesser Goldfinch *Carduelis psaltria* (Say)

Least Shrew *Cryptotis parva* (Say)

Short-tailed Shrew *Blarina brevicauda* (Say)

Small-footed Myotis *Myotis subulatus* (Say)

Coyote *Canis latrans* Say

Swift Fox *Vulpes velox* (Say)

Golden-mantled Ground Squirrel *Citellus lateralis* (Say)

Colorado Chipmunk *Eutamias quadrivittatus* (Say)

Thomas Say, from a miniature reproduced in Weiss and Zeigler's *Thomas Say* (1931).

Titian Ramsay Peale, in the uniform of a Long Expedition Naturalist. By C W Peale, 1819.

Golden-mantled Ground Squirrel - a Thomas Say discovery.

South Platte.

greetings. Our party stopped on the bank of the river, and the horses were unloaded and driven into the water. They swam beautifully, and with great regularity, and arrived safely on the opposite shore, where they were confined in a large lot, enclosed with a fence. After some difficulty, and considerable detention, we succeeded in procuring a large flat-bottomed boat, embarked ourselves and goods in it, and landed on the opposite side near our horse pen, where we encamped …

We remained in camp the whole of next day [4th], and traded with the Indians. … [Sunday 4th May 1834: "No regard paid by any of Capt. W's company to the Sabbath and but little in appearance by ours for we were obliged as we judged to do things which we should not have done under other circumstances" (Jason Lee 1916: 119).]

On the 20th [also *20th* in MS Journal = error for 5th] in the morning, we packed our horses and rode out of the Kaw settlement, leaving the river immediately, and making a N.W. by W. course – and the next day came to another village of the same tribe, consisting of about thirty lodges, and situated in the midst of a beautiful level prairie.

[**Harris's Sparrow** was heard again "on the 5th May, not far from the banks of Little Vermilion" (Nuttall 1840: 555).]

The Indians stopped our caravan almost by force, and evinced so much anxiety to trade with us, that we could not well avoid gratifying them. We remained with them about two hours, and bought corn, moccasins and leggings in abundance … We encamped in the evening on a small stream called Little Vermillion creek, where we found an abundance of excellent catfish, exactly similar to those of the Schuylkill river. Our people caught them in great numbers. Here we first saw the large ravens, (Corvus corax.) [**Common Raven**]. They hopped about the ground all around our camp; and as we left it, they came in, pell-mell, croaking, fighting, and scrambling for the few fragments that remained. [*The Raven (Corvus corax) is plentiful here, & the Yellow-headed Troopials, Cow Buntings & Grakles follow us from camp to camp* (MS Journal).]

8th. – This morning Mr. [Milton] Sublette left us to return to the settlements. He has been suffering for a considerable time with a fungus in one of his legs, and it has become so much worse since we started, in consequence of irritation caused by riding, that he finds it impossible to proceed. His departure has thrown a gloom over the whole camp. We all admired him for his amiable qualities, and his kind and obliging disposition. For myself, I had become so much attached to him, that I feel quite melancholy about his leaving us.

[9th] The weather is now very warm, and there has been a dead calm all day, which renders travelling most uncomfortable. We have frequently been favored with fresh breezes, which make it very agreeable, but the moment these fail us we are almost suffocated with intense heat. Our rate of travelling is about twenty miles per day, which, in this warm weather, and with heavily packed horses, is as much as we can accomplish with comfort to ourselves and animals.

On the afternoon of the next day [10th], we crossed a broad Indian trail, bearing northerly, supposed to be about five days old, and to have been made by a war party of Pawnees. We are now in the country traversed by these Indians, and are daily expecting to see them, but Captain W. seems very desirous to avoid them, on account of their well known thieving propensities, and quarrelsome disposition …

The little streams of this part of the country are fringed with a thick growth of pretty trees and bushes, and the buds are now swelling, and the leaves expanding, to "welcome back the spring." The birds, too, sing joyously amongst them, **grosbeaks, thrushes, and buntings**, a merry and musical band. I am particularly fond of sallying out early in the morning, and strolling around the camp. The light breeze just bends the tall tops of the grass on the boundless prairie, the birds are commencing their matin carollings, and all nature looks fresh and beautiful. The horses of the camp are lying comfortably on their sides, and seem, by the glances which they give me in passing, to know that their hour of toil is approaching, and the patient kine are ruminating in happy unconsciousness.

[*The beautiful White-crowned Bunting (Fringilla leucophrys)* [**White-crowned Sparrow**] *is numerous on these prairies. Inhabits the bushes near the water. The Black-throated & Savannah Finches (F. americana, & savanna)* [**Dickcissel** and **Savannah Sparrow**] *are also common, & the delightful notes of the* **Yellow-breasted Chat** *&* **Wood Thrush** *(Turdus melodius) are heard all around us* (MS Journal).]

11th. – We encountered some rather serious difficulties to-day in fording several wide and deep creeks, having muddy and miry bottoms. Many of our horses, (and particularly those that were packed,) fell into the water, and it was with the greatest difficulty and labor that they were extricated ... [*I saw this evening three or four* **Rose-breasted Grosbeaks** *– (Fringilla ludoviciana). They sang delightfully* (MS Journal).]

12th. – Our scouts came in this morning with the intelligence that they had found a large trail of white men, bearing N.W. We have no doubt that this is Wm. Sublette's party, and that it passed us last evening. They must have travelled very rapidly to overtake us so soon, and no doubt had men ahead watching our motions. It seems rather unfriendly, perhaps, to run by us in this furtive way, without even stopping to say good morning, but Sublette is attached to a rival company, and all stratagems are deemed allowable when interest is concerned. It is a matter of some moment to be the first at the mountain rendezvous, in order to obtain the furs brought every summer by the trappers ...

Pronghorns.

15th. – We saw to-day several large white wolves [**Gray Wolf**], and two herds of antelopes [**Pronghorn**]. The latter is one of the most beautiful animals I ever saw. When full grown, it is nearly as large as a deer. The horns are rather short, with a single prong near the top, and an abrupt backward curve at the summit like a hook. The ears are very delicate, almost as thin as paper, and hooked at the tip like the horns. The legs are remarkably light and beautifully formed, and as it bounds over the plain, it seems scarcely to touch the ground, so exceedingly light and agile are its motions. This animal is the *Antelope furcifer* of zoologists [now *Antilocapra americana*], and inhabits the western prairies of North America exclusively. [Jason Lee (1916: 123) reported that the hunters killed two Pronghorns on the 15th.] The ground here is strewn with great quantities of buffalo bones; the skulls of many of them in great perfection. I often thought of my friend Doctor M. [= Samuel Morton] and his *golgotha*, while we were kicking these fine specimens about the ground. We are now travelling along the banks of the Blue river, - a small fork of the Kanzas. The grass is very luxuriant and good, and we have excellent and beautiful camps every night.

Sandhill Cranes.

This morning [16th] a man was sent ahead to see W. Sublette's camp, and bear a message to him, who returned in the evening with the information that the company is only one day's journey beyond, and consists of about thirty-five men. We see his deserted camps every day, and, in some cases, the fires are not yet extinguished. It is sometimes amusing to see the wolves lurking like guilty things around these camps seeking for the fragments that may be left; as our party approaches, they sneak away with a mean, hang-dog air which often coaxes a whistling bullet out of the rifle of the wayfarer.

[*17th. – In about an hour's march from our encampment of last night we arrived at the head of the Blue, then deviated a little from our general course, making about N.N.W., & at 4 o'clock arrived on a branch of the Platte river, about a hundred yards in width. The stream is here not deep, perhaps four or five feet, the water dark & turbid like the Missouri, & having nearly the same taste. Wolves and Antelopes are in great abundance. Our hunters frequently kill the latter & bring the meat to camp. On the afternoon of the next day [18th], we encamped on the main Platte. It is from 1½ to 2 miles in width, very shoal, large sand flats & small verdant islands appearing in every part. We saw great numbers of* **Sandhill Cranes**, *Herons, (Ardea herodias)* [**Great Blue Heron**] *&* **Long-billed Curlews** *stalking about through the shallow water & searching for their aquatic food. The prairie is here as level as a race course, not the slightest undulation appearing throughout the whole extent of vision in a N. & Westerly direction, but to the Eastward of the river & about 5 miles from it, is seen a range of high bluffs or sand banks stretching away to the S.E. until they are lost in the far distance. The ground here is in many places encrusted with an impure salt, which by the taste appears to be a combination of the sulphate & muriate of Soda. We are now within about three days journey of the usual haunts of the Buffalo, & our men look forward to our arrival amongst them with considerable anxiety. The* **Cliff Swallow** *(Hirundo fulva) is numerous, & associates here with our three common species* [i.e. **Bank, Tree and Barn Swallow**. Northern Rough-winged Swallow probably also present.] *The* **Bobolink** *or Rice Bird (Icterus agripennis) is also in considerable numbers* (MS Journal).]

The next morning [19th], we perceived two men on horseback, at a great distance; and upon looking at them with our telescope, discovered them to be Indians … Captain W. rode out alone and joined them, while the party proceeded slowly on its way. In about fifteen minutes he returned with the information that they were of the tribe called Grand Pawnees. They told him that a war party of their people, consisting of fifteen hundred warriors, was encamped about thirty miles below; and the captain inferred that these men had been sent to watch our motions, and ascertain our place of encampment; he was therefore careful to impress upon them that we intended to go but a few miles further, and pitch our tents … we went on, and on, without slackening our pace, until 12 o'clock at night. We then called a halt on the bank of the river, made a hasty meal, threw ourselves down in our blankets, without pitching the tents, and slept soundly for three hours. We were then aroused, and off we went again, travelling steadily the whole day [20th], making about thirty-five miles, and so got quite clear of the Grand Pawnees. [Wyeth (1899: 72) says he rode about 50 miles on 20th May.]

The antelopes are very numerous here. There is not half an hour during the day in which they are not seen, and they frequently permit the party to approach very near them. This afternoon, two beautiful does came bounding after us, bleating precisely like sheep. The men imitated the call, and they came up to within fifty yards of us, and stood still; two of the hunters fired, and both the poor creatures fell dead. We can now procure as many of these animals as we wish, but their flesh is not equal to common venison, and is frequently rejected by our people. A number are, however, slaughtered every day, from mere wantonness and love of killing, the greenhorns glorying in the sport, like our striplings of the city, in their annual murdering of robins and sparrows.

20th. – This afternoon, we came in sight of a large *gang* of the long-coveted buffalo [**American Bison**]. They were grazing on the opposite side of the Platte, quietly as domestic cattle, but as we neared them, the foremost *winded* us, and started back, and the whole herd followed in the wildest confusion, and were soon out of sight. There must have been many thousands of them. Towards evening, a large band of **elk** came towards us at full gallop, and passed very near the party. The appearance of these animals produced a singular effect upon our horses, all of which became restive, and about half the loose ones broke away, and scoured over the plain in full chase after the elk. Captain W. and several of his men went immediately in pursuit of them, and returned late at night, bringing the greater number. Two have, however, been lost irrecoverably. [*I procured to day a handsome sparrow which is new to me. It is clearly allied to the Lark Finch* [Lark Sparrow] *(F. gramaca of Prince Bonaparte)* (MS Journal). It probably was a **Lark Sparrow**[4]. Townsend said: "This species inhabits several hundred miles of the Platte country in great numbers … It affects generally the low bushes of Wormwood (*Artemesia*), from the summit of which it pours forth a variety of pretty notes. It appears to be a very pugnacious species …" (Audubon 1840-44, 3:145, where this text is included in error under "Brown Song-Finch", see Stone 1906: 312).]

Elk.

The day following [21st], we saw several small herds of buffalo on our side of the river. Two of our hunters started out after a large bull that had separated himself from his companions, and gave him chase on fleet horses.

Away went the buffalo, and away went the men, hard as they could dash; now the hunters gained upon him, and pressed him hard; again the enormous creature had the advantage, plunging with all his might, his terrific horns often ploughing up the earth as he spurned it under him. Sometimes he would double, and rush so near the horses as almost to gore them with his horns, and in an instant would be off in a tangent, and throw his pursuers from the track. At length the poor animal came to bay, and made some unequivocal demonstrations of combat; raising and tossing his head furiously, and tearing up the ground with his feet. At this moment a shot was fired. The victim trembled like an aspen, and fell to his knees, but recovering himself in an instant, started again as fast as before. Again the determined hunters dashed after him, but the poor bull was nearly exhausted, he proceeded but a short distance and stopped again. The hunters approached, rode slowly by him, and shot two balls through his body with the most perfect coolness and precision … This is the most common mode of killing the buffalo, and is practiced very generally by the travelling hunters; many are also destroyed by approaching them on foot, when, if the bushes are sufficiently dense, or the grass high enough to afford concealment, the hunter, – by keeping carefully to leeward of his game, – may sometimes approach so near as almost to touch the animal …

Lark Sparrow (detail from plate CCCXC).

Towards evening, on rising a hill, we were suddenly greeted by a sight which seemed to astonish even the oldest amongst us. The whole plain, as far as the eye could discern, was covered by one enormous mass of buffalo. Our vision, at the very least computation, would certainly extend ten miles, and in the whole of this great space, including about eight miles in width from the bluffs to the river bank, there was apparently no vista in the incalculable multitude …

22nd. – On walking into our tent last night at eleven o'clock, after the expiration of the first watch, (in which I had served as a supernumerary, to prevent the desertion of the men,) and stooping to lay my gun in its usual situation near the head of my pallet, I was startled by seeing a pair of eyes, wild and bright as those of a tiger, gleaming from a dark corner of the lodge, and evidently directed upon me. My first impression, was that a wolf had been lurking around the camp, and had entered the tent in the prospect of finding meat. My gun was at my shoulder instinctively, my aim was directed between the eyes, and my finger pressed the trigger. At that moment a tall Indian sprang before me with a loud *wah!* seized the gun, and elevated the muzzle above my head; in another instant, a second Indian was by my side, and I saw his keen knife glitter as it left the scabbard. I had not time for thought, and was struggling with all my might with the first savage for the recovery of my weapon, when Captain W., and the other inmates of the tent were aroused, and the whole matter was explained, and set at rest in a moment. The Indians were chiefs of the tribe of Pawnee Loups, who had come with their young men to shoot buffalo: they had paid an evening visit to the captain, and as an act of courtesy had been invited to sleep in the tent … But the excitement of the scene through which I had just passed, effectually banished repose. I frequently directed my eyes towards the dark corner, and in the midst of the shapeless mass which occupied it, I could occasionally see the glittering orbs of our guest shining amidst the surrounding obscurity. At length fatigue conquered watchfulness, and I sank to sleep, dreaming of Indians, guns, daggers, and buffalo.

Upon rising the next morning, all had left the tent: the men were busied in cooking their morning meal; kettles were hanging upon the rude cranes, great ribs of meat were roasting before the fires, and loading the air with fragrance, and my dreams and midnight reveries, and apprehensions of evil, fled upon the wings of the bright morning …

23rd. – When we rose this morning, not a single buffalo, of the many thousands that yesterday strewed the plain, was to be seen. It seemed like magic. Where could they have gone? I asked myself this question again and again, but in vain. At length I applied to [Paul] Richardson, who stated that they had gone to the bluffs, but for what reason he could not tell … He and [Nicholas] Sandsbury (another hunter) were then about starting on a hunt to supply the camp, and I concluded to accompany them; Mr. Lee, the missionary, also joined us, and we all rode off together …We had not ridden a mile before we entered upon a plain of sand of great extent, and observed ahead vast clouds of dust rising and circling in the air as though a tornado or a whirlwind were sweeping over the earth …

For myself, I strolled on for nearly an hour, leading my horse, and peering over every hill, in the hope of finding a buffalo within range, but not one could I see that was sufficiently near; and when I attempted the stealthy approach which I had seen Richardson practice with so much success, I felt compelled to acknowledge my utter insufficiency. I had determined to kill a buffalo, and as I had seen it several times done with so much apparent ease, I considered it a mere moonshine matter, and thought I could compass it without difficulty; but now I had attempted it, and was grievously mistaken in my estimate of the required skill. I had several times heard the guns of the hunters, and felt satisfied that we should not go to camp without meat, and was on the point of altering my course to join them, when, as I wound around the base of a little hill, I saw about twenty buffalo lying quietly on the ground within thirty yards of me. Now was my time. I took my picket from my saddle, and fastened my horse to the ground as quietly as possible, but with hands that almost failed to do their office, from my excessive eagerness and trembling anxiety. When this was completed, I crawled around the hill again, almost suspending my breath from fear of alarming my intended victims, until I came again in full view of the unsuspecting herd. There were so many fine animals that I was at a loss which to select; those nearest me appeared small and poor, and I therefore settled my aim upon a huge bull on the outside. Just then I was attacked with the "*bull fever*" so dreadfully, that for several minutes I could not shoot. At length, however, I became firm and steady, and pulled my trigger at exactly the right instant. Up sprang the herd like lightning, and away they scoured, and my bull with them. I was vexed, angry, and discontented; I concluded that I could never kill a buffalo, and was about to mount my horse and ride off in despair, when I observed that one of the animals had stopped in the midst of his career. I rode towards him, and sure enough, there was my great bull trembling and swaying from side to side, and the clotted gore hanging like icicles from his nostrils. In a few minutes after, he fell heavily upon his side, and I dismounted and surveyed the unwieldy brute, as he panted and struggled in the death agony.

American Bison.

When the first ebullition of my triumph had subsided, I perceived that my prize was so excessively lean as to be worth nothing, and while I was exerting my whole strength in a vain endeavor to raise the head from the ground for the purpose of removing the tongue, the two hunters joined me, and laughed heartily at my achievement. Like all inexperienced hunters, I had been particular to select the largest bull in the gang, supposing it to be the best, (and it proved, as usual, the poorest,) while more than a dozen fat cows were nearer me, either of which I might have killed with as little trouble.

… It was now past mid-day; the weather was very warm, and the atmosphere was charged with minute particles of sand, which produced a dryness and stiffness of the mouth and tongue, that was exceedingly painful and distressing. Water was now the desideratum, but where was it to be found? … Soon afterwards, a bull was killed, and we all assembled around the carcass to assist in the manipulations. The animal was first raised from his side where he had lain, and supported upon his knees, with his hoofs turned under him; a longitudinal incision was then made from the nape, or anterior

"The extensive and apparently interminable green plains .."

base of the hump, and continued backward to the loins, and a large portion of the skin from each side removed; these pieces of skin were placed upon the ground, with the under surface uppermost, and the *fleeces*, or masses of meat, taken from along the back, were laid upon them. These fleeces, from a large animal, will weigh, perhaps, a hundred pounds each, and comprise the whole of the hump on each side of the vertical processes, (commonly called the *hump ribs*,) which are attached to the vertebra. The fleeces are considered the choice parts of the buffalo, and here, where the game is so abundant, nothing else is taken, if we except the tongue, and an occasional marrow bone.

This, it must be confessed, appears like a useless and unwarrantable waste of the goods of Providence; but when are men economical, unless compelled to be so by necessity? ...

But to return to ourselves. We were all suffering from excessive thirst, and so intolerable had it at length become, that Mr. Lee and myself proposed a gallop over to the Platte river, in order to appease it; but Richardson advised us not to go, as he had just thought of a means of relieving us, which he immediately proceeded to put in practice. He tumbled our mangled buffalo over upon his side, and with his knife opened the body, so as to expose to view the great stomach, and still crawling and twisting entrails. The good missionary and myself stood gaping with astonishment, and no little loathing, as we saw our hunter plunge his knife into the distended paunch, from which gushed the green and gelatinous juices, and then insinuate his tin pan into the opening, and by depressing its edge, strain off the water which was mingled with its contents.

Richardson always valued himself upon his politeness, and the cup was therefore first offered to Mr. Lee and myself, but it is almost needless to say that we declined ... [Jason Lee (1916: 125-126) also remembered the drink: "It consisted of water drawn from the paunch of the buffaloe ... but it was too thick with the excrement to please my fancy though they affirmed with oaths that it was very good".]

Before we left the spot, however, Richardson induced me to taste the blood which was still fluid in the heart, and immediately as it touched my lips, my burning thirst, aggravated by hunger, (for I had eaten nothing that day,) got the better of my abhorrence; I plunged my head into the reeking ventricles, and drank until forced to stop for breath. I felt somewhat ashamed of assimilating myself so nearly to the brutes, and turned my ensanguined countenance towards the missionary who stood by, but I saw no approval there: the good man was evidently attempting to control his risibility, and so I smiled to put him in countenance; the roar could no longer be restrained, and the missionary laughed until the tears rolled down his cheeks ...

When we arrived at the camp in the evening, and I enjoyed the luxury of a hearty draft of water, the effect upon my stomach was that of a powerful emetic: the blood was violently ejected without nausea, and I felt heartily glad to be rid of the disgusting encumbrance. I never drank blood from that day.

On the morning of the 24th of May we forded the Platte river, or rather its south fork, along which we had been travelling during the previous week. On the northern side, we found the country totally different in its aspect. Instead of the extensive and apparently interminable green plains, the monotony of which had become so wearisome to the eye, here was a great sandy waste, without a single green thing to vary and enliven the dreary scene. It was a change, however, and we were therefore enjoying it, and remarking to each other how particularly agreeable it was, when we were suddenly assailed by vast swarms of most ferocious little black gnats; the whole atmosphere seemed crowded with them, and they dashed into our faces, assaulted our eyes, ears, nostrils, and mouths, as though they were determined to bar our passage through their territory. These little creatures were so exceedingly minute that, singly, they were scarcely visible; and yet their sting caused such excessive pain, that for the rest of the day our men and horses were rendered almost frantic, the former bitterly imprecating,

North Platte.

and the latter stamping, and kicking, and rolling in the sand, in tremendous, yet vain, efforts to rid themselves of their pertinacious little foes. It was rather amusing to see the whole company with their handkerchiefs, shirts, and coats, thrown over their heads, stemming the animated torrent, and to hear the greenhorns cursing their tormentors, the country, and themselves, for their foolhardiness in venturing on the journey. When we encamped in the evening, we built fires at the mouths of the tents, the smoke from which kept our enemies at a distance, and we passed a night of tolerable comfort, after a day of most peculiar misery.

The next morning [25th] I observed that the faces of all the men were more or less swollen, some of them very severely, and poor Captain W. was totally blind for two days afterwards. [25th. – *Our course this morning lay along the edge of some very high bluffs abounding in limestone & swamp cedars & skirting the river – here I killed a very beautiful new bird* (JKT to C. Townsend, 27th June 1834). This was probably the **Lark Bunting**[5] - see pages 64 - 65.] We made a noon camp to-day on the north branch or fork of the river, and in the afternoon travelled along the bank of the stream. In about an hour's march, we came to rocks, precipices, and cedar trees, and although we anticipated some difficulty and toil in the passage of the heights, we felt glad to exchange them for the vast and wearisome prairies we had left behind. Soon after we commenced the ascent, we struck into an Indian path very much worn, occasionally mounting over rugged masses of rock, and leaping wide fissures in the soil, and sometimes picking our way over the jutting crags, directly above the river. On top of one of the stunted and broad spreading cedars, a **bald eagle** had built its enormous nest; and as we descended the mountain, we saw the callow young lying within it, while the anxious parents hovered over our heads, screaming their alarm.

In the evening we arrived upon the plain again; it was thickly covered with ragged and gnarled bushes of a species of wormwood, (*Artemesia,*) [commonly known as sagebrush] which perfumed the air, and at first was rather agreeable. The soil was poor and sandy, and the straggling blades of grass which found their way to the surface were brown and withered. Here was a poor prospect for our horses; a sad contrast indeed to the rich and luxuriant prairies we had left. On the edges of the little streams, however, we found some tolerable pasture, and we frequently stopped during the day to bait our poor animals in these pleasant places.

Lark Bunting male at nest

LARK BUNTING

Townsend wrote about his discovery in his *Narrative* preferring a different name:

"PRAIRIE FINCH *Fringilla bicolor*, (Townsend) … This very pretty and distinct species inhabits a portion of the platte country, east of the first range of the Rocky Mountains. It appears to be strictly gregarious. Feeds upon the ground, along which it runs swiftly, like the grass finch (F. graminea,) [Vesper Sparrow *Pooecetes gramineus*] to which it is somewhat allied. As the large flocks, (consisting often from sixty to a hundred,) were started from the ground by our caravan in passing, the piebald appearance of the males and females promiscuously intermingled, presented a curious, but by no means unpleasing effect. While the flock is engaged in feeding, the males are frequently observed to rise suddenly to a considerable height, and poising themselves over their companions, with their wings in constant and rapid motion, they become nearly stationary. In this situation, they pour forth a number of very lively and sweetly modulated notes, and at the expiration of about a minute, descend to the ground, and course about as before. I never observed this bird west of the Black Hills [Laramie Mountains, Wyoming]" (Townsend 1839c: 346-347).

The type specimens of Lark Bunting (ANSP 23951 male, ANSP 23952 female) collected by Townsend. These were almost certainly the specimens that Audubon used in the preparation of the above watercolour (NYHS 1863.17.390), that also contains a Song Sparrow (bottom left).

Lark Finch.
FRINGILLA GRAMMACA, Say.
Male 1.

Prairie Finch.
FRINGILLA BICOLOR, Townsend.
2. Male. 3. Female.

Brown Song Sparrow.
FRINGILLA CINEREA, Gmel.
4. Male.

Plate CCCXC from *The Birds of America*, closely matching the original watercolour but with the addition of a Lark Sparrow (at top) also collected by Townsend.

Western Meadowlark and flowers of short-grass prairie (at right).

We observed here several species of small marmots, (*Arctomys*,) [**ground squirrels** or **prairie dogs**] which burrowed in the sand, and were constantly skipping about the ground in front of our party. The short rattlesnake of the prairies was also abundant [**Western Rattlesnake** *Crotalus viridis*], and no doubt derived its chief sustenance from foraging among its playful little neighbors. Shortly before we halted this evening, being a considerable distance in advance of the caravan, I observed a dead gopher, (*Diplostoma*,) [**Plains Pocket Gopher**] – a small animal about the size of a rat, with large external cheek pouches, – lying upon the ground; and near it a full grown rattlesnake, also dead. The gopher was yet warm and pliant, and had evidently been killed but a few minutes previously; the snake also gave evidence of very recent death, by a muscular twitching of the tail, which occurs in most serpents, soon after life is extinct. It was a matter of interest to me to ascertain the mode by which these animals were deprived of life. I therefore dismounted from my horse, and examined them carefully, but could perceive nothing to furnish even a clue. Neither of them had any external or perceptible wound. The snake had doubtless killed the quadruped, but what had killed the snake? There being no wound upon its body was sufficient proof that the gopher had not used his teeth, and in no other way could he cause death.

I was unable to solve the problem to my satisfaction, so I pocketed the animal to prepare its skin, and rode on to the camp.

The birds thus far have been very abundant. There is a considerable variety, and many of them have not before been seen by naturalists. As to the plants, there seems to be no end to them, and Mr. N. is finding dozens of new species daily. In the other branches of science, our success has not been so great, partly on account of the rapidity and steadiness with which we travel, but chiefly from the difficulty, and almost impossibility, of carrying the subjects. Already we have cast away all our useless and superfluous clothing, and have been content to mortify our natural pride, to make room for our specimens. Such things as spare waistcoats, shaving boxes, soap, and stockings, have been ejected from our trunks, and we are content to dress, as we live, in a style of primitive simplicity. In fact, the whole appearance of our party is sufficiently primitive; many of the men are dressed entirely in deerskins, without a single article of civilized manufacture about them; the old trappers and hunters wear their hair flowing on their shoulders, and their large grizzled beards would scarcely disgrace a Bedouin of the desert …

[*26th. The Buffalo have disappeared for the present – none have been seen for two days although. our hunters have been out constantly – they have however killed a few Deer & Antelopes, which with our beef & bacon enable us to live sufficiently well* (MS Journal).]

During the whole day [27th] a most terrific gale was blowing directly in our faces, clouds of sand were driving and hurtling by us, often with such violence as nearly to stop our progress; and when we halted in the evening, we could scarcely recognise each other's faces beneath their odious mask of dust and dirt.

There have been no buffalo upon the plain to-day, all the game that we have seen, being a few elk and antelopes; but these of course we did not attempt to kill, as our whole and undivided attention was required to assist our progress.

28th. – We fell in with a new species of game to-day; – a large band of wild horses. They were very shy, scarcely permitting us to approach within rifle distance, and yet they kept within sight of us for some hours …

Chimney Rock

29th. [May 1834] … There is also a remarkable pillar, a kind of obelisk, standing at a considerable distance from the principal bluffs, upon a broad round pedestal. The pillar is perhaps 200ft. in height tapering to an obtuse point – & so small [thin] to appearance that the observer feels surprised that the powerful gales by which the Country is sometimes visited shd. not long ago have prostrated it. It is well known to the hunters & trappers who travel this route by the name of <u>Chimney</u> (JKT to C. Townsend, 27th June 1834).

Cottonwood grove on the North Platte.

In the afternoon, I committed an act of cruelty and wantonness, which distressed and troubled me beyond measure, and which I have ever since recollected with sorrow and compunction. A beautiful doe antelope came running and bleating after us, as though she wished to overtake the party; she continued following us for nearly an hour, at times approaching within thirty or forty yards, and standing to gaze at us as we moved slowly on our way. I several times raised my gun to fire at her, but my better nature as often gained the ascendancy, and I at last rode into the midst of the party to escape the temptation. Still the doe followed us, and I finally fell into the rear, but without intending it, and again looked at her as she trotted behind us. At that moment, my evil genius and love of sport triumphed; I slid down from my horse, aimed at the poor antelope, and shot a ball through her side. Under other circumstances, there would have been no cruelty in this; but here, where better meat was so abundant, and the camp was so plentifully supplied, it was unfeeling, heartless murder. It was under the influence of this too late impression, that I approached my poor victim. She was writhing in agony upon the ground, and exerting herself in vain efforts to draw her mangled body farther from her destroyer; and as I stood over her, and saw her cast her large, soft, black eyes upon me with an expression of the most touching sadness, while the great tears rolled over her face, I felt myself the meanest and most abhorrent thing in creation. But now a finishing blow would be mercy to her, and I threw my arm around her neck, averted my face, and drove my long knife through her bosom to the

CHESTNUT-COLLARED LONGSPUR

Some early texts and a variety of names for this Townsend discovery.

"CHESTNUT-COLLARED LARK-BUNTING … Mr. TOWNSEND procured a single male of this new species, respecting which he has sent me the following notice. "It is by no means a common bird; keeps in pairs, and appears to live exclusively on the ground; is remarkably shy, and although I saw the female several times, I was unable to procure it" (Audubon 1840-44, 3: 53).

"Chestnut-coloured [sic] Finch … Inhabits the prairies of the Platte River, where it is not uncommon" (Townsend 1837).

"CHESTNUT-COLORED [sic] FINCH … Inhabits the plains of the Platte river, near the first range of Rocky Mountains. It appears to live exclusively on the ground, and is a very rare and shy species. I procured but one specimen" (Townsend 1839c: 344-345).

"BROWN COLLARED LARK BUNTING … We met with this elegant species early in May, on the wide grassy plains of the Platte, soon after arriving at that stream [on 18th May]. They were now paired, probably for the season, were rather shy, and kept wholly on the ground. We heard it utter no note more than a chirp, as they kept busily foraging for sustenance" (Nuttall 1840: 537).

Chestnut-collared Longspur type specimen collected by Townsend (ANSP 24099). Audubon says that it was the only one procured and must therefore be the bird he painted opposite (second from top).

Chestnut-collared Longspur as it appeared in plate CCCXCIV of *The Birds of America*.

Audubon's watercolour (NYHS 1863.17.394) of six Townsend specimens, some inscribed with Townend's date of collection. Chestnut-collared Longspur (second from top) "May 28th 1834"; Lark Sparrow (third from top) "May 20th 1834". The three other species are from further west: Lazuli Bunting (top), Spotted Towhee male and female (at bottom), Golden-crowned Sparrow (above the towhees).

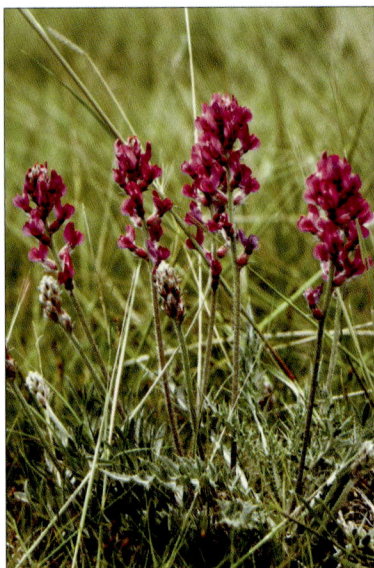

Loco Weed *Oxytropis sp.*

heart. I did not trust myself to look upon her afterwards, but mounted my horse, and galloped off to the party, with feelings such as I hope never to experience again. For several days the poor antelope haunted me, and I shall never forget its last look of pain and upbraiding.

The bluffs on the southern shore of the Platte, are, at this point, exceedingly rugged, and often quite picturesque; the formation appears to be simple clay, intermixed, occasionally, with a stratum of limestone, and one part of the bluff bears a striking and almost startling resemblance to a dilapidated feudal castle [Castle Rock. Jason Lee (1916: 128) says that on 28 May Nuttall travelled an extra five miles out to see it]. There is also a kind of obelisk, standing at a considerable distance from the bluffs, on a wide plain, towering to the height of about two hundred feet, and tapering to a small point at the top. This pillar is known to the hunters and trappers who traverse these regions, by the name of the "*chimney*." [On 29th May, Wyeth (1899: 73) says they "No[o]ned at the Chimney … after travelling this forenoon 11 miles."] Here we diverged from the usual course, leaving the bank of the river, and entered a large and deep ravine between the enormous bluffs. [**Blue Grosbeaks** were seen here: "Mr. Townsend and myself met with them in May on the borders of the Platte, near Scott's Bluffs where they were already mated and breeding" (Nuttall 1840: 624).]

Scott's Bluff.

The road was very uneven and difficult, winding from amongst innumerable mounds six to eight feet in height, the space between them frequently so narrow as scarcely to admit our horses, and some of the men rode for upwards of a mile kneeling upon their saddles. These mounds were of hard yellow clay, without a particle of rock of any kind, and along their bases, and in the narrow passages, flowers of every hue were growing. It was a most enchanting sight; even the men noticed it, and more than one of our matter-of-fact people exclaimed, beautiful, beautiful! Mr. N. was here in his glory. He rode on ahead of the company, and cleared the passages with a trembling and eager hand, looking anxiously back at the approaching party, as though he feared it would come ere he had finished, and tread his lovely prizes under foot.

The distance through the ravine is about three miles. We then crossed several beautiful grassy knolls, and descending to the plain, struck the Platte again, and travelled along its bank. Here one of our men caught a young antelope, which he brought to the camp upon his saddle. It was a beautiful and most delicate little creature, and in a few days became so tame as to remain with the camp without being tied, and to drink, from a tin cup, the milk which our good missionaries spared from their own scanty meals. The men christened it "*Zip Coon*" …

On the afternoon of the 31st, we came to green trees and bushes again, and the sight of them was more cheering than can be conceived, except by persons who have travelled for weeks without beholding a green thing, save the grass under their feet. We encamped in the evening in a beautiful grove of cottonwood trees, along the edge of which ran the Platte, dotted as usual with numerous islands.

In the morning, Mr. N. and myself were up before the dawn, strolling through the umbrageous forest, inhaling the fresh, bracing air, and making the echoes ring with the report of our gun, as the lovely tenants of the grove flew by dozens before us. I think I never before saw so great a variety of birds within the same space. All were beautiful, and many of them quite new to me; and after we had spent an hour amongst them, and my game bag was teeming with its precious freight, I was still loathe to leave the place, lest I should not have procured specimens of the whole.

None but a naturalist can appreciate a naturalist's feelings – his delight amounting to ecstasy – when a specimen such as he has never before seen, meets his eye, and the sorrow and grief which he feels when he is compelled to tear himself from a spot abounding with all that he has anxiously and unremittingly sought for.

This was peculiarly my case upon this occasion. We had been long travelling over a sterile and barren tract, where the lovely denizens of the forest could not exist, and I had been daily scanning the great extent of the desert, for some little *oasis* such as I had now found; here was my wish at length gratified, and yet the caravan would not halt for me; I must turn my back upon the *El Dorado* of my fond anticipations, and hurry forward over the dreary wilderness which lay beyond.

What valuable and highly interesting accessions to science might not be made by a party, composed exclusively of naturalists, on a journey through this rich and unexplored region! The botanist, the geologist, the mamalogist [sic], the ornithologist, and the entomologist, would find a rich and almost inexhaustible field for the prosecution of their enquiries, and the result of such an expedition would be to add most materially to our knowledge of the wealth and resources of our country, to furnish us with new and important facts relative to its structure, organization, and natural productions, and to complete the fine native collections in our already extensive museums.

1. One of the trappers counted fewer men at the start: "Our party consisted of forty men engaged in the service accompanied by Mess Nuttall and Townsend Botanists and Ornithologists with two attendants; likewise Rev's Jason and Daniel Lee Methodist Missionaries with four attendants on their way to establish a Mission in Oregon: which brot. our numbers (including six independent Trappers) to fifty Eight men" (Russell 1914: 1).

2. This is an early Kansas record for this elegant kite that was much more numerous in the early 1800s than it is now. Thomas Say also saw these kites in this vicinity (four, between Franklin and Fort Osage) in July 1819 (Evans 1997: 54). It is still occasionally recorded as a spring overshoot, north of its remaining US breeding sites in the south-east of the country.

3. This "Marmot" cannot have been any of the *Marmota* species that breed in the mountains to the west. In the 1830s the genus *Arctomys* included ground squirrels and prairie dogs. The specimen referred to by Townsend has not been traced but it could be either ANSP 348 Franklin's Ground Squirrel or ANSP 852 Thirteen-lined Ground Squirrel; both are referred to as being from "Missouri" in Townsend's letter to J.E. Gray, 13th January 1849. BMNH.

4. Audubon's original watercolour of the Lark Sparrow (NYHS 1863.17.394) is inscribed with the words "Platte River / Male May 20th 1834". So the specimen that Audubon painted is almost certainly the one that Townsend mentions in his journal.

5. The new bird collected on 25th May was either a Lark Bunting or Chestnut-collared Longspur. The type specimens of both species no longer have original labels. However, Audubon inscribed details from the labels onto his paintings of these two species. Inscribed on the original watercolour beside the female Lark Bunting (1863.17.390) are the words "Female May 28th 1834 / Platte River / J.K. Townsend". There is no inscription beside the male Lark Bunting so it may have been collected on 25th May (or on the preceding day, since that is when Nuttall (1840: 203) says the Lark Buntings were first seen).

On the original watercolour of the Chestnut-collared Longspur (NYHS 1863.17.394) are the words "Platte River / Male May 28th 1834". If the Chestnut-collared Longspur really was collected on 28th May as the inscription says, and because Townsend says that only one specimen was ever collected, then the type locality should be "Platte River, about thirty miles east of Chimney Rock", this distance being based on the fact that they were at Chimney Rock at noon on 29th May, having travelled eleven miles that day. However, since an error in transcription by Audubon is quite possible (misreading Townsend's "25th" as "28th") there is no need to change the current locality ("Forks of Platte River" some 125 miles to the eastward of Chimney Rock), as it may be just as accurate.

Townsend seems not to have preserved any prairie dog specimens yet he must have seen countless Black-tailed Prairie Dogs on the journey from Independence to Laramie's Fork. The White-tailed Prairie Dog (at right) lives in smaller colonies at high, cool elevations and if he saw them at all it would have been further west between Laramie's Fork and Fort Hall.

Laramie's Fork to Green River

"Although we have a supply of meat and it is the Sabbath of the Lord, yet the men have been chasing and killing ... two grisly bears and many Buffaloe."

Cyrus Shepard diary, Sunday 8th June 1834 ... Wyeth Journal, 8th June 1834.

The Laramie Mountains

On the 1st of June, we arrived at Laramie's fork of the Platte, and crossed it without much difficulty ...

On the 2d, we struck a range of high and stony mountains, called the Black Hills [Laramie Mountains[1]]. [Here Townsend first collected **Brewer's Sparrow**, his bird becoming the type specimen in 1856 when John Cassin described it as a new species – see p. 82.] The general aspect here, was dreary and forbidding; the soil was intersected by deep and craggy fissures; rock jutted over rock, and precipice frowned over precipice in frightful, and apparently endless, succession. Soon after we commenced the ascent, we experienced a change in the temperature of the air; and towards mid-day, when we had arrived near the summit, our large blanket *capeaus*, – which in the morning had been discarded as uncomfortable, – were drawn tightly around us, and every man was shivering in his saddle as though he had an ague fit. The soil here is of a deep reddish or ferruginous hue, intermixed with green sand; and on the heights, pebbles of chalcedony and agate are abundant.

Laramie's Fork, near the site of the original Fort Laramie that is mentioned by Wyeth (1899:73): *"June 1st ... At the crossing we found 13 of Sublettes men camped for the purpose of building a fort, he having gone ahead"* .

We crossed, in the afternoon, the last and steepest spur of this chain, winding around rough and stony precipices, and along the extreme verges of tremendous ravines, so dangerous looking that we were compelled to dismount and lead our horses.

On descending to the plain, we saw again the north fork of the Platte, and were glad of an opportunity of encamping. Our march to-day has been an unusually wearisome one, and many of our loose horses are bruised and lame.

Laramie Mountains.

Mountain Plover collected "*table lands of the Rocky Mountains...near the last streams of the Platte*" A new species Probably 14th June

Prairie Falcon specimen "*sources of the Platte*" No date

Red Buttes

North Platte

Brewer's Sparrow collected 'Black Hills' (Laramie Mountains) A new species No date

Sweetwater

Laramie Mountains

American Avocet "*in great numbers*" 9th June

South Pass 7,550ft

Independence Rock

Common Poorwill first seen by Nuttall 10th June

Horned Lark nest found 10th June

First **Bighorn Sheep** and **Grizzly Bear** encounters 10th June

First **Western Tanager** seen 4th June

Fort Williams (Fort Laramie)

Female **Evening Grosbeak** collected 3rd June later painted by Audubon

Laramie River

14th June	10th June	9th June	7th June	3rd June	1st June

1834

0	50	100 miles

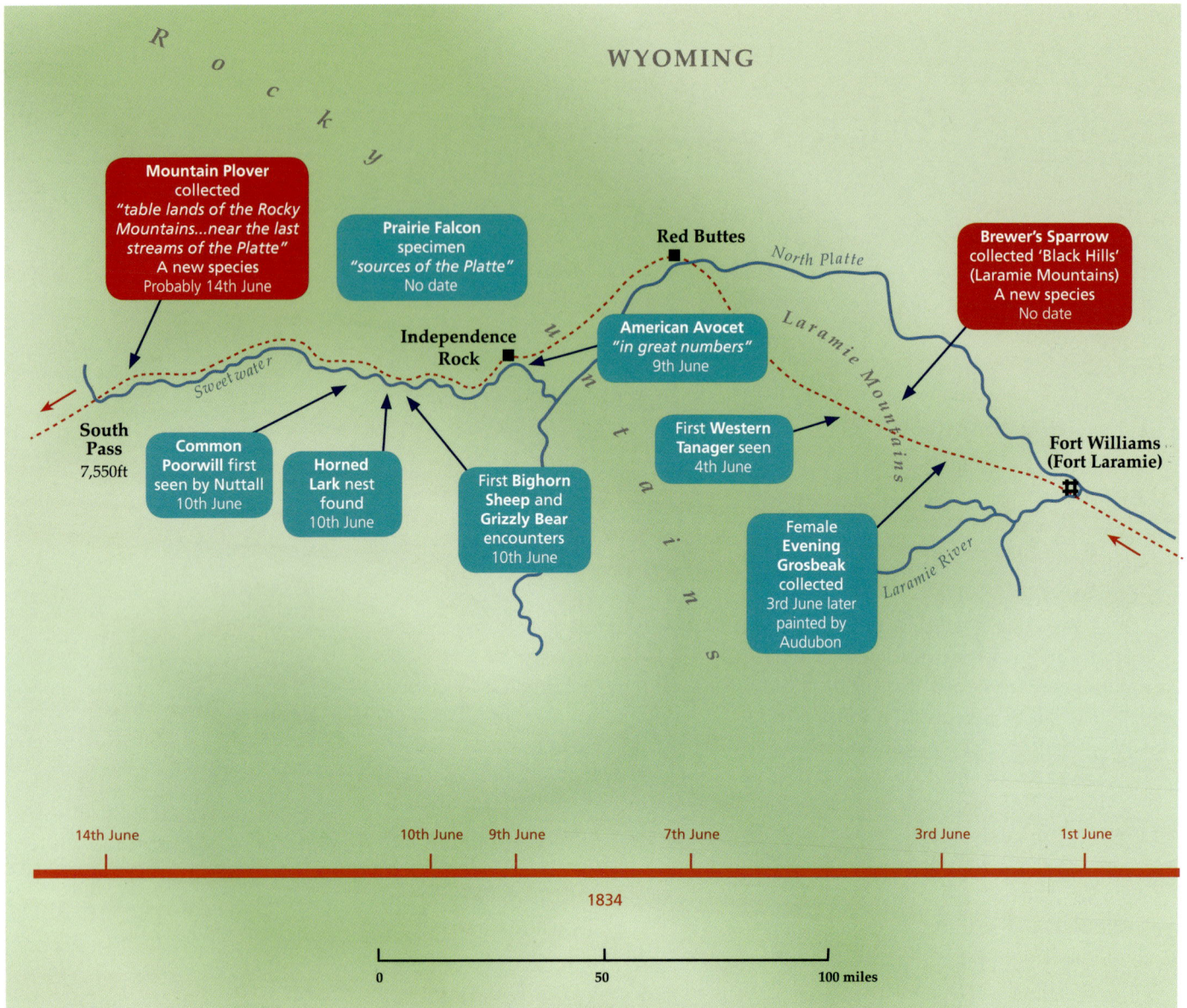

Map 4 - Laramie's Fork to South Pass, 1st June to 14th June 1834.

Red Buttes.

7th. – The country has now become more level, but the prairie is barren and inhospitable looking to the last degree. The twisted, aromatic wormwood covers and extracts the strength from the burnt and arid soil. The grass is dry and brown, and our horses are suffering extremely for want of food. Occasionally, however, a spot of lovely green appears, and here we allow our poor jaded friends to halt, and roam without their riders, and their satisfaction and pleasure is expressed by many a joyous neigh, and many a heart-felt roll upon the verdant sward.

In the afternoon, we arrived at the "Red Butes," two or three brown-red cliffs, about two thousand feet in height. This is a remarkable point in the mountain route. One of these cliffs terminates a long, lofty, wooded ridge, which has bounded our southern view for the past two days. The summits of the cliffs are covered with patches of snow, and the contrast of the dazzling white and brick-red produces a very pretty effect.

The next day [8th], we left the Platte river, and crossed a wide, sandy desert, dry and desolate; and on the 9th, encamped at noon on the banks of the Sweet-water. Here we found a large rounded mass of granite, about fifty feet high, called Rock Independence. Like the Red Butes, this rock is also a rather remarkable point in the route. On its smooth, perpendicular sides, we see carved the names of most of the mountain *bourgeois*, with the dates of their arrival. We observed those of the two Sublette's, Captains Bonneville, Serre [Cerré], Fontinelle, &c., and after leaving our own, and taking a hearty, but hasty lunch in the shade of a rock, and a draught from the pure and limpid stream at its base, we pursued our journey.

View from Independence Rock.

The river is here very narrow, often only twelve or fifteen feet wide, shallow, and winding so much, that during our march, to-day, we crossed it several times, in order to pursue a straight course. The banks of the stream are clothed with the most luxuriant pasture, and our invaluable dumb friends appear perfectly happy.

We saw here great numbers of a beautiful brown and white avocet, (the *Recurvirostra americana* **[American Avocet]** of ornithologists.) These fine birds were so tame as to allow a very near approach, running slowly before our party, and scarcely taking wing at the report of a gun. They frequent the marshy plains in the neighborhood of the river, and breed here.

On the 10th, about ninety miles to the west, we had a striking view of the Wind-river mountains. They are almost wholly of a dazzling whiteness, being covered thickly with snow, and the lofty peaks seem to blend themselves with the dark clouds which hang over them. This chain gives rise to the sources of the Missouri, the Colorado of the west [Green River], and Lewis' river of the Columbia [Snake River], and is the highest land on the continent of North America.

Independence Rock.

Bighorn Sheep.

We saw, to-day, a small flock of the hairy sheep of the Rocky Mountains [**Bighorn Sheep**], the big horn of the hunters, *(Ovis montana.)* We exerted ourselves in vain to shoot them. They darted from us, and hid themselves amongst the inaccessible cliffs, so that none but a chamois hunter might pretend to reach them. Richardson [one of their hunters] says that he has frequently killed them, but he admits that it is dangerous and wearisome sport; and when good beef is to be found upon the plains, men are not anxious to risk their necks for a meal of mutton.

In the afternoon, one of our men had a somewhat perilous adventure with a **grizzly bear**. He saw the animal crouching his huge frame in some willows which skirted the river, and approaching on horseback to within twenty yards, fired upon him. The bear was only slightly wounded by the shot, and with a fierce growl of angry malignity, rushed from his cover, and gave chase ... the man rode in among his fellows, pale and haggard from overwrought feelings, and was probably effectively cured of a propensity for meddling with grizzly bears.

A small striped rattlesnake [**Western Rattlesnake**] is abundant on these plains: – it is a different species from our common one at home, but is equally malignant and venomous. The horses are often startled by them, and dart aside with intuitive fear when their note of warning is sounded in the path.

12th. – The plains of the Sweet-water at this point, – latitude 43° 6´, longitude 110° 30´, – are covered with little salt pools, the edges of which are encrusted with alkaline efflorescences, looking like borders of snow. The rocks in the vicinity are a loose, fine grained sandstone, the strata nearly horizontal, and no organic remains have been discovered. We have still a view of the lofty Wind-river mountains on our right hand, and they have for some days served as a guide to determine our course ...

American Avocet at nest.

A SUCCESSION OF BIRDS

Although Townsend omitted much of the natural history from the *Narrative* we know from other sources that he was beginning to see western birds that he had never seen before. Some of the specimens from this period were used by Audubon in the preparation of *The Birds of America* – sometimes Audubon says so in the *Ornithological Biography* and sometimes his watercolours are marked up with the source and dates of specimens. At other times Audubon does not say or is rather vague as to where his birds originated but often the implication is that they came from Townsend. Occasionally we can match particular skins to particular plates because Audubon says he used a Townsend specimen and we know that only one was collected.

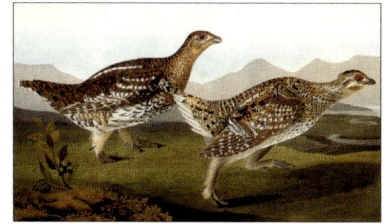

1st June – "Mr. TOWNSEND informs me that while crossing the north branch of the Platte (Lorimie's [sic] Fork), he found this species [**Sharp-tailed Grouse**] breeding, and that as an article of food it proved to be a very well-flavoured and plump bird, considerably superior to any of the other larger species that occur in the United States" (Audubon 1840-44, 5: 111; based on Nuttall 1840: 807).

1st June and following days – Nuttall saw the Red-shafted **Northern Flicker** "among the narrow belt of forest which borders Lorimie's Fork of the Platte" and often thereafter as they travelled westwards (Audubon 1840-44, 4: 290). Nuttall (1840: 311) also says[2] that they first saw **Say's Phoebe** "within the first range of the Rocky Mountains called the Black Hills, and in the vicinity of that northern branch of the Platte known by the name of Larimie's [sic] Fork".

Early June – Audubon's plate of the "Arkansaw Flycatcher" [**Western Kingbird**] is almost certainly prepared from specimens supplied by Townsend: "In my possession are some remarkably fine skins, from Mr. TOWNSEND's collection, which differ considerably from the figure given by BONAPARTE, who first described the species [first collected by Thomas Say] ... Mr. TOWNSEND's notice respecting it is as follows: "... It is numerous along the banks of the Platte, particularly in the vicinity of trees and bushes. It is found also, though not so abundantly, across the whole range of the Rocky mountains ... (Audubon 1840-44, 1: 199-200). Nuttall also mentions seeing the Western Kingbird: "In our journey westward, we first met with this bold and querulous species, early in July [= June], in the scanty woods which border the north-west branch of the Platte, within the range of the Rocky Mountains" (Nuttall 1840: 306).

3rd June – Townsend collected a **Black-headed Grosbeak**; his specimen was labelled "Black Hills/Male/June 3 – 34 / J.K. Townsend" and for many years resided in the Charleston Museum (Sass 1908). On the same day he collected a female **Evening Grosbeak** that was painted by Audubon, and according to Audubon was similarly labelled "Black Hills, Female, June 3, 1824 [sic]" (Audubon 1840-44, 3: 220).

4th June – The naturalists first saw **Western Tanagers**: "WILSON was the first ornithologist who figured this handsome bird. From his time until the return of Mr. TOWNSEND from the Columbia river no specimen seems to have been procured. That gentleman forwarded several males in much finer condition than those brought by LEWIS and CLARKE. Some of these I purchased, and, on his return to Philadelphia, I was presented with a female by my young friend Dr. TRUDEAU [error for Dr Townsend]. The account of this species is by THOMAS NUTTALL, who, however, was unacquainted with the female.

  "We first observed this fine bird in a thick belt of wood near Lorimer's [sic] Fork of the Platte, on the 4th of June, at a considerable distance to the east of the first chain of the Rocky Mountains (or Black Hills) [Laramie Mountains, Wyoming], so that the species in all probability continues some distance down the Platte ..."
  Mr. TOWNSEND says that "this handsome bird [**Western Tanager**]... is rare on the banks of the Platte, but rather common in the woods and oak groves of the Columbia. None were seen after leaving the Black Hills, until we reached the lower country ...'" (Audubon 1840-44, 3: 231-232). ▶

From top:

Sharp-tailed Grouse, plate CCCLXXXII, probably from Townsend specimen.

Western Kingbird, plate CCCLIX (detail), Evening Grosbeak female, plate CCCCXXIV (detail), Western Tanager, plate CCCLIV (detail), all from Townsend specimens.

Prairie Falcon, plate 16 from Cassin's *Illustrations of the Birds of California, Texas, Oregon, British and Russian America* (1856). Drawn from specimens at ANSP, perhaps including Townsend specimen at right (ANSP 2175).

About 4th June – Nuttall reported that they saw **Bullock's Oriole**: "About fifty or sixty miles to the north-west of the usual crossing-place of that branch of the La Platte called Larimie's Fork, we observed it making a nest quite similar to that of the Baltimore-bird [Baltimore Oriole]. Nuttall also noted that the **Mountain Bluebird** was "not uncommon" in this same vicinity (Audubon 1840-44, 4: 43; 2:178).

About 7th June – Somewhere in this rocky region Townsend collected a **Prairie Falcon** and seems to have thought that it was a Merlin. Eighteen years later John Cassin described it as a new species ("FALCO POLYAGRUS. – CASSIN. THE AMERICAN LANIER FALCON") in his *Illustrations of the Ornithology of California etc* (1856). Cassin used Townsend's specimen from "near the sources of the Platte river" as the type, not knowing that the falcon had already been described (in 1851, by Herman Schlegel, from a Mexican specimen.

"On June 10," says Mr. NUTTALL, "on the plains by the banks of the sweet water of the Platte, we started the Shore Lark **[Horned Lark]** from her nest in a small depression on the ground. It was made of bent grass, lined with coarse bison hair. The eggs were olive-white, minutely spotted all over with a darker tinge" (Audubon 1840-44, 3: 48; based on Nuttall 1840: 524).

Horned Lark and nestlings.

10th June – Nuttall reported that the **Common Poorwill** *Phalaenoptilus nuttalli* was "first seen by us on the 10th of June, amidst the naked granite hills of the upper branch of the Platte, called Sweet Water, about twilight, uttering from the clefts of the rocks at intervals a low wailing cry in the manner of the Whip-poor-Will" (Nuttall 1840: 747). Although no specimen was obtained, Audubon (1839, 5: 335) mentioned the bird briefly in the *Ornithological Biography*, calling it "NUTTALL'S WHIP-POOR-WILL *CAPRIMULGUS NUTTALLII*" with the observation that "It was frequently seen [in the Rocky Mountains] by [Nuttall], often within a few feet, but was not procured, probably because he is not in the habit of carrying a gun on his rambles." In 1844 Audubon was able to describe it properly from a specimen collected by J.G. Bell on Audubon's Missouri River Expedition (see e.g. Mearns and Mearns 1992: 344).

16th June – Townsend shot and prepared a **Say's Phoebe**: Audubon's original watercolour of Say's Phoebe (NYHS 1863.17.359) is inscribed "Male Bird. June 16th. / 1834. Townsend".

16th June – A specimen of **Clark's Nutcracker** is listed in the card catalogue of the Smithsonian Institution for 16th June 1834 (USNM 1929, card catalogue only, probably no surviving specimen).

16th June – Nuttall heard **Common Poorwill** again, noting that "near the banks of the Sandy river of the Colorado, I again heard the nocturnal cry of this bird something like *pevai, pevai, pevai*" (Nuttall 1840: 747). In this vicinity, no date specified, he also found a nest of the **Mountain Bluebird** on "a cliff of the Sandy River ... the female and male were both feeding their brood" (Audubon 1840-44, 2:178; based on Nuttall 1840: 515).

Lewisia flowers at Independence Rock.

Say's Phoebe, plate CCCLIX (detail). The bird at left was collected 16th June 1834.

Common Poorwill, plate 495 from Audubon's octavo edition of *The Birds of America*, painted from a specimen collected during Audubon's 1843 expedition up the Missouri. Townsend failed to collect any.

Mountain Plover, plate CCCL.
The finished plate was prepared from
Audubon's original watercolour at right
(NYHS 1863.17.350), that in turn was
prepared from Townsend's type specimen
beneath (ANSP 24,353). We know that
Audubon used this specimen because
there were no others then available, and
because of what Townsend says:

"ROCKY MOUNTAIN PLOVER ...
Inhabits the table land of the Rocky
mountains. I saw but one specimen of
this beautiful bird, and, as our party was
on the move, I was compelled to kill it
without delay" (Townsend 1839a: 349).

MOUNTAIN PLOVER

"ROCKY-MOUNTAIN PLOVER...

For the following brief account of this bird, I am indebted to my learned and obliging
friend THOMAS NUTTALL.

"This remarkable species, so much allied to the *Charadrius Wilsoni* [Wilson's Plover],
was scarcely seen by us for more than one or two days, and then on the central table-
land of the Rocky Mountains, in the plains near the last of the streams of the Platte,
pursued in our western and northern route. It being the month of July [= June] when
we saw it, there is little doubt but that it was breeding in this subalpine region. The only
individual shot, was seen skulking and running through the wormwood bushes which so
generally clothe those arid and dry wastes. After running some time, it would remain
perfectly still, as if conscious of the difficulty of distinguishing it from the colour of
the grey soil on which it stood. All that we saw were similar to the present individual,
and none, however flushed, took to the wing. We do not recollect hearing from it the
slightest complaint or note of any kind, being intent probably on concealing its young
or eggs by a perfect silence."

The skin from which I made my drawing was that of a female; and it is my opinion,
that the male, when found, will have as distinct markings as those exhibited by *Charadrius
melodus* [Piping Plover] or *Charadrius semipalmatus* [Semipalmated Plover]" (Audubon
1840-44, 5: 213).

Audubon's original watercolour of the Mountain Plover (NYHS 1863.17.350) is
inscribed "*Rocky Mountains / female, June 4, 1834 / J.K. Townsend*". If it had been collected
on 4th June it is more likely that Townsend would have labelled it "Black Hills" as he
did for other specimens collected on that date. Moreover both Townsend and Nuttall
say that it was seen on the "table-lands" a description that is more consistent with the
country they were traversing on 14th June, and more consistent with the habitat and
known range of the Mountain Plover.

SAGE THRASHER

Plate CCCLXIX (detail). This Sage Thrasher in *The Birds of America* was prepared from the Townsend specimen below (ANSP 23,728). There is no additional information about the specimen on the original watercolour as there is on the Mountain Plover watercolour.

"MOUNTAIN MOCKING-BIRD ...This interesting and hitherto unfigured species was procured on the Rocky Mountains by Mr. TOWNSEND, who forwarded a single specimen to Philadelphia, where I made a drawing of it. The following notice by Mr. NUTTALL shews that it is nearly allied in its habits to the Mocking-bird [Northern Mockingbird]:-

"On the arid plains of the central table-land, betwixt the northern sources of the Platte and the Colorado of the West [Green River], in the month of June, we frequently heard the cheering song of this delightful species, whose notes considerably resemble those of the Brown Thrush [Brown Thrasher], with some of the imitative powers of the Mocking-bird ... Just before arriving at Sandy Creek of the Colorado, while resting for refreshment at noon, I had the good fortune to find the nest in a wormwood bush, on the margin of a ravine, from whence the male was singing with its usual energy. It contained four almost emerald green eggs, spotted with dark olive of two shades, more numerous towards the greater end, the spots large and roundish. The nest itself was made of small twigs and rough stalks, lined with stripes of bark and bison wool ..." (Audubon 1840-44, 2: 194).

"Mountain Mocking-bird ...Rather smaller than the Mocking-bird. Inhabits the plains of the Rocky Mountains" (Townsend 1837: 192).

Sage Thrasher nest.

Brewer's Sparrow, plate CCCXCVIII (detail), and Townsend's type specimen (ANSP 24050).

a. Townsend's specimen catalogue (see appendix 6) was prepared chronologically, each additional species received a field number as it was encountered, the same number for all specimens of that species. In McEuen's list (see appendix 5) that featured some of Townsend's specimens, we see that "Cream colored sparrow" (Clay-colored Sparrow) appears as No.12, and that No.15 is referred to as "same as 12". This goes against Townsend's practice of giving different numbers to the same species so perhaps he realised that Clay-colored and Brewer's Sparrows were different species at the time of collecting, but was not so sure when he and McEuen compiled the list in 1839. We do not know the exact dates for the relevant entries in the catalogue so we do not know how many days between collection dates separated Nos.12 and 15. However, it is just possible that No.12 was a Clay-colored Sparrow, because it was taken further east (perhaps on 20th May 1834 in stream-side brush); and No.15 was instead a Brewer's Sparrow, because by then Townsend was further west within suitable habitat for the second species while in the Laramie Mountains, from 2nd to 10th June.

TOWNSEND COLLECTS THE FIRST BREWER'S SPARROWS

Audubon painted Brewer's Sparrow specimens collected by Townsend but incorrectly identified them as Clay-colored Sparrows. The two species are quite similar in appearance and there has been much confusion between them: many recent commentators on the plates for *The Birds of America* have persisted in misidentifying Audubon's painting of Brewer's Sparrow as the Clay-colored Sparrow, just as Audubon had done (Mearns and Mearns 1995).

Townsend's first specimens of Brewer's Sparrow were taken in the Laramie Mountains in early June 1834 and were later deposited in the Academy of Natural Sciences of Philadelphia. In 1856 John Cassin used one of these specimens when writing the first formal description of Brewer's Sparrow and explained how it differed from Clay-colored Sparrow. Cassin named the new species after Thomas Brewer of Boston for "an ardor in devotion to Ornithological science rarely paralleled" – but with no connection to the sparrow (Mearns and Mearns 1992: 116-121).

As Townsend travelled further westward to Big Sandy Creek, on the Pacific side of South Pass, he collected more Brewer's Sparrows and these were the ones that Audubon acquired. Audubon used them as models for his painting of what he called the "Clay-coloured Bunting", saying that: "The specimens from which the above descriptions have been taken, were procured on the 15th of June, 1834, on the Rocky Mountains, by Mr. TOWNSEND" (Audubon 1840-44, 3: 72). Audubon's misidentification arose because his painting pre-dated Cassin's work on the two sparrows. Neither Audubon, nor anyone else, had yet realized that there were two different species.

One of the ways in which they differ is in their preferred habitat. Brewer's Sparrow inhabits open areas of low brush, particularly sagebrush. The area near Big Sandy Creek is open, gently undulating tableland chiefly covered in grasses and sagebrush, indeed Townsend may well have collected the type of the Sage Thrasher on the same day as the Brewer's Sparrows. On 16th June Townsend reported "large herds of buffalo on the plains of Sandy river, grazing in every direction on the short and dry grass." In other words, it was exactly the sort of habitat for Brewer's Sparrows.

The Clay-colored Sparrow has a more northerly and easterly distribution than Brewer's Sparrow, and inhabits brushy hillsides, thickets, thorn scrub and deciduous forest edge. When Audubon got into this habitat during his expedition up the Missouri to Fort Union in 1843 and first saw Clay-colored Sparrows he thought that he had found a new species and promptly named it Shattuck's Bunting. This was because up until then the sparrows that he had always thought of as Clay-colored Sparrow had been Brewer's Sparrows!

So what did Townsend make of the sparrows? The answer is: perhaps not very much. Wyeth was in a big hurry to get to the Fur Trappers' Rendezvous near the Green River, so any birds that Townsend collected in June had to be snatched up and prepared in haste without much time to ponder their identities.[a]

South Pass

On the 14th, we left the Sweet-water, and proceeded in a south-westerly direction to Sandy river, a branch of the Colorado of the west. We arrived here at about 9 o'clock in the evening, after a hard and most toilsome march for both man and beast. We found no water on the route, and not a single blade of grass for our horses. Many of the poor animals stopped before night, and resolutely refused to proceed; and others with the remarkable sagacity, peculiar to them, left the track in defiance of those who drove and guided them, sought and found water, and spent the night in its vicinity. The band of missionaries, with their horses and horned cattle, halted by the way, and only about half the men of the party accompanied us to our encampment on Sandy. We were thus scattered along the route for several miles; and if a predatory band of Indians had then found us, we should have fallen an easy prey.

Sagebrush west of South Pass.

The next morning [15th] by about 10 o'clock all our men and horses had joined us, and, in spite of the fatigues of the previous day, we were all tolerably refreshed, and in good spirits. Towards noon we got under way, and proceeded seven or eight miles down the river to a spot where we found a little poor pasture for our horses. Here we remained until the next morning [16th?], to recruit. I found here a beautiful new species of mocking bird **[Sage Thrasher]**, which I shot and prepared (see p. 81). Birds are, however generally scarce, and there is here very little of interest in any department of natural history. [Townsend collected more **Brewer's Sparrows** on the 15th and it was these specimens that were painted by Audubon - see p. 82] We are also beginning to suffer somewhat for food: buffalo are rarely seen, the antelopes are unusually shy, and the life of our little favorite, "Zip," has been several times menaced. I believe, however, that his keeper, from sheer fondness, would witness much greater suffering in the camp, ere he would consent to the sacrifice of his playful little friend.

16th. – We observed a hoar frost and some thin ice, this morning at sunrise; but at mid-day, the thermometer stood at 82°. We halted at noon, after making about fifteen miles, and dined. Saw large herds of buffalo on the plains of Sandy river, grazing in every direction on the short and dry grass. Domestic cattle would certainly starve here, and yet the bison exists, and even becomes fat; a striking instance of the wonderful adaptation of Providence.

17th. – We had yesterday a cold rain, the first which has fallen in our track for several weeks. Our vicinity to the high mountains of the Wind river will perhaps account for it. To-day at noon, the mercury stood at 92° in the shade, but there being a strong breeze, we did not suffer from heat.

Our course was still down the Sandy river, and we are now looking forward with no little pleasure to a rest of two or more weeks at the mountain rendezvous on the Colorado. Here we expect to meet all the mountain companies who left the States last spring, and also the trappers who come in from various parts, with the furs collected by them during the previous year. All will be mirth and jollity, no doubt, but the grand desideratum with some of us, is to allow our horses to rest their tired limbs and exhausted strength on the rich and verdant plains of the Siskadee. At our camp this evening, our poor horses were compelled to fast as heretofore, there being absolutely nothing for them to eat. Some of the famished animals attempted to allay their insatiable cravings, by cropping the dry and bitter tops of the wormwood with which the prairie is strewed ...

Least Chipmunk.

LEAST CHIPMUNK

The type specimen of Least Chipmunk (ANSP 248) was collected "near mouth of Big Sandy". It was described by John Bachman in the *Journal of Academy of Natural Sciences of Philadelphia* (1839, 8: 71-73) where he quoted from a letter received from Townsend: "It is found very plentiful along the banks of the Rio Colorado [Green River], but, I think, does not inhabit a very extensive range, as I never saw it after leaving the river. It keeps almost constantly among heaps of stones, on the tops of which it often perches, extending its long tail over its back, and curving it over in front of its head. At such times it emits a lively, garrulous note, like the squeaking of a young puppy; but, if approached, darts off with astonishing swiftness, carrying the tail level with the ground, and almost eluding the eye by the activity of its motions, and conceals itself under some jutting rock, or in the interstices of a stone heap, until the intruder has passed."

Arrival at the Green River

June 19th. – We arrived to-day on the Green river, Siskadee or Colorado of the west, – a beautiful, clear, deep, and rapid stream, which receives the waters of Sandy, – and encamped upon its eastern bank. After making a hasty meal, as it was yet early in the day, I sallied forth with my gun, and roamed about the neighborhood for several hours in quest of birds. On returning, towards evening, I found that the whole company had left the spot, the place being occupied only by a few hungry wolves, ravens and magpies [**Gray Wolf, Common Raven** and **American Magpie**], the invariable gleaners of a forsaken camp.

I could not at first understand the meaning of all I saw. I thought the desertion strange, and was preparing to make the best of it, when a quick and joyful neigh sounded in the bushes near me, and I recognized the voice of my favorite horse. I found him carefully tied, with the saddle, &c., lying near him. I had not the least idea where the company had gone, but I knew that on the rich, alluvial banks of the river, the trail of the horses would be distinct enough, and I determined to place my dependence, in a great measure, upon the sagacity of my excellent dumb friend, satisfied that he would take me the right course. I accordingly mounted, and off we went at a speed which I found some difficulty in restraining. About half an hour's hard riding brought us to the edge of a large branch of the stream, and I observed that the horses had here entered. I noticed other tracks lower down, but supposed them to have been made by the wanderings of the loose animals. Here then seemed the proper fording place, and with some little hesitation, I allowed my nag to enter the water; we had proceeded but a few yards, however, when down he went off a steep bank, far beyond his depth. This was somewhat disconcerting; but there was but one thing to be done, so I turned my horse's head against the swift current, and we went snorting and blowing for the opposite shore. We arrived at length, though in a sadly wet and damaged state, and in a few minutes after, came in view of the new camp.

Captain W. explained to me that he had heard of good pasture here, and had concluded to move immediately, on account of the horses; he informed me, also, that he had crossed the stream about fifty yards below the point where I had entered, and had found an excellent ford. I did not regret my adventure, however, and was congratulating myself upon my good fortune in arriving so seasonably, when, upon looking to my saddle, I discovered that my coat was missing. I had felt uncomfortably warm when I mounted, and had removed the coat and attached it carelessly to the saddle; the rapidity of the current had disengaged it, and it was lost forever. The coat itself was not of much consequence after the service it had seen, but it contained the second volume of my journal, a pocket compass, and other articles of essential value to me. I would gladly have relinquished every thing the garment held, if I could have recovered the book; and although I returned to the river, and searched assiduously until night, and offered large rewards to the men, it could not be found.

The journal commenced with our arrival at the Black Hills, and contained some observations upon the natural productions of the country, which to me, at least, were of some importance; as well as descriptions of several new species of birds, and notes regarding their habits, &c., which cannot be replaced ...

In consequence of remaining several hours in wet clothes, after being heated by exercise, I rose the next morning with so much pain, and stiffness of the joints, that I could scarcely move. But notwithstanding this, I was compelled to mount my horse with the others, and to ride steadily and rapidly for eight hours. I suffered intensely during this ride; every step of my horse seemed to increase it, and induced constant sickness and retching[3].

When we halted, I was so completely exhausted, as to require assistance in dismounting, and shortly after, sank into a state of insensibility from which I did not recover for several hours. Then a violent fever commenced, alternating for two whole days, with sickness and pain. I think I never was more unwell in my life; and if I had been at home, lying on a feather bed instead of the cold ground, I should probably have fancied myself an invalid for weeks.

1. In the early 1800s the term Black Hills applied to all the outlying foothills east of the Rocky Mountains. The Black Hills that Townsend referred to are in Wyoming and are now called the Laramie Mountains.

2. Although Nuttall gives the date as 14th June (by which time they had crossed South Pass) he evidently meant 4th June.

3. Townsend was not the only person who was ill. Jason Lee (1916: 137) says that his cousin Daniel Lee was very sick on the evening of the 19th June and "in extreme pain" due to bathing in cold water and only just able to ride on the next day.

Rock Wren, plate CCCLX (detail).

While Townsend was sick, Nuttall was out and about and on his return to the East Coast presented an adult female Rock Wren to Audubon who quoted Nuttall's observations: "On the 21st of June, on the ledges of the bluffs which border the bottom of Hare's [= Ham's] Fork of the Siskadee (or Colorado of the West), I heard, and at length saw this curious Mountain Wren. Among these arid and bare hills of the central table-land they were quite common ... In the same rocky retreats they are commonly accompanied by a kind of small striped Ground Squirrel [**Least Chipmunk**], like that of the eastern coast in many respects, but much smaller. These little animals, which are numerous, the White-chinned Buzzard, *Buteo vulgaris* of RICHARDSON and SWAINSON [**Swainson's Hawk**], and the **Raven** frequently hover over and pounce upon. We met with this species as far west as the lowest falls of the Columbia, and within a few miles of Fort Van Couver, but among rocks and cliffs as usual." (Audubon 1840-44, 2: 113-114; based on Nuttall 1840: 491-492).

Nuttall reported a nest of Swainson's Hawk not mentioned by Townsend; and it is probable that Townsend took the opportunity to collect one or both of the adults since "White chin Buzzard" is listed as No. 34 in Townsend's catalogue of specimens (see appendix 6). Audubon used a Townsend specimen in preparation of plate CCCLXXII (at right) but it is not known where that specimen was obtained (see appendix 8). The rabbit is an eastern species with no connection with Townsend.

SWAINSON'S HAWK

"WHITE THROATED BUZZARD ... MR. TOWNSEND and myself observed this bird on the woody margins of the Rocky Mountain streams which pass into the Colorado of the West [Green River], about the [beginning of the] month of July, when they were breeding, in a tree where Mr. T. found the nest containing 2 white eggs. Its habits closely resembled those of the Red Tailed Buzzard" (Nuttall 1840:112).

Green River to Fort Hall

"Prof. Nuttall, of Cambridge, Massachusetts, who accompanied us for the purpose of making botanical enquiries, has met with the most flattering success. Mr. Townsend, the ornithologist, from Philadelphia, who also accompanied us, has surpassed his most sanguine expectations. Last evening he informed me that he has discovered fifteen new species of birds, six that were doubtful, and twenty that were lately found ..."

R.L. Edwards, 23rd June 1834.

The Trappers' Rendezvous

June 22nd – We are now lying at the rendezvous. W. Sublette, Captains Serre [Cerré], Fitzpatrick, and other leaders, with their companies, are encamped about a mile from us, on the same plain, and our own camp is crowded with a heterogeneous assemblage of visitors. The principal of these are Indians, of the Nez Percé, Banneck and Shoshoné tribes, who come with their furs and peltries which they have been collecting at the risk of their lives during the past winter and spring, to trade for ammunition, trinkets, and "fire water." There is, in addition to these, a great variety of personages amongst us; most of them calling themselves white men, French-Canadians, half-breeds, &c., their color nearly as dark, and their manners wholly as wild, as the Indians with whom they constantly associate. These people, with their obstreperous mirth, their whooping, and howling, and quarrelling, added to the mounted Indians, who are constantly dashing into and through our camp, yelling like fiends, the barking and baying of savage wolf-dogs, and the incessant cracking of rifles and carbines, render our camp a perfect bedlam. A more unpleasant situation for an invalid could scarcely be conceived. I am confined closely to the tent with illness, and am compelled all day to listen to the hiccoughing jargon of drunken traders, the *sacré* and *foutre* of Frenchmen run wild, and the swearing and screaming of our own men, who are scarcely less savage than the rest, being heated by the detestable liquor which circulates freely among them.

Green River.

The Trappers' Rendevous of 1837, by Alfred Jacob Miller.

Green River, near its junction with Big Sandy Creek.

It is very much to be regretted that at times like the present, there should be a positive necessity to allow the men as much rum as they can drink, but this course has been sanctioned by all leaders of parties who have hitherto visited these regions, and reform cannot be thought of now. The principal liquor in use here is alcohol diluted with water. It is sold to the men at *three dollars* the pint! Tobacco of very inferior quality, such as could be purchased in Philadelphia at about ten cents per pound, here brings two dollars! and everything else in proportion. There is no coin in circulation, and these articles are therefore paid for by the independent mountain-men, in beaver skins, buffalo robes, &c.; and those who are hired to the companies, have them charged against their wages ...

["I have skinned ~~about~~ 35 species of birds, about 14 of which are <u>new</u> – the remainder very rare & little known & some of them extremely beautiful. I might have added about 10 more to my list of species, but as I am certain of procuring them in abundance nearer our final stopping place I reserve the space for others. Among the plants our fr[d]. N. has succeeded to admiration – he has been completely in his element & his spirits have been uniformly good – his new species he does not attempt to compute – he has frequently found a doz. in the course of a day" (JKT to C. Townsend, 27th June 1834).]

30th. – Our camp here is a most lovely one in every respect, and as several days have elapsed since we came, and I am convalescent, I can roam about the country a little and enjoy it. The pasture is rich and very abundant, and it does our hearts good to witness the satisfaction and comfort of our poor jaded horses. Our tents are pitched in a pretty little valley or indentation in the plain, surrounded on all sides by low bluffs of yellow clay. Near us flows the clear deep water of the Siskadee [Green River], and beyond, on every side, is a wide and level prairie, interrupted only by some gigantic peaks of mountains and conical *butes* in the distance. The river, here, contains a great number of large trout, some grayling, and a small narrow-mouthed white fish, resembling a herring [**Mountain Whitefish** *Prosopium williamsoni*]. They are all frequently taken with the hook, and, the trout particularly, afford excellent sport to the lovers of angling. Old Izaac Walton would be in his glory here, and the precautionary measures which he so strongly recommends in approaching a trout stream, he would not need to practice, as the fish is not shy, and bites quickly and eagerly at a grasshopper or minnow.

Buffalo, antelopes, and elk are abundant in the vicinity, and we are therefore living well. We have seen also another kind of game, a beautiful bird, the size of a half grown turkey, called the cock of the plains, (*Tetrao urophasianus*) [**Greater Sage-Grouse**]. We first met with this noble bird on the plains[1], about two day's journey east of Green river [i.e. on the Sandy River], in flocks, or *packs*, of fifteen or twenty, and so exceedingly tame as to allow an approach to within a few feet, running before our horses like domestic fowls, and not unfrequently hopping under their bellies, while the men amused themselves by striking out their feathers with their riding whips. When we first saw them, the temptation to shoot was irresistible; the guns were cracking all around us, and the poor grouse falling in every direction; but what was our disappointment, when, upon roasting them nicely before the fire, we found them so strong and bitter as not to be eatable. From this time the cock of the plains was allowed to roam free and unmolested, and as he has failed to please our palates, we are content to admire the beauty of his plumage, and the grace and spirit of his attitudes.

Onwards to the Snake River

July 2d. – We bade adieu to the rendezvous this morning; packed up our movables, and journied along the bank of the river. Our horses are very much recruited by the long rest and good pasture which they have enjoyed, and, like their masters, are in excellent spirits ...

Greater Sage-Grouse, plate CCCLXXI (detail).

Violet-green Swallow, plate CCCLXXXV (detail).

Nuttall commented on the number of **Violet-green Swallows**: "We first met with this elegant species within the table-land of the Rocky Mountains, and they were particularly abundant around our encampment on Harris [Ham's] Fork, a branch of the Colorado of the west. They are nearly always associated with the **Cliff Swallow**, here likewise particularly numerous ... In the Rocky Mountains, near our camp, we observed them to go in and out of deserted nests of the Cliff Swallows, which they appeared to occupy in place of building nests of their own" (Audubon 1840-44, 1:186; based on Nuttall (1840: 725) who also quotes Townsend as saying: "It inhabits the neighborhood of the Colorado of the West, and breeds along the margin on bluffs of clay, where it attaches its nest[2], formed of mud and grasses, resembling in some measure that of the Cliff Swallow, but wanting the pendulous neck. The eggs are 4, of a dark clay color, with a few spots of reddish-brown at the larger end."

William Drummond Stewart. Sketch by Alfred Jacob Miller, 1837.

Ham's Fork.

Many of our men have left us, and joined the returning companies, but we have had an accession to our party of about thirty Indians; Flat-heads, Nez Percés, &., with their wives, children, and dogs. Without these our camp would be small; they will probably travel with us until we arrive on Snake River, and pass over the country where the most danger is to be apprehended from their enemies, the Black-feet.

Some of the women in this party, particularly those of the Nez Percé nation, are rather handsome, and their persons are decked off in truly savage taste. Their dresses of deer skin are profusely ornamented with beads and porcupine quills; huge strings of beads are hung around their necks, and their saddles are garnished with dozens of little hawk's bells, which jingle and make music for them as they travel along. Several of these women have little children tied to their backs, sewed up papoose fashion, only their heads being seen ...

We were joined at the rendezvous by a Captain Stewart[3], an English [= Scottish] gentleman of noble family, who is travelling for amusement, and in search of adventure. He has already been a year in the mountains, and is now desirous of visiting the lower country, from which he may probably take passage to England by sea. Another Englishman; a young man, named Ashworth[4], also attached himself to our party, for the same purpose.

Our course lay along the bank of Ham's fork, through a hilly and stony, but not a rocky country; the willow flourished on the margin of the stream, and occasionally the eye was relieved, on scanning the plain, by a pretty clump of cottonwood or poplar trees. The cock of the plains is very abundant here, and our pretty little summer yellow bird, (*Sylvia aestiva*,) [**Yellow Warbler**] one of our most common birds at home, is our constant companion. How natural sounds his little monotonous stave, and how it seems to carry us back to the dear scenes which we have exchanged for the wild and pathless wilderness!

[On this day they heard the familiar **Veery**, though it would have been from the darker more heavily spotted western population: "On the 3d of July we were serenaded by this old acquaintance, in the very central chain of the Rocky Mountains, on the borders of Ham's Fork of the Colorado, as well as in the thickets of Lewis River of the Shoshonee [Snake River]" (Nuttall 1840: 397).]

4th. – We left Ham's fork this morning, – now diminishing to a little purling brook, – and passed across the hills in a north-westerly direction for about twenty miles, when we struck Muddy creek. This is a branch of Bear river, which empties into the Salt lake, or "lake Bonneville," as it has been lately named, for what reason I know not. Our camp here, is a beautiful and most delightful one. A large plain, like a meadow, of rich, waving grass, with a lovely little stream running through the midst, high hills, capped with shapely cedars on two sides, and on the others an immense plain, with snow clad mountains in the distance. This being a memorial day, the liquor kegs were opened, and the men allowed in abundance. We, therefore, soon had a renewal of the coarse and brutal scenes of the rendezvous. Some of the bacchanals called for a volley in honor of the day, and in obedience to the order, some twenty or thirty "happy" ones reeled into line with their muzzles directed to every point of the compass, and when the word "fire" was given, we who were not "happy" had to lie flat upon the ground to avoid the bullets which were careering through the camp. ["This being the 4th of July the men must needs show their "Independance" and such another drunken crazy hooting quarrelling fighting frolic I seldom witnessed. Yes, even in this western world ardent spirits is the bane of poor infatuated men" (Jason Lee 1916: 142-143).]

In this little stream, the trout are more abundant than we have yet seen them. One of our *sober* men took, this afternoon, upwards of thirty pounds. These fish would probably average fifteen or sixteen inches in length, and weigh three-quarters of a pound; occasionally, however, a much larger one is seen.

WYOMING

Wind River Range

Townsend joins
hunting party
16th - 26th July

Grizzly Bear attack and first started to
see White-tailed Jackrabbit
10th July

Nuttall finds an American Pika.
His account written up by Bachman
in *Viviparous Quadrupeds*

Fort Hall

Blackfoot River

Ross's Creek

Green River

More Brewer's Sparrow collected
15th June
These specimens later painted by
Audubon

Whooping Cranes
White Pelicans
Canvasbacks
Black Ducks
N. Shovelers
9th July

Snake River

Portneuf River

IDAHO

Sage Thrasher collected.
A new species
15th June

White Clay Pits
(Soda Springs)

Sweetwater

Sheep
Rock

Lewis's Woodpecker
Clark's Nutcracker
8th July

Bison "in every direction"
Common Poorwill heard again
16th June

South
Pass
7,550ft

Green-tailed Towhee
collected. A new species
12th July

Ham's Fork

Bison
Pronghorn
Elk, Sage Grouse
"Abundant"

Sandy River

First Greater Sage Grouse seen
Two days east of Green River

Bear
Lake

Least Chipmunk
collected "near mouth of
Big Sandy". A new species
No date

"Muddy Creek"

Trappers'
Rendevous
1834

UTAH

Swainson's Hawk nest
found on wooded stream
leading into the Green River
No date

Black Fork

Great
Salt
Lake

Bear River

Green River

15th July - 6th August   10th July   8th July   5th July   22nd June - 2nd July   19th June   15th June   14th June

1834

0   50   100 miles

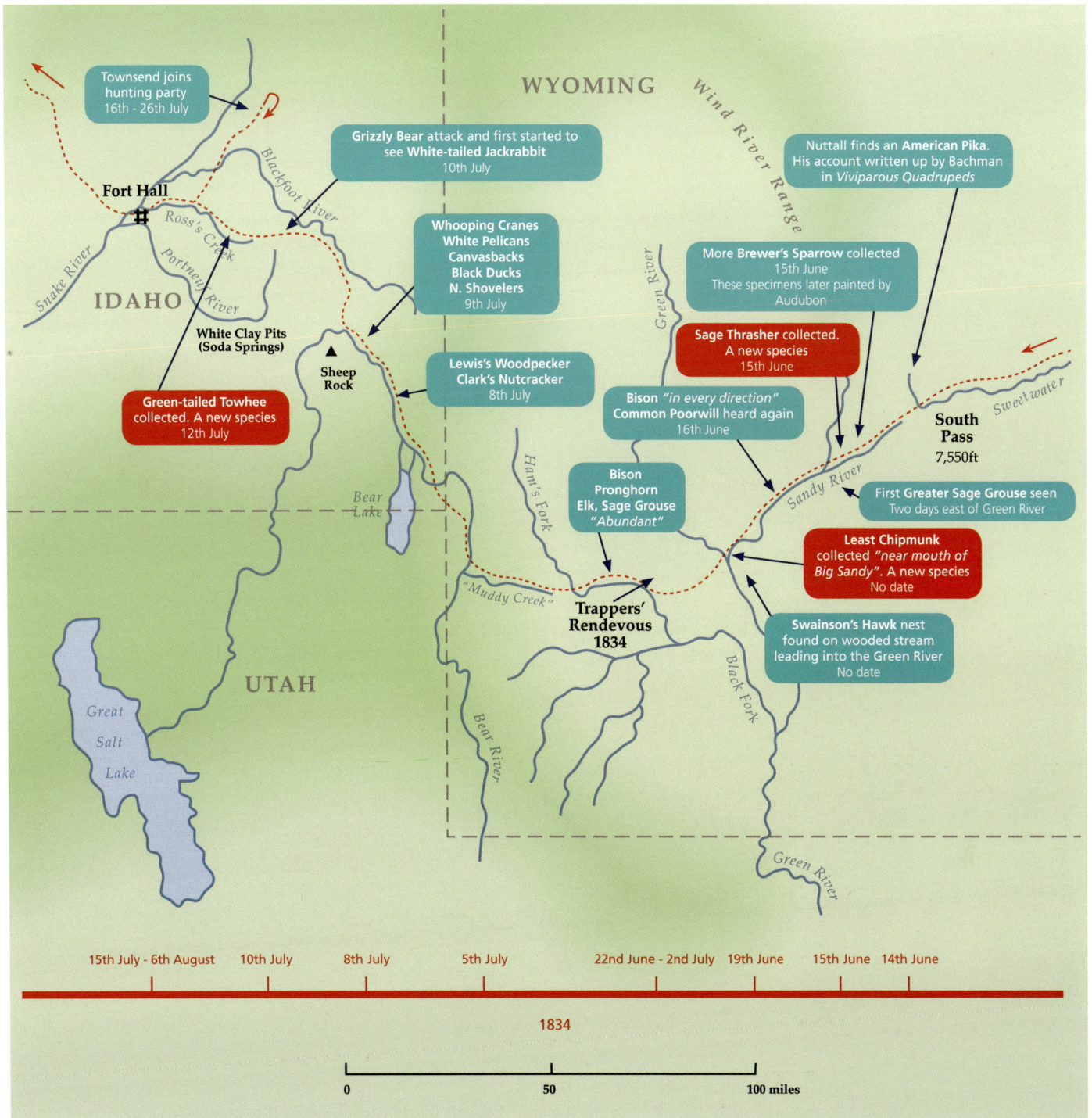

Map 5 - South Pass to Fort Hall, 14th June to 6th August 1834.

Lupins at Ham's Fork.

Flowers and conifers near the Bear River.

Aspens.

5th. – We travelled about twenty miles this day, over a country abounding in lofty hills, and early in the afternoon arrived on Bear river, and encamped. This is a fine stream of about one hundred and fifty feet in width, with a moveable sandy bottom. The grass is dry and poor, the willow abounds along the banks, and at a distance marks the course of the stream, which meanders through an alluvial plain of four to six miles in width ...

On the next day [6th] we crossed the river, which we immediately left, to avoid a great bend, and passed over some lofty ranges of hills and through rugged and stony valleys between them; the wind was blowing a gale right ahead, and clouds of dust were flying in our faces, so that at the end of the day, our countenances were disguised as they were on the plains of the Platte. The march to-day has been a most laborious and fatiguing one both for man and beast; we have travelled steadily from morning till night, not stopping at noon; our poor horses' feet are becoming very much worn and sore, and when at length we struck Bear river again and encamped, the wearied animals refused to eat, stretching themselves upon the ground and falling asleep from exhaustion.

Trout, grayling, and a kind of char are very abundant here – the first very large. The next day we travelled but twelve miles, it being impossible to urge our worn-out horses farther. Near our camp this evening we found some large gooseberries and currants, and made a hearty meal upon them. They were to us peculiarly delicious. We have lately been living entirely upon dried buffalo, without vegetables or bread; even this is now failing us, and we are upon short allowance. Game is very scarce, our hunters cannot find any, and our Indians have killed but two buffalo for several days. Of this small stock they would not spare us a mouthful, so it is probable we shall soon be hungry.

The alluvial plain here presents many unequivocal evidences of volcanic action, being thickly covered with masses of lava, and high walls and regular columns of basalt appear in many places. The surrounding country is composed, as usual, of high hills and narrow, stony valleys between them; the hills are thickly covered with a growth of small cedars, but on the plain, nothing flourishes but the everlasting wormwood, or *sage* as it here called.

["On the 6th of July we already saw flocks of young Shore Larks [Horned Larks] flying about, on the plains of Bear River in the Rocky Mountains (Nuttall 1840: 524).]

Our encampment on the 8th, was near what are called the "White-clay pits," still on Bear river. The soil is soft chalk, white and tenacious; and in the vicinity, are several springs of strong supercarbonated water, which bubble up with all the activity of artificial fountains. The taste was very agreeable and refreshing ...

In a thicket of common red cedars [Rocky Mountain Juniper *Juniperus scopulorum*[5]], near our camp, I found, and procured several specimens of two beautiful and rare birds which I had never before seen – the Lewis' woodpecker and Clark's crow, (*Picus torquatus* and *Corvus columbianus*) [**Lewis's Woodpecker** and **Clark's Nutcracker**[6]].

We remained the whole of the following day [9th] in camp to recruit our horses, and a good opportunity was thus afforded me of inspecting all the curiosities of this wonderful region, and of procuring some rare and valuable specimens of birds. Three of our hunters sallied forth in pursuit of several buffalo whose tracks had been observed by some of the men, and we were overjoyed to see them return in the evening loaded with meat and marrow bones of two animals which they had killed.

92

LEWIS'S WOODPECKER AND CLARK'S NUTCRACKER

Lewis's Woodpecker and Clark's Nutcracker as depicted by Audubon in plates CCCCXVI (detail) and CCCLXII (detail). Audubon probably used Townsend's specimen for both these species.

"LEWIS' WOODPECKER. Picus Torquatus ... Mr. Townsend says, "We first found them on Bear river ... They frequently perch crossways upon the smaller branches of trees, as well as against their trunks, climb with the usual ease and activity of other species, and are in the frequent habit of darting out from the tree on which they had stationed themselves, and after having performed a circular gyration in the air, returning immediately to the branch from which they had started; as they near the latter again, they spread their wings horizontally, and sail to their perch like some of the Hawks. Both sexes incubate" (Audubon 1840-44, 4: 280-281).

"CLARKE'S Crow, *Corvus columbianus*. First found on Bear river, and afterwards on the Blue Mountains, plentiful. Its flight is very unlike that of the Common Crow [American Crow], being performed by jerks, like that of the Woodpecker. When sitting, it is almost constantly screaming; its voice is very harsh and grating, and consists of one rather prolonged note. It breeds here in very high pine trees ... The *Corvus columbianus* is never seen within five hundred miles of the mouth of the Columbia. It appears generally to prefer a mountainous country and pine trees; and feeds chiefly on insects and their larvae. J.K.T." (quoted in Audubon 1840-44, 4: 127-128).

White Pelicans.

We saw here the **whooping crane**[7], and **white pelican**, numerous; and in the small streams near the bases of the hills, the common canvass-back duck [**Canvasback**], shoveller [**Northern Shoveler**], and **black duck**[8], (*Anas obscura*,) were feeding their young.

We were this evening visited by Mr. Thomas McKay, an Indian trader of some note in the mountains. He is a step-son of Dr. McLaughlin, the chief factor at Fort Vancouver, on the Columbia, and the leader of a party of Canadians and Indians, now on a hunt in the vicinity. This party is at present in our rear, and Mr. McKay has come ahead in order to join us, and keep us company until we reach Portneuf river, where we intend building a fort.

10th. – We were moving early this morning: our horses were very much recruited, and seemed as eager as their masters to travel on. It is astonishing how soon a horse revives, and overcomes the lassitude consequent upon fatigue, when he is allowed a day's rest upon tolerable pasture ...

Early in the afternoon we passed a large party of white men, encamped on the lava plain near one of the small streams. Horses were tethered all around, and men were lolling about playing games of cards, and loitering through the camp, as though at a loss for employment. We soon ascertained it to be Captain Bonneville's company resting after the fatigues of a long march. Mr. Wyeth and Captain Stewart visited the lodge of the "bald chief," and our party proceeded on its march. The difficulties of the route seemed to increase as we progressed, until at length we found ourselves wedged in among huge blocks of larva and columns of basalt, and were forced, most reluctantly, to retrace our steps for several miles, over the impediments which we had hoped we were leaving forever behind us. We had nearly reached Bonneville's camp again, when Captains Wyeth and Stewart joined us, and we struck into another path which proved more tolerable ...

Wyeth gave us a rather amusing account of his visit to the worthy captain. He and Captain Stewart were received very kindly by the veteran, and every delicacy that the lodge afforded was brought forth to do them honor. Among the rest, was some metheglen or diluted alcohol sweetened with honey, which the good host had concocted; this dainty beverage was set before them, and the thirsty guests were not slow in taking advantage of the invitation so obligingly given. Draught after draught

Blackfoot River.

of the precious liquor disappeared down the throats of the visitors. Towards evening, we struck Blackfoot river, a small sluggish, stagnant stream, heading with the waters of a rapid rivulet passed yesterday, which empties into the Bear river. This stream passes in a north-westerly direction through a valley of about six miles in width, covered with quagmires, through which we had great difficulty in making our way. [*It is the resort of great numbers of geese, ducks, cranes, &c., which breed along its margin. The large* **white pelican** *is also frequent* (MS Journal). [It may have been here, that they first encountered the **White-tailed Jackrabbit**: "We first saw it on the plains of the Black-foot River, west of the mountains, and observed it in all similar situations during our route to the Columbia. When first seen, which was in July, it was lean and unsavoury, having, like our common species, the larva of an insect imbedded in its neck" (JKT, quoted in Bachman 1839c: 92), see p. 126-127.] As we approached our encampment, near a small grove of willows, on the margin of the river, a tremendous **grizzly bear** rushed out upon us. Our horses ran wildly in every direction, snorting with terror, and became nearly unmanageable. Several balls were instantly fired into him, but they only seemed to increase his fury ... One of the pack horses was fairly fastened upon by the terrific claws of the brute, and in the terrified animal's efforts to escape the dreaded gripe, the pack and saddle were broken to pieces and disengaged. One of our mules also lent him a kick in the head while pursuing it up an adjacent hill, which sent him rolling to the bottom. Here he was finally brought to a stand.

The poor animal was so completely surrounded by enemies that he became bewildered. He raised himself upon his hind feet, standing almost erect, his mouth partly open, and from his protruding tongue the blood fell fast in drops. While in this position, he received about six more balls, each of which made him reel. At last, as in complete desperation, he dashed into the water, and swam several yards with astonishing strength and agility, the guns cracking at him constantly; but he was not to proceed far. Just then, Richardson, who had been absent, rode up, and fixing his deadly aim upon him, fired a ball into the back of his head, which killed him instantly. The strength of four men was required to drag the ferocious brute from the water ...

This evening, our pet antelope, poor little "Zip Coon," met with a serious accident. The mule on which he rode, got her feet fastened in some lava blocks, and, in the struggle to extricate herself, fell violently on the pointed fragments. One of the delicate legs of our favorite was broken, and he was otherwise so bruised and hurt, that, from sheer mercy, we ordered him killed ... We have sometimes taken young grizzly bears, but these little fellows, even when not larger than puppies, are so cross and snappish, that it is dangerous to handle them ... The young buffalo calf is also very often taken, and if removed from the mother, and out of sight of the herd, he will follow the camp as steadily as a dog ...

I had an adventure of this sort a few days before we arrived at the rendezvous. I captured a large bull calf, and with considerable difficulty, managed to drag him into the camp, by means of a rope noosed around his neck, and made fast to the high pommel of my saddle. Here I attached him firmly by a cord to a stake driven into the ground, and considered him secure. In a few minutes, however, he succeeded in breaking his fastenings, and away he scoured out of the camp. I lost no time in giving chase, and although I fell flat into a ditch, and afforded no little amusement to our people thereby, I soon overtook him, and was about seizing the stranded rope, which was still around his neck, when, to my surprise, the little animal showed fight; he came at me with all his force, and dashing his head into my breast, bore me to the ground in a twinkling. I, however, finally succeeded in recapturing him, and led and pushed him back into the camp; but I could make nothing of him; his stubbornness would neither yield to severity or kindness, and the next morning I loosed him and let him go.

White-tailed Jackrabbit.

11th. – On ascending a hill this morning, Captain Wyeth, who was at the head of the company, suddenly espied an Indian stealing cautiously along the summit ... The appearance of this Indian is a proof that others are lurking near; and if the party happens to be large, they may give us some trouble. We are now in a part of the country which is almost constantly infested by the Blackfeet; we have seen for several mornings past, the tracks of moccasins around our camp, and not unfrequently the prints of unshod horses, so that we know we are narrowly watched; and the slumbering of one of the guard, or the slightest appearance of carelessness in the conduct of the camp, may bring the savages whooping upon us like demons.

Our encampment this evening is one of the head branches of the Blackfoot river, from which we can see the three remarkable conic summits known by the name of the "*Three Butes*" or "*Tetons.*" Near these flows the Portneuf, or south branch of Snake or Lewis' river. Here is to be another place of rest, and we look forward to it with pleasure both on our own account and on that of our wearied horses.

Green-tailed Towhee. First described by Audubon from an immature bird collected by Townsend on 12th July 1834 near Ross's Creek. Audubon never painted this species - the first colour plate of an adult male (below) appeared in Cassin's *Illustrations of the Birds of California* (1856), depicted by G. G. White and W. E. Hitchcock from a specimen collected by William Gambel or Samuel Woodhouse.

GREEN-TAILED TOWHEE

"GREEN-TAILED SPARROW. *Fringilla Chlorura* ... The following notice respecting this bird is by Dr TOWNSEND: – "July 12. 1834. I shot this morning a new and singularly marked Sparrow[9]. The specimen is, however, unfortunately young, and the plumage is not fully developed. I feel in great hopes of finding the adult in similar situations on our route. It is a true *Fringilla*. The head is of a light brownish colour, spotted with dusky; back varied with a dusky and greenish-olive; rump brownish, spotted with dusky; wings plain dusky, the outer vanes, as well as the tail-feathers, greenish-yellow; axillaries yellow; throat white; a longitudinal line of black on either side; breast and flanks white, spotted or streaked with black; belly whitish; vent tinged with light brown, inclining to ochreous." "The measurements of this species I find I have not given. I probably omitted them from the supposition that I should at a future time find the perfect bird. In this I was, however, disappointed: I never saw it afterwards" (Audubon 1839, 5: 336).

12th. – In the afternoon we made camp on Ross's creek, a small branch of Snake river. The pasture is better than we have had for two weeks, and the stream contains an abundance of excellent trout. Some of these are enormous, and very fine eating. They bite eagerly at a grasshopper or minnow, but the largest fish are shy, and the sportsman requires to be carefully concealed in order to take them. We have here none of the fine tackle, jointed rods, reels, and silkworm gut of the accomplished city sportsman; we have only a piece of common cord, and a hook seized on with half-hitches, with a willow rod cut on the banks of the stream; but with this rough equipment we take as many trout as we wish, and who could do more, even with all the curious contrivances of old Izaac Walton or Christopher North?

The band of Indians which kept company with us from the rendezvous, left us yesterday, and fell back to join Captain Bonneville's party, which is travelling on behind. We do not regret their absence; for although they added strength to our band, and would have been useful in case of an attack from Blackfeet, yet they added very materially to our cares, and gave us some trouble by their noise, confusion, and singing at night.

[On 12th July 1834 David Douglas, a veteran botanist of the lower Columbia River, was killed in the Hawaiian Islands. Townsend and Nuttall did not hear the news until November.]

On the 14th, we travelled but about six miles, when a halt was called, and we pitched our tents upon the banks of the noble Shoshoné or Snake river. It seems now, as though we were really nearing the western extremity of our vast continent. We are now on a stream which pours its waters directly into the Columbia, and we can form some idea of the great Oregon river by the beauty and magnitude of its tributary. Soon after we stopped, Captain W., Richardson, and two others left us to seek for a suitable spot for building a fort, and in the evening they returned with the information that an excellent and convenient place had been pitched upon, about five miles from our present encampment ...

The next morning [15th] we moved early, and soon arrived at our destined camp. This is a fine large plain on the south side of the Portneuf, with an abundance of excellent grass and rich soil. The opposite side of the river is thickly covered with large timber of the cottonwood and willow, with a dense undergrowth of the same, intermixed with service-berry [*Amelanchier sp.*] and currant bushes [*Ribes sp.*].

Most of the men were immediately put to work, felling trees, making horse-pens, and preparing the various requisite materials for the building, while others were ordered to get themselves in readiness for a start on the back track, in order to make a hunt, and procure meat for the camp. To this party I have attached myself, and all my leisure time to-day is employed in preparing for it. [There is, no doubt, a number of interesting birds to be found here, but I have not time to procure any, having engaged to start in the morning with a hunting party (MS Journal extract, Hulbert 1934: 191).]

Our number will be twelve, and each man will lead a mule with a pack-saddle, in order to bring in the meat that we may kill. Richardson is the principal of this party, and Mr. Ashworth has also consented to join us, so that I hope we shall have an agreeable trip. There will be but little hard work to perform; our men are mostly of the best, and no rum or cards are allowed.

[Nuttall reported "On the 15th of July, arriving at the borders of the Shoshonee, or Snake river, we first met with the Common Magpie [**American Magpie**] on our route, mostly accompanied by the [**Common**] **Raven**, but there were no Crows. The young birds were so familiar and greedy, approaching the encampment in quest of food, as to be easily taken by the Indian boys, when they soon became reconciled to savage domesticity. The old birds were sufficiently shy, but the young ones were observed hopping and croaking around us, and tugging at any offal or flesh meat

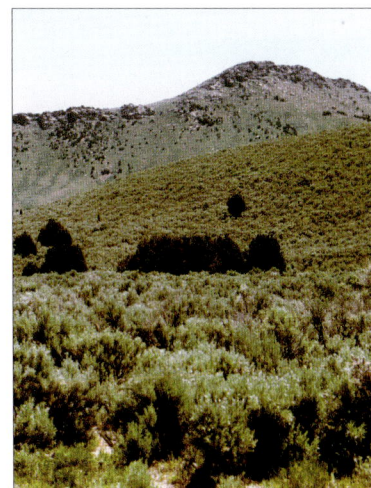

Sagebrush habitat near Ross's Creek (above). American Magpie and Common Ravens (below).

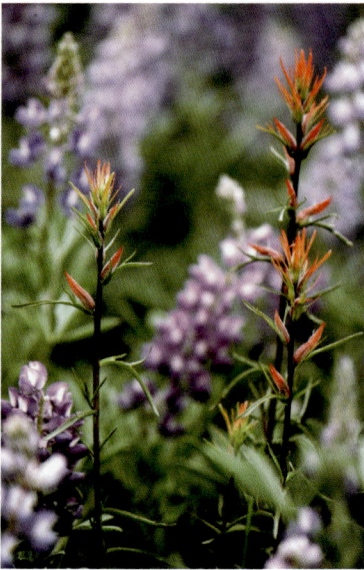

Indian Paintbrush *Castilleja sp.*

thrown out, like so many Vultures ... If chased off for an instant, they returned the next, and their monotonous and gluttonous croak was heard around us at all hours of the day" (Audubon 1840-44, 4: 100; based on Nuttall 1840: 232-233). Nuttall also reported seeing **Marsh Wrens**, of the western population, "in a marsh near Fort Hall. It is a remarkably active and quaint little species, skipping and diving about with great activity after its insect food and their larvae among the rank grass and rushes, near ponds, and the low banks of rivers ... neglected and seldom seen, it rears its young in security" (Nuttall 1840: 497). Presumably Townsend did not see this wren species as it is not on any of his lists.]

1. Nuttall (1840: 804) records Greater Sage-Grouse were first seen near Laramie's Fork but he seems to have become confused with the Sharp-tailed Grouse. Both he and Townsend say that Sharp-tailed Grouse were first seen at Laramie's Fork.

2. Townsend is mistaken, Nuttall is correct: the Violet-green Swallow does not build a mud nest.

3. Sir William Drummond Stewart (1795-1871) was from Perthshire, Scotland. A veteran of the Peninsular Wars and decorated for bravery at Waterloo, he afterwards developed a taste for the American West, making several hunting trips up the Missouri and into the Rocky Mountains between 1833 and 1838, making a final extravagant trip in 1843. In 1837 he hired the artist Alfred Jacob Miller who produced some of the most evocative paintings of western scenes, trappers and native tribes.

4. Little is known about Ashworth except that he was a scrounger and for that reason not liked by John McLoughlin at Fort Vancouver.

5. Townsend is referring to the Common or Eastern Red Cedar *Juniperus virginiana* but here it would be Rocky Mountain Juniper *Juniperus scopulorum* that may grow to 40 feet tall. This area is outside the range of Western Red Cedar *Thuja plicata*.

6. The USNM card catalogue of specimens records a Clark's Nutcracker collected by Townsend that is dated 16th June 1834, a few days before reaching the Trappers' Rendezvous.

7. Graustein (1967: 300) says that the Whooping Cranes seen by Townsend near Soda Springs must have been Sandhill Cranes. But Whooping Cranes may well have bred in that area at that time (nearby Gray's Lake was chosen for reintroduction of the species in the late 1900s). When Townsend saw Sandhill Cranes in Missouri he called them Sandhill Cranes. When he saw these cranes near Soda Springs he called them Whooping Cranes. In addition Nuttall had experience of Whooping Cranes in 1811 when he saw hundreds and would have kept Townsend right if need be. However, it is unlikely that Townsend saw Whooping Cranes on the Columbia River (Jobanek 1994).

8. Although Jobanek (1994) considered most Black Duck records for Oregon to be dubious, there seems to be no good reason not to accept Townsend's record for Idaho.

9. According to Audubon, a young male Golden-crowned Sparrow was also shot by Townsend on 12th July 1834 (Audubon 1840-44, 3: 163). Were there really two similar looking juvenile birds taken that day or has Audubon misidentified the Green-tailed Towhee as a Golden-crowned Sparrow? Interestingly, the type specimen of the Green-tailed Towhee (USNM1896) has an old label (not in JKT's handwriting) indicating that even the museum staff had at one time mistakenly identified it as "*Zonotrichia atricapilla*" [Golden-crowned Sparrow].

Fort Hall to Boise River

"I heard a sort of angry growl or grunt directly before me – and instantly after, saw a grizzly bear of the largest kind erect himself upon his hind feet within a dozen yards of me, his savage eyes glaring with horrible malignity, his mouth wide open, and his tremendous paws raised as though ready to descend upon me. For a moment, I thought my hour had come."

Townsend's *Narrative*, 23rd July 1834.

Eleven-day bison hunting excursion from Fort Hall

July 16th.– Our little hunting party of twelve men [not including Wyeth or Nuttall], rode out of the encampment this morning, at a brisk trot, which gait was continued until we arrived at our late encampment on Ross' creek, having gone about thirty miles. Here we came to a halt, and made a hearty meal on a buffalo which we had just killed ...

The next day [17th] we rode thirty-four miles, and encamped on a pretty little stream, fringed with willows, running through the midst of a large plain. Within a few miles, we saw a small herd of buffalo, and six of our company left the camp for a hunt. In an hour two of them returned, bringing the meat of one animal. We all commenced work immediately, cutting it in thin slices, and hanging it on the bushes to dry. By sundown, our work was finished, and soon after dark, the remaining hunters came in, bringing the best parts of three more. This will give us abundance of work for to-morrow, when the hunters will go out again ...

On the 20th, we moved our camp to a spot about twelve miles distant, where Richardson, with two other hunters, stopped yesterday and spent the night. They had killed several buffalo here, and were busily engaged in preparing the meat when we joined them ... [*This evening the roaring of the bulls in the gang near us is terrific, and these sounds are mingled with the howling of large packs of wolves, which regularly attend upon them, and the hoarse screaming of hundreds of ravens* [**Common Ravens**] *flying over head* (MS Journal, extracts in Hulbert 1934: 194).]

"Hunting the Buffalo" - one of only two illustrations in Townsend's *Narrative* (1839).

21st.– The buffalo appear even more numerous than when we came, and much less suspicious than common. The bulls frequently pass slowly along within a hundred yards of us, and toss their shaggy and frightful looking heads as though to warn us against attacking or approaching them ...

This afternoon [23rd] I observed a large flock of wild geese [**Canada Geese**] passing over; and upon watching them, perceived that they alighted about a mile and a half from us, where I knew there was a lake. Concluding that a little change of diet might be agreeable, I sallied forth with my gun across the plain in quest of the birds. I soon arrived at a thick copse of willow and currant bushes, which skirted the water, and was about entering, when I heard a sort of angry growl or grunt directly before me – and instantly after, saw a grizzly bear of the largest kind erect himself upon his hind feet within a dozen yards of me, his savage eyes glaring with horrible malignity, his mouth wide open, and his tremendous paws raised as though ready to descend upon me. For a moment, I thought my hour had come, and that I was fated to die an inglorious death away from my friends and my kindred; but after waiting a moment in agonizing suspense, and the bear showing no inclination to advance, my lagging courage returned, and cocking both barrels of my gun, and presenting it as steadily as my nerves would allow, full at the shaggy breast of the creature, I retreated slowly backwards ... when I had placed about a hundred yards between us, I wheeled about and flew, rather than ran, across the plain towards the camp. Several times during this run for life, (as I considered it,) did I fancy that I heard the bear at my heels; and not daring to look over my shoulder to ascertain the fact, I only increased my speed, until the camp was nearly gained, when, from sheer exhaustion I relaxed my efforts, fell flat upon the ground, and looked behind me. The whole space between me and the copse was untenanted, and I was forced to acknowledge, with a feeling strongly allied to shame, that my fears alone had represented the bear in chase of me ...

[24th. – We are visited every night by that fiend of the Rocky Mountains, the brown or grisly bear. The audacious fellows sometimes come grunting within a few yards of where we are lying, attracted by the meat which is piled around us; but they have never had the temerity to rob us; they generally retire peaceful and hungry as they came, after they have scented the stranger ... We have now been absent eight days, and we suppose that it will occupy three days to return, which will nearly consume the time allotted us. We are about sixty miles distant from the main party (MS Journal, extracts in Hulbert 1934: 195).]

On the morning of the 25th, we commenced baling up our meat in buffalo skins dried for the purpose. Each bale contains about a hundred pounds, of which a mule carries two; and when we had finished, our twelve long-eared friends were loaded. Our limited term of absence is now nearly expired, and we are anxious to return to the fort in order to prepare for the journey to the lower country ...

We travelled, this day, thirty miles, and the next afternoon, at 4 o'clock, arrived at the fort. On the route we met three hunters, whom Captain W. had sent to kill game for the camp. They informed us that all hands have been for several days on short allowance, and were very anxious for our return ...

Our people were all delighted to see us arrive, and I could perceive many a longing and eager gaze cast upon the well filled bales, as our mules swung their little bodies through the camp. My companion, Mr. N., had become so exceedingly thin that I should scarcely have known him; and upon my expressing surprise at the great change in his appearance, he heaved a sigh of inanity, and remarked that I "would have been as thin as he if I had lived on old *Ephraim* for two weeks, and short allowance at that." I found, in truth, that the whole camp had been subsisting, during our absence, on little else than two or three grizzly bears which had been killed in the neighborhood; and with a complacent glance at my own rotund and *cow-fed* person, I wished my *poor* friend better luck for the future ...

American Bison.

At the fort, affairs look prosperous: the stockade is finished; two bastions have been erected, and the work is singularly good, considering the scarcity of proper building tools. The house will now soon be habitable, and the structure can then be completed at leisure by men who will be left here in charge, while the party travels on to its destination, the Columbia. ["This Fort ... is on Lewis' Fork [Snake River] in an unpleasant situation being surrounded with sand which is sometimes driven before the wind in as great quantity as snow in the east" (Jason Lee 1916: 242).]

The next day [27th], being the Sabbath, our good missionary, Mr. Jason Lee, was requested to hold a meeting, with which he obligingly complied. A convenient, shady spot was selected in the forest adjacent, and the greater part of our men, as well as the whole of Mr. McKay's company, including the Indians, attended. The usual forms of the Methodist service, (to which Mr. L. is attached,) were gone through, and were followed by a brief, but excellent and appropriate exhortation by that gentleman. The people were remarkably quiet and attentive, and the Indians sat upon the ground like statues. Although not one of them could understand a word that was said, they nevertheless maintained the most strict and decorous silence, kneeling when the preacher kneeled, and rising when he rose, evidently with a view of paying him and us a suitable respect, however much their own notions as to the proper and most acceptable forms of worship, might have been opposed to ours.

A meeting for worship in the Rocky mountains is almost as unusual as the appearance of a herd of buffalo in the settlements. A sermon was perhaps never preached here before; but for myself, I really enjoyed the whole scene; it possessed the charm of novelty, to say nothing of the salutary effect which I sincerely hope it may produce.

Mr. Lee is a great favorite with the men, deservedly so, and there are probably few persons to whose preaching they would have listened with so much complaisance. I have often been amused and pleased by Mr. L.'s manner of reproving them for the coarseness and profanity of expression which is so universal amongst them ... [Jason Lee said "O, that I could address the Indians in their language. I did not attempt to preach, but gave a short exhortation from I Cor[inthians]. 10-21[= 10-31]. – "Whether therefore ye eat or drink" &c." (Lee 1916: 241).] .

["*Fort Hall*" *is situated on the left or south bank of the river, and about one hundred yards from it. The right bank is thickly covered with excellent cottonwood timber, and to the left, in the distance, rise high stony mountains ... The beaver appears numerous on this river and its tributaries; the otter* **[Northern River Otter]** *is also not uncommon: a number of the men have set their traps and catch several daily* (MS Journal, extracts in Hulbert 1934: 201).]

On the 30th of July, Mr. McKay and his party left us for Fort Vancouver, Captain Stewart and our band of missionaries accompanying them. The object of the latter in leaving us, is, that they may have an opportunity of travelling more slowly than we should do, on account, and for the benefit of the horned cattle which they are driving to the lower country. We feel quite sad in the prospect of parting from those with whom we have endured some toil and danger, and who have been to some of us as brothers, throughout our tedious journey; but, if no unforeseen accident occurs, we hope to meet them all again at Walla-Walla, the upper fort on the Columbia. As the party rode off, we fired three rounds, which were promptly answered, and three times three cheers wished the travellers success.

August 5th.– At sunrise this morning, the "star-spangled banner" was raised on the flag-staff at the fort, and a salute fired by the men, who, according to orders, assembled around it. All in camp were then allowed the free and uncontrolled use of liquor, and, as usual, the consequence was a scene of rioting, noise, and fighting, during the whole day; some became so drunk that their senses fled them entirely, and they were therefore harmless; but by far the greater number were just sufficiently under the influence of the vile trash, to render them in the conduct disgusting and tiger-like. We had

Jason Lee (1803 - 1845), Methodist missionary and Oregon pioneer.

"gouging," biting, fisticuffing, and "stamping" in the most "scientific" perfection; some even fired guns and pistols at each other, but these weapons were mostly harmless in the unsteady hands which employed them. Such scenes I hope never to witness again; they are absolutely sickening, and cause us to look upon our species with abhorrence and loathing ...

Townsend and Nuttall continue westwards from Fort Hall

The next morning [6th] we commenced packing, and at 11 o'clock bade adieu to "Fort Hall." Our company [led by Wyeth] now consists of but thirty men, several Indian women, and one hundred and sixteen horses. We crossed the main Snake or Shoshoné river, at a point about three miles from the fort. It is here as wide as the Missouri at Independence, but, beyond comparison, clearer and more beautiful.

Immediately on crossing the river, we entered upon a wide, sandy plain, thickly covered with wormwood, and early in the afternoon, encamped at the head of a delightful spring, about ten miles from our starting place.

On the route, our hunters killed a young **grizzly bear**, which, with a few grouse, made us an excellent dinner ... [therefore probably **Sharp-tailed Grouse**, rather than Greater Sage-Grouse].

We shall now, for about ten days, be travelling through the most dangerous country west of the mountains, the regular hunting ground of the Blackfeet Indians, who are said to be often seen here in parties of hundreds, or even thousands, scouring the plain in pursuit of the buffalo. Traders, therefore, seldom travel this route without meeting them, and being compelled to prove their valor upon them; the white men are, however, generally the victors, although their numbers are always vastly inferior.

7th.– We were moving this morning with the dawn, and travelled steadily the whole day, over one of the most arid plains we have seen [Eastern Snake River Plain], covered thickly with jagged masses of lava, and twisted wormwood bushes. Both horses and men were jaded to the last degree; the former from the rough, and at times almost impassable nature of the track, and the latter from the excessive heat and parching thirst. We saw not a drop of water during the day ...

["7th. Started at day light and traveled 10 hours as fast as possible N.W. by W. 30 miles to the Bute ... the most southerly one ... the day was hot and we suffered some for water and found but a small supply on the N. side of the Bute" (Wyeth 1899: 77).]

[7th. Nuttall reported seeing **Clark's Nutcrackers** once more: "We again saw a considerable flock of the young birds early in August, in a lofty ravine, near the summit of the 'Three Butes,' or isolated and very remarkable mountains, about 30 or 40 miles west of Lewis's river, of the Shoshonee. They appeared somewhat shy, and were scattered through a grove of aspens, flying with a slight chatter, scarcely a caw, from the tops of the bushes and trees, to the ground, probably in quest of insect food ...and as Mr. Townsend remarks, [Clark's Nutcrackers have] a constant predilection for the pine forests" (Nuttall 1840: 252).]

Clark's Nutcracker.

Soon after night-fall, some signs of water were seen in a small valley to our left, and, upon ascending it, the foremost of the party found a delightful little cold spring; but they soon exhausted it, and then commenced, with axes and knives, to dig it out and enlarge it. By the time that Mr. N., and myself arrived, they had excavated a large space which was filled to overflowing with muddy water. We did not wait for it to settle, however, but throwing ourselves flat upon the ground, drank until we were ready to burst ...

[As they settled down for the night, Nuttall reported another **Common Poorwill**: "On the 7th of August, while camped in the high ravine of the southernmost of the Three Butes or insulated mountains which are so conspicuous from Lewis's River [Snake River], this bird, in the evening, flew from under a stone near the summit of the mountain. It was smaller than any other species we have, had no white marks or blotches either on the tail or wings, was bright ferruginous on the head, back and wings. It flew about hawking for insects for two or three hours near our elevated camp, but was now silent, the breeding season being probably past" (Nuttall 1840: 747).]

The next morning [8th] we made an early start towards a range of willows which we could distinctly see, at the distance of fifteen or twenty miles, and which we knew indicated Goddin's creek, so called from a Canadian of that name who was killed in the vicinity by the Blackfeet. Goddin's son, a half-breed, is now with us as a trapper; he is a fine sturdy fellow, and of such strength of limb and wind, that he is said to be able to run down a buffalo on foot, and kill him with arrows.

Goddin's creek [now Big Lost River] was at length gained, and after travelling a few miles along its bank we encamped in some excellent pasture. Our poor horses seemed inclined to make up for lost time here, as yesterday their only food was the straggling blades of a little dry and parched grass growing among the wormwood on the hills ...

The next morning [9th] we left Goddin's creek, and travelled for ten miles over a plain, covered as usual with wormwood bushes and lava. Early in the day, the welcome cry of "a buffalo! A buffalo!" was heard from the head of the company, and was echoed joyfully down the whole line. At the moment, a fine large bull was seen to bound from the bushes in our front, and tear off with all his speed over the plain. Several hunters gave him chase immediately, and in a few minutes we heard the guns that proclaimed his death. The killing of this animal is a most fortunate circumstance for us: his meat will probably sustain us for three or four days ...

On leaving the plain this morning [=10th?], we struck into a defile between some of the highest mountains we have yet seen. In a short time we commenced ascending, and continued passing over them, until late in the afternoon, when we reached a plain about a mile in width, covered with excellent grass, and a delightful cool stream flowing through the middle of it. Here we encamped, having travelled twenty-seven miles.

Mountain Cottontail collected *"near mouth of Malheur"*. A new species
No date

"Townsend's Mocking Thrush" seen. **Red Squirrel** and **Yellow-pine Chipmunk** collected
12th August

Common Poorwill seen on the butte by Nuttall
7th August

Fort Boise (built by HBC late 1834)

"Imperial Woodpecker" seen
14th August

Sawtooth Mountains

Goddin's Creek (Big Lost River)

Hyndman Peak 12,078ft

Three Buttes

IDAHO

South Fork Boise River

"American Pheasants" seen
13th August

Big Southern Butte 7,559ft

Western Snake River Plain

Snake River

Townsend's Pocket Gopher perhaps collected near here. A new species
No date

Camas Creek

Malade River (Big Wood River)

Eastern Snake River Plain

Fort Hall

OREGON

Swainson's Hawk co-type possibly collected near here, from *"plains of the Snake River"*
No date

Snake River

Snake River

NEVADA

UTAH

| 24th August | 22nd August | 20th August | 17th August | 13th August | 12th August | 7th August | 6th August |

1834

| 0 | 50 | 100 miles |

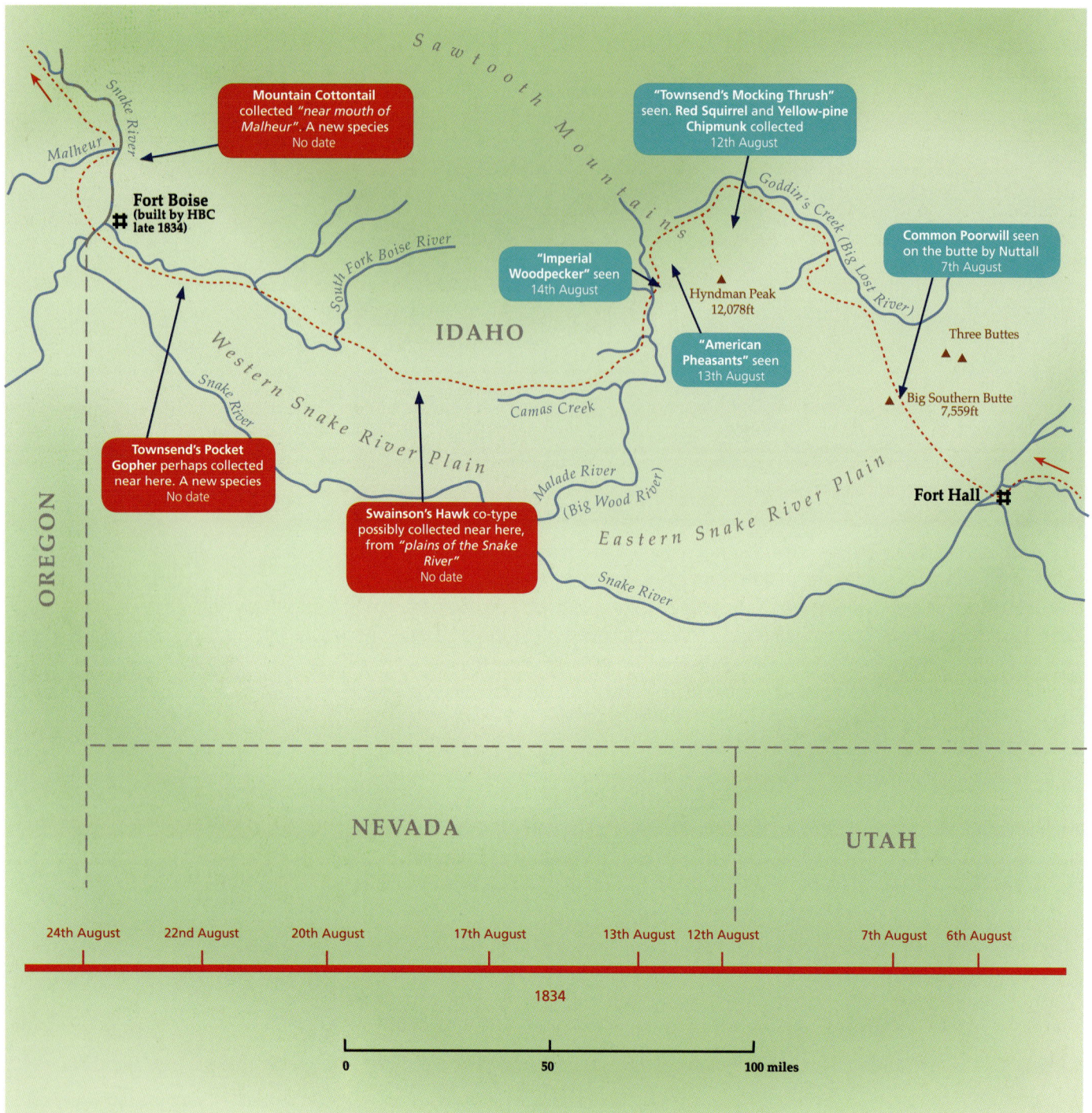

Map 6 - Across the Sawtooth Mountains, 6th - 24th August 1834

Eastern Snake River Plain looking towards the Big Lost River and Sawtooth Mountains.

104

Our journey, to-day, has been particularly laborious. We were engaged for several hours, constantly in ascending and descending enormous rocky hills, with scarcely the sign of a valley between them; and some of them so steep, that our horses were frequently in great danger of falling, by making a mis-step on the loose, rolling stones. I thought the Black Hills [Laramie Mountains], on the Platte, rugged and difficult of passage, but they sink into insignificance when compared with these.

We observed, on these mountains, large masses of greenstone, and beautiful pebbles of chalcedony and fine agate; the summits of the highest are covered with snow. In the mountain passes, we found an abundance of large, yellow currants [*Ribes aureum*], rather acid, but exceedingly palatable to men who have been long living on animal food exclusively. We all ate heartily of them; indeed, some of our people became so much attached to the bushes, that we had considerable difficulty to induce them to travel again.

10th.– We commenced our march at seven this morning, proceeding up a narrow valley, bordering our encampment in a north-easterly direction. The ravine soon widened, until it became a broad, level plain, covered by the eternal "sage" bushes, but was much less stony than usual. About mid-day, we left the plain, and shaped our course over a spur of one of the large mountains; then taking a ravine, in about an hour we came to the level land, and struck Goddin's creek again, late in the afternoon.

Our provision was all exhausted at breakfast, this morning, (most of our bull meat having been given to a band of ten trappers, who left us yesterday) ...[Townsend does not mention that this difficult day was his 25th birthday.]

On the day following [11th], Richardson killed two buffalo, and brought his horse heavily laden with meat to the camp. Our good hunter walked himself, that the animal might be able to bear the greater burthen. After depositing the meat in the camp, he took a fresh horse, and accompanied by three men, returned to the spot where the game had been killed, (about four miles distant,) and in the evening, brought in every pound of it, leaving only the heavier bones. The wolves [**Gray Wolves**] will be disappointed this evening; they are accustomed to dainty picking when they glean after the hunters, but we have now abandoned the "wasty ways" which so disgraced us when game was abundant; the despised leg bone, which was wont to be thrown aside with such contempt, is now polished of every tendon of its covering, and the savory hump is used as a kind of dessert after a meal of coarser meat.

Speaking of wolves, I have often been surprised at the perseverance and tenacity with which these animals will sometimes follow the hunter for a whole day, to feed upon the carcass he may leave behind him. When an animal is killed, they seem to mark the operation, and stand still at a most respectful distance, with drooping tail and ears, as though perfectly indifferent to the matter in progress. Thus will they stand until the game is butchered, the meat placed upon the saddle, and the hunter is mounted and on his way; then, if he glances behind him, he will see the wily forager stealthily crawling and prowling along towards the smoking remains, and pouncing upon it, and tearing it with tooth and nail, immediately as he gets out of reach.

During the day, the wolves are shy, and rarely permit an approach to within gun-shot; but at night, (where game is abundant,) they are so fearless as to come quite within the purlieus of the camp, and there sit, a dozen together, and howl hideously for hours. This kind of serenading, it may be supposed, is not the most agreeable; and many a time when on guard, have I observed the unquiet tossing of the bundles of blankets near me, and heard issue from them, the low, husky voice of some disturbed sleeper, denouncing heavy anathemas on the unseasonable music.

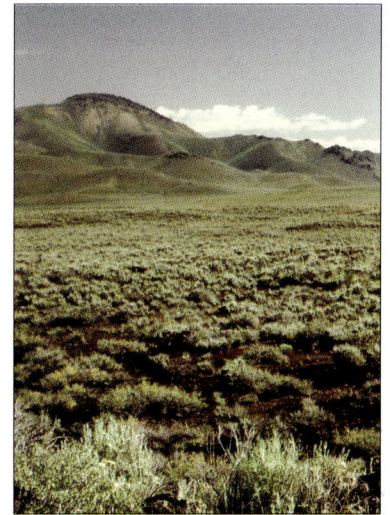

Sagebrush on the Snake River Plain.

From top: Audubon's original watercolour of Yellow-pine Chipmunks for plate XXIV of *The Viviparous Quadrupeds* (Note the date to the right of the chipmunks that Audubon has taken from Townsend's specimen labels). For a Yellow-pine Chipmunk specimen (ANSP 247) collected by Townsend and used by Audubon for the watercolour above see page 351.

Red Squirrel, plate V (detail) from *The Viviparous Quadrupeds*.

12th August. Townsend collected two **Yellow-pine Chipmunks** that were painted by Audubon. The original watercolour is inscribed "*White lined ground squirrel / Tamias quadrivittatus Say / August 12th 1834, / Rocky Mountains, / J.K.T / 1. Male. 2. Female.*" In *The Viviparous Quadrupeds* the two specimens appear in plate XXIV as "Four-striped Ground Squirrel", with two additional figures, but there is no mention of Townsend in the accompanying text. The Four-striped Ground Squirrel is now known as the Colorado Chipmunk *Tamias quadrivittatus* but Townsend's specimens were not that species because its range does not extend far enough northwards into Idaho and Wyoming to have been collected by him. The figures in the original watercolour have the olive-brown rumps and yellowish underside to the tail of Yellow-pine Chipmunk.

12th August. Townsend collected a squirrel that Bachman described as the "SCIURUS Richardsonii. Columbia Pine Squirrel". Audubon depicted it in plate V of *The Viviparous Quadrupeds* under the name of Richardson's Columbian Squirrel. It proved to be a form of **Red Squirrel**.

"This small species was first noticed by Lewis and Clarke, who deposited a specimen in the Philadelphia Museum [Peale's Museum], where it still exists. I have compared it with a specimen brought by Mr. Townsend, and find them identical. Richardson, who appears not to have seen it, supposes it to be a mere variety of the SCIURUS *hudsonicus* [Red Squirrel]. On the contrary, Mr. Townsend says in his notes: "It is evidently a distinct species; its habits are very different from those of the SCIURUS *hudsonicus*. It frequents the pine trees in the high range of the Rocky Mountains, west of the great chain, feeding upon the seeds contained in the cones.".. The specimen from which I have described is labelled "Rocky Mountains, August 12th, 1834" (Bachman 1839b: 64-67).

12th.– We shaped our course, this morning, towards what appeared to us a gap in a high and rugged mountain, about twenty miles ahead. After proceeding eight or ten miles, the character of the country underwent a remarkable and sudden change. Instead of the luxuriant sage bushes, by which the whole plains have hitherto been covered, and the compact and dense growth of willows which has uniformly fringed every stream and rivulet, the ground was completely denuded; not a single shrub was to be seen, nor the smallest appearance of vegetation, except in small patches near the water. The mountains, also, which had generally been rocky, and covered with low, tangled bushes, here abound in beautiful and shapely pine trees. Some of the higher peaks are, however, completely bare, and capped with enormous masses of snow.

After we had travelled about twelve miles, we entered a defile between the mountains, about five hundred yards wide, covered, like the surrounding country, with pines; and, as we proceeded, the timber grew so closely, added to a thick undergrowth of bushes, that it appeared almost impossible to proceed with our horses. The farther we advanced, the more our difficulties seemed to increase; obstacles of various kinds impeded our progress; – fallen trees, their branches tangled and matted together, large rocks and deep ravines, holes in the ground, into which our animals would be precipitated without the possibility of avoiding them, and an hundred other difficulties which beggar description. ["*12th* – walked barefoot to camp during the night through an infernal rough rocky prickly Bruisy swampy woody hole" (Wyeth 1899: 78).]

We travelled for six miles through such a region as I have attempted to describe, and at 2 o'clock encamped in a clear spot of ground, where we found excellent grass, and a cold, rapid stream. Soon after we stopped, Captain W. and Richardson left us, to look for a pass through the mountains ... [they] returned early next morning [13th in MS Journal], with the mortifying intelligence that no practicable pass through the mountain could be found. They ascended to the very summit of one of the highest peaks, above the snow and the reach of vegetation, and the only prospect which they had beyond, was a confused mass of huge angular rocks, over which even a wild goat could scarcely have made his way. Although they utterly failed in the object of their exploration, yet they were so fortunate as to kill a buffalo, (*the* buffalo,) the meat of which they brought on their horses ...

[13th.–] After a good breakfast, we packed our horses, and struck back on our trail of yesterday, in order to try another valley which we observed bearing parallel with this, at about three miles distant, and which we conclude must of course furnish a path through the mountain ... We have named this rugged valley, "Thornburg's *pass*," after one of our men of this name, (a tailor,) whom we have to thank for leading us into all these troubles. Thornburg crossed this mountain two years ago, and might therefore be expected to know something of the route, and as he was the only man in the company who had been here, Captain W. acted by his advice, in opposition to his own judgement ...

In the bushes, along the stream in this valley, the **black-tailed deer** (*Cervus macrourus*) is abundant. The beautiful creatures frequently bounded from their cover within a few yards of us, and trotted on before us like domestic animals; "they are so unacquainted with man" and his cruel arts, that they seem not to fear him.

We at length arrived on the open plain again, and in our route towards the other valley, we came to a large, recent Indian encampment, probably of Bannecks, who are travelling down to the fisheries on Snake river ... We travelled rapidly along the level land at the base of the mountain, for about three miles; we then began to ascend, and our progress was necessarily slow and tedious. The commencement of the Alpine path was, however, far better than we had expected, and we entertained the hope that the passage could be made without difficulty or much toil, but the farther we progressed, the more laborious the travelling became ...

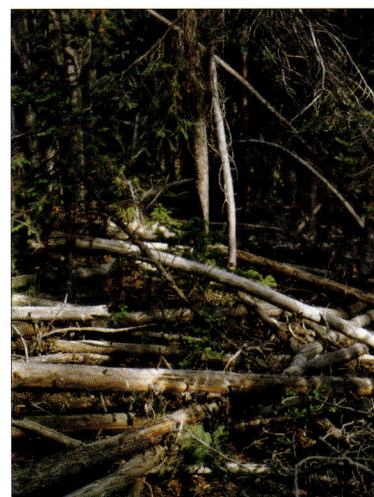

"fallen trees, their branches tangled and matted together... and an hundred other difficulties which beggar description..."

Black-tailed Deer.

American Pika. Known at that time as the Little Chief Hare, pikas are now classified in a separate family from hares and rabbits. Audubon does not say if he used any Townsend specimens in the preparation of plate LXXXIII (detail at right) but he may well have done.

As they struggled through the mountains Townsend collected an American Pika. According to the Rev Bachman, "It was a male, taken on the 13th of August, 1834 … The skull and teeth had been removed in the specimen we examined, except the upper incisors" (Bachman 1837; 354-355). The only surviving pika specimen attributable to Townsend must have been a different one, as it is a body mount with the skull still inside (ANSP 372); there is no precise date or collecting locality for this second specimen so it could have been taken here or acquired elsewhere.

Townsend seems not to have written anything about these curious mammals but Nuttall, in a letter to Bachman, wrote:

"I found its range to be in that latitude (42°) almost entirely alpine. I first discovered it by its peculiar cry, far up the mountain of the dividing ridge between the waters of the Columbia and Colorado, and the Missouri, hiding amongst loose piles of rocks, such as you generally see below broken cliffs. From this retreat I heard a slender, but very distinct bleat, so like that of a young kid or goat, that I at first concluded it to be such a call; but in vain trying to discover any large animal around me, at length I almost literally say, the mountain brought forth nothing much larger than a mouse, as I discovered that this little animal was the real author of this unexpected note" (Audubon and Bachman 1845-48: 258).

Late in the afternoon, we completed the passage across the mountain, and with thankful hearts, again trod the level land. We entered here a fine rich valley or plain, of about half a mile in width, between two ranges of the mountain. It was perfectly covered with willow, and through the middle of it, ran a rapid and turbulent mountain torrent, called Malade river [Rivière des Malades]. It contains a great abundance of **beaver**, their recent dams being seen in great numbers, and in the night, when all was quiet, we could hear the playful animals at their gambols, diving from the shore into the water, and striking the surface with their broad tails ...

[Audubon reported that Townsend and Nuttall had seen Tengmalm's Owl [**Boreal Owl**] "on the Malade River Mountains, where it was so tame and unsuspicious that Mr. NUTTALL was enabled to approach within a few feet of it, as it sat upon the bushes" (Audubon 1840-44, 1: 122).]

14th.– We travelled down the Mallade [sic] river, and followed the Indian trail through the valley. The path frequently passed along near the base of the mountain, and then wound its way a considerable distance up it, to avoid rocky impediments and thick tangled bushes below, so that we had some climbing to do; but the difficulties and perils of the route of yesterday are still so fresh in our memory, that all minor things are disregarded, at least by *us*. Our poor horses, however, no doubt feel differently, as they are tired and foot sore.

["*14th* ... Saw no game today **dusky grouse** plenty for three days past" (Wyeth 1899: 78).]

The next day [15th] we came to a close and almost impenetrable thicket of tangled willows, through which we had great difficulty in urging our horses. The breadth of the thicket was about one hundred yards, and a full hour was consumed in passing through it. We then entered immediately a rich and beautiful valley, covered profusely with a splendid blue Lupin. The mountains on either side are of much less height than those we have passed, and entirely bare, the pine trees which generally cover and ornament them, having disappeared. During the morning, we ascended and descended several high and stony hills, and early in the afternoon, emerged upon a large, level prairie, and struck a branch of Mallade river [now Big Wood River], where we encamped.

While we were unloading, we observed a number of Indians ahead, and not being aware of their character, stood with our horses saddled, while Captain W. and Richardson rode out to reconnoitre. In about half an hour they returned, and informed us that they were *Snakes* who were returning from the fisheries, and travelling towards the buffalo on the "big river," (Shoshoné [Snake River].) We therefore unsaddled our poor jaded horses and turned them out to feed upon the luxuriant pasture around the camp ...

The next morning [16th] we steered west across the wide prairie, crossing within every mile or two, a branch of the tortuous Mallade, near each of which good pasture was seen; but on the main prairie scarcely a blade of grass could be found, it having lately been fired by the Indians to improve the crops of next year. We have seen to-day some lava and basalt again on the sides of the hills, and on the mounds in the plain, but the level land was entirely free from it.

At noon on the 17th, we passed a deserted Indian camp, probably of the same people whose trail we have been following. There were many evident signs of the Indians having but recently left it, among which was that of several white wolves lurking around in the hope of finding remnants of meat, but, as a Scotchman would say, "I doubt they were mistaken," for meat is scarce here, and the frugal Indians rarely leave enough behind them to excite even the famished stomach of the lank and hungry wolf ...

In the afternoon we arrived at the "Kamas prairie," so called from a vast abundance of this esculent root which it produces, (the *Kamassa esculenta* of Nuttall [*Camassia quamash*].) The plain is a beautiful level one of about a mile over, hemmed in by low,

Cottonwood felled by Beavers.

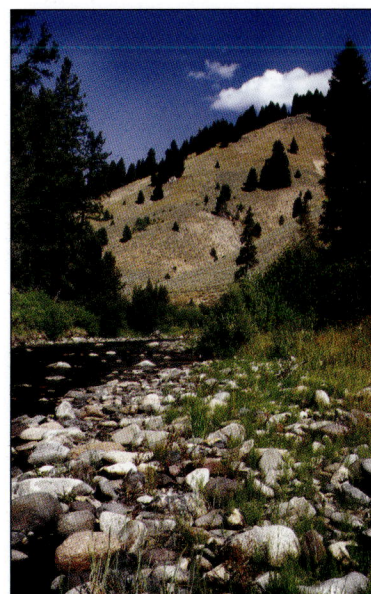
A tributary of Big Wood River.

THREE MYSTERY BIRDS SEEN IN THE SAWTOOTH MOUNTAINS

12th August. On this day, probably in the pines, Townsend had his first encounter with **Townsend's Solitaire**. "On the 12th of August 1834," says Dr TOWNSEND, "I saw a Thrush, cinereous brown above, whitish below, with a long rounded tail, every feather except the two middle ones largely tipped with white. About two months [= two weeks] subsequently, on the Shoshone River [Snake River], West of the Rocky Mountains, I killed one of these birds; but it was moulting, and I foolishly threw it away in the expectation of finding it on the Columbia. I never saw it afterwards."

Audubon, who supplied this quote, went on to designate it as a new species despite having ridiculously little information about it, naming the bird: "TOWNSEND'S MOCKING THRUSH *TURDUS TOWNSENDI*" (Audubon 1831-39, 5: 336).

Later, while on the Columbia River, Townsend was given a specimen of Townsend's Solitaire that Audubon named "Townsend's Ptilogonys *Ptilogonys Townsendi*" - see p. 208. It is curious that neither Townsend nor Audubon made any connection with the bird Townsend saw here in the Sawtooth Mountains and the Townsend's Solitaire specimen acquired later. The bird he saw in the mountains could possibly have been a worn adult Sage Thrasher, though Townsend does not say it was streaky and he would surely have been familiar with that species by then.

Franklin's Spruce Grouse.

13th August. Townsend saw some birds that he mistakenly thought were pheasants. Although they must have been **Franklin's Spruce Grouse**, Audubon went so far as to describe a new species: "AMERICAN PHEASANT *PHASIANUS AMERICANUS* ... The existence of a species of Pheasant in America appears to be proved by the following notice from Dr TOWNSEND. "On the 13th of August 1834, I saw two specimens of the long-tailed Pheasant. They inhabited the pine-trees in a deep and tangled dell, called Thornbury's Pass, near Mallade River, in the Rocky mountains. They appeared rather tame, flying for short distances before our party, and alighting near the summits of the tall pines. This bird in length appeared about equal to the English Pheasant, but not so heavy-bodied. The tail was as long, or longer. The general colour was dark brown or black, with some white below. The hunters mentioned having seen several of these birds during the day, and one of the rascals shot one with his rifle, and left it lying upon the ground. They had never met with such birds before" (Audubon 1831-39, 5: 335).

The long tail length is a bit of a puzzle. Franklin's Spruce Grouse does not have a particularly long tail (it has a shorter tail than the Dusky Grouse) but the dark colour "with some white below" strongly suggests Franklin's Spruce Grouse. Here in the Sawtooth Mountains the expedition had just entered the southernmost spur of the range of Spruce Grouse, and after leaving the mountains they moved out of its range and never saw them again. Graustein (1967: 303, 443) identifies the "pheasants" as Sharp-tailed Grouse but this cannot be right because the naturalists and hunters were all very familiar with that bird, and the plumage description does not fit well.

14th August. Audubon described a **Pileated Woodpecker** seen by Townsend on this day believing it to be the Imperial Woodpecker *Campephilus imperialis* from Mexico. Audubon quotes Townsend's notes in full: "On the 14th of August 1834, I saw several specimens of a large black Woodpecker, about the size of *Picus principalis* [Ivory-billed Woodpecker]. A broad band of white appeared to extend transversely across the wings and back. It inhabited the tall pine trees, and was very shy. The note was almost exactly that of the Red-headed Woodpecker, so much so that at first Mr NUTTALL and myself were both deceived by it. I lingered behind the party, which at that time was travelling rapidly, and at last got a shot at one of them with slugs, my large shot having been entirely expended. The bird fell wounded into a thicket at a considerable distance. I searched for an hour, without finding it, and was at last compelled to relinquish it and follow the party, which had been leaving me at a rapid trot, to find my way as I best could, and keep out of the reach of Indians, who were dogging us continually. Who can describe the chagrin and positive misery of a poor fellow in my then situation!" (Audubon 1831-39, 5: 313-314).

The call, habitat and range (but not the size) suggests a male Williamson's Sapsucker. The simpler explanation is that this was the larger Pileated Woodpecker, a species that is widespread in the eastern United States, is absent in the plains, but occurs in Idaho, Oregon and Washington. In the Sawtooth Mountains the expedition was just re-entering the range of this woodpecker that they had last seen in Missouri. In 1836 Townsend included the Pileated Woodpecker in the bird report for the Columbia River that he prepared for the Reverend Parker (see appendix 9) but never seems to have attempted to correct Audubon's claimed identification. This error was further compounded by John Cassin who included text and a colour plate of the Imperial Woodpecker in his *Illustration of the Birds of California, Texas, Oregon and Russian America* (1856) - based solely on those few words from Townsend.

rocky hills, and in spring, the pretty blue flowers of the Kamas are said to give it a peculiar, and very pleasing appearance. At this season, the flowers do not appear, the vegetable being indicated only by little dry stems which protrude all over the ground among the grass.

We encamped here, near a small branch of Mallade river; and soon after, all hands took their kettles and scattered over the prairie to dig a mess of kamas. We were, of course, eminently successful, and were furnished thereby with an excellent and wholesome meal. When boiled, this little root is palatable, and somewhat resembles the taste of the common potato; the Indian mode of preparing it, is, however, the best – that of fermenting it in pits under ground, into which hot stones have been placed. It is suffered to remain in these pits for several days; and when removed, is of a dark brown colour, about the consistence of softened glue, and sweet, like molasses. It is then often made into large cakes, by being mashed, and pressed together, and slightly baked in the sun. There are several other kinds of bulbous and tuberous roots, growing in these plains, which are eaten by the Indians, after undergoing a certain process of fermentation and baking. Among these, that which is most esteemed, is the white or biscuit root, the *Racine blanc* of the Canadians, – (*Eulophus ambiguus*, of Nuttall.) [Bread-root *Psoralea sp.*] This is dried, pulverized with stones, and after being moistened with water, is made into cakes and baked in the sun. The taste is not unlike that of a stale biscuit, and to a hungry man, or one who has long subsisted without vegetables of any kind, is rather palatable. ["*17th* ... Killed some **dusky grouse** and dug some kamas which assisted our living a little also found some choke cherries" (Wyeth 1899: 78).]

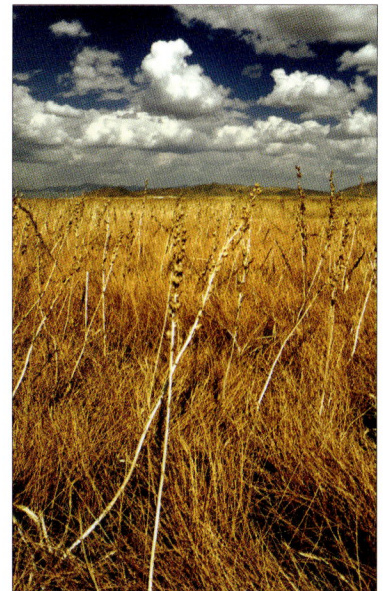

Kamas prairie in August.

On the morning of the 18th, we commenced ascending the hills again, and had a laborious and toilsome day's march ... We comfort ourselves, however, by supposing that we have now nearly passed the most rugged country on the route, and hope, before many days, to reach the valley of the Shoshoné [Snake], where the country will be level, and the pasture good. We are anxious, also, to fall in with the Snake Indians, in order to get a supply of salmon, as we have been living for several days on a short allowance of wretched, dry meat, and this poor pittance is now almost exhausted.

19th.– This morning was cold, the thermometer stood at 28o, and a thick skim of ice was in the camp kettles at sunrise. Another hard day's travel over the hills, during which we lost two of our largest and stoutest horses. Towards evening, we descended to a fine large plain, and struck *Boisée*, or Big Wood river [now Boisie River; not Big Wood River of today which was formerly Mallade River], on the borders of which we encamped. This is a beautiful stream, about one hundred yards in width, clear as crystal, and, in some parts, probably twenty feet deep. It is literally *crowded* with salmon, which are springing from the water almost constantly. Our mouths are watering most abundantly for some of them, but we are not provided with suitable implements for taking any, and must therefore depend for a supply on the Indians, whom we hope soon to meet.

We found, in the mountain passes, to-day, a considerable quantity of a small fruit called the choke-cherry, a species of prunus, growing on low bushes [*Prunus virginiana* or *P. emarginata*]. When ripe they are tolerable eating, somewhat astringent ... We have seen, also, large patches of service bushes [Serviceberry *Amelanchier sp.*], but no fruit. It seems to have failed this year, although ordinarily so abundant that it constitutes a large portion of the vegetable food of both Indians and white trappers who visit these regions.

August 20th.– ...The route, this morning, lay along the Boisée. For an hour, the travelling was toilsome and difficult, the Indian trail, leading along the high bank of the river, steep and rocky, making our progress very slow and laborious ... We have all been disappointed in the distance to this river, and the length of time required to reach it. Not a man in our camp has ever travelled this route before, and all we have known about it has been the general course.

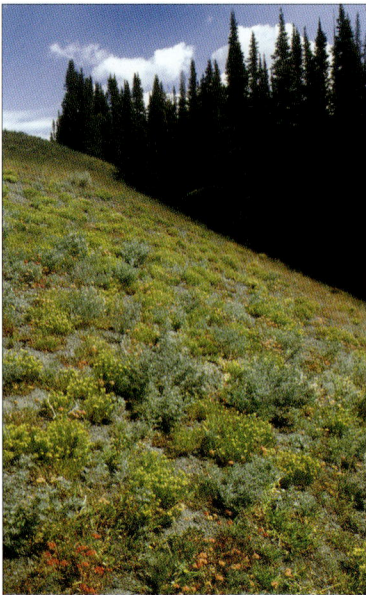

Hilly terrain west of Big Wood River.

In the afternoon, we observed a number of Indians on the opposite side of the river, engaged in fishing for salmon ...

21st.– The timber along the river banks is plentiful, and often attains a large size. It is chiefly of the species called balsam poplar, (*Populus balsamifera*.) [Black Cottonwood *Populus balsamifera*]

Towards noon to-day, we observed ahead several groups of Indians, perhaps twenty in each, and on the appearance of our cavalcade, they manifested their joy at seeing us, by the most extravagant and grotesque gestures, dancing and capering most ludicrously. Every individual of them was perfectly naked, with the exception of a small thong around the waist, to which was attached a square piece of flannel, skin, or canvass, depending half way to the knees. Their stature was rather below the middle height, but they were strongly built and very muscular. Each man carried his salmon spear, and these, with the knives stuck in their girdles, appeared to be their only weapons, not one of them having a gun ...

22d.– Last night during the second guard, while on my walk around the camp, I observed one of my men squatted on the ground, intently surveying some object which appeared to be moving among the horses. At his request, I stooped also, and could distinctly perceive something near us which was certainly not a horse, and yet was certainly a living object. I supposed it to be either a bear or a wolf, and at the earnest solicitation of the man, I gave the word "fire." The trigger was instantly pulled, the sparks flew from the flint, but the rifle was not exploded. At the sound, an Indian sprang from the grass where he had been crouching, and darted away towards the Snake camp. His object certainly was to appropriate one of our horses, and very fortunate for him was it that the gun missed fire, for the man was an unerring marksman. This little warning will probably check other similar attempts by these people.

Early in the morning I strolled into the Snake camp. It consists of about thirty lodges or wigwams, formed generally of branches of trees tied together in a conic summit, and covered with buffalo, deer, or elk skins. Men and little children were lolling about the ground all around the wigwams, together with a heterogeneous assemblance of dogs, cats, some tamed prairie wolves [**Coyotes**], and other "*varmints.*" The dogs growled and snapped when I approached, the wolves cowered and looked cross, and the cats ran away and hid themselves in dark corners. They had not been accustomed to the face of a white man, and all the quadrupeds seemed to regard me as some monstrous production, more to be feared than loved or courted. This dislike, however, did not appear to extend to the bipeds, for many of every age and sex gathered around me, and seemed to be examining me critically in all directions. The men looked complacently at me, the women, the dear creatures, smiled upon me, and the little naked, pot-bellied children crawled around my feet, examining the fashion of my hard shoes, and playing with the long fringes of my leathern inexpressibles. But I scarcely know how to commence a description of the *tout en semble* of the camp, or to frame a sentence which will give an adequate idea of the extreme filth, and most horrific nastiness of the whole vicinity ... Some of the women had little children clinging like bullfrogs to their backs, without being fastened, and in that situation extracting their lactiferous sustenance from the breast, which was thrown over the shoulders.

It is almost needless to say, that I did not remain long in the Snake camp ... When I returned to our camp, the trading was going on as briskly as yesterday. A large number of Indians were assembled around, all of whom had bundles of fish, which they were anxious to dispose of ...

[Bailey (1915) suggested that Townsend collected the type specimen of **Townsend's Pocket Gopher** on the 22nd August, see page 114.]

Boise River.

[23rd] ... Towards evening, we arrived on Snake river, crossed it at a ford[1], and encamped near a number of lodges along the shore. Shortly afterwards, Captain W., with three men, visited the Indians, carrying with them some small articles, to trade for fish. In about half an hour they returned, bringing only about ten salmon ... Being desirous to escape from the immediate vicinity of the village, we moved our camp about four miles further, and stopped for the night.

1. In this vicinity, later in 1834, Thomas McKay began the construction of Fort Boise for the Hudson's Bay Company. It was on the north side of the Boise River at its junction with the Snake River.

TOWNSEND'S POCKET GOPHER

When he visited Europe in 1838, Bachman took three pocket gophers with him to compare with those collected by the Scottish collectors Richardson, Drummond and Douglas, and other naturalists. Bachman showed the specimens to Dr Richardson who was preparing a paper on pocket gophers. Richardson correctly decided that one of Townsend's specimens was a new species and named it *Geomys Townsendii* but the paper was never published. Bachman acknowledged this when he published some of his findings in the *Journal of the Academy of Natural Sciences of Philadelphia*:

"GEOMYS *Townsendii*, (Richardson's Manuscripts.) …
The specimen was procured by Mr. Townsend on the Columbia river.
I am obliged to confess that I should not have ventured to publish this species as distinct from the preceding [GEOMYS *borealis*, (Richardson.)], on my own responsibility. The discriminating eye of Dr. Richardson, however, who has studied this genus more carefully than I have had an opportunity of doing, may have detected marks of difference which I have not been able to discover" (Bachman 1839c: 105).

But Bachman was not really convinced and later lumped *Geomys Townsendii* together with *Geomys borealis*, the Northern Pocket Gopher. His main description in the *Viviparous Quadrupeds* was taken "from three specimens of this pouched Sand-rat, obtained by the late Mr. TOWNSEND, on the Columbia river, two of which appeared to be in summer pelage, and the third in its autumnal coat". A fourth description referred to another specimen "sent by Mr. TOWNSEND, marked in RICHARDSON's MSS. as *Geomys Townsendii*" (Audubon and Bachman 1854: 389-390).

All the figures in plate CXLII of *The Viviparous Quadrupeds* (detail at right) are probably painted from Townsend specimens, and although said by Audubon and Bachman to be Northern Pocket Gophers, one of them may depict the type specimen of Townsend's Pocket Gopher (at right, ANSP 147). Townsend's Pocket Gopher was first described by Bachman from a specimen collected by Townsend, either in the area around Fort Hall or near the junction of the Boise and Snake Rivers. Bailey (1915) suggested that Townsend collected the type specimen on 22nd August 1834 (see appendix 7B).

Across the Blue Mountains to Fort Walla Walla

"I was surprised to find Mr. N. and Captain T. picking the last bones of a bird which they had cooked. Upon inquiry, I ascertained that the subject was an unfortunate owl which I had killed in the morning, and had intended to preserve, as a specimen. The temptation was too great to be resisted by the hungry Captain and naturalist, and the bird of wisdom lost the immortality which he might otherwise have acquired."

Townsend's *Narrative*, 2nd September 1834.

Western Snake River Plain to the Blue Mountains

We passed, this morning [24th August], over a flat country, very similar to that along the Platte, abounding in wormwood bushes, the pulpy-leaved thorn, and others, and deep with sand, and at noon stopped on a small stream called *Malheur's creek.* [Near the mouth of the Malheur, Townsend collected the type specimen of the **Mountain Cottontail**, see pp.117-118.]

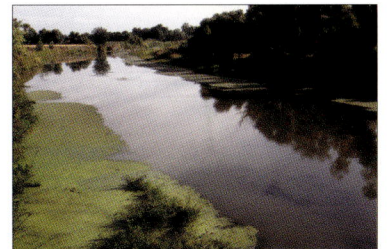

Malheur River, a few miles from its junction with the Snake River (below).

Here a party of nine men was equipped, and despatched up the river, and across the country, on a trapping expedition, with orders to join us early in the ensuing winter, at the fort on the Columbia. Richardson was the chief of this party, and when I grasped the hand of our worthy hunter, and bade him farewell, I felt as though I were taking leave of a friend. I had become particularly attached to him, from the great simplicity and kindness of his heart, and his universally correct and proper deportment. I had been accustomed to depend upon his knowledge and sagacity in every thing connected with the wild and roving life which I had led for some months past, and I felt that his absence would be a real loss, as well as to myself, as to the whole camp, which had profited so much by his dexterity and skill.

Our party will now consist of only seventeen men, but the number is amply sufficient, as we have passed over the country where danger is to be apprehended from Indians. We followed the course of the creek [downstream] during the afternoon, and in the evening encamped on Snake river, into which Malheur empties. The river is here nearly a mile wide, but deep and clear, and for a considerable distance, perfectly navigable for steamboats, or even larger craft, and it would not seem improbable, that at some distant day, these facilities, added to the excellence of the alluvial soil, should induce the stout and hardy adventurers of our country to make permanent settlements here.

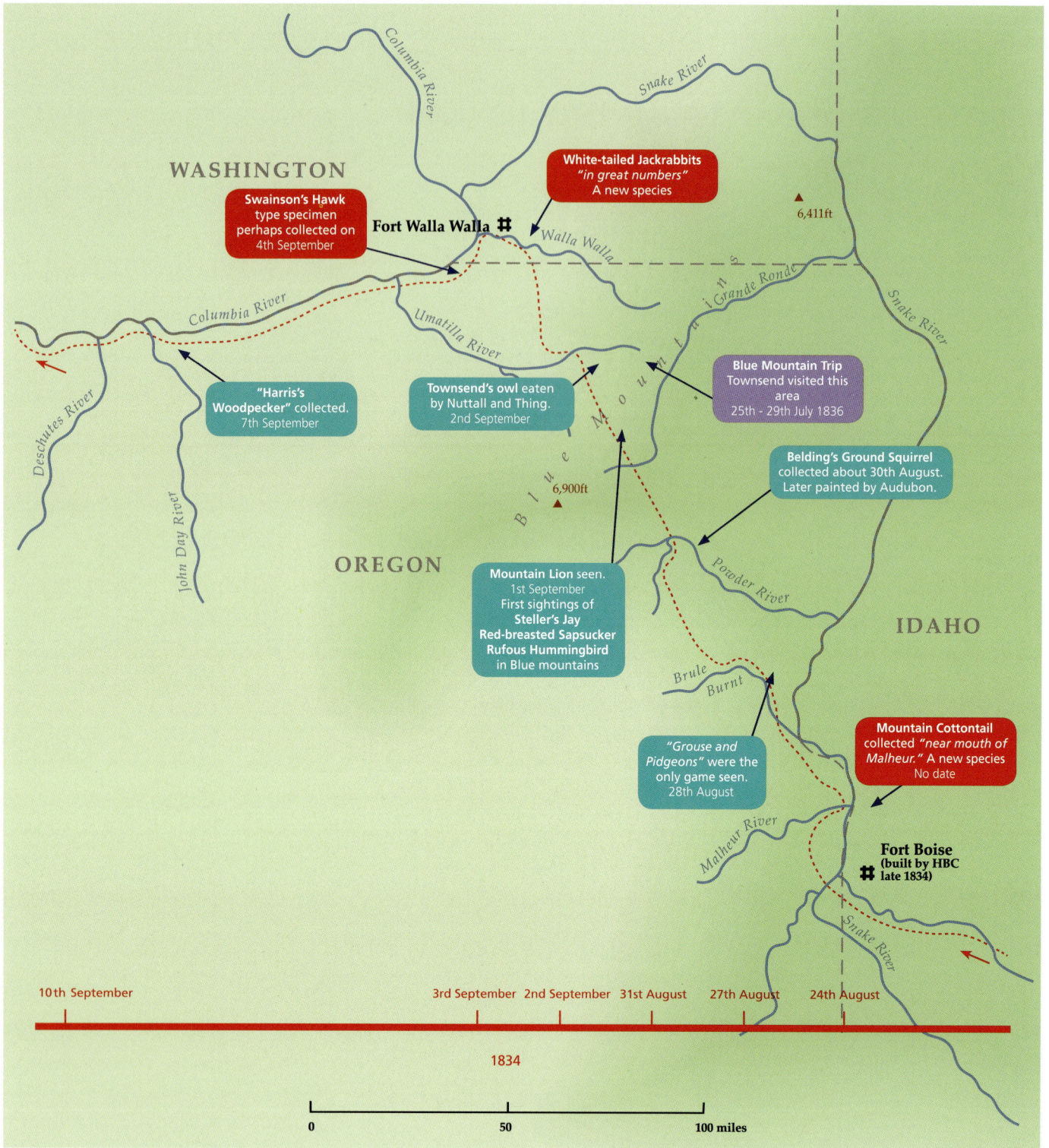

WASHINGTON

Swainson's Hawk type specimen perhaps collected on 4th September

White-tailed Jackrabbits *"in great numbers"* A new species

Fort Walla Walla #

Columbia River

Snake River

Walla Walla

Columbia River

Umatilla River

6,411ft

"Harris's Woodpecker" collected. 7th September

Townsend's owl eaten by Nuttall and Thing. 2nd September

Blue Mountain Trip Townsend visited this area 25th - 29th July 1836

Deschutes River

John Day River

OREGON

Blue Mountains

Grande Ronde

Snake River

6,900ft

Belding's Ground Squirrel collected about 30th August. Later painted by Audubon.

Powder River

Mountain Lion seen. 1st September First sightings of **Steller's Jay Red-breasted Sapsucker Rufous Hummingbird** in Blue mountains

IDAHO

Brule

Burnt

"Grouse and Pidgeons" were the only game seen. 28th August

Mountain Cottontail collected *"near mouth of Malheur."* A new species No date

Malheur River

Fort Boise (built by HBC late 1834) #

Snake River

10th September

3rd September 2nd September 31st August 27th August 24th August

1834

0   50   100 miles

Map 7 - Across the Blue Mountains and down the Columbia River, 24th August - 10th September 1834.

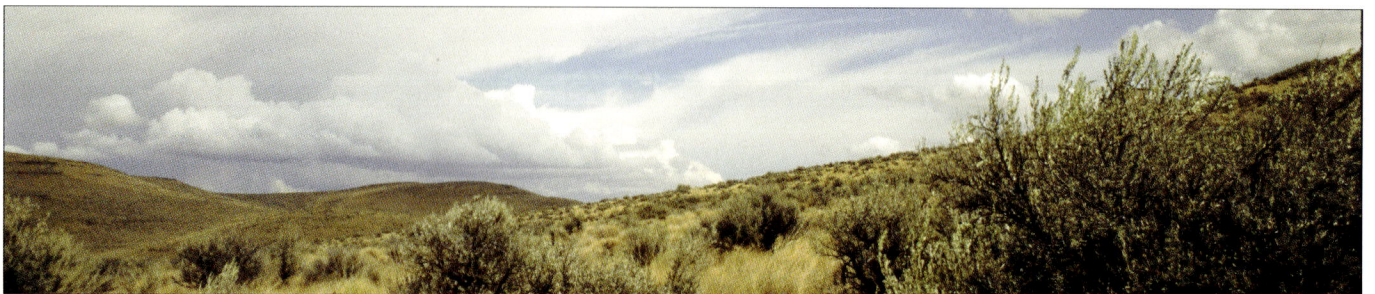

MOUNTAIN COTTONTAIL

Plate XCIV (detail) from Audubon's *Viviparous Quadrupeds.*

This small rabbit was discovered and named twice. John Bachman published the first description of the Mountain Cottontail in the *Journal of the Academy of Natural Sciences of Philadelphia* (1837) under the following name:

"LEPUS *Nuttallii*. NUTTALL'S Little Hare ...

To the kindness and liberality of THOMAS NUTTALL, Esq., whose repeated and perilous journeys over the prairies, the mountains, and through the forests of the far West, have raised him to a high rank among modern travellers; and whose researches, as a Botanist, are duly appreciated by all men of science – I am indebted for a new species of hare, which I am now about to describe:– the most diminutive of any species of true hare yet discovered; and in size and colour bearing a strong resemblance to L. LAGOMYS *princeps* of RICHARDSON [Pika] ...

The only information which I have been able to obtain of the habits of this little hare, is contained in the following note from Mr. NUTTALL, which accompanied the specimen:– "This little hare we met with west of the Rocky Mountains, inhabiting thickets by the banks of several small streams which flow into the Shoshonee and Columbia rivers[a]. It was frequently seen, in the evening, about our encampments, and appeared to possess all the habits of the LEPUS *Americanus* [Snowshoe Hare]" (Bachman 1837: 345-348).

The second Mountain Cottontail to be described was named "Worm-wood Hare" by Bachman because Townsend had found several specimens in the wormwood bushes around Fort Walla (see p.220 for more details). Bachman did not realise that both these hares were actually the same species, that "Nuttall's Little Hare" was simply a young one. The mistake was still not realised when Audubon and Bachman compiled *The Viviparous Quadrupeds* so the Mountain Cottontail was included twice: as "Worm-wood Hare" (plate LXXXVIII) and as "Nuttall's Little Hare" (plate XCIV).

So why all the confusion? Why did they even consider that there could have been two very small species of rabbit? The answer may lie in another article about "Nuttall's Little Hare" where Bachman added a few notes from Townsend who, besides incorrectly believing that their specimen was full grown, appears to have been influenced by the old hands in the expedition who had probably come across the Pygmy Rabbit, an even smaller rabbit than the Mountain Cottontail: ▶

"LEPUS *Nuttallii*, (nob.) Nuttall's Little Hare ... To the information already given on this species ... I subjoin a note which was kindly furnished me by Mr. Townsend. "The specimen from which you described, was doubtless that of an adult animal. We saw, perhaps twenty or thirty, and all the same size; several of the hunters and trappers attached to our party, who had perambulated the country for years, and were acquainted with, perhaps, every animal in it, said that it never attained a greater size. It is remarkably gentle, hopping before you like a domesticated animal"" (Bachman 1839c: 79-80).

Townsend and Nuttall may themselves have seen and collected this smaller species, the Pygmy Rabbit, or they may just have seen adult and half grown Mountain Cottontails. All we can be sure of is that all the specimens described by John Bachman were Mountain Cottontails[b]. The Pygmy Rabbit was not scientifically recognised until 1891[c]; it is largely nocturnal, often residing in empty Badger holes, and like the Mountain Cottontail has a range that includes the country between Fort Hall and Fort Walla Walla.

a. Townsend does not mention the Mountain Cottontail in his *Narrative* and in his scientific appendix we get no further clues as to where the type specimen (ANSP 382) was found because he only repeats what Bachman had written in 1837. The type locality has since been designated as "near the mouth of Malheur River" with a date of August 1834 (Nelson, 1909. *North American Fauna* 29: 201, 203), but the type might equally well have been collected in July, beside the Snake River, near Fort Hall, where Nuttall had ample leisure to roam around while the expedition waited for more than a week for a buffalo hunting party to return.

However, a donation of birds and mammals from Townsend to the National Institute includes "1 Lepus Nuttalii [sic], (unique specimen,) *Rocky Mountains*." (*Proceedings of the National Institution*, 1842, p. 147). A footnote goes on to explain: "This specimen was shot by Mr. J. K. Townsend, on the north branch of the Platte river, in the Rocky Mountains. Some twenty or thirty of the same species were seen by the party with which Mr. Townsend travelled, but, unfortunately, the present was the only specimen procured. Dr. Bachman, before writing his description of this new animal, visited the principal museums in Europe with the view of finding another example of the same species, but was unsuccessful. This therefore is an unique specimen."

Since Townsend must have supplied or written this information it seems that the Mountain Cottontail was first seen at the eastern edge of its range, *east* of South Pass. Bachman clearly says in his original description that the specimen was provided by Nuttall from the "west of the Rocky Mountains". Why this specimen should have been considered unique is puzzling (see note 2).

b. There seem to have been several specimens of "Nuttall's Little Hare": A. The specimen described by Bachman and painted by J.W. Audubon. B. Townsend donated "1 Lepus Nuttalii (unique specimen,) *Rocky Mountains*" to the National Institution in 1842. C. There was a "Lepus Nuttalli" (and 3 "Lepus artemesia") amongst the mammals offered to J.E. Gray in 1849 (that subsequently went to ANSP). D. There was a "small Lepus" in the list of duplicates that Audubon sold for Townsend in Britain; it was purchased by Lord Derby (LM D573a 20th August 1836, Fort Walla Walla). E. The type of *Sylvigalus nuttalli* resides at Philadelphia (ANSP 382) and could be A or C, or even B.

c. The Mountain Cottontail and Pygmy Rabbit are superficially similar, the latter being smaller with no white on the tail. When C. Hart Merriam described it in 1891 he said: "When I first saw this little Rabbit in the field the idea occurred to me that it might be *Lepus nuttalli* of Bachman, described in 1837 from a specimen collected on the Snake Plains by Townsend ... But on returning to the East and examining the type of *L. nuttalli* ... I find the latter to be a young Cotton-tail, with a "rather long, full tail," pure white beneath ... and I wholly concur in the opinion that it is the young of the species afterward described by Bachman (in 1839) as *L. Artemisia*" (Merriam 1891: 75).

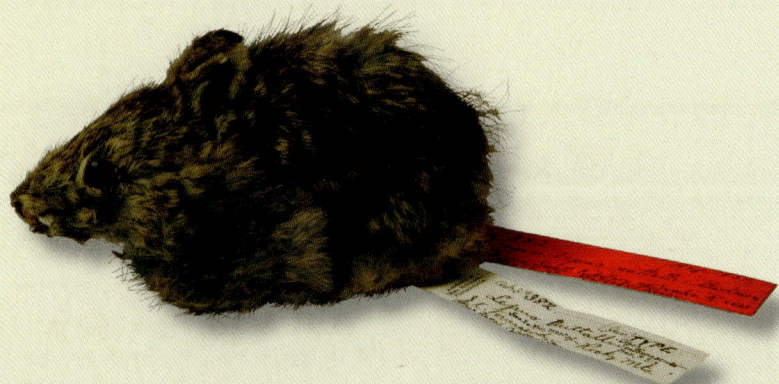

The type specimen of the Mountain Cottontail (ANSP 382).

I have not observed that the Indians often attempt fishing in the "big river," where it is wide and deep; they generally prefer the slues, creeks, &c. Across these, a net of closely woven willows is stretched, placed vertically, and extending from the bottom to several feet above the surface. A number of Indians enter the water about a hundred yards above the net, and, walking closely, drive the fish in a body against the wicker work. Here they frequently become entangled, and are always checked; the spear is then used dexterously, and they are thrown out, one by one, upon the shore. With industry, a vast number of salmon might be taken in this manner; but the Indians are generally so indolent and careless of the future, that it is rare to find an individual with provision enough to supply his lodge for a week.

"Spearing the Salmon" from Townsend's *Narrative*.

25th.– Early in the day the country assumed a more hilly aspect. The rich plains were gone. Instead of a dense growth of willow and the balsam poplar, low bushes of wormwood, &c., predominated, intermixed with tall, rank prairie grass.

Towards noon, we fell in with about ten lodges of Indians, (Snakes and Bannecks,) from whom we purchased eighty salmon. This has put us in excellent spirits. We feared that we had lost sight of the natives, and as we had not reserved half the requisite quantity of provisions for our support to the Columbia, (most of our stock having been given to Richardson's trapping party,) the prospect of several days abstinence seemed very clear before us.

In the afternoon, we deviated a little from our general course, to cut off a bend in the river, and crossed a short, high hill, a part of an extensive range which we have seen for two days ahead, and which we suppose to be in the vicinity of Powder river, and in the evening encamped in a narrow valley, on the borders of the Shoshoné.

26th.– Last night I had the misfortune to lose my favorite, and latterly my only riding horse, the other having been left at Fort Hall, in consequence of a sudden lameness, with which he became afflicted only the night before our departure. The animal was turned out as usual, with the others, in the evening, and as I have never known him to stray, I conclude that some lurking Indian has stolen him ... This is the most serious loss I have met with. The animal was particularly valuable to me, and no consideration would have induced me to part with it here ... Captain W. has kindly offered me the use of horses until we arrive at Columbia.

We commenced our march early, travelling up a broad, rich valley, in which we encamped last night, and at the head of it, on a creek called Brulé [Burnt River], we found one family, consisting of five Snake Indians, one man, two women, and two children ... We bought, of this family, a considerable quantity of dried choke-cherries [*Prunus virginianus* or *P. emarginata*], these being the only article of commerce which they

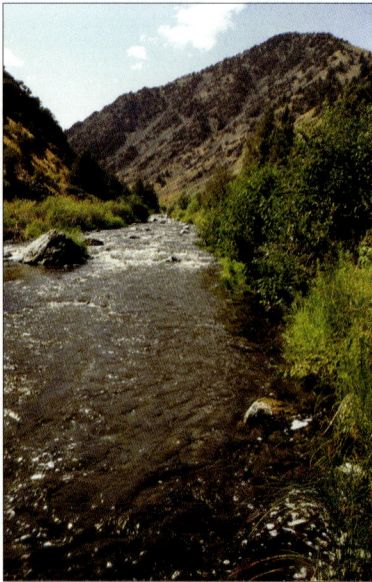
Burnt River.

possessed. This fruit they prepare by pounding it with stones, and drying it in masses in the sun. It is then good tasted, and somewhat nutritive, and it loses, by the process, the whole of the astringency which is so disagreeable in the recent fruit.

Leaving the valley, we proceeded over some high and stony hills, keeping pretty nearly the course of the creek. The travelling was, as usual in such places, difficult and laborious, and our progress necessarily slow and tedious. Throughout the day, there was no change in the character of the country, and the consequence was, that three of our poor horses gave up and stopped.

27th.– This morning, two men were left at the camp, for the purpose of collecting and bringing on, moderately, the horses left yesterday, and others that may hereafter fail. We were obliged to leave with them a stock of provision greater in proportion than our own limited allowance, and have thus somewhat diminished our chance of performing the remainder of the journey with satisfied appetites, but there is some small game to be found on the route, **grouse, ducks**, &c., and occasionally a **beaver** may be taken, if our necessities are pressing. We made a noon camp on Brulé, and stopped at night in a narrow valley, between the hills[1].

28th.– Towards noon to-day, we lost the trail among the hills, and although considerable search was made, we were not able to find it again. We then directed our course due north, and at 2 o'clock struck Powder river, a narrow and shallow stream, plentifully fringed with willows. We passed down this river for about five miles and encamped. Captain W. immediately left us to look for the lost trail, and returned in about two hours, with the information that no trace of it could be found ... Game has been exceedingly scarce, with the exception of a few grouse, pigeons [**Mourning Doves**], &c. We have not seen a deer, antelope [**Pronghorn**], or any other quadruped larger than a hare, since we left the confines of the buffalo country ...

29th.– We commenced our march early this morning, following the [Powder] river to a point about six miles above where we struck it yesterday. We then took to the hills, steering N.N.W., – it being impossible, from the broken state of the country, to keep the river bank.

Soon after we commenced the ascent, we met with difficulties in the shape of high, steep, banks, and deep ravines, the ground being thickly strewed with sharp, angular masses of lava and basalt. As we proceeded, these difficulties increased to such a degree, as to occasion a fear that our horses could never proceed. The hills at length became like a consolidated mass of irregular rock, and the small strips of earthy matter that occasionally appeared, were burst into wide fissures by the desiccation to which the country at this season is subject. Sometimes, as we approached the verges of the cliffs, we could see the river winding its devious course many hundred feet below, rushing and foaming in eddies and whirlpools, and fretting against the steep sides of the rocks, which hemmed it in. These are what are called the cut-rocks, the sides of which are in many places as smooth and regular as though they had been worked with the chisel, and the opening between them, through which the river flows, is frequently so narrow that a biscuit might be thrown across it ...

The next day [30th] we were still travelling over the high and steep hills, which, fortunately for our poor horses, were far less stony than hitherto. ["*30th*. Made 8 miles up the creek through Cut Rocks during which time killed one **Salmon** and Two **Otter**" (Wyeth 1899: 80).] At about noon we descended to the plain, and struck the river in the midst of a large level prairie. We proceeded up stream for an hour, and to our great joy suddenly came in sight of a broad, open trail stretching away to the S.W. [N.W ?] ... Towards evening we crossed a single range of low hills and came to a small round prairie, with good water and excellent pasture. Here we found a family of *Kayouse* Indians, and encamped within sight of them. Two squaws from this family, visited us soon after, bringing some large kamas cakes and fermented roots, which we purchased of them.

Richardson's Ground Squirrel painted by Audubon, now re-identified as Belding's Ground Squirrel

Belding's Ground Squirrel was first described and named by C. Hart Merriam in 1888 from specimens from Donner, Placer County, California. Yet we believe this species was collected by Townsend in 1834, painted by J.J. Audubon and included by him in *The Viviparous Quadrupeds* as plate L (Audubon and Bachman 1845-48: 159-60, text to plate L). It has long been identified as Richardson's Ground Squirrel but there have been numerous revisions in the classification of the ground squirrels since then, and, according to the current understanding of their distribution, Townsend did not pass through the range of Richardson's Ground Squirrel. Which species then did he collect – and is there any supporting evidence that Audubon painted a different species?

Examination of plate L, supposedly of Richardson's Ground Squirrel, painted from two Townsend specimens, shows that one figure has a dark tip to the tail, while the other has lost the hair off its tail. Audubon and Bachman's text for "Spermophilus Richardsonii. Richardson's Spermophile" says that "The tail is the colour of the back; the hairs on the margins, near the end, are dark-brown tipped with yellowish-white". The only closely related species that is plainly marked on the body and has a dark tip to the tail is Belding's Ground Squirrel.

This identification is further supported elsewhere in the text where they say that: "The specimens from which our figures and descriptions were made, were obtained by Mr. TOWNSEND ... in the Rocky Mountains, (about latitude 45° [N],)". Townsend's route crossed the 45° N line of latitude only once, in the upper reaches of the Powder River, Oregon, on about 30th August 1834. Although Townsend says that it was *about* 45° N, he would have had a good idea of his position because Captain Thing was taking accurate readings of their position each day. This approximation of place (and time) clearly places the location where the specimens were found within the range of Belding's Ground Squirrel. Most Belding's Ground Squirrels enter aestivation in August and this continues into winter hibernation. These ones may have been late in retreating underground, or Townsend could have dug them up. Unfortunately, the specimens appear to have been lost or destroyed after Audubon had painted them.

"high and steep hills which, fortunately for our horses, were far less stoney than hitherto."

Belding's Ground Squirrels in plate L of Audubon's *Viviparous Quadrupeds* where they are included under the name of Richardson's Ground Squirrel.

Steller's Jay, first seen by the naturalists in the Blue Mountains (below).

31st.– Our route this morning, was over a country generally level and free from rocks; we crossed, however, one short, and very steep mountain range, thickly covered with tall and heavy pine trees, and came to a large and beautiful prairie, called the *Grande ronde*. Here we found Captain Bonneville's company, which has been lying here several days, waiting the arrival of its trapping parties. We made a noon camp near it, and were visited by Captain Bonneville. This was the first time I had seen this gentleman. His manners were affable and pleasing, and he seemed possessed of a large share of bold, adventurous, and to a certain extent, romantic spirit, without which no man can expect to thrive as a mountain leader. He stated that he preferred the "free and easy' life of a mountain hunter and trapper, to the comfortable and luxurious indolence of a dweller in civilized lands, and would not exchange his homely, but wholesome mountain fare, and his buffalo lodge, for the most piquant dishes of the French *artiste*, and the finest palace in the land. This came well from him, and I was pleased with it, although I confess I had become somewhat weary of rough travelling and rough fare, and looked forward with no little pleasure to a long rest under a Christian roof, and a general participation in Christian living.

With the captain, came a whole troop of Indians, Kayouse, Nez Percés, &c. They were very friendly towards us, each of the chiefs taking us by the hand with great cordiality, appearing pleased to see us, and anxious to point out to us the easiest and most expeditious route to the lower country. These Indians are, almost universally, fine looking, robust men, with strong aqualine features, and a much more cheerful cast of countenance than is usual among the race. Some of the women might almost be called beautiful, and none that I have ever seen are homely. Their dresses are generally of thin deer or antelope skin, with occasionally a bodice of some linen stuffs, purchased from the whites, and their whole appearance is neat and cleanly, forming a very striking contrast to the greasy, filthy, and disgusting Snake females ...

After making a hasty meal, and bidding adieu to the captain, and our friendly Indian visitors, we mounted our horses, and rode off. About half an hour's brisk trotting brought us to the foot of a steep and high mountain, called the *Blue*. This is said to be the most extensive chain west of the dividing ridge, and, with one exception perhaps the most difficult of passage. The whole mountain is densely covered with tall pine trees, with an undergrowth of service bushes and other shrubs, and the path is strewed, to a very inconvenient degree, with volcanic rocks. [They first saw **Steller's Jay** "in the Blue Mountains of the Oregon, east of Walla-Walla" where they were "scarce and shy"

– though they later found them more numerous on the Columbia. **Rufous-necked Hummingbirds** were also first seen "near the Blue Mountains of the Columbia, in the autumn, as we proceeded to the West" (Nuttall 1840: 243, 714).] In some of the ravines we find small springs of water; they are, however, rather rare, and the grass has been lately consumed, and many of the trees blasted by the ravaging fires of the Indians. ["*1 Sept.* ... from the Three Butes every day has been thick smoke like fog enveloping the whole country last night we camped at 10 ock [o'clock] having found no water and the whole country burnt as black as my Hat affording as poor a prospect for a poor sett of Horses as need be" (Wyeth 1899: 81).] These fires are yet smouldering, and the smoke from them effectually prevents our viewing the surrounding country, and completely obscures the beams of the sun. We travelled this evening until after dark, and encamped on a small stream in a gorge, where we found a plot of grass that had escaped the burning. ["*31st.* Killed 5 Hens today" (Wyeth 1899: 81). These were probably **Dusky Grouse**. Townsend says that this species was "first found in the Blue Mountains near Wallah Wallah, in large flocks, in September", though Wyeth mentions "dusky grouse" seen two weeks earlier, on 14th August, in the Sawtooth Mountains (Audubon 1840-44, 5: 90; Wyeth 1899: 78).]

September 1st.– Last evening, as we were about retiring to our beds, we heard, distinctly, as we thought, a loud haloo, several times repeated, and in a tone like that of a man in great distress. Supposing it to be a person who had lost his way in the darkness, and was searching for us, we fired several guns at regular intervals, but as they elicited no reply, after waiting a considerable time, we built a large fire, as a guide, and lay down to sleep.

Early this morning, a large panther [**Mountain Lion**] was seen prowling around our camp, and the halooing of last night was explained. It was the dismal, distressing yell by which this animal entices its prey, until pity or curiosity induces it to approach to its destruction. The panther is said to inhabit these forests in considerable numbers, and has not infrequently been known to kill the horses of a camp. He has seldom the temerity to attack a man, unless sorely pressed by hunger, or infuriated by wounds.

The path through the valley, in which we encamped last night, was level and smooth for about a mile; we then mounted a short, steep hill, and began immediately to descend. The road down the mountain wound constantly, and we travelled in short, zig-zag lines, in order to avoid the extremely abrupt declivities; but occasionally, we were compelled to descend in places that made us pause before making the attempt: they were, some of them, almost perpendicular, and our horses would frequently slide several yards, before they could recover. To this must be added enormous jagged masses of rock, obstructing the road in many places, and pine trees projecting their horizontal branches across the path.

The road continued, as I have described it, to the valley in the plain, and a full hour was consumed before we reached it. The country then became comparatively level again to the next range, where a mountain was to be ascended to the same height as the last. Here we dismounted and led our horses, it being impracticable, in their present state, to ride them. It was the most toilsome march I ever made, and we were all so much fatigued, when we arrived at the summit, that rest was indispensable to us as to our poor jaded horses. Here we made a noon camp, with a handful of grass and no water ... The route, in the afternoon, was over the top of the mountain, the road tolerably level, but crowded with stones ... We travelled steadily until 9 o'clock, when we saw ahead the dark outline of a high mountain, and soon after heard the men who rode in front, cry out, joyously, at the top of their voices, "*water! water!*" It was truly a cheering sound, and the words were echoed loudly by every man in the company. We had not tasted water since morning, and both horses and men have been suffering considerably for the want of it.

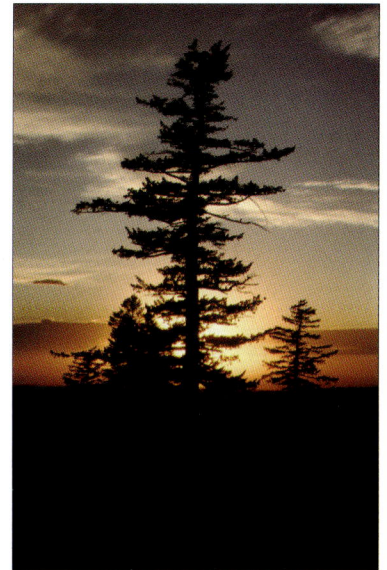

Sunset in the Blue Mountains.

2d.– Captain W. and two men, left us early this morning for Walla-walla, where they expect to arrive this evening, and send us some provision, of which we shall be in need, to-morrow.

Our camp moved soon after, under the direction of Captain Thing, and in about four miles reached *Utalla river* [Umatilla River], where it stopped, and remained until 12 o'clock.

As we were approaching so near the abode of those in whose eyes we wished to appear like fellow Christians, we concluded that there would be a propriety in attempting to remove at least one of the heathenish badges which we had worn throughout the journey; so Mr. N.'s razor was fished out from its hiding place in the bottom of his trunk, and in a few minutes our encumbered chins lost their long-cherished ornaments; we performed our ablutions in the river, arrayed ourselves in clean linen, trimmed our long hair, and then arranged our toilet before a mirror, with great self-complacence and satisfaction. I admired my own appearance considerably, (and this is, probably, an acknowledgement that few would make,) but I could not refrain from laughing at the strange, party-colored appearance of my physiognomy, the lower portion being fair, like a woman's, and the upper, brown and swarthy as an Indian.

Having nothing prepared for dinner to-day I strolled along the stream above the camp, and made a meal on rose buds, of which I collected an abundance; and on returning, I was surprised to find Mr. N. and Captain T. picking the last bones of a bird which they had cooked. Upon inquiry, I ascertained that the subject was an unfortunate **owl**[2] which I had killed in the morning, and had intended to preserve, as a specimen. The temptation was too great to be resisted by the hungry Captain and naturalist, and the bird of wisdom lost the immortality which he might otherwise have acquired.

In the afternoon, soon after leaving the Utalla, we ascended a high and very steep hill, and came immediately in view of a beautiful, and regularly undulating country of great extent. We have now probably done with high, rugged mountains; the sun shines clear, the air is bracing and elastic, and we are all in fine spirits.

Arrival at Fort Walla Walla

The next day [3rd], the road being generally level, and tolerably free from stones, we were enabled to keep our horses at the swiftest gait to which we dare urge them. We have been somewhat disappointed in not receiving the expected supplies from Walla-walla, but have not suffered for provision, as the grouse and hares are very abundant here, and we have shot as many as we wished.

At about noon we struck the Walla-walla river, a very pretty stream of fifty or sixty yards in width, fringed with tall willows, and containing a number of salmon, which we can see frequently leaping from the water. The pasture here, being good, we allowed our horses an hour's rest to feed, and then travelled on over the plain, until near dark, when, on rising a sandy hill, the noble Columbia burst at once upon our view. I could scarcely repress a loud exclamation of delight and pleasure, as I gazed upon the magnificent river, flowing silently and majestically on, and reflected that I had actually crossed the vast American continent, and now stood upon a stream that pours its waters directly into the Pacific. This, then, was the great Oregon, the first appearance of which gave Lewis and Clark so many emotions of joy and pleasure, and on this stream our indefatigable countrymen wintered, after the toils and privations of a long, and protracted journey through the wilderness. My reverie was suddenly interrupted by one of the men exclaiming from his position in advance, "there is the fort." We had, in truth approached very near, without being conscious of it. There stood the fort on the bank of the river; horses and horned cattle were roaming about the vicinity, and on the borders of the little Walla-walla, we recognised the white tent of our long lost missionaries. These we soon joined, and were met and received by them like brethren. Mr. N. and myself were invited to sup with them upon a dish of stewed hares [**White-tailed Jackrabbits**] which they had just prepared, and it is almost needless to say that we did full justice to the good men's cookery. [Daniel Lee remembered the moment well: "Mr. Nuttall, and Mr. Townsend, made their appearance at our camp, hungry as wolves. Part of a kettle of stewed rabit [sic] was standing by, which being given to the two naturalists, they showed off the way to get a living to perfection." Lee and Frost 1884: 124).] [The type specimen of the White-tailed Jackrabbit is from Walla Walla.] They told us that they had travelled comfortably from Fort Hall, without any unusual fatigue, and like ourselves, had no particular stirring adventures. Their route, although somewhat longer, was a much less toilsome and difficult one, and they suffered but little for food, being well provided with dried buffalo meat, which had been prepared near Fort Hall ...

WHITE-TAILED JACKRABBIT

"LEPUS *Townsendii* Townsend's Hare ...

This species, which is another of the discoveries of Mr. Townsend, and of which no specimen exists in any museum that I have had an opportunity of examining, is one of the most singular hares that has ever fallen under my notice ... The specimen from which the above description and drawing were taken, was a female, procured by Mr. Townsend on the Walla-walla, one of the sources of the Columbia river. The following note, by the discoverer of this hare, reached me at Edinburgh, after the description was made. It will afford valuable information in regard to its habits.

"The specimen is that of a female – and the species is common on the Rocky Mountains. I made particular inquiries, both of the Indians and British traders, as to the changes it undergoes at different seasons, and they all agreed that it never was lighter colored. We first saw it on the plains of the Black-foot River, west of the mountains, and observed it in all similar situations during our route to the Columbia. When first seen, which was in July, it was lean and unsavoury, having, like our common species, the larva of an insect imbedded in its neck; but when we arrived at Walla-Walla, in September, we found the Indians, and the persons attached to the fort, using them as a common article of food. Immediately after we arrived, we were regaled with a dish of hares, and I thought I had never eaten any thing more delicious. They are found here in great numbers on the plains covered with wormwood, (Artemesia.) It is so exceedingly fleet that no ordinary dog can catch it. I have frequently surprised it in its form, and shot it as it leapt away, but I found it necessary to be very expeditious, and to pull the trigger at a particular instant, or the game was off among the wormwood, and I never saw it again. The Indians kill them with arrows, by approaching them stealthily, as they lie concealed under the bushes, and in the winter take them with nets. To do this, some one or two hundred Indians, men, women, and children, collect, and enclose a large space with a slight net, about five feet wide, made of hemp; the net is kept in a vertical position by pointed sticks attached to it, and driven into the ground. These sticks are placed about five or six feet apart, and at each one an Indian is stationed, with a short club in his hand. After these arrangements are completed, a large number of Indians enter the circle, and beat the bushes in every direction. The frightened hares dart off towards the net, and, in attempting to pass, are knocked on the head and secured. Mr. Pambrun, the superintendent of Fort Walla-walla. From whom I obtained this account,

White-tailed Jackrabbit *Lepus townsendi* collected by Townsend and described by Bachman (Holotype, ANSP 378).

says that he has often participated in this sport with the Indians, and has known several hundred to be thus taken in a day. When captured alive, it does not scream like the common gray rabbit, (LEPUS *sylvaticus*) [Eastern Cottontail *Sylvilagus floridanus*].

This hare inhabits the plains exclusively, and seems particularly fond of the vicinity of the aromatic wormwood.* [footnote: *The specimen was stuffed with this article.] Immediately as you leave these bushes, in journeying towards the sea, you lose sight of this hare. The specimen I have sent is not the largest I have seen, some individuals measuring several inches more in length" (Bachman 1839c: 90-94).

The missionaries informed us that they had engaged a large barge to convey themselves and baggage to Fort Vancouver, and that Captain Stewart and Mr. Ashworth were to be of the party. Mr. N. and myself were very anxious to take a seat with them, but to our disappointment, were told that the boat would scarcely accommodate those already engaged. We had therefore to relinquish it, and prepare for a journey on horseback to the *Dalles*, about eighty miles below, to which place Captain W. would precede us in the barge, and engage canoes to convey us to the lower fort.

This evening, we purchased a large bag of Indian meal, of which we made a kettle of mush, and mixed with it a considerable quantity of horse tallow and salt. This was, I think, one of the best meals I ever made. We all ate heartily of it, and pronounced it princely food. We had been long without bread stuff of any kind, and the coarsest farinaceous substance, with a proper allowance of grease, would have been highly prized.

The next morning [4th], we visited Walla-walla Fort, and were introduced, by Captain W., to Lieutenant Pierre S. Pambrun, the superintendent. Wyeth and Mr. Pambrun had met before, and were well acquainted; they had, therefore, many reminiscences of bygone days to recount, and long conversations, relative to the variety of incidents which had occurred to each, since last they parted.

The fort is built of drift logs, and surrounded by a stockade of the same, with two bastions, and a gallery around the inside. It stands about a hundred yards from the river, on the south bank, in a bleak and unprotected situation, surrounded on every side by a great, sandy plain, which supports little vegetation, except the wormwood and thorn-bushes. On the banks of the little river [Walla Walla River], however, there are narrow strips of rich soil, and here Mr. Pambrun raises the few garden vegetables necessary for the support of his family. Potatoes, turnips, carrots, &c., thrive well, and Indian corn produces eighty bushels to the acre.

At about 10 o'clock, the barge got under way, and soon after, our company with its baggage, crossed the [Walla Walla?] river in canoes, and encamped on the opposite shore.

There is a considerable number of Indians resident here, Kayouses and a collateral band of the same tribe, called Walla-wallas. They live along the bank of the river, in shantys or wigwams of drift wood, covered with buffalo or deer skins. They are a

Fort Walla Walla by J. M. Stanley, 1847.

miserable, squalid looking people, are constantly lolling around and in the fort, and annoy visitors by the importunate manner in which they endeavor to force them into some petty trade for a pipe, a hare, or a grouse. All the industrious and enterprising men of this tribe are away trading salmon, kamas root, &c. to the mountain companies.

Notwithstanding the truly wretched plight in which these poor people live, and the privations which they must necessarily have to suffer, they are said to be remarkably honest and upright in their dealings, and generally correct in their moral deportment. Although they doubtless have the acquisitive qualities so characteristic of the race, they are rarely known to violate the principles of common honesty. A man may leave his tent unguarded, and richly stored with every thing which ordinarily excites the cupidity of the Indian, yet, on returning after a long absence, he may find all safe. What a commentary is this on the habits and conduct of our *Christian* communities!

The [Columbia] river is here about three-fourths of a mile in width, – a clear, deep, and rapid stream, the current being generally from three to four miles an hour. It is the noblest looking river I have seen since leaving our Delaware. The banks are in many places high and rocky, occasionally interrupted by broad, level sandy beaches. The only vegetation along the margin, is the wormwood, and other low, arid plants, but some of the bottoms are covered with heavy, rank grass, affording excellent pasture for horses.

[Slipp (1947) has suggested 4th September 1834 as the day that the type specimen of **Swainson's Hawk** was collected: "shot by Mr. TOWNSEND on a rock near the Columbia River, on which it had its nest" (Audubon 1840-44, 1: 30). See appendix 8 for discussion regarding the true type locality.]

1. They followed the Burnt River further upstream than later travellers, many with wagons, who usually cut off sooner and took a more north-easterly route via Flagstaff Hill.

2. Unidentified owl: it could have been one of several species. Because they were so hungry even something as small as a Western Screech-Owl could have been eaten, though at about 5oz (150gms) this species would not make much of a meal.

Mount Hood and
Columbia River.

Fort Walla Walla to Fort Vancouver

"How delightful, I have often exclaimed, must have been the feelings of those enthusiastic naturalists, NUTTALL and TOWNSEND, while traversing the ridges of the Rocky Mountains! How grand and impressive the scenery presented to their admiring gaze ... while on wide-spread wings the Great Vulture [California Condor] sailed overhead watching the departure of the travellers ... But now I see them, brother-like, with lighter steps, descending toward the head waters of the famed Oregon. They have reached the great stream, and seating themselves in a canoe, shoot down the current ..."

J.J. Audubon (1840-44) *Ornithological Biography* 4: 27.

[September] 5th.– This morning we commenced our march down the Columbia. We have no provision with us except flour and horse tallow, but we have little doubt of meeting Indians daily, with whom we can trade for fish. Our road will now be a rather monotonous one along the bank of the river, tolerably level, but often rocky, so that very rapid travelling is inadmissible. The **mallard** duck, the widgeon [**American Wigeon**], and the **green-winged teal** are tolerably abundant in the little estuaries of the river. Our men have killed several, but they are poor, and not good.

6th.– We have observed to-day several high, conical stacks of drift-wood near the river. These are the graves of the Indians. Some of these cemeteries are of considerable extent, and probably contain a great number of bodies. I had the curiosity to peep into several of them, and even to remove some of the coverings, but found nothing to compensate for the trouble.

We bought some salmon from Indians whom we met to-day, which, with our flour and tallow, enable us to live very comfortably.

7th.– We frequently fall in with large bands of Indian horses. There are among them some very beautiful animals, but they are generally almost as wild as deer, seldom permitting an approach to within a hundred yards or more. They generally have owners, as we observe upon many of them strange hieroglyphic looking characters, but there are no doubt some that have never known the bit, and will probably always roam the prairie uncontrolled. When the Indians wish to catch a horse from one of these bands, they adopt the same plan pursued by the South Americans for taking the wild animal.

The Dalles by J. M. Stanley, 1847.

7th September 1834 - Townsend collected a female Hairy Woodpecker, and later acquired a male that also went to Audubon. Although depicted in plate CCCCXVII (detail above) as Harris's Woodpecker it was not a new species. There is no surviving woodpecker specimen for 7th September 1834 but there is one for 27th September 1834: Hairy Woodpecker female, USNM 1869.

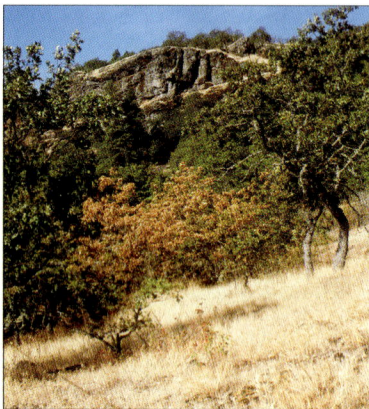

Oak Chaparral. After the endless sagebrush, there is a narrow belt of oak chaparral and grassland near the Dalles that continues for only about 25 miles downstream. Near the mouth of the Hood River the dense conifer forests suddenly begin and stretch all the way to the Pacific, interspersed with some oak woodland at lower elevations.

HAIRY WOODPECKER

Audubon described the dark and heavily marked Hairy Woodpeckers of the Pacific coastal regions as a new species, naming it Harris's Woodpecker:

"It is to Mr. TOWNSEND that we are indebted for the discovery of this singularly marked species, of which he has sent me a pair of specimens in excellent preservation, both shot on the Columbia river, the male on the 18th of January 1836, the female on the 7th of September 1834. Having been left at liberty to give names to whatever new species might occur among the birds transmitted to me by that zealous naturalist, I have honoured the present Woodpecker with the name of my friend EDWARD HARRIS, Esq., ... he merits this tribute as an ardent and successful cultivator of ornithology, and an admirer of the works of Him whose good providence gave me so noble-hearted a friend" (Audubon 1840-44, 4: 242).

"This species, so much resembling the common *P. villosus* [Hairy Woodpecker], is abundant in the forests on the Columbia river. Its habits are very similar to those of its near relative. Builds a loose and unsubstantial nest, in the hollow of a decayed tree, and lays four white eggs" (Townsend 1839: 348).

8th.– Our road to-day has been less monotonous, and much more hilly than hitherto. Along the bank of the river, are high mountains, composed of basaltic rock and sand ... At our camp in the evening, eight Walla-walla's came to see us. The chief was a remarkably fine looking man, but he, as well as several of his party, was suffering from a severe purulent opthalmia which had almost deprived him of sight. He pointed to his eyes, and contorting his features to indicate the pain he suffered, asked me by signs to give him medicine to cure him. I was very sorry that my small stock of simples did not contain anything suited to his complaint, and I endeavoured to tell him so. I have observed that this disease is rather prevalent among the Indians residing on the river, and I understand from the chief's signs that most of the Indians towards the lower country were similarly affected.

9th.– The character of the country has changed considerably since we left Walla-walla. The river has become gradually more narrow, until it is now but about two hundred yards in width, and completely hemmed in by enormous rocks on both sides. Many of these extend for considerable distances into the stream in perpendicular columns, and the water dashes and breaks against them until all around is foam. The current is here very swift, probably six or seven miles to the hour; and the Indian canoes in passing down, seem literally to *fly* along its surface. The road to-day has been rugged to the very last degree. We have passed over continuous masses of sharp rock for hours together, sometimes picking our way along the very edge of the river, several hundred feet above it; again, gaining the back land, by passing through any casual chasm or opening in the rocks, where we were compelled to dismount, and lead our horses.

This evening, we are surrounded by a large company of Chinook Indians, of both sexes, whose temporary wig-wams are on the bank of the river. Many of the squaws have young children sewed up in the usual Indian fashion, wrapped in a skin, and tied firmly to a board, so that nothing but the head of the little individual is seen.

These Indians are very peaceable and friendly. They have no weapons except bows, and these are used more for amusement and exercise, than as a means of procuring them sustenance, their sole dependence being fish and beaver, with perhaps a few hares and grouse, which are taken in traps. We traded with these people for a few fish and beaver skins, and some roots, and before we retired for the night, arranged the men in a circle, and gave them a smoke in token of our friendship.

10th.– This afternoon we reached the *Dalles*. The entire water of the river here flows through channels of about fifteen feet in width, and between high, perpendicular rocks; there are several of these channels at distances of from half a mile to a mile apart, and the water foams and boils through them like an enormous cauldron.

On the opposite side of the river there is a large Indian village ... We were disappointed in not seeing Captain W. here, as this was the spot where we expected to meet him; the chief, however, told us that we should find him about twelve miles below, at the next village. We were accordingly soon on the move again, and urging our horses to their fastest gait, we arrived about sunset. The captain, the chief of the village, and several other Indians, came out to meet us and make us welcome. Captain W. has been here two days, and we were pleased to learn that he had completed all the necessary arrangements for transporting ourselves and baggage to [Fort] Vancouver in canoes. The route by land is said to be a very tedious and difficult one, and, in some places, almost impassable, but even were it otherwise, I believe we should all much prefer the water conveyance, as we have become very tired of riding ...

We have concluded to leave our horses here, in charge of the chief of the village, who has promised to attend to them during the winter, and deliver them to our order in the spring. Captain W. having been acquainted with this man before, is willing to trust him.

Conifer forest on the banks of the Columbia River.

11th.– Early this morning, we launched our three [Western Red Cedar dugout] canoes, and each being provided with an Indian, as helmsman, we applied ourselves to our paddles, and were soon moving briskly down the river. In about an hour after, the wind came out dead ahead, and although the current was in favor, our progress was sensibly checked. As we proceeded, the wind rose to a heavy gale, and the waves ran to a prodigious height. At one moment our frail bark[1] danced upon the crest of a wave, and at the next, fell with a surge into the trough of the sea, and as we looked at the swell before us, it seemed that in an instant we must inevitably be engulphed. At such times, the canoe ahead of us was entirely hidden from view, but she was observed to rise again like a seagull, and hurry on into the same danger ... In the mean time the boat had become half full of water, shipping a part of every surf that struck her, and as we gained the shallows every man sprang overboard, breast deep, and began hauling the canoe to shore ... The goods had suffered considerably by the wetting; they were all unbaled and dried by a large fire, which we built on the shore ...

12th.– The gale continues with the same violence as yesterday, and we do not therefore think it expedient to leave our camp. Mr. N.'s large and beautiful collection of new and rare plants was considerably injured by the wetting it received; he has been

133

constantly engaged since we landed yesterday, in opening and drying them. In this task he exhibits a degree of patience and perseverance which is truly astonishing; sitting on the ground, and steaming over the enormous fire, for hours together, drying the papers, and re-arranging the whole collection, specimen by specimen, while the great drops of perspiration roll unheeded from his brow. Throughout the whole of our long journey, I have had constantly to admire the ardor and perfect indefatigability with which he has devoted himself to the grand object of his tour. No difficulty, no danger, no fatigue has ever daunted him, and he finds his rich reward in the addition of nearly a *thousand* new species of American plants, which he has been enabled to make to the already teaming flora of our vast continent. My bale of birds, which was equally exposed to the action of the water, escaped without any material injury.

In the afternoon, the gale not having abated, Captain W. became impatient to proceed, as he feared his business at [Fort] Vancouver would suffer by delay; he accordingly proposed taking one canoe, and braving the fury of the elements, saying that he wished five men, who were not afraid of water, to accompany him ... The more sedate amongst us did not much approve of this somewhat hasty measure of our principal; it appeared like a useless and daring exposure of human life, not warranted by the exigencies of the case. Mr. N. remarked that he would rather lose all his plants than venture his life in that canoe.

On the 13th the wind shifted to due north, and was blowing somewhat less furiously than on the previous day. At about noon we loaded our canoes, and embarked ...

14th. – Before sunrise, a light rain commenced, which increased towards mid-day to a heavy shower, and continued steadily during the afternoon and night ... We made an early start, and proceeded on very expeditiously until about noon, when we arrived at the "cascades," and came to a halt above them, near a small Indian village. These cascades, or cataracts are formed by a collection of large rocks, in the bed of the river, which extend, for perhaps half a mile. The current for a short distance above them, is exceedingly rapid, and there is said to be a gradual fall, or declivity of the river, of about twenty feet in the mile. Over these rocks, and across the whole river, the water dashes and foams most furiously, and with a roar which we heard distinctly at the distance of several miles.

It is wholly impossible for any craft to make its way through these difficulties, and our light canoes would not live an instant in them. It is, therefore, necessary to make a portage, either by carrying the canoes over land to the opposite side of the cataracts, or by wading in the water near the shore, where the surges are lightest, and dragging the unloaded boat through them by a cable. Our people chose the latter method ... In the meantime, Mr. N., and myself were sent ahead to take the best care of ourselves that our situation and the surrounding circumstances permitted. We found a small Indian trail on the river bank, which we followed in all its devious windings, up and down hills, over enormous piles of rough flinty rocks, through brier bushes, and pools of water, &c. &c., for about a mile, and descending near the edge of the river, we observed a number of white men who had just succeeded in forcing a large barge through the torrent, and were then warping her into still water near the shore. Upon approaching them more closely, we recognised, to our astonishment, our old friend Captain Stewart, with the good missionaries, and all the rest who left us at Walla-walla on the 4th. Poor fellows! Every man of them had been over breast deep in water, and the rain, which was still falling in torrents, was more than sufficient to drench what the waves did not cover, so that they were most abundantly soaked and bedraggled ... They informed us that Captain W.'s canoe had been dashed to pieces on the rocks above, and that he and all his crew were thrown into the water, and forced to swim for their lives. They all escaped and proceeded down the river, this morning, in a canoe, hired of the Indians here, one of whom accompanied them, as pilot.

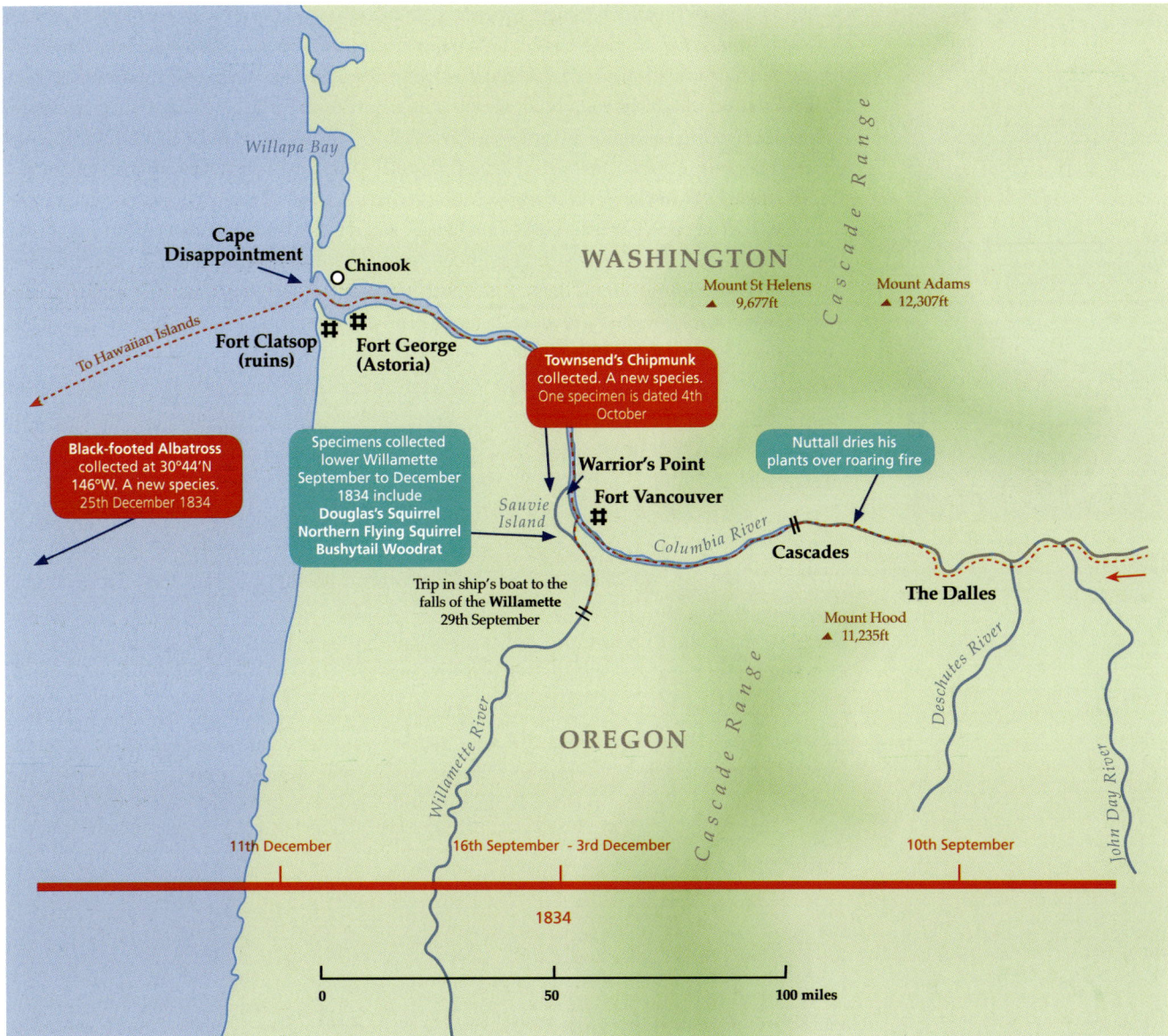

Map 8 - The Lower Columbia River, 10th September - 11th December 1834.

Cape Disappointment

Chinook

WASHINGTON

Willapa Bay

Mount St Helens
▲ 9,677ft

Mount Adams
▲ 12,307ft

Cascade Range

To Hawaiian Islands

Fort Clatsop (ruins)

Fort George (Astoria)

Townsend's Chipmunk collected. A new species. One specimen is dated 4th October

Black-footed Albatross collected at 30°44'N 146°W. A new species. 25th December 1834

Specimens collected lower Willamette September to December 1834 include **Douglas's Squirrel Northern Flying Squirrel Bushytail Woodrat**

Sauvie Island

Warrior's Point

Fort Vancouver

Columbia River

Nuttall dries his plants over roaring fire

Cascades

The Dalles

Trip in ship's boat to the falls of the **Willamette** 29th September

Mount Hood
▲ 11,235ft

Cascade Range

Willamette River

OREGON

Deschutes River

John Day River

11th December

16th September - 3rd December

10th September

1834

0   50   100 miles

Columbia River.

Beacon Rock, 33 miles upriver from Fort Vancouver.

After a hasty meal of fish, purchased on the spot, our friends reloaded their boat and got under way, hoping to reach [Fort] Vancouver by next morning ... In the mean time, we had all to walk back along the circuitous and almost impassable Indian trail, and carry our wet and heavy baggage from the spot where the boats had been unloaded ... Over this most miserable of all roads, with the cold rain dashing and pelting upon us during the whole time, until we felt as though we were frozen to the very marrow, did we all have to travel and return four separate times, before our baggage was properly deposited. It was by far the most fatiguing, cheerless, and uncomfortable business in which I was ever engaged, and truly glad was I to lie down at night on the cold, wet ground, wrapped in my blankets, out of which I had just wrung the water, and I think I never slept more soundly or comfortably than that night ...

15th.– ... The man whom we sent yesterday to the village, returned this morning; he stated that one canoe only could be had, but that three Indians, accustomed to the navigation, would accompany us; that they would soon be with us, and endeavor to repair our damaged boat. In an hour they came, and after the necessary clamping and caulking of our leaky vessel, we loaded, and were soon moving rapidly down the river. The rain ceased about noon, but the sun did not appear during the day.

[Wyeth could not contain his impatience and decided to press on to Fort Vancouver: "*15th. Early in the morning having hired another canoe put ahead and in a rainy day at about 12 ock. met the Bg [Brig] May Dacre in full sail up the River boarded her and found all well she had put into Valparaiso having been struck by Lightning and much damaged. Capt. Lambert was well and brot me 20 Sandwich Islanders and 2 Coopers 2 Smiths and a Clerk*" (Wyeth 1899: 82).]

16th.– The day was a delightful one; the sky was robed in a large flaky cumulus, the glorious sun occasionally bursting through among the clouds, with dazzling splendor. We rose in the morning in fine spirits, our Indians assuring as that "King George," as they called the fort, was but a short distance from us. At about 11 o'clock, we arrived, and stepped on shore at the *end of our journey*.

It is now three days over six months since I left my beloved home. I, as well as the rest, have been in some situations of danger, of trial, and of difficulty, but I have passed through them all unharmed, with a constitution strengthened, and invigorated by healthful exercise, and a heart which I trust can feel deeply, sincerely thankful to that kind and overruling Providence who has watched over and protected me [... *thankful to that excellent Being who has said, "Although ye forsake me, I will not cast you off for ever"* (MS Journal).]

On the beach in front of the fort, we were met by Mr. Lee, the missionary, and Dr. John McLoughlin, the chief factor, and Governor of the Hudson's Bay posts in this vicinity. The Dr. is a large, dignified and very noble looking man, with a fine expressive countenance, and remarkably bland and pleasing manners. The missionary introduced Mr. N. and myself in due form, and we were greeted and received with a frank and unassuming politeness which was most peculiarly grateful to our feelings. He requested us to consider his house our home, provided a separate room for our use, a servant to wait upon us, and furnished us with every convenience which we could possibly wish for. I shall never cease to feel grateful to him for his disinterested kindness to the poor houseless and travel-worn strangers.

1. Bark is an alternative spelling of barque (small boat). It is unlikely to have been a birch bark canoe; on the lower Columbia River, large canoes, sometimes 12 paces in length, were made from hollowed trunks of Western Red Cedar.

136

ZOOLOGICAL DISCOVERY IN THE PACIFIC NORTHWEST BY 1834

When Townsend arrived on the Columbia River in late 1834 he was by no means the first naturalist to work in the area. The majority of the birds and larger mammals, whether on the estuary, rocky shore or in the coastal forests, had been described in Europe by a variety of Russian, German and British naturalists long before he arrived, indeed before Lewis and Clark built Fort Clatsop at the mouth of the Columbia for the winter of 1805-06, and before Alexander Wilson and Audubon had even set foot in the United States. So, while it is true that Townsend was the first zoologist to work in lower Idaho, and in a few parts of Oregon and Washington, when it came to the coastal fauna he had been preceded by many shipboard naturalists, a few itinerant overland botanical collectors, and the fur trappers of the Hudson's Bay Company.

The literature is replete with references to the birds encountered along the Pacific coast by the early Spanish, French, Russian and British explorers, though many of them did not describe the birds and mammals scientifically, nor manage to bring back specimens (Pearse 1968). The best of the French and Spanish explorers had the misfortune to either lose their ships, or else their specimens languished in museums without being properly assessed. The Russians were more successful because they had settlements in western Alaska, southwards down the coast to Sitka and almost to San Francisco, so there had been many opportunities to send birds and mammals back to St Petersburg where such species as Short-tailed Albatross, Tufted Puffin, Rhinoceros Auklet and Pelagic Cormorant were described by P.S. Pallas, and Brandt's Cormorant and Spectacled Eider by J.F. von Brandt. The British expeditions to the area included Captain Cook's last voyage (1777-79), the specimens from which were described by John Latham (but not given scientific names until worked through by J.F. Gmelin in 1788-89). These new birds included Surfbird, Marbled Murrelet, Ancient Murrelet, Whiskered Auklet, Rufous Hummingbird, Varied Thrush and Golden-crowned Sparrow, which can all occur on or near the Columbia but were first obtained to the north, some of them from various Bering Sea locations. In comparison, the later British voyages of George Vancouver and Captain Beechey yielded only a few new animals.

Lewis and Clark, the first Americans to cross the continent, wrote good accounts of a range of birds and mammals but none of them constituted a formal description and they have lost much of the credit for their zoological work. They collected pitifully few specimens, and most of those were from the Rocky Mountains or the plains to the east: Lewis's Woodpecker, Clark's Nutcracker, Western Tanager (all three species being described and painted from their specimens by Alexander Wilson) and a few other birds and mammals, for example, Greater Sage-Grouse, Pronghorn, Mountain Goat, Mountain Beaver and Columbian Ground Squirrel. President Jefferson also received from them one live American Magpie and one live Black-tailed Prairie Dog.

By contrast, John Franklin's overland expeditions to arctic Canada (1819-22 and 1825-27) have long been recognised by zoologists for their valuable achievements. Success came largely through the separate efforts of Dr John Richardson and Thomas Drummond. Neither of them travelled near the lower Columbia River but Richardson preserved, and afterwards received from Drummond, David Douglas and Meredith Gairdner, a quantity of specimens, some of them common to the Columbia River area or collected there by Douglas and Gairdner. They were described by Richardson and William Swainson in the *Fauna Boreali-Americana* (1829-37) and included such novelties as Trumpeter Swan, White-tailed Ptarmigan, Gray-crowned Rosy-Finch, American Pika, Rainbow Trout, White Sturgeon and Pacific Lamprey, as well as a great many new subspecies (Houston 1998).

▶

David Douglas (1798 - 1834), Scottish plant collector who acquired a small number of birds from the Columbia River in 1825 - 27.

So, even at Fort Vancouver itself, Townsend had been preceded by David Douglas, as well as by the Hudson's Bay Company surgeons John Scouler, W.F. Tolmie and Gairdner, though Townsend was able to collaborate with the latter two by exchanging specimens because they were still working in the area when he arrived. Douglas, an energetic and restless Scottish botanical collector, had first arrived at the mouth of the Columbia River by ship in 1825 and explored the area around Cape Disappointment, Fort Astoria and Fort Vancouver, sometimes with Scouler, and later Douglas went up the Columbia to the cascades and sixty miles southwards along the Willamette River. He left the area in the spring of 1827, heading overland to Hudson Bay where he took ship to England. He was back at Fort Vancouver in May 1830 and made even more extensive plant collections than on his previous trip, this time wandering much further southwards and northwards, but losing all his specimens on the Fraser River, and losing his life in the Hawaiian Islands in 1834. Although Douglas collected birds and mammals he lacked either the skill or the preservatives to make a complete success of it, or may have just been unlucky in their transportation since many perished before reaching Britain, and others suffered from neglect by the Horticultural Society in London. Those few that went to the museums of Glasgow, Edinburgh and London were made use of by Richardson and Audubon but have now largely disappeared. Douglas seems to have concentrated on the bigger species of birds (Hall 1934), being one of the few early collectors to acquire both male and female California Condors (Harris 1941). He described the Mountain Quail and Franklin's Grouse as new but the latter is now regarded as a race of Spruce Grouse.

After all this collecting activity in the region would there be anything new for Townsend to find? There certainly was. He collected small birds and rodents in the oak woodlands and dark conifer forests, shorebirds at the mouth of the Columbia, voles, moles and pocket gophers from underground, bats from the storehouses of Fort Vancouver, as well as butterflies, moths and beetles from a variety of habitats and found new species in almost all these groups. He continued to collect reptiles but these were destroyed in July 1835 (see p.194). He also helped Nuttall to gather terrestrial, marine, and freshwater molluscs, a diversion they both enjoyed when bird and plant hunting was slow.

Townsend became the first to make a proper job of collecting birds and mammals in the Pacific Northwest. He had the identification skills and preservation techniques, and the good fortune, to safely relay hundreds of birds and a smaller number of mammals to Philadelphia. Many of Townsend's specimens were either new or exceedingly rare in collections and allowed Audubon to give a western balance to *The Birds of America* and *The Viviparous Quadrupeds*.

Douglas Fir *Pseudotsuga menziesii.*

138

Fort Vancouver and the lower Columbia River

"Mr. N. and myself are now residing on board the brig, and pursuing with considerable success our scientific researches through the neighborhood. I have shot and prepared here several new species of birds, and two or three undescribed quadrupeds, besides procuring a considerable number, which, though known to naturalists, are rare, and therefore valuable."

Townsend's *Narrative*, 5th November 1834.

[17th – 24th September 1834] FORT VANCOUVER is situated on the north bank of the Columbia on a large level plain, about a quarter of a mile from the shore. The space comprised within the stoccade is an oblong square, of about one hundred, by two hundred and fifty feet. The houses built of logs and frame-work, to the number of ten or twelve, are ranged around in a quadrangular form, the one occupied by the doctor being in the middle. In front, and enclosed on three sides by the buildings, is a large open space, where all the in-door work of the establishment is done. Here the Indians assemble with their multifarious articles of trade, beaver, otter, venison, and various other game, and here, once a week, several scores of Canadians are employed, beating the furs which have been collected, in order to free them from dust and vermin.

Mr. N. and myself walked over the farm with the doctor, to inspect the various improvements which he has made. He has already several hundred acres fenced in, and under cultivation, and like our own western prairie land, it produces abundant crops, particularly of grain, without requiring any manure. Wheat thrives astonishingly; I never saw better in any country, and the various culinary vegetables, potatoes, carrots, parsnips, &c., are in great profusion, and of the first quality. Indian corn does not flourish so well as at Walla-walla, the soil not being so well adapted to it; melons are well flavoured, but small; the greatest curiosity, however, is the apples, which grow on small trees, the branches of which would be broken without the support of props. So profuse is the quantity of fruit that the limbs are covered with it, and it is actually *packed*

Fort Vancouver by Henry Warre, 1845.

Dr John McLoughlin and a reconstruction of his house at Fort Vancouver.

together precisely in the same manner that onions are attached to ropes when they are exposed for sale in our markets.

On the farm is a grist mill, a threshing mill, and a saw mill, the two first, by horse, and the last, by water power; besides many minor improvements in agricultural and other matters, which cannot but astonish the stranger from a civilized land, and which reflect great credit upon the liberal and enlightened chief factor.

In the propagation of domestic cattle, the doctor has been particularly successful. Ten years ago a few head of neat cattle were brought to the fort by some fur traders from California; these have now increased to near seven hundred. They are a large framed, long horned breed, inferior in their milch qualities to those of the United States, but the beef is excellent, and in consequence of the mildness of the climate, it is never necessary to provide them with fodder during the winter, an abundant supply of excellent pasture being always found.

On the farm, in the vicinity of the fort, are thirty or forty log huts, which are occupied by the Canadians, and others attached to the establishment. These huts are placed in rows, with broad lanes or streets between them, and the whole looks like a very neat and beautiful village. The most fastidious cleanliness appears to be observed; the women may be seen sweeping the streets and scrubbing the door-sills as regularly as in our own proverbially cleanly city.

Lewis's Woodpecker specimen collected by Townsend on 22nd September 1834 (OUM 3523).

TOWNSEND'S BIG-EARED BAT

Townsend's Big-eared Bat (top left) in a rarely seen Audubon watercolour (NYHS Z3333 [S-30]), painted from a Townsend specimen from the Columbia River (see appendix 17). None of Audubon's bat paintings were included in *The Viviparous Quadrupeds*.

"TOWNSEND'S GREAT-EARED BAT ... Inhabits the Columbia river district, rather common. Frequents the store houses attached to the forts, seldom emerging from them even at night. This, and a species of *Verpertilio* [*sic*], (*V. subulatus,*) [**Western Small-footed Myotis**] which is even more numerous, are protected by the gentlemen of the Hudson's Bay Company, for their services in destroying the *dermestes* [beetles] which abound in their fur establishments" (Townsend 1839a: 324-325).

Sunday, September 25th.– Divine service was performed in the fort this morning by Mr. Jason Lee. This gentleman and his nephew had been absent some days in search of a suitable place to establish themselves, in order to fulfil the object of their mission. They returned yesterday, and intend leaving us to-morrow with their suite for the station selected, which is upon the Wallammet [Willamette] river, about sixty miles south of the fort.

In the evening we were gratified by the arrival of Captain Wyeth from below, who informed us that the brig from Boston, which was sent out by the company to which Wyeth is attached, had entered the river, and was anchored about twenty miles below, at a spot called Warrior's point, near the western entrance to the Wallammet.

Captain W. mentioned his intention to visit the Wallammet country, and seek out a convenient location for a fort which he wishes to establish without delay, and Mr. N. and myself accepted an invitation to accompany him in the morning. He has brought with him one of the brig's boats, and eight oarsmen, five of whom are Sandwich [Hawaiian] Islanders.

We have experienced for several days past, gloomy, lowering, and showery weather; indeed the sun has scarcely been seen for a week past. This is said to indicate the near approach of the rainy season, which usually sets in about the middle of October, or even earlier. After this time, until December, there is very little clear weather, showers or heavy clouds almost constantly prevailing.

SPECIMENS AND OBSERVATIONS, SEPTEMBER-NOVEMBER 1834

September specimens were taken at or near Fort Vancouver. From October to at least mid November the two naturalists based themselves on the brig at Warrior's Point, at the north end of Sauvie Island.

22nd September 1834, **Lewis's Woodpecker**, OUM 3523.

22nd September 1834, **Lewis's Woodpecker**, USNM 1909, card catalogue.

26th September 1834, **Sooty Grouse**. Audubon 1840-44, 5: 93).

26th September [1834?], **Douglas's Squirrel**. Inscription on Audubon watercolour (Tyler 2000: 124); probably painted from ANSP 286

27th September 1834, **Hairy Woodpecker**, USNM 1869.

4th October 1834, **Townsend's Chipmunk**, USNM 92/38797.

5th October 1834, **Oregon Dark-eyed Junco**, OUM 13517.

11th October 1834, **Spotted Towhee**, (Audubon 1840-44, 3: 166).

11th October 1834, **Western Scrub-Jay**, USNM 2841, card catalogue.

16th October 1834, **Oregon Dark-eyed Junco**, USNM 1948.

18th October 1834, **Song Sparrow**, USNM 83644, card catalogue.

18th October 1834, **Northern Flicker**, USNM 1886, card catalogue.

29th October 1834, **Bushy-tailed Woodrat**, LM D353a.

[no day] October 1834, **Downy Woodpecker**, OUM 3658.

October [no day or year stated, = 1834?], **Ruby-crowned Kinglet**, ANSP 16848.

7th November [1834?], **Downy Woodpecker**, OUM 3659.

- November [1834 or 35], **Northern Pygmy Owl**, (Audubon 1840 - 44, 1: 117).

DOUGLAS'S SQUIRREL

Plate XLVIII (detail) from Audubon's *Viviparous Quadrupeds.*

Audubon's original watercolour of Douglas's Squirrel for plate XLVIII of *The Viviparous Quadrupeds* is inscribed: "Columbia River Sep[r] 26th (no year) Sciurus Douglassii / John K. Townsend Douglass' Squirrel" (Tyler 2000: 124). This is a common species on the lower Columbia River and Townsend may well have collected it during his first few weeks there.

DUSKY GROUSE AND SOOTY GROUSE

Audubon wrote in his *Ornithological Biography* that "Mr. Townsend, in the notes with which he has favoured me, has the following observations:– Dusky Grouse, *Tetrao obscurus*. *Qul-al-lalleun* of the Chinooks. First found in the Blue Mountains, near Wallah Wallah, in large flocks, in September. Keep in pine woods altogether, never found on the plains; they perch on the trees. Afterwards found on the Columbia river in pairs in May [Sooty Grouse]. The eggs are numerous, of a cinereous-brown colour, blunt at both ends, and small for the size of the bird. The actions of the female, when the young are following her, are precisely the same as the Ruffed Grouse, using all the arts of that bird in counterfeiting lameness, &c. Female smaller than the male, lighter coloured, and wants the yellow warty skin upon the sides of the neck" (Audubon 1840-44, 5: 90-91, 93). Audubon had in his possession a Dusky Grouse "killed by Mr. Townsend on the Columbia river, Sept. 26. 1834" – now Sooty Grouse.

Plate CCCLXI of *The Birds of America* was captioned Dusky Grouse. Although long known as Blue Grouse the earlier name is now back in favour following the separation of Blue Grouse into Dusky and Sooty Grouse in 2006 (47th Supplement to the *AOU Check-list*). Townsend would have seen Dusky Grouse in the Blue Mountains and Sooty Grouse beside the Columbia River nearer to the coast

Excursion to the falls on the Willamette River

On the 29th, Captain Wyeth[1], Mr. N., and myself, embarked in the ship's boat for our exploring excursion [of about three days]. We had a good crew of fine robust sailors, and the copper colored islanders, – or *Kanakas*, as they are called, – did their duty with great alacrity and good will. At about five miles below the fort, we entered the upper mouth of the Wallammet. This river is here about half the width of the Columbia, a clear and beautiful stream, and navigable for large vessels to the distance of twenty-five miles. It is covered with numerous islands, the largest of which is that called *Wappatoo Island*, about twenty miles in length [now Sauvie Island[2]]. The vegetation on the main land is good, the timber generally pine and post oak [Garry Oak *Quercus garryana*], and the river is margined in many places with a beautiful species of willow with large ob-lanceolate leaves like those of the peach, and white on their under surface [perhaps Arroyo Willow *Salix lasiolepis*]. The timber on the islands is chiefly oak, no pine growing there.

At about 10 o'clock we overtook three men whom Captain W, had sent ahead in a canoe and we all landed soon after on the beach and dined on a mess of salmon and peas which we had provided. We were under way again in the afternoon, and encamped at about sunset. We have as yet seen no suitable place for an establishment, and to-morrow we proceed to the falls of the river, about fifteen miles further. Almost all the land in the vicinity is excellent and well calculated for cultivation, and several spots which we have visited, would be admirably adapted to the captain's views, but

Wappatoo or Arrowhead *Sagittaria latifolia*.

Garry Oak *Quercus garryana*.

that there is not a sufficient extent unincumbered, or which could be fitted for the purposes of tillage in a space of time short enough to be serviceable; others are at some seasons inundated, which is an insurmountable objection.

We embarked early the next morning [30th], and at 11 o'clock arrived at the falls, after encountering some difficulties from rapids, through which we had to warp our boat. There are here three falls on a line of rocks extending across the river, which forms the bed of the upper channel. The water is precipitated through deep abrazed gorges, and falls perhaps forty feet at an angle of about twenty degrees. It was a beautiful sight when viewed from a distance, but it became grand and almost sublime as we approached it nearer. I mounted the rocks and stood over the highest fall, and although the roar of the cataract was almost deafening, and the rays of the bright sun reflected from the white and glittering foam threatened to deprive me of sight, yet I became so absorbed in the contemplation of the scene, and the reflections which were involuntary excited, as to forget every thing else for the time, and was only aroused by Captain W. tapping me on the shoulder, and telling me that every thing was arranged for our return. While I visited the falls, the captain and his men had found what they sought for; and the object of our voyage being accomplished, we got on board immediately and shaped our course down the river with a fair wind, and the current in favor.

About two miles below the cataract is a small village of Klikatat Indians. Their situation does not appear different from what we had been accustomed to see in the neighborhood of the fort. They live in the same sort of miserable loose hovels, and are the same wretched, squalid looking people. Although enjoying far more advantages, and having in a much greater degree the means of rendering themselves comfortable, yet their mode of living, their garments, their wigwams, and every thing connected with them, is not much better than the Snakes and Bannecks, and very far inferior to that fine, noble-looking race, the Kayouse, whom we met on the *Grande ronde*.

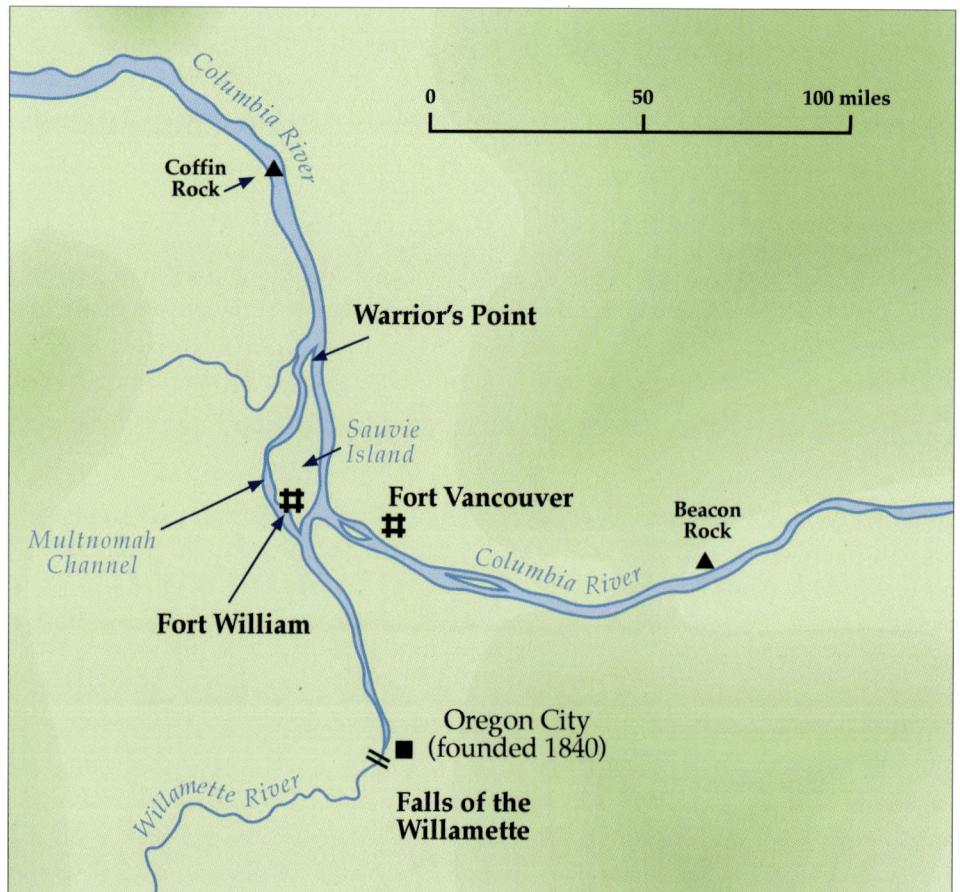

Map 9. Fort Vancouver and vicinity.

144

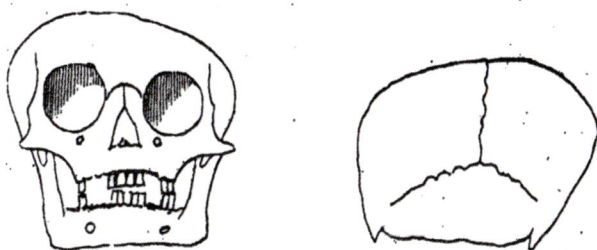

COWALITSK.

A custom prevalent, and almost universal amongst these Indians, is that of flattening, or mashing in the whole front of the skull, from the superciliary ridge to the crown. The appearance produced by this unnatural operation is almost hideous, and one would suppose that the intellect would be materially affected by it. This, however, does not appear to be the case … It is even considered among them a degradation to possess a round head, and one whose *caput* has happened to be neglected in his infancy, can never become even a subordinate chief in his tribe, and is treated with indifference and disdain, as one who is unworthy a place amongst them … The mode by which the flattening is effected, varies considerably with the different tribes. The Wallammet Indians place the infant, soon after birth, upon a board, to the edges of which are attached little loops of hempen cord or leather, and other similar cords are passed across and back, in a zig-zag manner , through these loops, enclosing the child, and binding it firmly down. To the upper edge of this board, in which is a depression to receive the back part of the head, another smaller one is attached by hinges of leather, and made to lie obliquely upon the forehead, the force of the pressure being regulated by several strings attached to its edge, which are passed through holes in the board upon which the infant is lying, and secured there.

I saw, to-day a young child from whose head the board had just been removed. It was, without exception, the most frightful and disgusting looking object that I ever beheld …

Warrior's Point and Fort Vancouver[3]

On the 1st of November [= October?] we arrived at the brig. She was moored, head and stern, to a large rock near the lower mouth of the Wallammet [at Warrior's Point]. Captain Lambert with his ship's company, and our mountain men, were all actively engaged at various employments; carpenters, smiths, coopers, and other artisans were busy in their several vocations; domestic animals, pigs, sheep, goats, poultry, &c., were roaming about as if perfectly at home, and the whole scene looked so like the entrance to a country village, that it was difficult to fancy oneself in a howling wilderness inhabited only by the wild and improvident Indian, and his scarcely more free and fearless neighbors, the bear and the wolf. An excellent temporary storehouse of twigs, thatched with grass, has been erected, in which has been deposited the extensive assortment of goods necessary for the settlement, as well as a number of smaller ones, in which the men reside. It is intended as soon as practicable, to build a large and permanent dwelling of logs, which will also include the store and trading establishment, and form the groundwork for an *American fort* [Fort William[4]] on the river Columbia.

5th.– Mr. N. and myself are now residing on board the brig, and pursuing with considerable success our scientific researches through the neighborhood. I have shot and prepared here several new species of birds, and two or three undescribed quadrupeds besides procuring a considerable number, which, though known to naturalists, are rare, and therefore valuable. My companion is of course in his element; the forest, the plain, the rocky hill, and the mossy bank yield him rich and most abundant supply …

Mount Hood rising above the Columbia River.

Townsend's Chipmunk. Plate XX from Audubon's *Viviparous Quadrupeds* was prepared from specimens collected by Townsend near the banks of the lower Columbia River. Because of recent taxonomic changes, the range of this species overlaps with the Sooty Grouse, not the Dusky Grouse (see text at right).

"Mr. Townsend, who procured the specimens from which we have drawn up our description, observes, "This pretty little fellow, so much resembling our common *T. striatus, (Lysteri,)* [Eastern Chipmunk] is quite common; it lives in holes in the ground; running over your foot as you traverse the woods. It frequently perches itself upon a log or stump, and keeps up a continual clucking, which is usually answered by another at some distance, for a considerable time. Their note so much resembles that of the dusky grouse, (*Tetrao obscurus,*) that I have more than once been deceived by it" (Audubon and Bachman 1845-48: 72, 74, plate XX, based on Bachman 1839b: 68-71). One Townsend specimen is dated 4th October 1834 (USNM 92/38797).

Townsend's Chipmunk, collected by Townsend "Oregon, 1834" (ANSP 241, lectotype, original label lost).

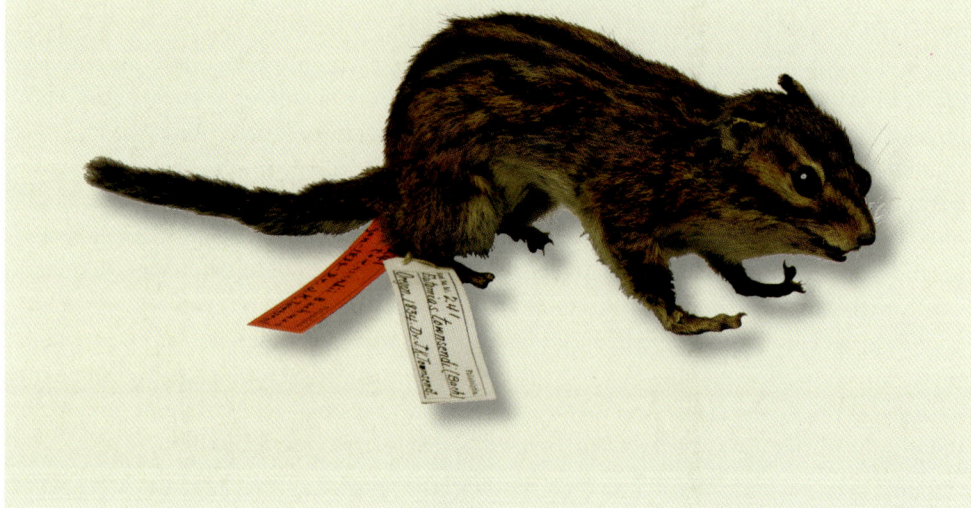

146

OREGON DARK-EYED JUNCO

Oregon Dark-eyed Junco. Plate
CCCXCVIII of *The Birds of America*
(detail).

5th October. Townsend collected an Oregon Dark-eyed Junco that was painted by
Audubon.

"OREGON SNOW-BIRD ... This species, which is so nearly allied to our Common
Snow-bird [Slate-colored Dark-eyed Junco], is another of those recently added to our
Fauna by Mr. TOWNSEND, from whom I purchased several specimens. All that I know
of its habits is derived from the following notice given me by Mr. NUTTALL. "It was first
seen by us in the woods of the Columbia, in the autumn and winter, flitting about in
small flocks, always in the forest, never in the open fields, or on the way-sides. At this
time they rarely utter an occasional chirp, or remain wholly silent. We afterwards saw
them inhabiting the same woods throughout the summer, in diminished numbers, or in
pairs, but I do not recollect hearing them utter any song, though they are probably not
silent in the season of breeding. With the nest, eggs, and young I am not acquainted."

I have represented the male and female from specimens procured by Mr. Townsend
on the Columbia river, on the 5th of October, 1834" (Audubon 1840-44, 3: 91).

"OREGON SNOW FINCH ... Common on the
Columbia river in winter. Gregarious. Voice, and
general habits similar to *F. hyemalis* [Slate-colored
Dark-eyed Junco]"
(Townsend 1839a:
345-346).

Oregon Dark-eyed Junco specimens
collected by Townsend on 5th October
1834 and in October 1835 (OUM 13517
and OUM 13516).

WESTERN SCRUB-JAY

The Western Scrub-Jay breeds all along the wooded banks of the Columbia River from the Dalles to Fort Vancouer, and in October the two naturalists must have often seen these wonderful jays, though rather strangely, Nuttall is the only one to put anything in print, under the name of Ultramarine Jay: "Early in October, on arriving in the forests of the Columbia, near Fort Vancouver, an establishment of the Hudson's Bay Company, we saw in the same situations with Steller's Jay the present species. Its habits are much like those of the Common Jay [Blue Jay]... It is a graceful, active, and rather shy bird, flying out straight from tree to tree, remarkable by its long tail and rather short wings; and its note is much less harsh and loud than that of Steller's Jay" (Nuttall 1840: 245). Back in mid-September, in poor weather and too preoccupied by the difficult descent of the Columbia River, Townsend and Nuttall had failed to notice the Western Scrub-Jay.

DOWNY WOODPECKER

Two Downy Woodpecker specimens collected by Townsend, dated "7th November [1834?]" and "October 1834" (OUM 3659, OUM 3658). Audubon described a male Downy Woodpecker from the Columbia River, understandably believing it to be a new species because of the way it differed from the eastern race, but he gave no sources or date for his specimen; he named his supposed new species after Meredith Gairdner, a surgeon at Fort Vancouver: "GAIRDNER'S WOODPECKER Picus Gairdnerii" (Audubon 1840-44, 4: 252-254). Apart from the description he gave little additional information and did not paint it.

BUSHY-TAILED WOODRAT

"Neotoma Drummondii.– Richardson. Rocky Mountain Neotoma ... Mr. TOWNSEND has kindly furnished us with some remarks on this species, from which we make the following extracts:– "I never saw it in the Rocky Mountains, but it is very common near the Columbia river. It is found in the store-houses of the inhabitants, where it supplies the place of the common rat, which is not found here. It is a remarkably mischievous animal, destroying every thing which comes in its way – papers, books, goods, &c. It has been known not unfrequently to eat entirely through the middle of a bale of blankets, rendering the whole utterly useless; and like a pet crow carries away every thing it can lay its *hands* on. Even candle-sticks, porter-bottles, and large iron axes, being sometimes found in its burrows" (Audubon and Bachman 1845-48: 98-100, plate XXIX).

Bushy-tailed Woodrat collected by Townsend on 29th October 1836. It was one of the specimens sold by Audubon, on Townsend's behalf, to Lord Derby (LM D353a).

NORTHERN PYGMY OWL

Northern Pygmy Owl: "Of this pretty little Owl I can only say that the single specimen from which I made the two figures in the plate before you, was sent to me by Mr. TOWNSEND, along with the following notice respecting it: – "I shot this bird on the Columbia River, near Fort Vancouver, in the month of November [1834 or 1835]. I first saw it on wing about mid-day, and its curious jerking or undulating flight struck me as extremely peculiar, and induced me to follow and secure it. It soon alighted upon a high branch of a pine tree, and I shot it with my rifle, the only gun I had with me, as I was at that time engaged in shooting cranes [Sandhill Cranes] along the banks of the river. The specimen is somewhat mutilated, in consequence of having lost one wing by the ball. The stomach contained nearly the whole body of a Ruby-crowned Wren, with a few small remnants of beetles and worms. It was a male …"

I have [subsequently] seen several specimens of this Owl in the Edinburgh Museum, which had also been sent from Fort Vancouver by Dr. MEREDITH GAIRDNER" (Audubon 1840-44, 1: 117-118).

Garry Oaks on Sauvie Island.

Audubon painted the Northern Flying Squirrel in two plates as three different species. The Oregon Flying Squirrel (Plate XV at right) was almost certainly drawn from a Townsend specimen, though there is no text to confirm it. Bachman merely notes that, "Mr. Townsend remarks, in regard to this species, that it inhabits the pine woods of the Columbia, near the sea; very rare. Habits of the P. *volucella* [Southern Flying Squirrel]" (Bachman 1839c: 101-103). Rhoads (1897: 324) suggested that this was one of the quadrupeds collected while Townsend was based on the brig at the mouth of the Willamette and gave it the probable type locality of near St Helens, Columbia County, Oregon – though there seems to be no evidence for such a precise locality as the only date is a museum catalogue date of '1834'.

Plate CXLIII (detail at left) shows the Northern Flying Squirrel under the names of Severn River Flying Squirrel and Rocky Mountain Flying Squirrel – the latter is the upper figure in the plate, drawn either from a Townsend or Drummond specimen, though once again the text does not make it clear.

[5th November continued] Mr. N. and myself have been anxious to escape the wet and disagreeable winter of this region, and visit some other portion of the country, where the inclemency of the season will not interfere with the prosecution of our respective pursuits. After some reflection and consultation, we concluded to take passage in the brig, which will sail in a few weeks for the Sandwich Islands. We shall remain there about three months, and return to the river in time to commence our peregrinations in the spring.

23d [November].– At Fort Vancouver. A letter was received yesterday by Dr. McLoughlin, from Captain Wyeth, dated Walla-walla, stating that the twelve Sandwich Islanders whom he took with him a week since[5] for a journey to Fort Hall, had deserted, each taking a horse … the consequence will then be, that the expedition must be abandoned, and the captain return to the fort to spend the winter.

Lower Columbia River.

December 3d.– Yesterday Mr. N. and myself went down the river to the brig, and this morning early the vessel left her moorings, and with her sails unloosed stood out into the channel way. The weather was overcast, and we had but little wind, so that our progress during the morning was necessarily slow. In the afternoon we ran aground in one and a half fathoms water, but as the tide was low, we were enabled to get her clear in the evening. The navigation of this river is particularly difficult in consequence of numerous shoals and sand bars, and good pilots are scarce, the Indians alone officiating in that capacity. Towards noon the next day, a Kowalitsk Indian with but one eye, who said his name was *George*, boarded us, and showed a letter which he carried, written by Captain McNeall, in the Hudson's Bay service, recommending said George as a capable and experienced pilot. We accepted his services gladly …

On the afternoon of the 4th, we passed along a bold precipitous shore, near which we observed a large isolated rock, and on it a great number of canoes, deposited above the reach of the tides. This spot is called *Mount Coffin*, and the canoes contain the dead bodies of Indians. They are carefully wrapped in blankets, and all the personal property of the deceased, bows and arrows, guns, salmon spears, ornaments, &., are placed within, and around his canoe. The vicinity of this, and all other cemeteries, is held so sacred by the Indians, that they never approach it, except to make similar deposites; they will often even travel a considerable distance out of their course, in order to avoid intruding upon the sanctuary of their dead.

We came to anchor near this rock in the evening, and Captain Lambert, Mr. N., and myself visited the tombs. We were especially careful not to touch or disarrange any of the fabrics, and it was well we were so, for as we turned to leave the place, we found that we had been narrowly watched …I have been very anxious to procure the skulls of some of these Indians …

6th.– The weather is almost constantly rainy and squally, making it unpleasant to be on deck; we are therefore confined closely to the cabin, and are anxious to get out to sea as soon as possible, if only to escape this.

In the afternoon, the captain and myself went ashore in the long-boat, and visited several Indian houses upon the beach. These are built of roughly hewn boards and logs, usually covered with pine bark, or matting of their own manufacture, and open at the top, to allow the smoke to escape. In one of these houses we found men, women, and children, to the number of fifty-two, seated as usual, upon the ground, around numerous fires, the smoke from which filled every cranny of the building, and to us was almost stifling, although the Indians did not appear to suffer any inconvenience from it. Although living in a state of the most abject poverty, deprived of most of the absolute necessities of life, and frequently enduring the pangs of protracted starvation, yet these poor people appear happy and contented. They are scarcely qualified to enjoy the common comforts of life, even if their indolence did not prevent the attempt to procure them.

Fort George

On the afternoon of the 8th, we anchored off *Fort George*, as it is called, although perhaps it scarcely deserves the name of a fort, being composed of but one principal house of hewn boards, and a number of small Indian huts surrounding it, presenting the appearance, from a distance, of an ordinary small farm house with its appropriate out-buildings … This is the spot where once stood the fort established by the direction of our honored countryman, John Jacob Astor. One of the chimneys of old Fort Astoria is still standing, a melancholy monument of American enterprise and domestic misrule. The spot where once the fine parterre overlooked the river, and the bold stoccade enclosed the neat and substantial fort, is now overgrown with weeds and bushes, and can scarce be distinguished from the primeval forest which surrounds it on every side …

ASTORIA, AS IT WAS IN 1813.

The next morning [9th?], we ran down into Baker's bay, and anchored within gunshot of the cape, when Captain Lambert and myself went on shore in the boat, to examine the channel, and decide upon the prospect of getting out to sea. This passage is a very dangerous one, and is with reason dreaded by mariners. A wide bar of sand extends from Cape Disappointment to the opposite shore, – called Point Adams, – and with the exception of a space, comprehending about half a mile, the sea at all times breaks furiously, the surges dashing to the height of the mast head of a ship, and with the most terrific roaring. Sometimes the water in the channel is agitated equally with that which covers the whole length of the bar, and it is then a matter of imminent risk to attempt a passage. Vessels have occasionally been compelled to lie in under the cape for several weeks, in momentary expectation of the subsidence of the dangerous breakers, and they have not unfrequently been required to stand off shore, from without, until the crews have suffered extremely for food and water. This circumstance must ever form a barrier to a permanent settlement here; the sands, which compose the bar, are constantly shifting, and changing the course and depth of the channel, so that none but the small coasting vessels in the service of the company can, with much safety, pass back and forth.

Mr. N. and myself visited the sea beach, outside the cape, in the hope of finding peculiar marine shells, but although we searched assiduously during the morning, we had but little success. We saw several deer in the thick forest on the side of the cape, and a great number of black shags, or cormorants, flying over the breakers, and resting upon the surf-washed rocks [**Brandt's, Pelagic** and **Double-crested Cormorants** all occur here.]

On the morning of the 11th, Mr. Hanson, the mate, returned from the shore, and reported that the channel was smooth; it was therefore deemed safe to attempt the passage immediately …

When we entered the channel, the water which had before been so smooth, became suddenly very much agitated, swelling, and roaring, and foaming around us, as if the surges were upheaved from the very bottom, and as if our vessel would fall in the trough of the sea, pitching down like a huge leviathan seeking its native depths, I could not but feel positive, that the enormous wave, which hung like a judgement over our heads, would inevitably engulph us … although I was aware of our imminent peril, and the tales that I had frequently heard of vessels perishing in this very spot, and in precisely such a sea, recurred to my mind with some force, yet I could not but

Brandt's Cormorants.

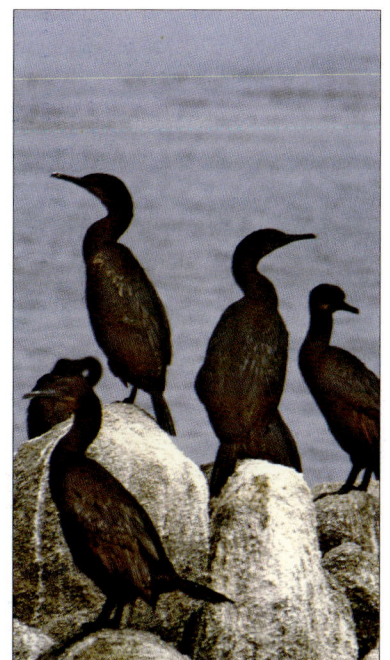

"At this spot several vessels belonging to the Hudson's Bay Company have been lost, and it was here that our noble Peacock, when attached to the United States Exploring Expedition foundered [18th July 1841], carrying with her the products of many months of labor and scientific toil performed by our energetic and indefatigable countrymen.

We were, as may be supposed, heartily glad to leave this frightful place, and in a few minutes were booming along over a beautiful placid sea, at the rate of eight knots an hour" (Townsend 1847: 88-89).

Titian Ramsay Peale was a naturalist on the *Peacock* and, as a result of the wreck, lost an unknown number of specimens from the Hawaiian Islands and elsewhere.

feel a kind of secret and wild joy at finding myself in a situation of such awful and magnificent grandeur. I thought of the lines of Shelly, and repeated them to myself in a kind of ecstasy.

> *"And see'st thou, and hear'st thou,*
> *And fear'st thou, and fear'st thou,*
> *And ride we not free*
> *O'er the terrible sea,*
> *I and thou?"*

In about twenty minutes we had escaped all the danger, and found ourselves riding easily in a beautiful placid sea. We set the sails, which had been shortened on the bar, and the gallant vessel feeling the impulse of the wind, rushed ahead as if exulting in the victory she had achieved.

1. Wyeth (1899: 83) simply says: "*28th*. Up the Wallamut with Mr. Nuttall and Townsend and Mr. Stout."

2. Wappatoo or Wapato Island was named because of the abundance there of Arrowhead *Sagittaria latifolia*; the roots being a staple food of the Indians. It is now called Sauvie Island, after Baptiste Sauvé who ran the Hudson's Bay Company dairy farm there.

3. There is no entry in the *Narrative* for any date in October, nor is there any Townsend MS material for the period to clarify his movements. Wyeth (1899: 83) is characteristically blunt and of no help in telling us where the naturalists went after returning from the Willamette Falls:

"*30th*. [September]. Returning down the rivers.

31st. [= 1st October]. At night reached the vessell at Carneans [Casineau's village, on or near Sauvie Island] from this time until the 13th. Oct. making preparation for a campaign into the Snake country and arrived on the 13th at [Fort] Vancouver and was received with great attention by all there."

Although the *Narrative* starts up again with 1st November it seems likely that this is a mistake for 1st October. This makes more sense, and ties in better with Wyeth's Journal, but then means that there are few entries for November. At the end of September the two naturalists would have returned from the Willamette, gone with Wyeth to the *May Dacre* at Warrior's Point. They may have stayed there until the 13th October when Wyeth went up river to Fort Vancouver, or they may have stayed on the brig for the rest of the month or more. Another possibility, though there seems little reason for it, is that they moved about frequently. Weather permitting, it would have been simple enough for them to go by canoe between Warrior's Point, Fort William and Fort Vancouver; the distance from Warrior's Point to Fort Vancouver being only about fifteen miles (see map). By February 1836, if not already, there was a little schooner running twice a week between the two forts.

Townsend's movements at this time are zoologically significant because of the need to designate localities for the new species he found, many of them collected in October 1834 rather than November 1834. Although the sites are close together, Fort Vancouver lies on the north bank of the Columbia, in Washington, while Warrior's Point and Fort William are on Sauvie Island, in Oregon. Most localities for his new species have the designation "Fort Vancouver" but many of them were no doubt from Sauvie Island.

From existing specimens we know that Townsend was adding many common resident mammals and birds to his collection, but none of them have localities any more specific than "Columbia River" and so do not help to determine his movements.

4. It was named Fort Williams in honour of a partner of Wyeth's in Boston, but for most of its brief existence was known as Fort William.

5. Wyeth (1899: 83-84) started out with the Sandwich Islanders on 13th October; they deserted about 10th November.

Part 2 - Chapter 9

First visit to the
Hawaiian Islands, 1835

"While at the Islands I made a considerable collection of the native shells & a few other small matters, but no birds to speak of. They are very scarce & very difficult of access."

J.K. Townsend to C. Townsend, 2nd May 1835.

The voyage to the Hawaiian Islands

[12th December 1834] - We saw, outside the bar, a great number of birds, of various kinds – ducks of several species, two or three kinds of guillemots, (*Uria* [**murrelets**],) – shags, (*Phalacrocorax* [**cormorants**],) among which was a splendid new species, brown albatross, (*Diomedea fusca* [**Black-footed Albatross**],) the common dusky pelican, (*P. fuscus* [**Brown Pelican**],) and numerous *Procellariae* [**fulmars, shearwaters, storm-petrels**], – also, the beautiful marine animal, called *Medusa* [**jellyfish**]. It is a cartilaginous or gummy substance, flattish, and about the size of a man's hand, with a tube projecting from it, expanded or flared out like the end of a clarionet. Within the body, near the posterior part is a large ovate ball, of a bright orange color, resembling the yolk of an egg.

14th.– There is to-day a heavy sea running, and we landsmen are affording some merriment to the seasoned crew, by our "lubberly" manner of "fetching away" in our attempts to walk the deck. I find, for myself, that I must for the present consent to relinquish an erect and dignified carriage, and adopt the less graceful, but safer method of clinging to the rails, &c., to assist locomotion. One thing, however, I cannot but feel thankful for, which is, that I have never felt in the least degree sea sick; and having so far escaped, I have no apprehension for the future.

Saw, in the afternoon, a large **sperm whale**, lazily rolling about a quarter of a mile ahead of the vessel. It occasionally spouted up a stream of water to the height of six or seven feet, but was perfectly quiescent until we approached near it, when it suddenly sank away and was lost to sight.

20th.– We observe constantly around us several species of dark **albatross, puffins** [= **shearwaters**?], petrels, &c. They follow closely in our wake, sailing over the surges with astonishing ease and grace, frequently skimming so near the surface that the eye loses them for an instant between the swells, but at such times they never touch the water, although we not unfrequently see them resting upon it.

23d.– The weather has become very mild, the thermometer ranging from 65° to 75°, indicating our approach to the tropics; and as a further proof of it, we saw, this morning, a beautiful tropic bird, (*Phaeton* [sic]) [**Red-tailed Tropicbird** *Phaethon rubricauda*[1]]. It sails around the vessel with an easy, graceful sweep, its long train being very conspicuous, and sufficiently distinguishing it from a tern, which, in other respects, it closely resembles. Its voice is very much like that of the great tern, (*Sterna hirundo*,) [Royal Tern *Sterna maxima* rather than Common Tern *Sterna hirundo*] being a harsh, loud, and gutteral croak, emitted while sailing high, and with its head curved downwards, examining the surface of the sea in search of its finny prey. [It is remarkable for having two central tail-feathers of a brilliant crimson, and about twice the length of the whole body. I procured several specimens, but unfortunately lost more than I obtained, from their falling into the sea after being shot flying over the ship (Townsend 1847: 90).]

Northern Fulmar.

On one of Townsend's passages between the Columbia and Hawaii he collected a Northern Fulmar at 38° N, long 134° W. (ANSP 5190, no date). There are two more specimens at Washington (USNM 2750 & USNM 2751, no dates or localities). Audubon may have used all or some of these Northern Fulmar skins when he vainly attempted to describe two new species of fulmar in his *Ornithological Biography*:

"PACIFIC FULMAR PROCELLARIA PACIFICA, Aud. (NOT FIGURED.) … Three skins transmitted to me by Mr. TOWNSEND appear to belong to two species of the Fulmar genus, distinct from that of the Atlantic seas. The first of these species I have named as above … North-west coast of America. Abundant" (Audubon 1840-44, 7: 208-209).

"SLENDER-BILLED FULMAR PROCELLARIA TENUIROSTRIS, Aud. (NOT FIGURED.) … The following note from Mr. TOWNSEND was appended to this specimen: - "Within a day's sail from the mouth of the Columbia river. … they are so tame as almost to allow themselves to be taken with the hand. The stomachs of most of those that I captured were found to contain a species of sepia [cuttlefish] and grease …" (Audubon 1840-44, 7: 210).

For a discussion about Townsend's controversial seabirds see appendix 10.

"Sketches of a Voyage, and residence in the South Sea islands"

Ten years after returning to Philadelphia, Townsend published a fresh account of his time in the Pacific that contained several interesting snippets that were not included in his *Narrative* of 1839. His article appeared in the *Literary Record and Journal of the Linnaean Association of Pennsylvania College* (1847) under the title of "Sketches of a Voyage, and residence in the South Sea islands". An almost identical version was later printed in *Philadelphia's Saturday Evening Post* (1849). The following extract from the *Literary Record* refers to the seabirds seen after leaving the Oregon coast:

"Off Cape Disappointment, and for many hundred miles out at sea, we observed great numbers of Sea Birds of various kinds, several of which I have myself described and published as new species. The little Guillemots, (Uria) [**Cassin's** and **Rhinoceros Auklets** are the most likely] were tumbling and rolling along on the surface, half swimming, half flying, and looking almost precisely like large eggs as they sported across our bows. From this resemblance, which is very striking, they have universally, among the sailors, obtained the name of "Egg Birds." Large Cormorants, (Phalacrocorax) of several species [**Double-crested, Pelagic** and **Brandt's Cormorants**], were very abundant, as were also various kinds of *Petrels*, (Procellaria) [**Northern Fulmar** and **various shearwaters**] and *Mother Carey's Chickens*, (Thalassidroma.) [**Fork-tailed** and **Leach's Storm-Petrel** are the most likely]. The last named birds are so called by the sailors from the superstition too well known to be repeated here. In connection with this superstition, Jack has also the credit of believing, not only that the appearance of this pretty and harmless bird is always indicative of the near approach of a storm, but that if any one has the temerity to catch and kill one, the vessel in which he sails will surely be overtaken by a tempest and destroyed …

A large brown Albatross (which I have named *Diomedea fusca*) inhabits these seas [**Black-footed Albatross**]. It differs considerably from the common white species which is so abundant around the two great Capes. Though smaller than the Cape Bird, it yet measures from twelve to fourteen feet across the wings. No one who has not seen this noble bird in flight can form any idea of the extreme ease and grace with which it skims over the foaming billows. Its long, falcate wing seems never to tire … On, on, he flies over the wide and wasteful ocean, without ever appearing to rest, except when he alights to pick up something floating on its surface (Townsend 1847: 88-113)."

BLACK-FOOTED ALBATROSS

On Christmas Day 1834 Townsend collected a **Black-footed Albatross**, a new species that was first described and named by Audubon, in volume V of the *Ornithological Biography* (1839): "For a specimen of this Albatross, I am indebted to Mr. TOWNSEND, who procured it on the 25th December, 1834, on the Pacific Ocean, in lat. 30°, 44′, N. long. 146° [W.]. It is clearly distinct from the other two described in this work, namely the Dusky and the Yellow-nosed; but I have received no information respecting its habits. Not finding any of the meagre notices or descriptions to which I can refer to agree with this bird, I have taken the liberty of giving it a name, being well assured that, should it prove to have been described, some person will kindly correct my mistake" (Audubon 1840-44, 7: 198-199). There was no accompanying painting.

The plumage of this species is quite variable: juveniles are uniformly brown, adults are slightly paler (and a rare light phase adult also occurs). It breeds in the western Hawaiian Islands and some winter off the Oregon coast.

Black-footed Albatross.

157

30th.– For the last four or five days we have been making but little headway, having been, occasionally, almost becalmed, and not going at any time more than two, or two and a half knots. The weather is so warm that our upper garments have become uncomfortable, the mean of the thermometer being about 77°, but we have, for several days past, been favored with cool, refreshing showers in the evening, which tend very much to our comfort.

Saw, this afternoon, in latitude 23° 20′ N., longitude 149° 30′ W., a "school" of eight or ten **sperm whales**. Several passed within twenty yards of the vessel, and we had an excellent opportunity of observing them. They were so near that we could distinctly see the expansion of the nostrils as they spouted the brine before them, with a noise like the blowing of a mighty forge. There were among them several calves, which were sporting around their dams, sometimes dashing against them head foremost, and gliding half out of water upon the backs of the old ones.

We were followed to-day [30th?] by several large dolphins [**Mahi-Mahi** or **Dolphin Fish** *Coryphaena hippurus*]. I had often heard of the surpassing beauty of this fish, but my imagination had never pictured any thing half so splendid as I then witnessed. We were going at about three knots, and the fish easily kept up with us, swimming in the wake without any apparent exertion, or even motion of their fins. At one moment they appeared to be of a clear, uniform grass-green, glistening and sparkling in the waves like emeralds, and in the next, they had changed their color wholly, appearing of an iridescent purple, with large oval spots of green and shining red; again, they were speckled and striped with all the colors of the rainbow, but without any one appearing predominant, and these changes were going on every minute while they remained near us, which was for the space of half an hour. When caught, and taken from the water; it is said that these changes occur precisely as when in their native element, with scarcely any diminution of brilliancy; and as vitality becomes less active, the variations are less frequent, until the colors finally settle into a dark greenish hue, and the animal is dead.

January 2d, 1835.– This evening at 5 o'clock, we made distinctly, the head land of three of the Sandwich group, Hawaii, Maui, and Morokai [Moloka'i], being within about eighteen miles of the nearest …

On the afternoon of the 4th [3rd?], we ran by several islands, and all within five miles. We could distinctly see the lofty and precipitous rocks of the coast, the deep ravines between them, and, by the assistance of our glasses, the green and rich looking vegetation of the interrupted plains.

The north coast of Oahu.

158

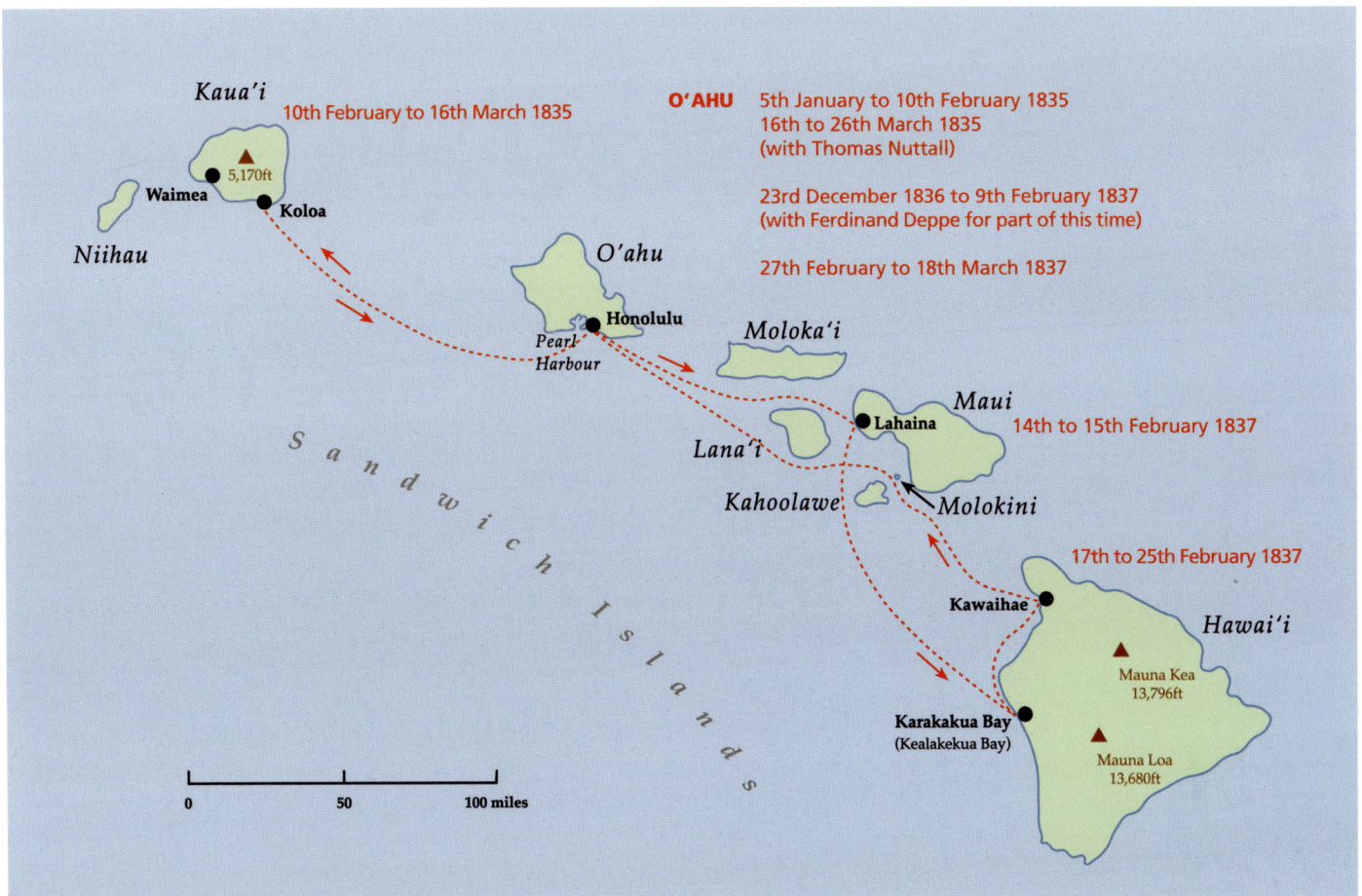

Map 10 – The Hawaiian Islands, showing Townsend's travels in 1835 and 1837.

O'ahu

At noon next day [4th], we made the island of Oahu, our destination, distant about forty miles. In the evening we were enabled to run, the moon shining brightly, and the atmosphere being unusually free from haze. At 10 o'clock we were within a few miles of the island, so that we could distinctly see a number of lights from the huts on the beach; we let go our anchor off a point called Diamond hill [Diamond Head]; and soon after, the mountain ranges, and the quiet valleys echoed the report of our pilot gun.

As I leaned over the rail this evening, gazing at the shore on our quarter, with its lofty peaks, and lovely sleeping vales, clearly defined by the light of the full orbed moon, I thought I never had witnessed any thing so perfectly enchanting, The warm breeze which came in gentle puffs from the land, seemed to bear fragrance on its wings, and to discourse of the rich and sunny climes from which it came. The whole scene was to me like fairy land. I thought of Captain Cook, and fancied his having been here, and gazing with delighted eyes upon the very prospect before me, little dreaming, that after all he had endured, he should here be sacrificed by the very people to whom he hoped to prove a benefactor and friend. The noise and bustle on deck, sailors running to and fro making the ship "snug" for harbor, and all the preparations for an arrival, effectually banished my meditations, and I descended to my state room, to sleep away the tedious hours, 'till the morrow should reveal all the new and strange features of the land to which we had come.

Diamond Head.

Early on the morning of the 5th, Mr. Reynolds, the deputy pilot, boarded us in a whale boat manned by natives, and accompanied by two American gentlemen, residents of the town of Honoruru [Honolulu], – Captain William S. Hinckley and P.A. Brinsmade, Esq. Our anchor was soon weighed, and with a fine, free wind, we rounded Diamond hill, and passed along a beautiful indentation in the shore, called Waititi [Waikiki] bay, within sight of a large coral reef, by which the whole island is surrounded. We very soon came in view of the lovely, sylvan looking village of Honoruru. The shore below the town from Waititi, to a considerable distance above, is fringed with graceful cocoanut trees, with here and there a pretty little grass cottage, reposing under their shade. As we approached the harbor, these cottages became more numerous, until at last they appeared thickly grouped together, with occasionally a pretty garden dividing them. The fort, too, which fronts the ocean, with its clean, white washed walls, and cannon frowning from the embrazures, adds very much to the effect of the scene; while behind, the noble hills and fertile valleys between, clothed with the richest verdure, soften down and mellow the whole, and render the prospect indescribably beautiful …

Our brig soon entered the narrow channel, opposite the harbor, and with a light, but steady breeze, stood in close to the town and let go her anchor within a hundred yards of the shore. As we were about leaving the vessel, Captain Charlton, H.B.M. consul, and Captain W. Darby of the H.B. Co.'s brig Eagle came on board, and gave Mr. N. and myself a passage to the shore in their boat. They walked with us to the house of Mr. Jones, the American consul, to whom I had a letter from my friend Doctor M. Burrough, of Philadelphia. We were received by this gentleman in a manner calculated to make us feel perfectly at home; a good and comfortable house was immediately provided for us, and every assistance was offered in forwarding our views. We dined at the sumptuous table of W. French, Esq., an American gentleman, and one of the most thriving merchants of the town, and were here introduced to several highly respectable foreign residents … In the afternoon we strolled out with two or three gentlemen to view the village and its environs.

The south-eastern part of Honolulu in 1826 (artist unknown).

The town of Honoruru contains about three hundred houses, the great majority of which are composed of grass exclusively, and those occupied by the natives consist of a single room. Others, in which many of the foreigners reside, are partitioned with boards, and form as comfortable and agreeable residences as could be desired in a climate always warm. There are some few houses of frame, and several of coral rock, built by the resident merchants and missionaries; but they are certainly not superior, except in being more durable, to those of grass, and probably not so comfortable in the intensly hot seasons. The houses are scattered about without any regard to regularity, the hard, clay passage-ways winding amongst them in every direction; but an air of neatness and simple elegance pervades the whole, which cannot fail to make a favorable impression on the stranger.

The natives are generally remarkably well formed, of a dark copper color, with pleasant and rather intellectual countenances, and many of the women are handsome.

The dress of the men, not in the employment of the whites, consists of a large piece of native cloth, called a *Tapa*, or a robe of calico thrown loosely round the body, somewhat like the Roman toga, and knotted on the left shoulder. The women wear a loose gown of calico, or native cloth, fastened tightly round the neck, but not bound at the waist, and often with the addition of several yards of cotton cloth tied above the hips.

Their hair is generally of a beautiful glossy black, and of unusual fineness … They display much taste in the arrangement of wild flowers amongst their hair, and a common ornament for the forehead is the Re of beautiful yellow feathers [from **mamo** and **o'o** species] …

8th.– Mr. N. and myself are now fairly domiciliated. We occupy a large and commodious room, in a building called the Pagoda, which is in a central part of the town; from our front windows we have a fine view of the harbor and the shipping, and from a balcony in the rear, we can see almost the entire length of the lovely valley of the Nuano [Nu'uanu], with its bold and rugged rocks, and the luxuriant verdure on their sides; while nearer, the little square taro patches, crowded together over the intermediate plain, look like pretty garden plots, as the broad green leaves of the plant are tinted by the sunbeams.

In the afternoon, a gentleman somewhat past middle age, in a plain, but neat garb, called upon me, and introduced himself as the Rev. Hiram Bingham, one of the missionaries resident upon the island. He gave me a very interesting account of the first landing and establishment of the missionaries at the Sandwich Islands, and discoursed very pleasantly on ordinary topics for half an hour. As Mr. N. was absent on a conchological excursion, I had the good man all to myself, and I may truly say, I have

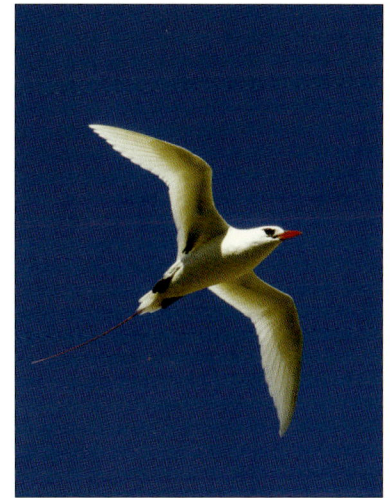

Red-tailed Tropicbird: "The native boys of the Sandwich and Society Islands adopt a singular mode of obtaining the long lanceolate tail-feathers, which are sometimes used as head ornaments by the natives, and are also sold in bundles as curiosities to strangers. The birds, at certain seasons, resort in immense numbers to the high and precipitous rocks of the coast, to breed. The boys visit these communities at the time when the birds are known to be sitting, and silently approaching the nest, quickly and adroitly pluck the two long feathers from the tail without doing other injury to the anxious parent. By this mode so many feathers are procured, that on almost any day, in the Island of Oahu, at least fifty bunches of the size of a man's arm might be purchased from the boys, who hawk them about for sale" (Townsend 1847: 90-91).

A selection of Townsend's specimens from O'ahu, from top: 'Akepa (ANSP 19841), O'ahu Creeper (ANSP 18582, ANSP 3375), O'ahu Amakihi (ANSP 3381, ANSP 3378).

King Kamehameha III.

rarely spent a half hour more agreeably. I was particularly interested in Mr. Bingham, from the circumstance of his being one of the oldest missionaries at these islands, and from the knowledge (which I had before acquired) of the very great influence he had exerted in the establishment of the missions, and of the excellent condition to which they had attained under his supervision and guardianship ...

10th.– This morning I saw the king for the first time. He is a very young man, only about twenty years of age, of ordinary size, and rather ordinary appearance. He was dressed in a little blue jacket, such as is worn by sailors when ashore, white pantaloons, and common black hat. He was walking in the street at a rapid, and not very dignified gate, and was followed closely by about twenty natives ...

The natives have very generally become acquainted with the pursuits of my companion and myself, and at almost all hours of the day, our mansion is besieged by men, women and children. Some bring shells, pearls, living birds, cocoanuts, bananas, &c., to sell, and others are attracted by curiosity to see us, which is no doubt much excited in regard to the use which we intend making of all the strange things they bring us.

["On the islands where few white men reside, and these composed almost exclusively of missionaries, the natives are simple, gentle and virtuous. I have never been more kindly or hospitably treated than in the houses of these primitive people residing on islands possessing but little to tempt foreigners to form settlements. They have always been ready to aid me in collecting birds, shells, &c. in many cases, not even expecting remuneration for their trouble.

They are most valuable adjuncts to the naturalist and collector. Being so truly amphibious in their habits, they serve the conchologist admirably in place of a dredge; diving into the sea, among the rocks, and searching the bottom for shells with wonderful pertinacity and success. They are also very successful bird-catchers. In the island of Oahu they procure the gum of a tree which they call "*Tu-tui*," [Kukuʻi *Aleurites moluccana*] and make of it a tenacious paste by moistening it with water. They smear little sticks with this paste and plunge them into the large pods of the banana, which contain a sweet juice of which the bird is extremely fond. The bird alights upon the gummed stick, and his feet are in an instant so firmly glued to it, that he cannot escape. By this mode, dozens of beautiful birds were brought to me almost daily, all alive and uninjured" (Townsend 1847: 115).]

Sunday, 11th.– Mr. Jones, the consul, called for us this morning, and we accompanied him to the Seamen's chapel in our neighborhood, the only church in the town in which English service is performed ... In the afternoon, Mr. N. and myself walked with the consul to the native church, at the lower extremity of the town ... The sermon, in the native language, by Mr. Bingham, was delivered in an easy and fluent manner, and in the whole of the great concourse there was scarcely a movement during the service. All seemed deeply engaged in the business for which they had assembled; and as I looked around upon the quiet and attentive multitude, a comparison with the wild, and idolatrous scenes which their assemblies exhibited in times past was irresistibly forced upon me.

A few days after this I was introduced by Captain Charlton, his Britannic majesty's consul, to the king KAUIKEAOULI, or TAMEHAMEHA [KAMEHAMEHA] III., as he is sometimes called. He was accompanied by John Young [junior[2]], one of his prime favorites, a fine, noble looking young man, who I thought looked much more like a king than his master. His majesty was very condescending and kind. He conversed easily and freely, though in broken English, and having understood that I had been somewhat of a traveller, was very curious to hear my adventures through the wild regions of the west. The stories of buffalo and grizzly bear hunting pleased him particularly, and his dark eye actually glittered as I recounted to him the stirring and thrilling incidents of the wild buffalo chase, and the no less moving perils of the encounter with the fierce bear of the prairies. He remarked that he should enjoy such hunting; that here there was nothing for his amusement but the chasing of wild cattle, and the common athletic exercise of quoits, bar-heaving, &c …

17th.– Mr. N. and myself were invited to participate in a *lu-au* dinner, to be given in the valley of Nuano this afternoon. At about 2 o'clock, Mr. Jones called for us, and furnished us with good horses, upon which we mounted, and galloped off to the valley. After a delightful ride of about five miles, over a good, though rather stony road, between the hills which enclose the valley, we arrived at a pretty little temporary cottage, formed entirely of the broad green leaves of the *ti* plant [Ti *Cordyline fruticosa*], and perched on a picturesque hill, overlooking the whole extent of our ride. Here we found a number of the foreign gentlemen; others soon joined us, and our company consisted of fifty or sixty persons, the king, John Young, and several other distinguished natives being of the party.

As the collation was not yet ready to be served up, Mr. Jones, Captain Hinckley, Mr. N. and myself remounted our horses for a visit to the great *pari* [pali], or precipice, two miles above. We found the road somewhat rough, and very hilly, in some places extremely narrow, and the path wound constantly through bushes and tall ferns to the elevated land which we were approaching. When within a few hundred yards of the

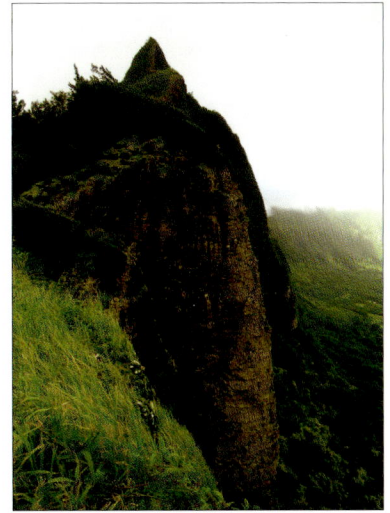

The Pali at the head of the Nu'uanu valley near Honolulu was depicted by Fisquet in plate 46 of Vaillant's *Voyage au tour du monde* (1845-52).

precipice, we left our horses in charge of several native boys, who had followed us for the purpose, and ascended to the edge of the pari. The wind was blowing a gale, so that it was necessary to remove our hats and bind up our heads with handkerchiefs, and when we stood upon the cliff, some care was required to keep our footing, and to brace ourselves against the furious blast which was eddying around the summit.

The pari is an almost perpendicular precipice, of about six hundred feet, composed of basaltic rock, with occasional strata of hard white clay.

On the north is seen the fertile and beautiful valley of Kolau [Kailua], with its neat little cottages, taro-patches and fields of sugar cane, spread out before you like a picture; and beyond, is the indented shore, with its high and pointed cliffs, margining the ocean as far as the eye can discern. Down this precipice, on the north side, is a sort of rude path, which the natives have constructed, and up this we saw a number of them toiling, clinging with their hands to the jutting crags above, to raise and support their bodies in the ascent. As they approached nearer to us, I was surprised to perceive that every man bore a burthen on his shoulder; some had large calabashes of poe [poi], suspended one on each end of a long pole, and others carried living pigs, similarly suspended, by having their feet tied together, and the pole passed between them. The porkers, although hanging back downwards, in a position not the most comfortable, did not complain of the treatment, until they were deposited on the terra firma of the summit, when they tuned their pipes to a lusty squeal, and made amends for their former silence.

This spot is the scene of the last great battle of King Tamehameha [Kamehameha 1], by which he acquired the sole and absolute sovereignty of the whole Sandwich group. The routed army of the petty island king was driven to take refuge among the wild crags of the pari, and hither it was followed by the conquering forces of the invader. No quarter was shown. The fugitives were hunted like savage beasts, and, almost to a man, were hurled from the giddy height and dashed to pieces on the frightful rocks below.

On returning to the cottage, we found that the dinner had been *dished* up, and that the guests were about taking their seats. Our table was the green grass, upon which had been arranged, with native taste, a circular *table cloth*, composed of ti leaves, placed one above another. On this the viands were laid. They consisted of fat pigs, and fat dogs, turkies, chickens, boiled ham, and fish, with vegetables of various kinds, taro, sweet potatoes, &c.,– all cooked in the native manner …

Kaua'i

February 10th.– Mr. Nuttall and myself were kindly invited by Captain Hinckley, to take a trip with him to the island of Kauai, in the brig Avon, owned by him. We embarked this morning, and with the pilot on board sailed out of the harbor. This being one of the leeward islands, towards which the trade-winds always blow, we made the passage quickly; and in the evening ran in to the harbor of Koloa, and anchored within half a mile of the shore.

The next morning we landed in our boat, and took possession of a large untenanted native house, near the beach, belonging to Captain H. Our servant busied himself in making our new residence comfortable, arranging the bedding and mosquitoe curtains, procuring mats for partitions, and, finally, in setting before us a good breakfast, cooked on the spot. We had scarcely finished our meal, when two horsemen rode up, and dismounting at the door, entered the house. Our visitors proved to be the king and John Young, who had mounted their horses this morning for a cattle hunt on the hills, but hearing of the arrival of the foreigners, had hastened to pay their respects to us, and assure us of their protection. The king, and his train came a few days since to this island, and propose remaining two or three weeks longer; his majesty's object appears

to be to inspect the condition of his people here, and to give them an opportunity of evincing their loyalty and affection. He was so kind as to express much interest for Mr. N. and myself, gave us one of his own body servants for our attendant, while we remain, and promised that in the afternoon he would send us some provisions.

Shortly after the king left us, we were visited by the Rev. P. J. Gulick, the missionary of this station, to whom I had a letter from my kind friend, Mr. Deill [Diell], and after sitting half an hour, we accompanied him to his house, about a mile distant.

This part of the island of Kauai exhibits no particularly interesting features: from the beach to the mission station there is a good road made by the natives over a gentle ascent of about two miles, on each side of which taro patches, yam and maize fields abound. Back from the ocean and at right angles with it, are seen several ranges of long, high hills, with narrow valleys between; the hills are covered with low trees of *Tu-tui* and *Pandanus* [Hala *Pandanus tectorius*], and the valleys with dense bushes, tall ferns, and broad leaved bananas.

The good missionary introduced us to his wife, a very intelligent and agreeable woman, and to his three pretty children, and we spent an hour with them very pleasantly. In the afternoon we returned to our cottage near the beach, where we found a native waiting for us with a hand cart filled with provisions of various kinds, which he said the king had sent to us as a present. There was a very large hog, three pigs, three or four turkies, and several pairs of chickens, all living; with vegetables in great abundance, taro, sweet-potatoes, melons, &c. I thought the man must certainly have made a mistake, but he assured me that it was right: "the king had sent them to the '*hauris*' (foreigners) who had just arrived, and wished him to say that in *tree* days he would send as much more" …

The next day we paid another visit to Mr. Gulick, and accepted a pressing invitation to make our home with him, his house being much more convenient to the valleys, which we wished to explore in search of birds and other natural objects. Here we had another interview with the king, who insisted upon our occupying a fine large house in the neighborhood of Mr. Gulick's residence, as a depository for our collections. We accepted this offer with pleasure, and the missionary's accommodation being somewhat contracted, we used the king's house as our study and sleeping apartment, taking our meals with the family of our kind friend.

Koloa and its sugar mill by Peter Hurd (1841). Surveying for a site for the sugar mill began in 1834.

We made here several long excursions over the hills and through the deep valleys, without much success. The birds are the same as those we found and collected at Oahu, but are not so numerous. They are principally creepers (*Certhia*) [*Drepanidinae* **Hawaiian Honeycreepers**] and honey-suckers (*Nectarinia*) [*Meliphagidae* **Honeyeaters**]; feed chiefly upon flowers, and the sweet juice of the banana, and some species are very abundant. The native boys here have adopted a singular mode of catching the honey sucking birds. They lay themselves flat upon their backs on the ground, and cover their whole bodies with bushes, and the campanulate flowers of which the birds are in search.

One of these flowers is then held by the lower portion of the tube between the finger and thumb; the little bird inserts his long, curved bill to the base of the flower, when it is immediately seized by the fingers of the boy, and the little flutterer disappears beneath the mass of bushes. In this way, dozens of beautiful birds are taken, and they are brought to us living and uninjured.

I'iwi preserved by Townsend (LM D511a).

166

I'iwi

TOWNSEND'S SPECIMENS FROM KAUA'I

Captain Cook's ships stopped at Kaua'i in 1778 and specimens of 'I'iwi were received when bartering with the natives. No other species from Kaua'i are known to have made their way to Europe from that visit (Olson and James 1994). Townsend was therefore the first to collect there (between 10th February – 16th March 1834, with Thomas Nuttall). Although the currently surviving endemic forest birds are now restricted to high altitude parts of the island the disastrous effects of introduced ground predators and mosquitoes had not then taken effect and Townsend did not have to travel far from the coast to find native forest birds. Unfortunately, few birds were preserved and the following quote by Townsend explains the disappointing results for the five week period that he was there: "After we had been here about four days, however, a heavy S.W. wind sprung up, blowing steadily towards Oahu, in consequence of which the Avon could not leave her anchorage, and we were compelled to remain where we were. Under some other circumstances this detention would not have been irksome; but we had made provision for only a few days' residence, and in a very short time all our ammunition, poison for preserving specimens, &c. were exhausted, and it was impossible to obtain even substitutes for these indispensable articles at this place. So we had nothing for it, but to yield to our fate with what grace we could, and spend the remainder of our forced sojourn in collecting plants, shells, and such other matters as the "moth and rust would not corrupt" (Townsend 1847: 122). Worse still, his specimens did not afterwards receive the attention they deserved and became widely scattered.

Two of the endemic bird species known with certainty to have been preserved by Townsend on Kaua'i are the '**Anianiau** (**Lesser 'Amakihi**) and the **Kaua'i 'O'o**. The 'Anianiau was described as a new species in 1887, from a specimen collected on Kaua'i by Valdemar Knudsen in 1866. Leonhard Stejneger described it in the *Proceedings of the US National Museum* 10: 94-95; using Knudsen's specimen in the US National Museum, even though an earlier Townsend specimen was in the same museum (USNM 14686) and another Townsend-collected specimen was at Philadelphia (ANSP 18648).

'Anianiau collected on Kaua'i by Townsend (USNM 14686).

KAUA'I 'O'O

Kaua'i 'O'o by F.W. Frohawk, detail from *Aves Hawaiiensis* (Wilson, S.B. & Evans, A.H. 1890-99). Townsend's type specimen of the Kaua'i 'O'o (ANSP 18581) may have been used by Frohawk for the painting – it was certainly sent to England to be examined by S.B. Wilson and Alfred Newton (see label on specimen).

In 1855
John Cassin wrote: "The only specimen that I have ever seen of this species was brought by Dr. Townsend from the Sandwich Islands and is marked as a male. It has heretofore in the collection of the Academy [of Natural Sciences of Philadelphia] been mistaken for *Certhia pacifica* Gmelin [Hawai'i Mamo *Drepanis pacifica*], but is clearly not that bird and but little resembles it as far as can be determined from the plate in Audebert, or from the descriptions by Gmelin and Latham" (Cassin 1855: 440).

Other specimens collected by Townsend on Kaua'i

Skins of the 'I'iwi, 'Amakihi, 'Elepaio, 'Akialoa and 'O'u were also in Townsend's collection but at that time these birds all occured also on O'ahu and Hawai'i and may or may not have come from Kaua'i. The 'I'iwi and 'O'u have not been separated into subspecies and the subspecies designations of the other birds are not always given in the museum databases, though we do know that the Kaua'i race of **Nukupu'u** *Hemignathus lucidus hanappe* was collected (FMNH 308799).

We know too that the two naturalists saw the **Short-eared Owl** during their stay here because Nuttall (1840: 141) says "we have observed it at Atooi [Kaua'i] one of the Sandwich islands" – but there is no evidence to link Townsend's two known owl specimens with this island, indeed one of them seems to have come from the big island, Hawai'i.

Nuttall found seventeen new species of shells on the Hawaiian Islands, including two on Kaua'i that are not found elsewhere. Nuttall's shells were described by T.A. Conrad in 'Descriptions of new Marine Shells from Upper California' in the *Journal of the Academy of Natural Sciences of Philadelphia* (1837). A fairly typical entry, for one of the Kaua'i shells, reads: "*Inhabits* the island of Atooi, on rocks bare at low water; not common." Collection dates of the specimens are not given.

Townsend specimens from the Hawaiian Islands, from top: I'iwi (ANSP 3364), Kaua'i 'Akialoa (ANSP 3360), 'O'u (ANSP 3358, ANSP 3357, ANSP 3356). The labels do not specify from which island they came but it is possible that Townsend collected them all on Kaua'i.

20th.– We expected to have left the island several days ago, but the Avon has not returned, and she would not now be able to come, in consequence of a steady S.W. wind which has prevailed for the last week. Our ammunition, and materials for the preparation of birds, are entirely exhausted, and we cannot here obtain a supply, so we amuse ourselves in collecting shells on the beaches, plants, fish, &c. We are living very comfortably in the house furnished us by the king, and we have become completely domesticated in the agreeable family of Mr. Gulick. We sometimes spend whole days wandering along the rocky coast in search of shells, and in these journeys we are always accompanied by a troop of boys and girls, and sometimes men and women, often to the number of twenty or thirty. They are indefatigable shell-hunters, and prove of great service to us, being compensated for each one that they bring us, with pins or needles. In their habits they are perfectly amphibious, diving into the sea, and through the dashing surf without the least hesitation, and exploring the bottom for an almost incredible time without rising to breathe …

… [about 12th March] as I was strolling near the palace, which, being on a hill, commanded a fine view of the ocean, the cry of "sail, O!" was uttered in a joyful tone, by a bevy of urchins, who were on the look out, and was echoed all round the neighborhood. The king, who had of late become unusually dull and spiritless, seemed suddenly to have acquired new life. He was seen rushing out of the house, like one distraught, and jumping and capering all about in a perfect agony of joy. Seeing me near him, he grasped my hand in the most cordial manner, while his eyes filled with tears: "We shall go back to Oahu, hauri, my people want me again; the wind has changed, and this ship is sent to take me away."

I participated in the anxiety of the king to return; for, in addition to the *ennui* which is always the accompaniment of a forced detention, even in pleasant places, I feared that our brig would leave the islands for the Columbia without us, and we should thus lose the opportunity of hailing the opening of spring, in our western world, the season which, of all others, is the most interesting to us.

170

The rugged coast of Kaua'i, a few miles east of Koloa. Townsend and Nuttall, and the king and his retinue, were delayed on the island because of bad weather. Townsend summed it up years later when he said "Instead of five *days*, we remained five *weeks* on this island; and, barring the impossibility of preparing birds (my favorite avocation,) we were very happy" (Townsend 1847: 122).

Next morning a messenger came to the king from Waimea, a port about fourteen miles distant, with information that the schooner which we had seen, had put in there, and only waited the embarkation of his majesty to steer direct for Oahu. No time was lost in taking advantage of this opportunity; and at noon, the spot, which the day before had swarmed with hundreds of dingy natives, was silent and untenanted … as we knew the vessel would be crowded, and therefore uncomfortable; and as the wind now blew steadily in its accustomed quarter, we had little doubt of the early arrival of the Avon …

Back to O'ahu

15th.– A sail was descried at daylight this morning, bearing towards our island; and while we were at breakfast, two fishermen called, to tell us that it was the Avon come at last. She was standing towards the harbor of Koloa, with a fair wind, and when Mr. N. and myself arrived at the beach, she was hauling in under the land. The Captain and Mr. Smith, a resident of Honoruru, came on shore to meet us. They informed us that the king arrived at Oahu last evening, after an unusually boisterous and uncomfortable passage, and that his majesty was almost beside himself with joy to receive once more the warm and affectionate greetings of his people at home.

171

The brig was detained here a day, in order to take in a cargo of live stock, pigs, goats, &c.; and the next morning we bade adieu to the kind and affectionate family of the missionary, and went on board. In the afternoon our anchor was weighed, and we were soon ploughing the wide ocean, while the rugged, iron-bound coast of Kauai rapidly receded from our view. We had on board several distinguished natives, as passengers, the principal of whom was Kekeoeva [Kekuanao'a], the governor, and at each meal which we took on board, the old gentleman asked an audible blessing on the viands, and regularly returned thanks at its conclusion.

After a pleasant passage of two days we arrived at Oahu, and were warmly greeted by our friends, who sympathized with us, and thought that our long tarriance must have been peculiarly irksome. They knew but little of the resources of the naturalist; they knew not that the wild forest, the deep glen, and the rugged mountain-top possess charms for him which he would not exchange for gilded palaces; and that to acquaint himself with nature, he gladly escapes from the restraints of civilization, and buries himself from the world which cannot appreciate his enjoyment.

22d.– I joined a party of ladies and gentlemen this morning, in an excursion to Pearl river[3], on the west side of the island. We embarked in several small schooners and barges, and had a delightful trip of two hours. The king, who was with us, with a number of his favorites ... procured for our accommodation several native houses, in which we slung our cots and hammocks, and slept at night. We took our meals under a large shaded ranai, and the amusements of the party were riding, shooting, and a variety of sylvan games, which rendered our pic-nic of three days a constant scene of pleasant festivity ...

Return to the Columbia River

On the 26th of March we embarked on board the brig May Dacre, upon our return to the Columbia. As we sailed out of the harbor, and the lovely shore of the island became more and more indistinct, I felt sad and melancholy in the prospect of parting, perhaps for ever, from the excellent friends who had treated me with such uniform kindness and hospitality.

We have had an accession to our crew of thirty Sandwich Islanders, who are to be engaged in the salmon fishery on the Columbia, and six of these have been allowed the unusual privilege of taking their wives with them. Some six or eight natives, of both sexes, friends and relatives of the crew, came on board when we weighed anchor, and their parting words were prolonged until the brig cleared the reef, and her sails filled with the fresh trade wind. They thought it then time to withdraw, and putting their noses together after their fashion, they bade their friends an affectionate farewell, and without hesitation dashed into the sea, and made directly for the shore. I thought of blue sharks, tiger sharks, and shovel-noses, and would not have run such a risk for all the wealth of all the islands.

April 5th.– Yesterday we had an *inkling* of a storm. Some rain fell in the morning early; and at noon, while sitting in the cabin, I was startled by hearing a flapping, as if all the sails in the ship were being torn to pieces, and a roaring not unlike the escape of steam from the boiler of a boat. Upon ascending to the deck, I observed the whole ocean covered with glittering white foam, the surges boiling, and dashing, and breaking over our vessel as she labored heavily amongst them. The sails were flapping about most unmanageably. The studding-sails, – all of which had been set before the storm, – were immediately taken in, and hands were sent aloft to furl the remaining canvass, until in a very short time we were scudding under bare poles, and defying the utmost fury of the elements. In about half an hour, the storm abated, and, soon after, entirely ceased; the wind became steady and fresh; the white folds of our canvass were again let loose, and away we went before a "smashing breeze" at the rate of ten knots an hour.

12th.– The mate has several times spoken of having seen large flocks of a small species of land bird sailing around the vessel. As it is scarcely possible they could have emigrated from the distant continent, Captain L. [Lambert] supposes that they are residents of a small uninhabited island, somewhere in these latitudes, which has long been supposed to exist, but has never been found. He who may hereafter be so fortunate as to discover this land, will probably be rendered a rich man for the remainder of his days, for it doubtless abounds in seal, which never having been interrupted, will be easily taken. I have not seen the birds spoken off, but suppose them to be some of the small *grallae*[4].

On the 15th, the wind, which had for several days been light, began steadily to increase, until we were running ten knots by the log. In the afternoon, the atmosphere became thick and hazy, indicating our approach to the shores of the continent. In a short time, a number of the small Auks [**auklets** and **murrelets**], – of which we saw a few immediately after leaving the Columbia, – were observed sporting in the waves, close under our bows; then several **gulls** of the species common on the river [mainly **Western Gulls** and **California Gulls**], and soon after large flocks of **geese** [several species are possible] and **canvass-back ducks** [**Canvasbacks** and perhaps **Redheads**].

The sea gradually lost its legitimate deep blue color, and assumed a dirty, green appearance, indicating soundings. Upon heaving the lead here, we got only eleven fathoms, and found that we had approached nearer than was prudent, having been misled by the haze. Wore ship immediately, and soon saw land, bearing east, which we ascertained to be south of Cape Disappointment. Stood off during the night, and the next morning at 4 o'clock, the wind favoring us, we bore up for the cape, and at 7 crossed the dangerous bar safely, and ran direct for the river.

1. When writing of tropicbirds Townsend only ever refers to the Red-tailed Tropicbird. The White-tailed Tropicbird *Phaethon lepturus* also occurs in Hawaiian waters and breeds on the islands but, in common with some other observers of that period, he probably thought that they were female Red-tailed Tropicbirds. Although Townsend says that he collected specimens of Red-tailed Tropicbirds none can now be traced.

2. John Young (1744-1835), an English sailor marooned on the island of Hawai'i in 1790, became a trusted advisor to King Kamehameha I. Young's second wife, Kaonaeha, a niece of Kamehameha, bore four children including Keoni Ana or John Young junior (1810-1857) (Cahill 1999). Townsend got to know the younger man well; he does not mention meeting the father but he met his widow on 23rd February 1837 while at Kawaihae on Hawaii.

3. At some time, perhaps during these three days, Nuttall went to the muddy marshes of the Pearl River and collected five new species of shells (Conrad 1837).

4. Jobanek (1999: 286) suggests that *grallae* refers to shorebirds. There is no island in this area: many species of shorebird and passerine are now known to fly out over the sea to shorten their migration. Despite the higher risk of not being able to land and feed it saves them time and energy.

All the birds in plate CCCXCV of *The Birds of America* were painted by Audubon from Townsend's specimens. From top: "Audubon's" Yellow-rumped Warbler (male and female); Hermit Warbler (male and female); and Black-throated Gray Warbler (male). Compare this plate with Audubon's original watercolour on p. 183.

N.º 79.

PLATE CCCXCV.

Drawn from Nature by J. J. Audubon, F. R. S. F. L. S.

Engraved, Printed and Coloured by R. Havell, 1837.

Audubons Warbler.
SYLVIA AUDUBONI, *Townsend.*
1. Male, 2. Female.

Hermit Warbler.
SYLVIA OCCIDENTALIS, *Townsend.*
3. Male, 4. Female.
Plant. Strawberry Tree.
EUONYMUS AMERICANA.

Black-throated gray Warbler.
SYLVIA NIGRESCENS, *Townsend.*
5 and 6. Males.

Return to the Columbia River

"I was in a state of absolute nudity, and at that moment, the inhabitants of an Indian village near, consisting of men, women, children and dogs, startled by the sound of my gun, were flocking out to see what was the matter."

(Townsend 1848: 266).

Upstream to Warrior's Point

On the 16th [April 1835], we anchored abreast of Oak point [forty miles upriver on the south bank, about thirty miles short of Warrior's Point]. Our decks were almost immediately crowded with Indians to welcome us, and among them we recognised many faces with which we were familiar. *Chinamus*, the Chinook chief, was the principal of these, who, with his wife, *Aillapust*, or *Sally*, as she is called at the fort, paid us an early visit, and brought us red deer [both Whitetail and Blacktail Deer are reddish in spring and summer] and sturgeon [*Acipenser sp.*] to regale upon after our voyage.

On the afternoon of the next day [17th], we ran up to Warrior's point, the brig's old mooring ground [at the north end of Sauvie Island]. The people here had been anxious to see us; extensive preparations had been made to prosecute the salmon fishery, and the coopers have been engaged the whole winter in making barrels to accommodate them. Mr. Walker, the missionaries' quondam associate, was in charge of the post, and he informed us that Captain Wyeth had returned only a few weeks since from the upper country, where he had been spending the winter, engaged in the arduous business of trapping, in the prosecution of which he had endured great and various hardships.

[Nuttall at once saw **Rufous Hummingbirds**: "On the 16th [= 17th?] of April, in the forests at the outlet of the Wahlamet, we first saw the males in numbers, darting, burring, and squeaking in the usual manner of their tribe; but when engaged in collecting its accustomed sweets in all the energy of life, it seemed like a breathing gem, or magic carbuncle of glowing fire, stretching out of its gorgeous ruff, as if to emulate the sun itself in splendor" (Nuttall 1840: 715).]

May 12th.– The rainy season is not yet over; we have had almost constant showers since we arrived, but now the weather appears settled. Birds are numerous, particularly the warblers, (Sylvia) [**Wood-Warblers** *Parulidae*]. Many of these are migratory, remaining but a few weeks: others breed here, and reside during the greater part of the summer. I have already procured several new species. [*Narrative* continues on p. 190.]

Sauvie Island near Warrior's Point.

A letter home, written by Townsend on 2nd May 1835 while on board Captain Wyeth's ship the *May Dacre*, anchored off Sauvie Island.

No. 9

On board Brig May Dacre
Columbia River
5th Mo. 2nd. 1835.

I hasten with great pleasure to take advantage of an opportunity which offers of sending a line to you, but from the very little time allowed me, it must be almost literally but a line. Three of Capt Wyeth's men have just arrived at this station from the farm 40 miles above and state their intention of starting for the States immediately. The people here are busily preparing them a small outfit & the moment it is ready they will commence their journey. I of course am not willing to neglect any chance of sending you an acct of my getting along, but in the present instance I feel very doubtful of your even receiving it. The three men travel entirely alone & the chances are very much against their ever arriving in their native country. They are however bold, fear-nothing fellows, & such often overcome obstacles & escape dangers that would prove insurmountable to the more timid.

N. & myself arrived here a fortnight ago after a very short & delightful passage of 20 days from Oahu. Since we have since spent our time in roamings about the neighbourhood, collecting Birds, Plants &c of which we find a plentiful supply. Have not visited Vancouver since our return but expect to do so in about 2 weeks more. Our visit at the Sandwich Islands was a very pleasant one. The residents generally were singularly kind to us & seemed to do all in their power to further our prospects & render us comfortable. To the Am. Consul particularly I shall never cease to feel grateful. While at the Islands I made a considerable collection of the native shells, & a few other small matters, but no birds to speak of; they are very scarce & very difficult of access. During our residence there, four or five vessels arrived from U. States - & the last brought news to the 10th of October. Oh how I wish that you had known of my intention to visit the Islands or that you could have anticipated it & I might then have heard from you several times - as it was I could not resist the desire of enquiring for letters of every vessel & altho' I certainly could not reasonably expect them, yet I felt always sad & disappointed when my anxious questions were answered in the negative. And now that I am without the remotest prospect of hearing one word from you - It sometimes makes me feel very unhappy, as you may well suppose, but I endeavour to reason it away. It is unavoidable & regrets cannot alter it. One effect however it invariably produces - viz. a desire to hasten my return to you as much as possible. But here again I am in the hands of fate as it were. I had at one time almost concluded to take passage in the vessel for home, but I found this would be inconvenient & inexpedient (as uncle John would say,) indeed almost

176

BIRDS AND MAMMALS OF THE LOWER COLUMBIA RIVER, MAY AND JUNE 1835

The spring of 1835 was Townsend and Nuttall's most frenetic collecting period. Townsend was in his element, collecting new and scarcely known species, as every day seemed to bring more migrants from the south. The following letter to his father, written "*On board Brig May Dacre / Columbia River*", is of special significance because it tells us that specimens obtained during the first two weeks of May were collected on or near Sauvie Island (Oregon) rather than at Fort Vancouver (Washington):

"N & myself arrived here a fortnight ago after a very short & delightful passage of 20 days from Oahu. We have since spent our time in roaming about the neighborhood, collecting birds, plants &c of which we find a plentiful supply. I have not visited [Fort] Vancouver since our return but expect to do so in about 2 weeks more … I shall write to you again by the Brig which will sail from here about the 1st of 10th mo next. Say to Dr Michener he shall also have a letter & something for his cabinet at the same time. Will somebody call upon Dr Morton, perhaps one of my brothers will do it, tell him of my existence, say that I have been tolerably successful in collecting, request him to remember me to the members of the Academy generally & that he shall have a letter from me by the Brig.

Most affectionately John

(JKT to C. Townsend, 2nd May 1835)

So, for the first two weeks of May, Townsend was in the vicinity of Warrior's Point and Fort William on Sauvie Island. Neither Townsend's *Narrative* nor Wyeth's journal throw any light on when Townsend first went up to Fort Vancouver after his return from the Hawaiian Islands. The *Narrative* for 20th May merely states that Captain Wyeth came down from Walla Walla the previous day and that Townsend immediately joined him on a trip to the Willamette Falls – either Townsend was already at Fort Vancouver or, more likely, he was still based on Wyeth's ship at Warrior's Point.

Cassin's Vireo. "About the beginning of May, in the oaks already almost wholly in leaf, on the banks of the Columbia, we heard around us the plaintive deliberate warble of this species [which they believed to be the "Solitary Vireo, or Flycatcher"] (Nuttall 1840: 346).

Wilson's Warbler. "Mr. Townsend and myself had the pleasure of observing the arrival of this little cheerful songster in the wilds of Oregon about the first week of May … by the 12th of May, they were already feeding their full-fledged young, though I also found a nest on the 16th of the same month, containing 4 eggs, and just commencing incubation" (Nuttall 1840: 335-336).

Wilson's Warbler collected in May 1835 (OUM 12721).

8th May 1835, **Bush-tit**. NYHS 1863.17.353.

9th May [1835], **Bush-tit**, NYHS 1863.17.353.

9th May 1835, **Pileated Woodpecker**, USNM 2792.

9th May 1835, **Townsend's Mole**. ANSP 449. Bachman's original description gives the locality as "the banks of the Columbia river". True (1896) restricted the type locality to "Fort Vancouver, or the immediate neighborhood" i.e. in Clark Co., Washington, but Townsend's letter of 2nd May 1835 indicates that the locality should be "on or near Sauvie Island, Oregon" because he had not yet gone to Fort Vancouver at that date.

Townsend's Mole, plate CXLV from Audubon's *Viviparous Quadrupeds*. Townsend's type specimen is dated 9th May 1835 (ANSP 449).

TOWNSEND'S MOLE

"Townsend's Shrew Mole … A specimen of this quadruped was kindly presented to me, by Mr. Nuttall, who requested that, in case it should prove a distinct species, it might be given under the above name. I subsequently received from Mr. Townsend another specimen, a little larger in size, which I presume to be a mere variety, although very singularly marked … The specimen brought by Mr. Townsend, is thicker, and about an inch longer. It has a white stripe about two lines wide, commencing under the chin, and running in a somewhat irregular line along the under surface of the body to within an inch and a half of the insertion of the tail; there is also a white streak commencing on the forehead, and extending along the snout.

The specimen of Mr. Townsend is labelled "Banks of the Columbia river, May 9th, 1835"" (Bachman 1839b: 58-60).

12th May 1835, **Golden-crowned Sparrow**, NYHS 1863.17.394.

12th May 1835, **Chestnut-backed Chickadee**, OUM 11571.

14th May 1835, **Black-throated Gray Warbler**, USNM 2915.

14th May 1835, **Spotted Towhee**, (Audubon 1840-44, 3: 166).

15th May 1835, **Orange-crowned Warbler**, USNM 2929.

16th May 1835, **Band-tailed Pigeon**, USNM 2825, card catalogue.

16th May 1835, **Orange-crowned Warbler**, USNM 1912.

The remaining specimens and observations are perhaps mostly from near Fort Vancouver, but will include some from the vicinity of Warrior's Point and Fort William, as Townsend seems to have moved between these sites fairly frequently. There was also a short trip of just a few days to the Willamette Falls (circa 20th-25th May). From 10th September (or even as early as 20th August) he based himself at Fort Vancouver and remained there most of the winter.

[24th? May 1835], **California Condor**, USNM 8005. Willamette Falls (Townsend 1848: 265-267).

27th May 1835, **Spotted Towhee**, USNM 2867.

28th May 1835, **Hermit Warbler**, ANSP 162369. Townsend (1839a: 341) says he shot a pair on this date, the only specimens for this species. If ANSP 162369 really was collected by Townsend it should have a red label indicating that it is a co-type (see photo).

29th May 1835, **Rufous Hummingbird**, USNM 1943.

29th May 1835, **Rufous Hummingbird**. Nuttall found a nest that was probably the one later painted by Audubon (Nuttall 1840: 715-716; Audubon 1840-44, 4: 200-201).

30th May 1835, **Rufous Hummingbird**. Townsend collected two specimens and another on the preceding day (Audubon 1840-44, 4: 203).

31st May 1835, **"Audubon's" Yellow-rumped Warbler**, USNM 2909.

31st May 1835, **MacGillivray's Warbler**, MCZ 35008.

"spring of 1835", **Townsend's Warbler**, "on the Columbia river" (Townsend 1839a: 342).

-- May 1835, **Rufous Hummingbird**, OUM 04978.

-- May 1835, **MacGillivray's Warbler**, USNM 1910.

-- May 1835, **MacGillivray's Warbler**, USNM 2907.

-- May 1835, **Wilson's Warbler**, OUM 12071.

-- May 1835, **Cassin's Vireo**, OUM 12721.

8th June [1835], **"Audubon's" Yellow-rumped Warbler**. Young already in family parties (Nuttall 1840: 415).

10th June 1835, **Rufous Hummingbird**, young female shot by Townsend (Audubon 1840-44, 4: 203).

10th June [no year = 1835?], **Swainson's Thrush**, first described by Nuttall (1840: 400-401) from a female "procured on the banks of the Columbia River on the 10th of June by my friend Mr. Townsend."

12th June [1835], **MacGillivray's Warbler**. A nest brought in to Nuttall (1840: 461).

14th June [1835], **Spotted Towhee**. Nuttall found nest and eggs (Audubon 1840-44, 3:165; Nuttall 1840: 611).

"About the middle of June", **Western Kingbird**, heard "in the dark swampy forests of the Wahlamet" (Nuttall 1840: 306).

16th June 1835, **Black-throated Gray Warbler**, USNM 1908.

19th June 1835, **Red-shafted Northern Flicker**, OUM 3299.

21st June [1835], **Bush-tit**. Flock of twelve seen "in the dark woods near Fort Vancouver" (Nuttall 1840: 270).

22nd June 1835, **Lazuli Bunting** male, NYHS 1863.17.394. Audubon's original watercolour is inscribed "Male [illegible] River / June 22nd 1835", taken from the specimen label.

22nd June 1835, **California Ground Squirrel**. Audubon's original watercolour for plate 49 of *The Viviparous Quadrupeds* is inscribed: "4 specimens from / J.K. Townsend all differing / vary greatly in size / from the Columbia River / June 22nd 1835" (Tyler 2000: 127).

26th June 1835, **Mountain Bluebird**, AMNH 39327

-- June 1835, **Mountain Bluebird**, (Audubon 1840-44, 2: 180).

-- June 1835, **bluebird species**? MCZ 16394.

-- June 1835, **Yellow-bellied Flycatcher [= Pacific-slope Flycatcher]**, OUM 14179.

Two warblers collected by Townsend on the lower Columbia River. From top: MacGillivray's Warbler (ANSP 23765); "Audubon's" Yellow-rumped Warbler (ANSP 23826).

Pacific-slope Flycatcher collected in June 1835 (OUM 14179).

BUSH-TIT

"BROWN-HEADED TITMOUSE …I first observed this little species on the Columbia river in May, 1835, and procured a pair[a]. They hopped through the bushes, and hung from the twigs in the manner of other titmice[b], twittering all the time, with a rapid enunciation, resembling the words, *tsish-tsish-tsee-tsee*. Upon my return, I found that Mr. Nuttall had observed the same birds a few hours previously in another place. He said that they frequently flew to the ground from the bushes, where they appeared to institute a rapid search for insects, and quickly returned to the perch, emitting their weak, querulous note the whole time without intermission. The stomachs of these birds contained fragments of minute coleopterous insects, and in the ovary of the female was an egg nearly ready for expulsion.

The nest, which Mr. N. found a few days afterwards, is a very curious and beautiful fabric, somewhat like that of the bottle tit [Long-tailed Tit *Aegithalos caudatus*] of Europe, being from eight to nine inches in length, formed of fine bent, lined with hair, and covered externally with mosses, the hole for entrance near the top. It was suspended from a low bush, and contained seven eggs, very small and beautifully shaped, and pure white" (Townsend 1839a: 337-338).

According to Mr. TOWNSEND, "the Chinooks name it *a-ha-ke-lok*. It is a constant resident about the Columbia river; hops about in the bushes, and frequently hangs from the twigs in the manner of other Titmice, twittering all the while with a rapid enunciation resembling the words *thshish, tshist, tsee, twee*. The irides are bright yellow" (Audubon 1840-44, 2:160).

a. All Townsend's Bush-tit specimens have been lost. Audubon's original watercolour (NYHS 1863.17.353) is inscribed "male May 8th 1835 / [female] May 9th 18[35]", the exact year for the female is lost at the page edge.

b. The Bush-tit *Psaltriparus minimus* is not now in the same genus as titmice (*Baeolophus*) or chickadees (*Poecile*).

CHESTNUT-BACKED CHICKADEE

"Chestnut-backed Titmouse … Inhabits the forests of the Columbia river.

It is nearly allied to P. *Hudsonius* [Boreal Chickadee *Poecile hudsonica*], but is distinguished by its smaller size and chestnut-coloured back" (Townsend 1837: 190).

"CHESTNUT-BACKED TITMOUSE … Mr. NUTTALL'S notice respecting the present species is as follows:–

"The Chestnut-backed Titmouse is seen throughout the year in the forests of the Columbia, and as far south as Upper California, in all of which tract it breeds, forming, as I have some reason to believe, a pendulous, or at least an exposed nest, like some of the European species … When the gun thins their ranks, it is surprising to see the courage, anxiety, and solicitude of these little creatures: they follow you with their wailing scold, and entreat for their companions in a manner that impresses you with a favourable idea of their social feelings and sympathy."

Mr. TOWNSEND says, that "the Chinook Indians call this species *a kul*. It inhabits the forests of the Columbia river, where it breeds and goes in flocks in the autumn, more or less gregarious through the season. The legs and feet are light blue"" (Audubon 1840-44, 2: 158).

Chestnut-backed Chickadee, possible co-type, collected on 12th May 1835 (OUM 11571).

Chesnut-backed Titmouse. 1. male. 2. female
Parus rufescens. Townsend.

Black-capt Titmouse 3. male & 9.
Parus atricapillus, Wils.
Willow Oak. Quercus phellos, L.

Chesnut crowned Titmouse. 5 m. 6. f
Parus minimus. Townsend. —
and nest.

No 71

Plate 353

Townsend's Warbler, type specimen collected in the spring of 1835 (USNM 2918).

TOWNSEND'S WARBLER

TOWNSEND'S WARBLER ... I procured but one specimen of this beautiful bird, on the Columbia river, in the spring of 1835. Early in autumn of the same year, I shot another male, in somewhat plainer livery.

It does not breed there, and I know nothing of its habits" (Townsend 1839a: 341-342).

"TOWNSEND'S WOOD-WARBLER ... Mr. NUTTALL has honoured this beautiful Warbler with the name of his friend and companion Mr. TOWNSEND. It was procured about the Columbia river. All the information respecting it that I possess is contained in the following brief notice by the former of these celebrated naturalists. "Of this fine species, we know very little, it being one of those transient visitors, which, on their way to the north, merely stop a few days to feed and recruit ... As this species frequents the upper parts of the lofty firs, it was almost an accident to obtain it at all. The female remains unknown" (Audubon 1840-44, 2: 59).

BLACK-THROATED GRAY WARBLER

"BLACK-THROATED GRAY WOOD-WARBLER ... This is another of the interesting species discovered and named by Mr. TOWNSEND, who informs me that ... it is abundant in the forests of the Columbia, where it breeds and remains until winter; and that the nest, formed externally of fibrous green moss, is generally placed on the upper branches of the oak, suspended between two small twigs. Mr. NUTTALL's notice respecting it is as follows:– "This curious species, so much resembling *Sylvia striata* [Blackpoll Warbler], was seen to arrive in early May; and from its song more regularly delivered at intervals, in the tops of deciduous-leaved trees, we have little doubt but that they breed in the forests of the Columbia. On the 23d May [1835] I had the satisfaction of harkening to the delicate but monotonous song of this bird, as he busily and intently searched every leafy bough and expanding bud for larvae and insects in a spreading oak, from whence he delivered his solitary note" (Audubon 1840-44, 2: 62).

Black-throated Gray Warbler, co-type collected on 14th May 1835 (USNM 2915).

No 76. 1, 2. Western blue Bird.
Plate 380. Sialia occidentalis, Townsend.

3, 4. Arctic blue bird.
Sylvia arctica.

5, 6. Black throated Gray W.
S. nigrescens, Townsend.
Plant.
Calycanthus floridi

7, 8. Hermit W.
Sylvia occidentalis, Townsend.

Carolina allspice

9. Townsend's W.
Sylvia Townsendii, Aud.

10, 11. Audubon's W.
Sylvia Audubonii, Nuttall.

"AUDUBON'S" YELLOW-RUMPED WARBLER

"AUDUBON'S WOOD-WARBLER ... This species, so very intimately allied to *Sylvia coronata* [Myrtle Warbler], that an observer might readily mistake the one for the other, was discovered by Mr. TOWNSEND, who has done me the honour of naming it after me. He states, that "the Chinook Indians know it by the name of '*Fout-sah*,' and that it is very numerous about the Columbia River, arriving there in the middle of March, and remaining to breed, but disappearing in the end of June. In the beginning of October it is again seen, with its plumage renewed ... I have given figures of the male and female, taken from specimens obtained by Mr. TOWNSEND on the Columbia" (Audubon 1840-44, 2: 26-27).

"Audubon's Warbler" the western race of Yellow-rumped Warbler, collected on 31st May 1835 (USNM 2909).

HERMIT WARBLER

"HERMIT WARBLER ... I shot a single pair of these birds in a pine forest on the Columbia river, on the 28th of May 1835. They were flitting about among the pine trees, very actively engaged in searching for insects, and frequently hanging from the boughs like titmice [chickadees, in this instance]" (Townsend 1839a: 340-341).

"HERMIT WOOD-WARBLER ... Mr. TOWNSEND's note is as follows:– I shot this pair of birds near Fort Vancouver, on the 28th of May, 1835. I found them flitting among the pine trees in the depth of a forest. They were actively engaged in searching for insects, and were frequently seen hanging from the twigs like Titmice. Their note was uttered at distant intervals, and resembled very much that of the Black-throated Blue Warbler, *Sylvia canadensis*" (Audubon 1840-44, 2: 60-61).

Only two specimens were procured, both depicted by Audubon (see page 183).

Hermit Warbler, type specimen collected on 28th May 1835 (ANSP 162369).

WESTERN BLUEBIRD

"WESTERN BLUE BIRD ... Common on the Columbia river in the spring. It arrives from the south early in April, and about the first week in May commences building. The nest is placed in the hollow of a decayed tree, and is very loose and unsubstantial. The eggs, four to five, are light blue, somewhat larger than those of the common blue bird, (*S. Wilsonii*.) [Eastern Bluebird] (Townsend 1839a: 343-344).

MOUNTAIN BLUEBIRD

For his written description of this species Audubon used a male shot by Townsend in "June 1835" and a female "shot on the 26th of the same month and year". Audubon misplaced Townsend's observations on the Western Bluebird under this species (Audubon 1840-44, 2: 178-180).

SPOTTED TOWHEE

"ARCTIC GROUND-FINCH ... My friend Mr. Nuttall has furnished me with the following account of it:–

"We found this familiar bird entirely confined to the western side of the Rocky Mountains ... On the 14th of June, I saw the nest of this species, situated in the shelter of a low undershrub, in a depression scratched out for its reception ... As usual, the pair shewed a great solicitude about their nest, the male in particular approaching boldly to scold and lament at the dangerous intrusion. This species extends into Upper California, and is occasionally seen there with the brown species of Swanson, Pipilo fuscus [= California Towhee *Pipilo crissalis*, recently separated from Canyon Towhee *Pipilo fuscus*]."

Mr. Townsend informs me that it is called "Chlawa-th'l" by the Chinook Indians, and is abundant on the banks of the Columbia, where it is found mostly on the ground, or on bushes near the ground, rarely ascending trees ... The male above described was shot by Mr. Townsend on the Columbia river, on the 14th of May, 1835; the female on the 11th of October, 1834" (Audubon 1840-44, 3: 164-166).

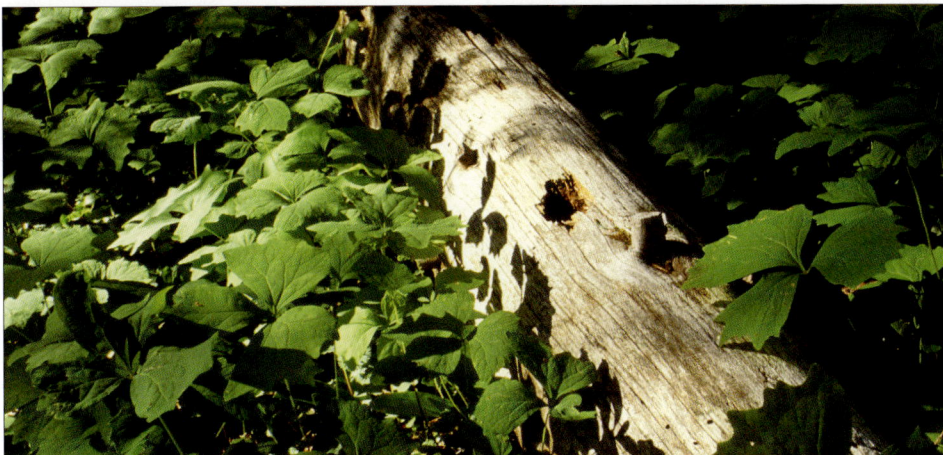

Vanilla Leaf *Achlys triphylla*.

185

MacGillivray's Warbler, plate
CCCXCIX (detail) and type specimens
collected in May 1835 (USNM 1910,
USNM 2907).

MACGILLIVRAY'S WARBLER

Townsend wrote the first published description of this warbler in the Scientific
Appendix to the *Narrative*, but it has become known as MacGillivray's Warbler (see
page 283).

"TOLMIE'S WARBLER *Sylvia Tolmiei* ... This pretty species, so much resembling the
curious *S. philadelphia* [Mourning Warbler] of Wilson, is common in spring on the
Columbia. It is mostly solitary, and extremely wary, keeping chiefly in the densest and
most impenetrable thickets, and gliding through them in a very cautious and suspicious
manner. It may, however, sometimes be seen towards mid-day, perched upon a dead
twig over its favorite place of concealment, and at such times it warbles a very sprightly
and pleasant little song, raising its head until the bill is almost vertical, and swelling its
throat in the manner of many of its relatives.

I dedicate the species to my friend W.F. Tolmie, Esq., of Fort Vancouver" (Townsend
1839a: 343).

RUFOUS HUMMINGBIRD

Rufous Hummingbird, plate
CCCLXXIX (detail) and a Townsend
specimen collected in May 1835 (OUM
04978).

"RUFF-NECKED HUMMING-BIRD ... Mr. TOWNSEND's note is as follows:- Nootka Sound Humming-bird ... On a clear day the male may be seen to rise to a great height in the air, and descend instantly near the earth, then mount again to the same altitude as at first, performing in the evolution the half of a large circle. During the descent it emits a strange and astonishingly loud note, which can be compared to nothing but the rubbing together of the limbs of trees during a high wind. I heard this singular note repeatedly last spring and summer, but did not then discover to what it belonged. I did not suppose it to be a bird at all, and least of all a Humming-bird. The observer thinks it almost impossible that so small a creature can be capable of producing so much sound ..."

The above descriptions are from two individuals shot by Mr. TOWNSEND on the "Columbia river, 30th May, 1835." A "young male, Columbia river, 29th May, 1835," resembles the female as above described, differing only in having the metallic spots on the throat larger. A "young female, Columbia river, June 10th, 1835," differs from the adult only in wanting the metallic spots on the throat, which is spotted with greenish-brown" (Audubon 1840-44, 4: 200-203).

Nuttall's Dogwood *Cornus nuttalli* one of Nuttall's best-known botanical discoveries from the Lower Columbia. He also gathered shells of about twenty-three new species of freshwater molluscs in Oregon in 1834 and 1835. The most productive site was "Wahlamet, near junction with the Columbia River." Isaac Lea described the shells in the *Transactions of the Academy of Natural Sciences of Philadelphia* (1839), naming several species after Nuttall.

(Opposite) Band-tailed Pigeon, plate CCCLXVII from *The Birds of America*, drawn from Townsend specimens, though not necessarily from these ones (ANSP 13264, ANSP 175964). The branch that the birds are perched on was painted by Maria Martin from a sample brought back by Nuttall. Audubon published the first description of this magnificent tree in his *Ornithological Biography*, naming it "Nuttall's Dog-wood *Cornus Nuttalli*".

BAND-TAILED PIGEON

"BAND-TAILED DOVE, OR PIGEON … In the course of Colonel Long's expedition to the Rocky Mountains, a single specimen of this large and handsome Pigeon was procured [in Colorado]. This individual was afterwards figured in the continuation of Wilson's American Ornithology. Many specimens, however, have more recently been obtained by Mr. Townsend, from whom I have procured three pairs of adult and some young birds … In my plate are represented two adult birds, placed on the branch of a superb species of *dogwood*, discovered by my learned friend Thomas Nuttall, Esq., when on his march toward the shores of the Pacific Ocean, and which I have graced with his name! … Mr. Townsend's notice respecting the bird here spoken of is as follows:–

"The Band-tailed Pigeon is called by the Chinook Indians 'akoigh homin.' It ranges from the eastern spurs of the Rocky Mountains across to the Columbia river, where it is abundant. It arrived in 1836 in very great numbers, on the 17th April, and continued in large flocks while breeding. Their breeding places are on the banks of the river. The eggs are placed on the ground, under small bushes, without a nest, where numbers congregate together. The eggs are two, of a yellowish-white colour … When sitting in the trees, they huddle very close together in the manner of the Carolina Parrot [Carolina Parakeet], and many may be killed at a single discharge of the fowling-piece. The flesh is tender and juicy, and therefore fine eating."

Mr. Nuttall has favoured me with an equally interesting notice. "This large and fine Pigeon, always moving about in flocks, keeps in Oregon only in the thick forests of the Columbia and the Wahlamet, and during the summer is more particularly abundant in the alluvial groves of the latter river, where throughout that season we constantly heard their cooing, or witnessed the swarming flocks feeding on the berries of the elder tree, those of the Great Cornel (*Cornus Nuttalli*) [Nuttall's Dogwood *Cornus nuttalli*], or, before the ripening of berries, on the seed-germs or the young pods of the balsam poplar … They are said to breed on the ground, or in low bushes, but I did not find the nest, although I saw the birds feeding around every day near Watpatoo [Wapatoo] Island [Sauvie Island]" " (Audubon 1840-44, 4: 312-315).

Band-tailed Pigeon, 1. Male 2. Female
COLUMBA FASCIATA. Say.
Plant Nuttall Cornel.
Genus Nuttall Aud.

Another trip to the Willamette Falls

20th [June].– Mr Wyeth, came down from Walla-walla yesterday, and this morning I embarked with him in a large canoe, manned by Kanakas, for a trip to the Wallammet falls in order to procure salmon. We visited fort William, (Wyeth's new settlement upon Wappatoo [Sauvie] island,) which is about fifteen miles from the lower mouth of the Wallammet. We found here the missionaries, Messrs. Lee and Edwards, who arrived to-day from their station, sixty miles above. They give flattering accounts of their prospects here; they are surrounded by a considerable number of Indians who are friendly to the introduction of civilization and religious light, and who treat them with the greatest hospitality and kindness. They have built several comfortable log houses, and the soil in their vicinity they represent as unusually rich and productive. They have, I think, a good prospect of being serviceable to this miserable and degraded people; and if they commence their operations judiciously, and pursue a steady, unwavering course, the Indians in this section of the country may yet be redeemed from the thraldom of vice, superstition, and indolence, to which they have so long submitted, and above which their energies have not enabled them to rise.

The spot chosen by Captain W. for his fort is on a high piece of land, which will probably not be overflown by the periodical freshets, and the soil is the rich black loam so plentifully distributed through this section of country. The men now live in tents and temporary huts, but several log houses are constructing which, when finished, will vie in durability and comfort with [Fort] Vancouver itself[1].

21st.– The large band-tail pigeon (*Colomba fasciata*) [**Band-tailed Pigeon**] is very abundant near the river, found in flocks of from fifty to sixty, and perching upon the dead trees along the margin of the stream. They are feeding upon the buds of the balsam poplar; are very fat, and excellent eating. In the course of the morning, and without leaving the canoe, I killed enough to supply our people with provision for two days.

24th.– We visited the falls to-day, and while Captain W. was inspecting the vicinity to decide upon the practicability of drawing his seine here, I strolled into the Indian Lodges on the bank of the river. The poor creatures were all living miserably, and some appeared to be suffering absolute want. Those who were the best supplied, had nothing

Fishing at the Willamette Falls, sketch by Drayton, circa 1841.

more than the fragments of a few **sturgeons** [*Acipenser sp.*] and lamprey eels [**Pacific Lamprey** *Lampreta tridentate*], kamas bread, &c. To the roofs of the lodges were hung a number of crooked bladders, filled with rancid seal oil, used as a sort of condiment with the dry and unsavory sturgeon.

On the Klakamas river, about a mile below, we found a few lodges belonging to Indians of the Kalapooyah tribe. We addressed them in Chinook, (the language spoken by all those inhabiting the Columbia below the cascades,) but they evidently did not comprehend a word, answering in a peculiarly harsh and gutteral language, with which we were entirely unacquainted. However, we easily made them understand by signs that we wanted salmon, and being assured in the same significant manner that they had none to sell, we decamped as soon as possible, to escape the fleas and other vermin with which the interior of their wretched habitations were plentifully supplied. We saw here a large Indian cemetery. The bodies had been buried under the ground, and each tomb had a board at its head, upon which was rudely painted some strange, uncouth figure. The pans, kettles, clothing, &c., of the deceased, were all suspended upon sticks, driven into the ground near the head board. [The *Narrative* continues on page 194]

CALIFORNIA CONDOR[a]

At the Willamette Falls, about the 24th May 1835, Townsend managed to secure his only specimen of a California Condor (USNM 8005). The condor was the first species that he wrote about in his uncompleted *Ornithology of the United States of North America* (1839) – see first extract below. Only many years later, in an obscure journal, did he ever publish an account of how he came to kill the bird – see second extract below.

California Condor, plate 1 from Townsend's *Ornithology of the United States of North America* (1839).

"CALIFORNIAN VULTURE CATHARTES CALIFORNIANUS ... The Californian Vulture inhabits the region of the Columbia river to the distance of five hundred miles from its mouth, and is most abundant in spring, at which season it feeds on the dead salmon that are thrown in great numbers upon the shores. It is also often met with near the Indian villages, being attracted by the offal of the fish thrown around the habitations. It associates with the Turkey Buzzard, (*C. aura*,) but is easily distinguished from that species in flight, both by its greater size and the more abrupt curvature of the wing ...

Their food, while on the Columbia, appears to consist almost exclusively of fish, as in the neighbourhood of the falls and rapids of the river it is always in abundance. ▶

191

The salmon, in their attempts to leap over the obstruction, become exhausted, and are cast up on the beaches in great numbers. Thither therefore resort all the unclean birds of the country – the present species, the Turkey Buzzard, Raven, &c. The Californian Vulture cannot, however, be called a plentiful species, as even in the situations mentioned it is rare to see more than two or three at a time, and these so shy as not to allow an approach to within a hundred yards, except by stratagem" (Townsend 1839g: 6-7).

"CALIFORNIAN VULTURE (*Cathartes Californianus*) … In my journey across the Rocky Mountains to the Oregon in 1834, I kept a sharp look-out for this rare and interesting bird in all situations on the route, which I thought likely to afford it a congenial dwelling place; but not one did I see. It was not indeed until my return to the coast in the spring [of 1835] from the Sandwich Islands, where I spent the winter, that I was gratified by a sight of the great Vulture.

In a journey of exploration which I made to the Willammet, in the month of April[b], when the river was crowded with Salmon, making their way up against the stream, urged by an abortive instinct to pass the barriers of the thirty feet fall, I observed dozens of Turkey Vultures constantly sailing over the boiling surges, with their bare heads curved downwards as if in search of prey. As I gazed upon them, interested in their graceful and easy motions, I heard a loud rustling sound over my head, which induced me to look upward; and there, to my inexpressible joy, soared the great Californian, seemingly intent upon watching the motions of his puny relatives below. Suddenly, while I watched, I saw him wheel, and down like an arrow he plunged, alighting upon an unfortunate Salmon which had just been cast, exhausted with his attempts to leap the falls, on the shore within a short distance. At that moment I fired, and the poor Vulture fell wounded, beside his still palpitating quarry. My prize being on the opposite side of the river, I lost no time in removing my clothing and plunging into the stream. A few vigorous strokes carried me across; I sprang upon the shore, and ran, with delighted haste, to secure the much coveted and valuable specimen. But I soon discovered that I had still something to do before the operation of skinning him was to commence. The huge creature had been only wing-broken, and as I approached him, seemed determined not to yield himself a willing captive. My gun had been left behind; I was in a state of absolute nudity, and at that moment, the inhabitants of an Indian village near, consisting of men, women, children and dogs, startled by the sound of my gun, were flocking out to see what was the matter. I looked about in vain for a stick; none was to be found, and my only weapons were stones, with which I continued, for a considerable time, to pelt the Vulture, who sometimes hobbled awkwardly away, when attacked, and at others dashed furiously at me, hissing like an angry serpent, and compelled me likewise to run. It must have been an amusing scene for the Indians

California Condor. The specimen is the immature bird (USNM 8005) that Townsend shot at the Willamette Falls in May 1835 – see text.

looking on, and I heard more than once, the loud, obstreperous laugh of the women, when the Vulture was flapping after me and I throwing sand in his eyes with my naked feet. After perhaps half an hour spent in this way, I was fortunate enough to hit him fairly on the head with a large stone, which stunned him and he fell. In an instant I alighted upon him, sitting upon his body; and firmly grasping his neck with my hands. One of the Indians, at my request, brought me a knife, and I soon despatched him by severing the spine. I hired one of the boys to cross the river in a canoe to bring over my clothes and gun, and when dressed, skinned my prize with the Indians crowding around me, curious to see the operation.

This was the only specimen I was ever able to procure[c]; for although, during the spring, I constantly saw the Vultures at all points where the Salmon was cast upon the shores, their extreme shyness uniformly prevented an approach to within gun-shot.

The Californian Vulture has never been seen east of the Rocky Mountains. It is said to be abundant on the coast of California, whence I have recently received two specimens, sent me by Lieut. Col. Fremont, who procured them during his late exploration of that remote country. I have no doubt, it would be found on the N.W. coast of N. America, at least as high up as the Russian Settlements. Its winter residence is unknown, but it is probably in the dreary and inaccessible sierras of Upper California. It is reputed to breed in the Umptqua country, about fifty or sixty miles south of the Oregon Rion, and it is said to lay two eggs, which are entirely black[d]!"

Townsend, J.K. 1848. POPULAR MONOGRAPH OF THE ACCIPITRINE BIRDS OF N.A. *Literary Record and Journal of the Linnaean Association of Pennsylvania College* 4: 265-267.

a. The type specimen of the California Condor was obtained by Archibald Menzies near Monterey, California, in December 1792. The description of it by George Shaw was published in 1798, and it seems to have been this bird in the British Museum (Natural History) that Audubon used for his plate in *The Birds of America*. Audubon resorted to Townsend's specimen only for his written description of an immature vulture.

b. From the text of the *Narrative*, even though the condor incident is not mentioned, it would seem that the condor was killed on 24th May 1835, or within a day or two of that date.

c. Townsend's specimen was acquired by Audubon, before passing to S.F. Baird and to the Smithsonian Institution (USNM 78005, immature female). Menzies, Douglas, Townsend, Deppe and Frémont were among the few early travellers to secure specimens of the elusive California Condor. Lewis and Clark, for ease of transportation, reduced their specimen to just a skull and a primary feather. Those that failed entirely included such famous naturalists and explorers as La Pérouse in 1786, Baron von Langsdorff in 1806, Dr Alex Collie, G.T. Lay and Lt Belcher from 1826 to 1827, Dr Botta from 1827 to 1828, Dr Néboux in 1837, Thomas Nuttall while near Monterey in 1836, William Gambel from 1842 to 1845, Titian Peale in 1840, Col Andrew Grayson from 1846 to 1857, William Hutton from 1847 to 1851, J.W. Audubon in 1849, J.G. Bell in 1849, Adolphus Heermann from 1849 to 1854, Dr Samuel Woodhouse in 1851, James Hepburn from 1852 to 1869, Dr J.S. Newberry in 1854, and even John Xántus from 1857 to 1858 (Harris 1941).

d. California Condors only lay a single whitish egg, though it may become soiled during incubation. The origin of this tale of black eggs seems to be David Douglas (1829) who reported "egg very large (fully larger than a goose-egg) nearly a perfect circle and of a uniform jet black" but in fact the story can be traced to Etienne Lucien who told Douglas that during the summer, condors "are seen in great numbers in the woody parts of the Columbia, from the ocean to the mountains of Lewis and Clark's River, four hundred miles in the interior" and that they laid two black eggs in stick nests lined with grass in pine trees (Douglas 1914, in Harris 1941: 20).

Nuttall's Cockle. At an unknown date in the muddy salt marshes near Point Adams, Nuttall gathered Nuttall's Cockles *Clinocardium nuttallii*, a common species ranging from the shores of the Bering Sea to southern California. This cockle and two other new marine shells from the Columbia, as well as others from California and elsewhere, were written up in the *Journal of the Academy of Natural Sciences of Philadelphia* (1837) by T.A. Conrad. Unfortunately no collection dates are given in Conrad's text, and there are no dates with the surviving specimens.

4th July. On this day, Nuttall was at "Point Chinhook, near the estuary of the Columbia" collecting plants, listening to the song of Swainson's Thrush, and watching the western race of Bewick's Wren in the thick conifer forests. This may have been the day he saw a small colony of Cliff Swallow nests on the "face of Pillar Rock, an isolated columnar mass of basalt, near Chinhook" (Nuttall 1840: 401, 492, 730).

SPECIMENS AND OBSERVATIONS, JULY 1835

We know, from the specimen record and from the texts of Townsend, Nuttall and Audubon, that both Townsend and Nuttall were still busy collecting in July but they were finding less that was new to them. Unfortunately Townsend's *Narrative* gives few details of his activities and movements for the summertime, probably because he had an established collecting routine, many days were similar, and he never went very far afield. There is only a single entry for June 1835 – on 6th June he discusses the superstitions and beliefs of the local tribes but does not say where he was staying. The next entry in the *Narrative* is not until Independence Day, when Townsend says he was back on the brig at Sauvie Island.

3rd July 1835, **Western Bluebird**, USNM 1931, card catalogue.

8th July 1835, **Red-breasted Sapsucker**, USNM 1938. Audubon (1840-44, 4: 261) noted that "Several specimens of this Woodpecker, which were procured by Mr. TOWNSEND on the Columbia river, are in my possession" and that it was also seen by Nuttall "in the forests of the Columbia and the Blue Mountains of the same country" and in "Upper California".

12th July 1835, **Violet-green Swallow**, USNM 1895, card catalogue.

12th July 1835, **Violet-green Swallow**, USNM 1945, card catalogue.

14th July 1838 [= 1835], **Violet-green Swallow**, AMNH 40224.

21st July 1835, **Townsend's Vole**. No surviving specimen – date mentioned in Bachman's original description (JANSP 1839 8: 60).

28th July 1835, **Black-headed Grosbeak**, USNM 2873.

30th July 1835, **Band-tailed Pigeon**, USNM 1933, card catalogue.

"About the close of July [1835]", **Swainson's Thrush**, nest found "in the prairies of Wahlamet" (Nuttall 1840: 401).

July 4th.– This morning was ushered in by the firing of cannon on board our brig, and we had made preparations for spending the day in festivity, when, at about 9 o'clock, a letter was received from Mr. Walker, who has charge of the fort on Wappatoo island [Fort William], stating that the tailor, Thornburg, had been killed this morning, by Hubbard, the gunsmith, and requesting our presence immediately, to investigate the case, and direct him how to act.

Our boat was manned without loss of time, and Captain L. and myself repaired to the fort, where we found every thing in confusion. Poor Thornburg, whom I had seen but two days previously, full of health and vigor, was now a lifeless corpse … Upon examining the body, we found that the two balls from the pistol had entered the arm below the shoulder, and escaping the bone, had passed into the cavity of the chest. The verdict of the jury was "justifiable homicide," [because he had acted in self defence] and a properly attested certificate, containing a full account of the proceedings, was given to Hubbard, as well for his satisfaction, as to prevent future difficulty, if the subject should ever be investigated by a judicial tribunal.

This Thornburg was an unusually bold and determined man, fruitful in inventing mischief, as he was reckless and daring in its prosecution. His appetite for ardent spirits was of the most inordinate kind. During the journey across the country, I constantly carried a large two-gallon bottle of whiskey, in which I deposited various kinds of lizards and serpents and when we arrived at the Columbia the vessel was almost full of these crawling creatures. I left the bottle on board the brig when I paid my first visit to the Wallammet falls, and on my return found that Thornburg had decanted the liquor from the precious **reptiles** which I had destined for immortality, and he and one of his pot companions had been "happy" upon it for a whole day … I did not discover the theft until too late to save my specimens, which were in consequence all destroyed.

SWAINSON'S THRUSH

Nuttall is given the credit for being the first to properly describe Swainson's Thrush. He included it, under the name "Western Thrush", in his *Manual of Ornithology* (1840), after carefully re-examining a Townsend specimen. The thrush is part of a difficult-to-identify group and Audubon seems not to have been convinced that it was a new species: he never included it in any edition of *The Birds of America*. Two specimens are currently claimed as the types (ANSP 23,644 and USNM 2040). Part of Nuttall's text reads as follows:

"WESTERN THRUSH … About the commencement of May, Mr. Townsend and myself observed the arrival of this species in the shady forests of the Oregon, where, shy and retiring, it was flitting through the low bushes or gathering insects on the ground. At intervals, on the commencement of the breeding period, we heard their notes, bearing indeed, some resemblance to the quaint warble of the Veery or Wilson's Thrush, though quite distinct, and easily recognisable from the notes of that Atlantic species … About the close of July, I found the nest of this species in the prairies of the Wahlamet; it was suspended in the forks of a stout stalk of a large fern … The only specimen from which I am now able to describe the species is that of a female procured on the banks of the Columbia on the 10th of June [1835 or 1836] by my friend Mr. Townsend. This neglect arose from the too hasty conclusion that it was no other than the well known Wilson's Thrush [Veery]" (Nuttall 1840: 400-401).

TOWNSEND'S VOLE

Townsend's Vole (fig 1 at far left) in plate CXLIV from Audubon's *Viviparous Quadrupeds*. Townsend's type specimen is lost or missing.

The Rev John Bachman published the original description of this new vole in the *Journal of the Academy of Natural Sciences of Philadelphia*: "ARVICOLA *Townsendii* Townsend's Meadow Mouse … Of this species, which was a male, obtained by Mr. Townsend on the Columbia river, on the 21st July, 1835, I can find no description in Richardson" (Bachman 1839b: 60).

Years later, he repeated these basic facts when *The Viviparous Quadrupeds* was nearing completion: "ARVICOLA TOWNSENDII Townsend's Arvicola … The late Mr. TOWNSEND, who captured this animal under an old log on the banks of the Columbia River, gave us no account of its habits … the specimen here described was obtained on the 21st of July, 1835, by Mr. TOWNSEND, on the shores of the Columbia river. It no doubt is widely distributed on the western side of the Rocky Mountains …" (Audubon 1845-48: 393, plate CXLIV, fig 1).

Townsend's rather vague collecting locality of "Columbia Riiver" was restricted by Bailey (1900) to "near the mouth of the Willamettte, on or near Wappatoo (or Sauvie) Island" – probably correctly, as Townsend was very likely at that locality on that date.

SPECIMENS AND OBSERVATIONS, AUGUST TO DECEMBER 1835

This was apparently a very quiet spell for Townsend with just a few specimens known to have been collected and few observations that can be positively dated to this period. This is partly because he was sick during the latter part of August and early September and because he later took over the surgeon's duties at Fort Vancouver following the departure of the ailing Dr Gairdner.

11th August 1835, **Western Small-footed Myotis**, ANSP 1126.

18th August 1835, **California Gull**, ANSP 4810.

8th October 1835, **American Dipper**, (Audubon 1831-39, 5: 304; NYHS 1863.17.435).

23rd October 1835, **Tundra Swan**, USNM 2765, card catalogue.

28th October 1835, **Red-breasted Sapsucker**, USNM 2796, card catalogue.

28th October 1835, **Townsend's Warbler**, USNM 2918.

-- October 1835, **Oregon Dark-eyed Junco**, OUM 13574.

11th [August].– Mr. Nuttall, who has just returned from the dalles, where he has been spending some weeks, brings distressing intelligence from above. It really seems that the "Columbia River Fishing and Trading Company" is devoted to destruction; disasters meet them at every turn, and as yet none of their schemes have prospered. This has not been for want of energy or exertion. Captain W. has pursued the plans which seemed to him best adapted for insuring success, with the most indefatigable perseverance and industry, and has endured hardships without murmuring, which would have prostrated many a more robust man; nevertheless, he has not succeeded in making the business of fishing and trapping productive, and as we cannot divine the cause, we must attribute it to the Providence that rules the destinies of men and controls all human enterprises.

Two evenings since, eight Sandwich Islanders, a white man and an Indian woman, left the cascades in a large canoe laden with salmon, for the brig. The river was as usual rough and tempestuous, the wind blew a heavy gale, the canoe was capsized, and eight out of the ten sank to rise no more ... Intelligence has also been received of the murder of one of Wyeth's principal trappers, named Abbot, and another white man who accompanied him, by the Banneck Indians ... it is always unsafe to travel among Indians, as no one knows at what moment a tribe which has always been friendly, may receive ill treatment from thoughtless, or evil-designing men, and the innocent suffer for the deeds of the guilty.

August 19th.– This morning, Captain Thing (Wyeth's partner) arrived from the interior. Poor man! He looks very much worn by fatigue and hardships, and seven years older than when I last saw him. He passed through the Snake country from Fort Hall, without knowing of the hostile disposition of the Bannecks ... Captain Thing lost every thing he had with him ... had two men severely, but not mortally, wounded. The Indians had seven killed, and a considerable number wounded.

20th.– [A severely wounded man arrives and "was immediately put in the hospital here ... under the excellent and skilful care of Doctor Gairdner." i.e. Townsend was at Fort Vancouver.]

The Indians of the Columbia were once a numerous and powerful people; the shore of the river, for scores of miles, was lined with their villages ... The depopulation here has been truly frightful. A gentleman told me, that only four years ago, as he wandered near what had formerly been a thickly peopled village, he counted no less than sixteen dead, men and women, lying unburied and festering in the sun in front of their habitations. Within the houses all were sick; not one had escaped the contagion [an intermittent fever]; upwards of a hundred individuals, men, women, and children, were writhing in agony on the floors of the houses, with no one to render them any assistance … Probably there does not now exist one, where, five years ago, there were a hundred Indians … About two hundred miles southward, the Indians are said to be in a much more flourishing condition, and their hostility to the white people to be most deadly. They believe that we brought with us the fatal fever which has ravaged this portion of the country, and the consequence is, that they kill without mercy every white man who trusts himself amongst them.

LETTERS TO PHILADELPHIA

On 10th September 1835 Townsend began a letter to his father from Fort Vancouver, completing it about a fortnight later with instructions about the distribution of the contents of the box that accompanied the letter:

"… allow me to mention … a Capt. Stewart, an english [= Scottish] gentleman of noble family, younger brother to the Earl of Bute, whom I met in the mountains. He travelled with our party from the rendezvous to [Fort] Vancouver, & left here last spring for the States. He is what may be called a genuine good fellow, cheerful, hilarious, & a first rate hunter, a great recommendation I assure you in the mountains. He has none of that exclusiveness & hauteur which is so apt to attach to Englishmen as a garment, but is perfectly free & easy, with a high opinion of Americans & American institutions, and an evident leaning to republican principles. He expects to reach Philadᵃ·, & promised to call upon you to inform you of my welfare & what he knew about me. You will be pleased with him I know … I do not intend to remain another year in this country …

… by the by I had almost forgotten to tell you – I have been sick, I even I. I should not have thought that possible, but indeed its true. I was "laid by the heels" by this miserable fever – the same disease that has been so shockingly fatal among the poor indians; all the people have it here, they consider it a thing of course, but it never attacks you until you become acclimated, strangers are never troubled with it … The disease is very easily managed by medicine, (Quinine) & I have never heard of a fatal case except among indians.

… do you know that I'm now in my 27ᵗʰ year – only think of it – in a very short time I shall be numbered upon the list of the sour ones, but I'll take care I don't stay there long, if I can find anyone to say yes at me. I have no stomach for the Copper beauties here, so must perforce go home to choose.

… I think of changing my route home. Captain W. intends to return by Santa Fee & has invited N. and myself to accompany him. We shall probably do so, but it is not yet decided … Dr. Morton will send you a small box when it arrives from Boston containing two or three little articles – the chief part of my Collection of the Curiosities of this Country not appertaining to Nat. history has been stolen from me by one of Capt. W's men who deserted. The things in your box marked for Dr. Morton you will please send as pr. direction …

affectionately John   (JKT to C. Townsend, 10th & 26th September 1835)

Townsend also wrote to Samuel Morton concerning skull shapes of the "Chinouks and Chickitats" (JKT to S.G. Morton, 26th September 1835. Letter untraced, but mentioned in Morton's *Crania Americana* (1839: 207)).

October 1st.– Doctor Gairdner, the surgeon of Fort Vancouver, took passage a few days ago to the Sandwich Islands, in one of the Company's vessels [the *Ganymede*]. He has been suffering for several months, with a pulmonary affection, and is anxious to escape to a milder and more salubrious climate[2]. In his absence, the charge of the hospital will devolve on me, and my time will thus be employed through the coming winter. There are at present but few cases of sickness, mostly ague and fever, so prevalent at this season. My companion, Mr. Nuttall, was also a passenger in the same vessel. From the islands, he will probably visit California, and either return to the Columbia by the next ship, and take the route across the mountains, or double Cape Horn to reach his home[3].

31st August. At this time Nuttall was staying with Jason and Daniel Lee on the Willamette. "At the close of August, in the plains 60 miles up the Wahlamet, flocks chiefly of young [**Lewis's Woodpeckers**], from 12 to 20 or more together were to be seen shifting backwards and forwards in the trees near the river, playing about like so many sportive [Eurasian] Jackdaws [*Corvus monedula*], which the young so much resemble in color" (Nuttall 1840: 679).

AMERICAN DIPPER

"MORTON'S WATER OUZEL ... I have honored this species with the name of my excellent friend, Doctor Samuel George Morton, of Philadelphia. It was shot by Captain W. Brotchie, near Fort McLoughlin [Bella Bella, B.C.], on the N.W. coast of America, in latitude about 49° N. He stated that it was common there, and inhabited, like the rest of its tribe, the rapid fresh water streams. He procured but one specimen" (Townsend 1839a: 339). Audubon says this specimen was a female and obtained in February 1836 (Audubon 1831-39, 5: 303). Fig 1 of plate CCCCXXXV.

"COLUMBIAN WATER OUZEL ... This fine bird inhabits the swiftly running streams of fresh water in the vicinity of Fort Vancouver. It is a very scarce species, as in all my peregrinations I have met with but two individuals, only one of which I was enabled to procure.

This I observed swimming about among the rapids of the stream, occasionally flying for short distances over the surface, and then diving into it, and reappearing after a long interval. Occasionally it would alight on the stones, and at such times jerked the tail in the manner of some of the sandpipers. I did not hear it utter any note" (Townsend 1839a: 339-340). Audubon says that this specimen was a male shot by Townsend on 8th October 1835 (Audubon 1831-39, 5: 304). Fig. 2 of plate CCCXXXV.

This was not a Townsend discovery: the American Dipper *Cinclus mexicanus* was first described by William Swainson in 1827 from a Mexican specimen. However, Townsend provided Audubon with specimens and field observations.

Audubon published two plates of dippers in the folio edition of *The Birds of America* and named the four figured birds as three different species: *Cinclus Americanus*, *Cinclus Mortoni* and *Cinclus Townsendi*.

In 1838, in volume IV of the *Ornithological Biography* (p. 501), Audubon provided the text for *Cinclus Americanus* but by the following year when volume V was completed, he had realised that they were all the same species and called them all *Cinclus Americanus* (pp. 303-304).

In the octavo edition of *The Birds of America* (1840-44) Audubon omitted the colour plate and all the text that referred to *Cinclus Mortoni* and *Cinclus Townsendi*.

American Dipper, plate CCCCXXXV, the very last bird to be included in the folio edition of *The Birds of America*. It was supposed to depict two new species brought back from the west by Townsend but both are American Dippers. Fig. 1, at left, *Cinclus Mortoni* painted from the specimen opposite, at top (USNM 2862). Fig. 2, at right, *Cinclus Townsendi* painted from the specimen opposite, below (USNM 2861).

American Dipper, plate CCCLXX. This plate was prepared earlier than the one above, from specimens "procured on the Rocky Mountains, on the 15th of June" – but no year is given. If the specimens came from Townsend then they were most likely collected in 1835 or 1836 because in mid June 1834 he was not in the correct habitat for American Dipper.

Rev Samuel Parker, 1835.

16th.– Several days since, the Rev. Samuel Parker, of Ithaca, N. York, arrived at the fort. He left his home last May, travelled to the rendezvous on the Colorado [Green River, Wyoming], with the fur company of Mr. Fontinelle, and performed the remainder of the journey with the Nez Percé or Cheaptin Indians. His object is to examine the country in respect to its agricultural and other facilities, with a view to the establishment of missions among the Indians. He will probably return to the States next spring, and report the result of his observations to the board of commissioners, by whose advice his pioneer journey has been undertaken.

On the 17th, I embarked with this gentleman in a canoe, for a visit to the lower part of the river. We arrived at the American brig in the afternoon, on board of which we quartered for the night, and the next morning early, the vessel cast off from the shore. She has her cargo of furs and salmon on board, and is bound to Boston, via the Sandwich and Society Islands. Mr. Parker took passage in her to Fort George, and in the afternoon I returned in my canoe to [Fort] Vancouver[4].

December 1st.– The weather is now unusually fine. Instead of the drenching rains which generally prevail during the winter months, it has been for some weeks clear and cool, the thermometer ranging from 35° to 45°.

The ducks and geese, which have swarmed throughout the country during the latter part of the autumn, are leaving us, and the swans [**Tundra Swans**, and perhaps some **Trumpeter Swans**] are arriving in great numbers. These are here, as in all other places, very shy; it is difficult to approach them without cover; but the Indians have adopted a mode of killing them which is very successful; that of drifting upon the flocks at night, in a canoe, in the bow of which a large fire of pitch pine has been kindled. The swans are dazzled, and apparently stupefied by the bright light, and fall easy victims to the craft of the sportsman.

20th.– Yesterday one of the Canadians took an enormous wolf in a beaver-trap. It is probably a distinct species from the common one, (*lupus,*) [**Gray Wolf**] much larger and stronger, and of a yellowish cinereous color. The man states that he found considerable difficulty in capturing him, even after the trap had been fastened on his foot. Unlike the lupus, (which is cowardly and cringing when made prisoner,) he showed fight, and seizing the pole in his teeth, with which the man attempted to despatch him, with one backward jerk, threw his assailant to the ground, and darted at him, until checked by the trap chain. He was finally shot, and I obtained his skin, which I have preserved.

1. After Wyeth's return to Boston in 1836 the fort was soon abandoned. Wyeth says the fort was eight miles from Fort Vancouver on the south-west side of Sauvie Island. This tallies with Townsend's location, "about fifteen miles from the lower mouth of the Wallammet", placing the fort site beside the Multnomah Channel, near the present junction of Reeder Road and Sauvie Island Road.

2. Meredith Gairdner died at Honolulu on 26th March 1837, aged 29. There is no known portrait.

3. Nuttall and Townsend discussed several overland return routes but both returned by sea, at separate times. Nuttall spent four months in the Hawaiian Islands, took a ship to California, and returned to Boston via Cape Horn. A major consignment of Townsend's specimens (and his letters of 26th September 1835) went with Nuttall to Hawaii but must soon have been shipped on from there because the specimens reached Dr. Morton in Philadelphia before Nuttall got back to the east coast.

4. Parker says that they were paddled down to Sauvie Island by Hawaiian Islanders, arrived at 5pm at the May Dacre that was moored alongside the natural wharf of basalt that is Warrior's Point, and were welcomed on board by Captain Lambert. The brig sailed down river with the tide on 19th October and in the evening Parker saw Halley's Comet, a phenomenon not mentioned by Townsend (Parker 1990: 148-151).

American Dipper habitat near the Columbia River.

Audubon stated that "The only original observations respecting the habits of the American Dipper ... have been favoured by Mr TOWNSEND:– "This bird inhabits the clear mountain streams in the vicinity of the Columbia river. When observed it was swimming among the rapids, occasionally flying short distances over the surface of the water, and then diving into it, and reappearing after a long interval. Sometimes it will alight along the margin, and jerk its tail upwards like a Wren. I did not hear it utter any note. The stomach was found to contain fragments of fresh-water snails. I observed that this bird did not alight on the surface of the water, but dived immediately from the wing" (Audubon 1840-44, pp. 184-185).

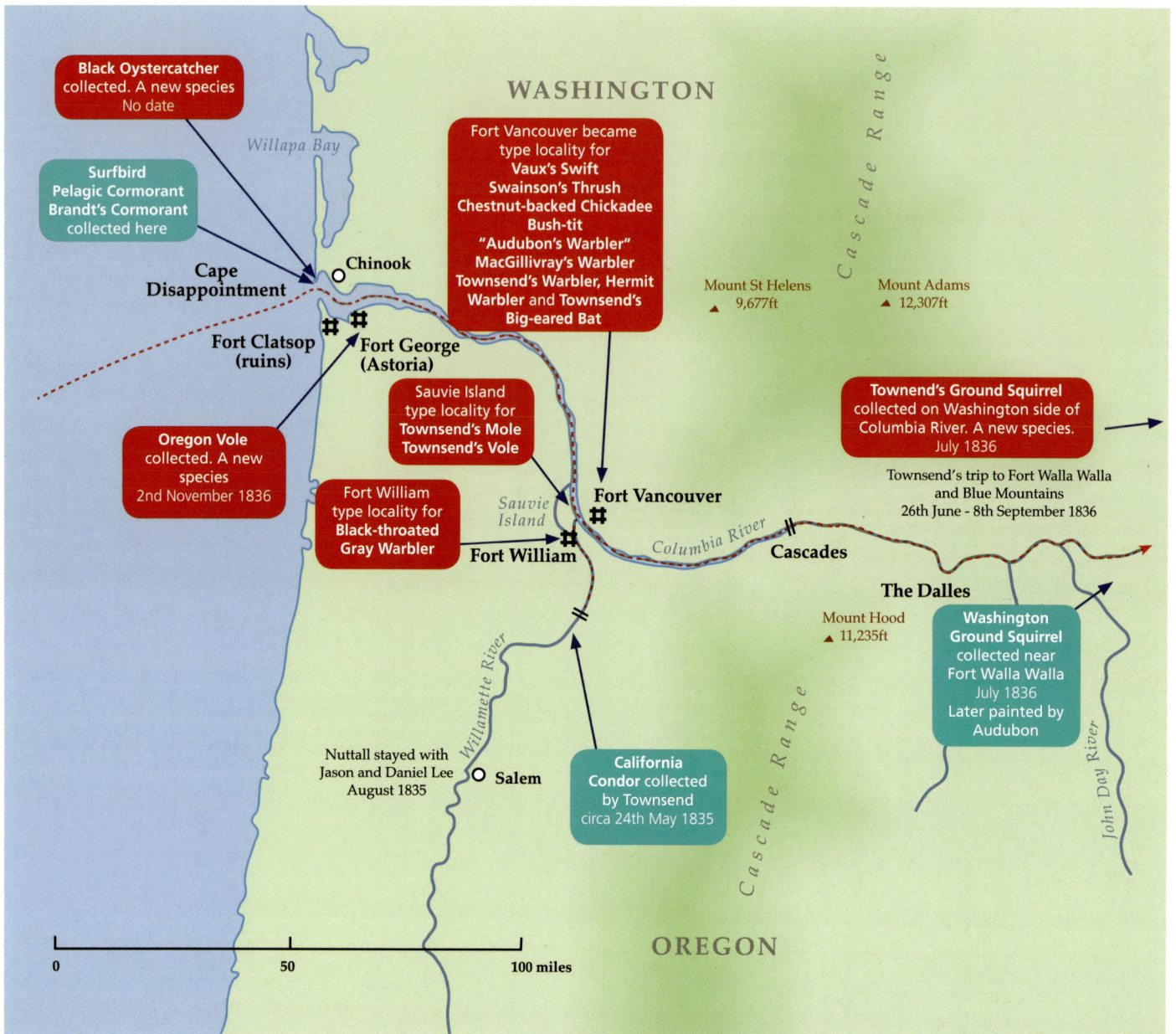

Black Oystercatcher collected. A new species
No date

Surfbird
Pelagic Cormorant
Brandt's Cormorant
collected here

WASHINGTON

Willapa Bay

Fort Vancouver became type locality for
Vaux's Swift
Swainson's Thrush
Chestnut-backed Chickadee
Bush-tit
"Audubon's Warbler"
MacGillivray's Warbler
Townsend's Warbler, Hermit Warbler and **Townsend's Big-eared Bat**

Cape
Disappointment

Chinook

Fort Clatsop
(ruins)

Fort George
(Astoria)

Sauvie Island type locality for
Townsend's Mole
Townsend's Vole

Sauvie Island

Fort Vancouver

Mount St Helens
▲ 9,677ft

Mount Adams
▲ 12,307ft

Cascade Range

Townend's Ground Squirrel collected on Washington side of Columbia River. A new species.
July 1836

Townsend's trip to Fort Walla Walla and Blue Mountains
26th June - 8th September 1836

Oregon Vole collected. A new species
2nd November 1836

Fort William type locality for
Black-throated Gray Warbler

Fort William

Columbia River

Cascades

The Dalles

Mount Hood
▲ 11,235ft

Washington Ground Squirrel collected near Fort Walla Walla
July 1836
Later painted by Audubon

Willamette River

Nuttall stayed with Jason and Daniel Lee August 1835

Salem

California Condor collected by Townsend
circa 24th May 1835

Cascade Range

John Day River

OREGON

0   50   100 miles

Map 11 - The lower Columbia River, 16th April 1835 - 30th November 1836

Beacon Rock, Columbia River.

The Columbia River, 1836

"Mr Townsend (a traveling naturalist) … is from Philadelphia, and has been in the mountains two years. He is sent here by a society to collect the different species of bipeds, and quadrupeds, peculiar to this country. We brought a parcel of letters to him, the first he had received since he had left home"

Narcissa Whitman's journal, entries for 1st & 12th September 1836.

Winter at Fort Vancouver

February 3d, 1836. – During a visit to Fort William, last week, I saw, as I wandered through the forest, about three miles from the house, a canoe, deposited, as is usual, in the branches of a tree, some fourteen feet from the ground. Knowing that it contained the body of an Indian, I ascended to it for the purpose of abstracting the skull; but upon examination, what was my surprise to find a perfect, embalmed body of a young female, in a state of preservation equal to any which I had seen from the catacombs of Thebes. I determined to obtain possession of it, but as this was not the proper time to carry it away, I returned to the fort, and said nothing of the discovery which I had made.

That night, at the witching hour of twelve, I furnished myself with a rope, and launched a small canoe, which I paddled up against the current to a point opposite the mummy tree. Here I ran my canoe ashore, and removing my shoes and stockings, proceeded to the tree, which was about a hundred yards from the river. I ascended, and making the rope fast around the body, lowered it gently to the ground; then arranging the fabric which had been displaced, as neatly as the darkness allowed, I descended, and taking the body upon my shoulders, bore it to my canoe, and pushed off into the stream. On arriving at the fort, I deposited my prize in the store house, and sewed around it a large Indian mat, to give it the appearance of a bale of guns. Being on a visit to the fort, with Indians whom I had engaged to paddle my canoe, I thought it unsafe to take the mummy on board when I returned to [Fort] Vancouver the next day, but left directions with Mr. Walker to stow it away under the hatches of a little schooner, which was running twice a week between the two forts.

On the arrival of this vessel, several days after, I received, instead of the body, a note from Mr. Walker, stating that an Indian had called at the fort, and demanded the corpse. He was the brother of the deceased, and had been in the habit of visiting the tomb of his sister every year. He had now come for that purpose, from his residence near the *"tum-water,"* (cascades,) and his keen eye had detected the intrusion of a stranger on the spot hallowed to him by many successive pilgrimages. The canoe of his sister was tenantless, and he knew the spoiler to have been a white man, by the tracks upon the beach, which did not incline inward like those of an Indian.

The case was so clearly made out, that Mr. W. could not deny the fact of the body being in the house, and it was accordingly delivered to him, with a present of several blankets, to prevent the circumstance from operating upon his mind to the prejudice of the white people. The poor Indian took the body of his sister upon his shoulders, and as he walked away, grief got the better of his stoicism, and the sound of his weeping was heard long after he had entered the forest.

Fort Vancouver (reconstruction).

SPECIMENS AND OBSERVATIONS, JANUARY AND FEBRUARY 1836

Townsend remained at Fort Vancouver throughout the winter and spring, confined to barracks, as it were, because he had taken on the surgeon's duties for the fort, following Dr Gairdner's departure for the Hawaiian Island at the end of the previous September. He therefore had much less time for collecting, and this is reflected in the specimen record and the paucity of published observations.

15th January 1836, **Hermit Thrush**, NYHS 1863.17.419.

18th January 1836, **Varied Thrush**, USNM 1844.

18th January 1836, **Song Sparrow**, USNM 1860.

18th January 1836, **Hairy Woodpecker**, (Audubon 1840-44, 4: 242).

25th January 1835 [= 1836], **Varied Thrush**, LM D1704 [Specimen labelled 1835 but Townsend was in the Hawaiian Islands in January of both 1835 and 1834.]

-- January 1836, **Varied Thrush**, OUM 11091.

10th February 1836, **Tundra Swan**, ANSP 6106. Nearly 100 years later, Wharton Huber of the Academy of Natural Sciences of Philadelphia reported on a method of preparing water bird skins: "Mr. F.W. Wood, one of our Taxidermists, who has made a specialty of remodelling old specimens, has recently degreased and remade, by the above method, the skin of a Whistling Swan [Tundra Swan], one of the dirtiest skins I ever saw. This Swan was shot on the Columbia River in 1836 by J.K. Townsend. It is now as clean and almost as white as a freshly collected specimen" (Huber 1930: 408-411, with photographs taken before and after treatment).

15th February 1836, **Fox Sparrow**, USNM 2874. Type of *Plectrophanes townsendi* (Audubon 1840-44, 3: 143-14).

20th February 1836, **bird species not named**, USNM 2746, card catalogue.

Song Sparrow, plate CCCXC (detail).

SONG SPARROW

"BROWN SONG-FINCH ... Of this bird I have received the following account from Mr. NUTTALL:– "This species, so much allied to *Fringilla iliaca* [Fox Sparrow] by its brown colour, inhabits the woody districts of the Columbia, very generally as far as the sea-coast, and continues as far south as Upper California ... We heard their cheerful notes throughout the summer; and every fine day in winter till the month of November, particularly in the morning, their song was still continued ...

Mr. TOWNSEND speaks of it as follows:- [A handwritten note by Townsend in a copy of the *Ornithological Biography* at the Academy of Natural Sciences of Philadelphia says that the text that follows is for the wrong species, the Lark Sparrow (Stone 1906: 312)]" (Audubon 1840-44, 3: 145-146).

VARIED THRUSH

"Mr. TOWNSEND ... informs me that he "first found this Thrush on the Columbia river in the month of October, and that it becomes more numerous in winter, which it spends in that region, though some remove farther south. It there associates with the Common [American] Robin, *Turdus migratorius*, but possesses a very different note, it being louder, sharper, and quicker than those of the latter, and in the spring, before it sets out for its yet unascertained breeding-place, it warbles very sweetly. It is called *Ammeskuk* by the Chinooks."

Mr. NUTTALL's notice respecting it is as follows:– "On the Columbia they are only winter birds of passage, arriving about October, and continuing more or less frequently throughout the winter. At this time they flit through the forest in small flocks, frequenting usually low trees, on which they perch in perfect silence, and are at times very timorous and difficult of approach ...""

The figures in my plate were taken from adult males and a fine female shot in spring" (Audubon 1840-44, 3: 22-23).

FOX SPARROW

Townsend may have been seeing more than one race of Fox Sparrow, e.g. *Passerella iliaca fuliginosa* (the dark coastal race of Pacific North West) and *P. i. unalaschensis* (from south-west Alaska, wintering in Oregon). The Fox Sparrow that Audubon mentions (below) as being collected on 15th February 1836 is catalogued as *Passerella iliaca townsendi*. (USNM 2874).

"TOWNSEND'S FINCH ... This species was discovered on the shores of the Columbia river, by Mr. Townsend, who sent me a perfect specimen, ticketed "Female, February 15th, 1836," together with the following notice. [A handwritten note by Townsend in a copy of the *Ornithological Biography* at the Academy of Natural Sciences of Philadelphia says that the text that follows is for the wrong species, as it describes Brewer's Sparrow (Stone 1906: 312)]" (Audubon 1840-44, 3: 143).

"TOWNSEND'S GROUND FINCH, OR LONGSPUR ... This species is common in the neighbourhood of Fort Vancouver on the Columbia. It inhabits the dense bushes chiefly in the vicinity of low, marshy places, and feeds upon coleopterous insects and worms, for which it searches in the ground by scratching up the earth with its feet. It is observed only in the autumn and winter" (Townsend 1839a: 345).

TOWNSEND'S PARENTS WANT HIM TO COME HOME

"My ardent desire and prayer for thee has been, that this long and perilous journey may suffice the remainder of thy life, that hereafter thee may feel thy mind settled and willing to remain in thy native City ... if thy inclination continued as it was in early life to be a Physician there never was a more propitious time for thee ...

(Charles Townsend to JKT, 3rd March 1836)

"We shall soon be in daily expectation of the arrival of your vessel, in which we cannot help hoping thou may have embarked, for surely by this time, methinks thou must be tired of dwelling in distant lands, or roaming the Forrests in search of the feathered tribe and the natural curiosities of the Earth, and quite ready to return to civilized life, the comforts of thy own home, and the embrace of those who love thee. Yes, and that this may satisfy thee for the remainder of thy life ... I must commend thee to "God and to the Word of his Grace which is nigh thee even in thy heart and in thy mouth" as thou art attentive to its dictates

thy warmly attached Mother"   (Priscilla Townsend to JKT, 3rd March 1836)

March 16th.– Doctor W.F. Tolmie, one of the surgeons of the Hudson's Bay Company, has just arrived from Fort Langley, on the coast, and has relieved me of the charge of the hospital, which will afford me the opportunity of peregrinating again in pursuit of specimens. The spring is just opening, the birds are arriving, the plants are starting from the ground, and in a few weeks, the wide prairies of the Columbia will appear like the richest flower gardens.

SPECIMENS AND OBSERVATIONS, APRIL-JUNE 1836

Townsend was now no longer employed as the surgeon at Fort Vancouver. While his medical obligations may have been satisfying during the winter months, the arrival of spring meant that he yearned to be out and about exploring once more. The fact that this was his second spring on the Columbia River reduced the number of specimens that he was preparing since most of the birds returning from the south were now familiar to him. In the middle of May he went downstream beyond Sauvie Island but not until 25th June did he manage to get away from the area.

1st April 1836, **Red-shafted Northern Flicker**, (Audubon 1840-44, 4: 291-292).

-- April 1836, **"Audubons" Yellow-rumped Warbler**, USNM 2911.

-- April 1836. **Townsend's Solitaire**, NYHS 1863.17.419.

17th April 1836, **Band-tailed Pigeon**, first arrival (Audubon 1840-44, 4: 312-315); Nuttall (1840: 754) also reports Townsend seeing very great numbers on this date.

4th May 1836, **MacGillivray's Warbler**, USNM 1861.

21st May 1836, **Western Bluebird**, LM D1454.

27th May 1836, **Evening Grosbeak**. "Columbia river, May 27, 1836.– The Evening Grosbeak, *Fringilla vespertina*, is very numerous in the pine-woods at this time. You can scarcely enter a grove of pines at any hour in the day without seeing numbers of them. They are very unsuspicious and tame, and I have, in consequence, been enabled to procure a fine suite of specimens" (JKT, in Audubon 1840-44, 3: 218).

29th May 1836, **American Magpie**, ANSP 3304 (see Jobanek 1994:13).

3rd June 1836, **Bullock's Oriole**, ANSP 3501.

3rd June 1836, **Lazuli Bunting**, pair shot (Audubon 1840-44, 3: 100).

10th June [no year = 1835 or 1836], **Swainson's Thrush**, first described by Nuttall 1840: 400-401) from a female "procured on the banks of the Columbia on the 10th of June by my friend Mr. Townsend."

16th June 1836, **Lewis's Woodpecker**, ANSP 24204.

21st June 1836, **Bullock's Oriole**, male and female (Audubon 1840-44, 4: 43-45).

-- June 1836, **flycatcher** [*Empidonax sp?*], MCZ 17286.

-- June 1836, **Lazuli Bunting**, OUM 13364.

-- June 1836, **Black Oystercatcher**, male (Audubon 1840-44, 5: 243). This was the "Bachman's Oystercatcher" specimen given to Townsend by Dr Tolmie – Townsend was not at the coast at this time.

Dr William Fraser Tolmie (1812-1886), a Scottish surgeon employed by the Hudson's Bay Company. He arrived in the Pacific Northwest in 1832 (on the same ship as Dr Meridith Gairdner), serving on Puget Sound and further north on the Canadian coast before returning to the Columbia River in the spring of 1836. In 1859 he took up farming on Vancouver Island. One of his sons, Simon Fraser Tolmie, was Premier of British Columbia from 1928-33.

Dr Tolmie exchanged both plants and birds with Townsend. It is probable that Tolmie sent some of Townsend's birds to the Inverness and Glasgow museums but the institutions there do not now possesses any of Tolmie's or Townsend's skins. MacGillivray's Warbler *Oporornis tolmiei* was named after him by Townsend (Mearns & Mearns 1992, pp. 436-440, 545).

Plate CCCCXIX from *The Birds of
America*.

Hermit Thrush (at top). Audubon used the name Little Tawny Thrush, and inscribed his original watercolour "Jany 15^t 1836 – Columbia River."

Townsend's Solitaire (middle). Townsend confessed that "Of this singular bird I know nothing, but that it was shot by my friend Captain W. Brotchie, of the Honourable Hudson's Bay Company, in a pine forest near Fort George, (Astoria.) It was the only specimen seen" (Townsend 1839a: 338-339). Audubon named the new species after Townsend (who had in fact probably seen the species in the Sawtooth Mountains on 12th August 1834, but had never managed to collect one for himself). There is no trace of the type specimen at ANSP or USNM. Audubon's original watercolour is inscribed "Columbia River April 1836".

Gray Jay (bottom). When Audubon wrote about the Gray Jay he said: "Its range is very extensive, as I have specimens procured by Mr. TOWNSEND on the Columbia river, and it has been observed by Dr. RICHARDSON as far northward as lat. 65°. The former of these naturalists states that he found "these birds at the site of Old Fort Astoria, on the Columbia river. They were very noisy and active; the voice is strong and harsh. The Indians however say, that they are rarely seen, and that they do not breed hearabouts" (Audubon 1840-44, 4: 124.)

AN ESSAY ON BIRDS AND A LETTER HOME

Townsend's essay appears in appendix 9, with a new commentary.

On 11th April 1836, Townsend wrote a short essay entitled "Description of the Birds of the Columbia River Region" for the Reverend Parker (a part of which was much modified and incorporated into Parker's *Journal of an Exploring Tour Beyond the Rocky Mountains* (1846)). The original manuscript has now been lost but a typescript copy exists and was reprinted and discussed by Jobanek and Marshall (1992).

On the same day, Townsend grabbed the opportunity to hastily pen a letter to his father, since the Rev Parker was about to return to the east coast. The letter reveals that Townsend is still undecided about his own return route and that he is feeling rather peeved that Nuttall had decided not to come back from the Hawaiian Islands after all:

" I regret that I cannot yet speak with certainty regarding the mode or time of my return. I am still favorable to take passage to England & think it most probable I shall embrace this opportunity. I shall however not recross the Rocky Mountains – that's decided, & if I choose a land route at all it will be through Mexico, after landing from the Sandwich Islands at Monterey, St. Francisco or some other port of California. The only objection to this is the considerable expense that must necessarily be incurred, but were I sure of the prompt assistance of the Acad. in these pecuniary matters I should not hesitate one moment, but choose this route & sail for the islands next month. I am however not sure of this & although I should certainly very much augment my collections in Nat. History by pursuing this course I fear that these prudential considerations will operate against the plan …

I think I told you in my late letter about Mr. Nuttall's desertion of me. He went to the [Hawaiian] islands last fall to get rid of a disagreeable winter & promised me on the word of an honest man to return early in the Spring but instead of his corporal presence in the ship on her return, I found only a paltry note (I can't call it a letter) saying that he was off, but not entering into any particulars of any kind nor doing as my sister Hannah & all other people of the right sort would have done, telling me all about it & withholding nothing ... He stated his determination of going 'round Cape Horn to Boston in a whaleship ..."

The bearer, the Rev^d. Saml. Parker is a clergyman "in good esteem with us." He crossed the Rocky Mount^ns last year on an appointment by the board of missions of Boston. His object has been, not to settle among the indians, but to visit them & their country for the purpose of deciding upon the expediency of establishing stations amongst them. He now returns to report & prepared to advise his friends to dispatch their heralds among these poor creatures without delay ...

You will think perhaps that I have apostacized & become a presbyterian. No. I am a quaker yet. I glory in it & I mean always to continue to be a quaker…

Your most affectionate John.   (JKT to C. Townsend, 11th April 1836)

May 13th.– [While camped on a plain downstream from Warrior's Point, Townsend witnessed a "medicine man" causing a sick girl so much distress that he angrily declared that he would cure the girl in three days if the "medicine man" was sent away. With nothing to lose, the father reluctantly agreed. Townsend gave the girl a cathartic followed by sulphate of quinine, and she recovered completely in two days.]

In consequence of my success in this case, I had an application to administer medicine to two other children similarly affected. My stock of quinine being exhausted, I determined to substitute an extract of the bark of the dogwood, (*Cornus Nuttalli,*) [Nuttall's or Pacific Dogwood *Cornus nuttallii*] and taking one of the parents into the wood with his blanket, I soon chipped off a plentiful supply, returned, boiled it in his own kettle, and completed the preparation in his lodge, with most of the Indians standing by, and staring at me, to comprehend the process. This was exactly what I wished; and as I proceeded, I took some pains to explain the whole matter to them, in order that they might at a future time be enabled to make use of a really valuable medicine, which grows abundantly every where throughout the country …

I administered to each of the children about a scruple of the extract per day. The second day they escaped the paroxysm, and on the third were entirely well.

EVENING GROSBEAK

Evening Grosbeak, plate CCCCXXIV (detail).

"The female remained utterly unknown until it was obtained by Mr. TOWNSEND, who found this Grosbeak abundant about the Columbia river, and procured a great number of specimens, several of which are in my possession. The following note from him contains all the information respecting its habits that I can lay before you.

"Columbia river, May 27, 1836. – The Evening Grosbeak, *Fringilla vespertina*, is very numerous in the pine-woods at this time. You can scarcely enter a grove of pines at any hour in the day without seeing numbers of them. They are very unsuspicious and tame, and I have, in consequence, been enabled to procure a fine suite of specimens. The accounts that have been published respecting them by the only two authors to whom I have access, Mr. NUTTALL and Prince BONAPARTE, are, I think, in many respects incorrect[1]. In the first place, it is stated that they are retiring and silent during the day, and sing only on the approach of evening. Here they are remarkably noisy during the whole of the day, from sunrise to sunset … Another error in the books is this, – they both state that the female is similar to the male in plumage. Now, this is entirely a mistake: she is so very different in colour and markings, that were it not for the size and colour of the bill, and its peculiar physiognomy, one might be induced to suppose it another species" (Audubon 1840-44, 3: 217-221).

BULLOCK'S ORIOLE

Bullock's Oriole male and female (at left) with Baltimore Oriole (at right), plate CCCCXXXIII (detail).

"BULLOCK'S TROOPIAL, ORIOLE, OR HANG-NEST … Mr. TOWNSEND… says … "it inhabits the Rocky Mountains near the Black Hills and the forests of the Columbia river. In the latter place it is a rather plentiful species. Its usual note consists of a single quavering call somewhat like one of the notes of the Scarlet Tanager, *Tanagra rubra*. At other times it warbles a little, but not with half the sweetness or compass of its near relative the Baltimore. It is a very active species, so much so that it is very difficult to get a shot at it while sitting, but it is easily killed on the wing. It evidently breeds here, and has probably now a nest (June 16th), but I have not been able to find it. The female is rarely seen, and is particularly shy and noiseless."

… The female is smaller and differs greatly in colouring … [The] description is taken from an individual killed on the 21st of June, 1836, on the Columbia river … A young male, killed on the Columbia river, on the 21st of June, 1836, and in its first plumage, resembles the female …" (Audubon 1840-44, 4: 43-45).

210

LAZULI BUNTING

"The Lazuli Finch, one of the handsomest of its tribe, was added to our Fauna by THOMAS SAY, who procured it in the course of LONG's expedition already mentioned. A figure of the only specimen then obtained was given in the continuation of WILSON's American Ornithology by the Prince of MUSIGNANO [Charles Bonaparte]. It has been my good fortune to procure a fine pair from Mr. TOWNSEND, who shot them on the Columbia river, on the 3d of June, 1836, so that I have been enabled to represent the female, which has not hitherto been figured, as well as the male. That enterprising naturalist has informed me, that "the Chinook Indians name this species *Tilkonapaooks*, and that it is rather a common bird on the Columbia, but is always shy and retiring in its habits, the female being very rarely seen. It possesses lively and pleasing powers of song, which it pours forth from the top branches of moderate-sized trees. Its nest, which is usually placed in the willows along the margins of streams, is composed of small sticks, fine grasses, and cow or buffalo hair."

A nest of this species [was] presented to me by Mr. Nuttall, who found it on the Columbia river ..." (Audubon 1840-44, 3: 100).

Lazuli Bunting, plates CCCXCVIII and CCCCXXIV (details, of male and female); and a Townsend specimen collected in June 1836 (OUM 13364). Despite the text at left from the *Ornithological Biography,* Audubon's original watercolour for this bird is inscribed "Male Columbia River / June 22nd 1835".

211

Peter Skene Ogden (1794-1854) was one of the most important fur-trader explorers. He worked for the North West Company and then the Hudson's Bay Company, making explorations in Oregon, the Salt Lake area, Humboldt River and California.

Upriver to Fort Walla Walla

June 26th.– I left [Fort] Vancouver yesterday, with the summer brigade, for a visit to Walla-walla, and its vicinity. The gentlemen of the party are, Peter Ogden, Esq., chief factor, bound to New Caledonia, Archibald McDonald, esq., for Colville, and Samuel Black, Esq., for Thompson's river, and the brigade consists of sixty men, with nine boats.

SPECIMENS AND OBSERVATIONS FROM THE TRIP TO FORT WALLA WALLA AND THE BLUE MOUNTAINS, 26th JUNE – 8th SEPTEMBER 1836.

This journey was not as productive as Townsend had hoped. He got away from Fort Vancouver later in the season than he would have liked and he was passing through well-worked habitats that he had traversed, albeit rather rapidly, on his arrival in the region in 1834. Once again, his MS Journal is a more productive source of wildlife encounters than the published *Narrative*. It is a pity he did not go instead to the timberline and alpine slopes of Mt Adams or Mt Hood where he would have come across Hoary Marmot, Mountain Chickadee, and also perhaps White-tailed Ptarmigan and Gray-crowned Rosy-Finch – but it was dangerous to travel away from the river without an escort.

30th September 1836, **California Ground Squirrel**, Townsend's *Narrative*.

3rd July 1836, **Snowy Egret**, ANSP 162372, Walla Walla.

-- July 1836, **Black Tern**, USNM 2020.

7th July 1836, **Sharp-tailed Grouse**, Townsend's *Narrative*.

[7th?] July [1836], **Townsend's Ground Squirrel**, ANSP 344.

no date, **Washington Ground Squirrel**, Audubon's *Viviparous Quadrupeds* plate CXLVII.

16th July 1836, **American Badger**, MS Journal.

12th August 1836, **Greater Sage-Grouse**, MS Journal.

12th August 1836, **Mountain Cottontail**, MS Journal.

13th August 1836, **Snowy Egret**, Fort Walla Walla (ex Harris Collection; specimen mentioned by Street (1948) but current location not known).

16th August 1836, **Greater Sage-Grouse**, MS Journal.

20th August 1836, **Mountain Cottontail**, LM D573a, Fort Walla Walla.

27th.– We arrived yesterday at the upper cascades, and made in the course of the day three portages. As is usual in this place, it rained almost constantly, and the poor men engaged in carrying the goods, were completely drenched. A considerable number of Indians are employed here in fishing, and they supply us with an abundance of salmon. Among them I recognise many of my old friends from below. [*In the neighbourhood of the cascades are many high mountains & enormous rocks of basalt many of them in the form of obelisks & pillars, & some at least 2000 feet in height. On the smooth sides of many of them I observed numbers of the* **Cliff Swallow** *(Hirundo rufa) & the birds themselves were circling around us in every direction. At our encampment in the evening the musquitoes were in swarms & made it very uncomfortable* (MS Journal).]

29th.– … In the afternoon, we passed the bold, basaltic point, known to the voyageurs by the name of "Cape Horn." The wind here blew a perfect hurricane, and but for the consummate skill of those who managed our boats, we must have had no little difficulty. [In the evening they camped "*9 miles below the Dalls*" (MS Journal).]

Cliff Swallows and nests.

30th.– We were engaged almost the whole of this day in making portages, and I had, in consequence, some opportunity of prosecuting my researches on the land. We have now passed the range of vegetation; there are no trees or even shrubs; nothing but huge, jagged rocks of basalt, and interminable sand heaps. I found here a large and beautiful species of marmot, (the *Arctomys Richardsonii*,) several of which I shot [= **California Ground Squirrel**[2] *Spermophilus beecheyi*]. Encamped in the evening at the village of the Indian chief, *Tilki*. I had often heard of this man, but now I saw him for the first time. His person is rather below the middle size, but his features are good, with a Roman cast, and his eye is deep black, and unusually fine. He appears to be remarkably intelligent, and half a century before the generality of his people in civilization.

CALIFORNIA GROUND SQUIRREL

Audubon's original watercolour of the California Ground Squirrel used for plate XLIX of his *Viviparous Quadrupeds*. Audubon inscribed the watercolour with the words "4 Specimens from J.K. Townsend all differing very greatly in sizes, from the Columbia River, June 22[d] – 1835". In Audubon's published text he says that one of the Townsend specimens is marked "Falls of the Columbia River" and another "Wall-walla"; the latter must therefore have been collected on a different date.

July 3d.– This morning we came to the open prairies, covered with wormwood bushes. The appearance, and strong odor of these, forcibly reminded me of my journey across the mountains, when we frequently saw no vegetation for weeks, except this dry and barren looking shrub.

The Indians here are numerous, and are now engaged in catching salmon, lamprey eels [**Pacific Lamprey**], &c. They take thousands of the latter, and they are seen hanging in great numbers in their lodges to dry in the smoke. As soon as the Indians see us approach, they leave their wigwams, and run out towards us, frequently wading to their breasts in the water, to get near the boats. Their constant cry is *pi-pi, pi-pi,* (tobacco, tobacco,) and they bring a great variety of matters to trade for this desirable article; fish, living birds of various kinds, young wolves, foxes, minks, &c.

[*A Collector of a Menagerie would do well amongst these people* (MS Journal, 4th July 1836).]

On the evening of the 6th, we arrived at Walla-walla or Nez Percés fort, where I was kindly received by Mr. Pambrun, the superintendent.

The next day [7th] the brigade left us for the interior, and I shouldered my gun for an excursion through the neighborhood. On the west side of the little Walla-walla river, I saw, during a walk of two miles, at least thirty rattlesnakes [**Western Rattlesnake** *Crotalus viridis*], and killed five that would not get out of my way. They all seemed willing to dispute the ground with me, shaking their rattles, coiling and darting at me with great fury. I returned to the fort in the afternoon with twenty-two **sharp-tailed grouse**, (*Tetrao phasianellus*,) the product of my day's shooting.

[7th?] July [1836], **Townsend's Ground Squirrel**, ANSP 344. Obtained on the "prairies near Walla-walla" (Bachman 1839b: 61-62). [Scheffer (1946) suggests that the type of Townsend's Ground Squirrel was taken on the day following his arrival at Fort Walla Walla [= 7th July]. In the MS Journal the encounter with the rattlesnakes on the west side of the Walla Walla River took place on 16th July, a more likely date for him to have taken the ground squirrel. Only on 16th August does Townsend enter the range of Townsend's Ground Squirrel when he specifically says he crossed the *Columbia* River.]

[*9th. At Fort Walla walla – Mr. Pambrun & myself rode out about two miles over the prairie for the purpose of shooting Grous. We soon came upon a large <u>pack</u> of birds & I killed nine of them in about an hour. These birds are about the size of the domestic hen & afford a most attractive & beautiful mark for the sportsman. They lie well, flush easily & freely & fly with great vigour & straight away making as much <u>whirring</u> with their wings as the Common pheasant [Ruffed Grouse]. It reminded me strongly of my old sporting days & I pursued it for a time with considerable zest. The flesh of this species (the* **Sharp-tailed Grous***, Tetrao phasianellus) is white like that of the Pheasant [Ruffed Grouse] & is excellent eating.*

16th. This afternoon a small camp of Nez Perces, or <u>Pierced Nose</u> indians arrived at the fort. They come from the <u>Grande ronde</u> where they have been catching Salmon, digging Kamas &c. One of them brought me a **Badger***, a very pretty animal, which I prepared* [there are three specimens at ANSP] (MS Journal).].

12th July. Unknown to Townsend, it was announced in Philadelphia that in the previous week his first consignment of specimens from the Columbia River had arrived safely at the Academy, and was said to consist of 300 birds and 50 quadrupeds (*Waldie's Circulating Library*, 12th July 1836).

TOWNSEND'S GROUND SQUIRREL AND WASHINGTON GROUND SQUIRREL

These two ground squirrels inhabit opposite sides of the Columbia River, in the region around Fort Walla Walla where the river flows down from the north and makes a great bends westward. One species is dappled with whitish spots, the other is a uniform grey-colour. For one species, there is a specimen and no Audubon painting. For the other, there is a painting but no surviving specimen.

Townsend's Ground Squirrel

Townsend's Ground Squirrel inhabits a relatively small area in southeast Washington on the western/northern side of the Columbia River. It is plainly marked, being undappled and lacking any black tip to the tail. There is a single Townsend specimen (ANSP 344) that lacks any accompanying data. However, Townsend tells us that it was perhaps[a] found on his way upriver on his way to the Blue Mountains:

"TOWNSEND'S MARMOT ... I procured a single specimen of this animal on the Columbia river, about three hundred miles above its mouth, in July [1836]. It was said to be common there at that season, but as I was travelling in boats to the interior, had but little time to search for it. I know but little of its habits. Disappears in August, and emerges in the spring in a very attenuated state" (Townsend 1839a: 316-317).

Scheffer (1946) says that Townsend's *Narrative* indicates that collecting took place around Fort Walla Walla on the eastern/southern side of the Columbia and does not mention any trips to the western/northern bank of the river. But Scheffer did not examine Townsend's MS version of the journal that mentions a single crossing of the Columbia River, on 16th August 1836, in order to collect specimens. Townsend could in any case have collected it on the western/northern bank of the Columbia as he made his way upriver with the summer brigade, a few weeks earlier.

The single surviving specimen of Townsend's Ground Squirrel collected by Townsend[b] cannot be the specimen that went to Bachman and then to Audubon and his son J.W. Audubon. The specimen that was portrayed by Audubon is a different species and not a Townsend's Ground Squirrel after all. It is a Washington Ground Squirrel.

▶

Washington Ground Squirrel depicted in *The Viviparous Quadrupeds*

Townsend also collected a Washington Ground Squirrel, a closely related species that inhabits only the eastern/southern side of the Columbia River, including the area around Fort Walla Walla. It has a dappled coat and a black tip to the tail and it sounds as though Bachman was referring to this species when he first described Townsend's Ground Squirrel. Why else would he say it was singularly marked when Townsend's Ground Squirrel is particularly uniform in colour?

"Townsend's Marmot [Townsend's Ground Squirrel] … This singularly marked and, I conceive, new species, is another of the discoveries of our indefatigable countryman, Mr. Townsend. In a letter to me he states, "That it inhabits, in the summer, the prairies near the Walla-walla; is rather common. It becomes excessively fat, and is eaten by the Indians. It disappears in August, and reappears early in spring, in a very attenuated state" (Bachman 1839b: 61).

Townsend's Ground Squirrel was named "American Souslik" on plate CXLVII (detail at left, fig 1) of Audubon's *Viviparous Quadrupeds*. The specimen is a Washington Ground Squirrel (ANSP 344) – see text.

215

Examination of figure 1 in plate CXLVII, painted by J.W. Audubon and said to be "Townsend's Ground Squirrel" clearly shows that *the specimen is dappled* and it must therefore be a Washington Ground Squirrel, the only dappled ground squirrel of similar appearance inhabiting that area. It is true that the figure in the painting should have a black tip to the tail but the specimen may have lost the hairs at the tip. The text that accompanies the plate describes the specimen as "speckled with white all over the back". Moreover, Bachman says in his general remarks that this specimen had "white and irregular specks".

To summarise, Townsend appears to have collected both undappled Townsend's Ground Squirrel (for which one specimen survives) and the dappled Washington Ground Squirrel (for which no specimen can now be traced). It was the latter species that was painted by J.W. Audubon in *The Viviparous Quadrupeds* and described in the text by Bachman.

a. The portion of Townsend's text that follows may refer to the Washington Ground Squirrel.

b. Townsend's Ground Squirrel has recently been split into three species: *townsendii* referring to the population in south central Washington (i.e. only on the western/northern side of the Columbia River), *canus* the Merriam's or Columbia Plateau Ground Squirrel, and *mollis* the Piute or Great Basin Ground Squirrel. Specimen ANSP 344 is labelled as Townsend's Ground Squirrel. It has not been identified either to race or any of the proposed new species. It is possible that it is not the nominate race and might have been collected as far eastwards as Idaho.

A trip to the Blue Mountains

25th.– I mounted my horse this morning for a journey to the Blue mountains. I am accompanied by a young half breed named Baptiste Dorion, who acts as guide, groom, interpreter, &c., and I have a pack horse to carry my little *nick-nackeries*. We shaped our course about N.E. [= S.E; there are no mountains 30-50 miles to north east.] over the sandy prairie, and in the evening encamped on the Morro river, having made about thirty miles. On our way we met two Walla-walla Indians driving down a large band of horses. They inform us that the Snakes have crossed the mountain to commence their annual thieving of horses, and they are taking them away to have them secure. I shall need to keep a sharp look out to my own small caravan, or I shall be under the necessity of turning pedestrian.

Sitka Mountain Ash *Sorbus sitchensis*.

July 26th.– At noon, to-day, we arrived at the Utalla, or Emmitilly [Umatilla] river, where we found a large village of Kayouse Indians, engaged in preparing kamas. Large quantities of this root were strewed about on mats and buffalo robes; some in a crude state, and a vast quantity pounded, to be made into cakes for winter store. There are of the Indians, about twelve or fifteen lodges. A very large one, about sixty feet long by fifteen broad, is occupied by the chief, and his immediate family. This man I saw when I arrived at Walla-walla, and I have accepted an invitation to make my home in his lodge while I remain here. The house is really a very comfortable one; the rays of the sun are completely excluded, and the ground is covered with buffalo robes. There are in the chief's lodge about twenty women, all busy as usual; some pounding kamas, others making leathern dresses, moccasins, &c. Several of the younger of these are very good looking, – I might almost say handsome. Their heads are of the natural form, – not flattened and contorted in the horrible manner of the Chinooks …

The next morning [27th in MS Journal], my friend the chief furnished me with fresh horses, and I and my attendant, with two Indian guides, started for a trip to the mountain. [*As we passed up one of the gorges we saw great numbers of the* **Sharp-tailed Grous** *lurking among the willows & occasionally alighting, - when disturbed, - in the Cottonwood trees* (MS Journal).] We passed up one of the narrow valleys or gorges which here run at right angles from the alpine land, and as we ascended, the scenery became more and more wild, and the ground rough and difficult of passage, but I had under me one of the finest horses I ever rode; he seemed perfectly acquainted with the country; I had but to give him his head, and not attempt to direct him, and he carried me triumphantly through every difficulty. Immediately as we reached the upper land, and the pine trees, we saw large flocks of the **dusky grouse**, (*Tetrao obscurus,*) a number of which we killed. Other birds were, however, very scarce. I am at least two months too late, and I cannot too much regret the circumstance. Here is a rich field for the ornithologist at the proper season. [*We started a* **Black-tailed Deer** *on the mountain but were not prepared to kill it; also two Foxes, one of them, the beautifully & highly esteemed variety called* Silver [**Red Fox, black phase**] (MS Journal).] We returned to our lodge in the evening loaded with grouse, but with very few specimens to increase my collection[3].

29th.– Early this morning our Indians struck their lodges, and commenced making all their numerous movables into bales for packing on the horses. I admired the facility and despatch with which this was done; the women alone worked at it, the men lolling around, smoking and talking, and not even once directing their fair partners in their task. The whole camp travelled with me to Walla-walla, where we arrived the next day.

Return to Fort Walla Walla

[There is no entry in the *Narrative* for any date in August 1836 despite several long entries in his journal:

[*Aug. 10th. I am this day 27 years of age. This is in all probability the last birth day I shall pass in this Country. When my next arrives I hope to be in my own happy & blessed land surrounded by our beloved family & friends. It is now 2 years & 5 months since I left home & nearly two years since I arrived at the Columbia* (MS Journal).

12th. This morning I crossed the river with an indian to hunt for the large Grous (T. urophasianus) [**Greater Sage-Grouse**]*. After walking about 3 miles I found a small pack of 6 birds from which I killed one. I soon after saw several more & shot in all 4 birds besides several which escaped wounded. This was in consequence of my shot being too small. It is a remarkably strong bird & is able to carry away a large quantity of small, - or even large, - shot, & so poorly was I provided that unless the bird flushed very close to me I was not able to kill it. As an instance of their uncommon tenacity of life Mr. Pambrun says that he has frequently shot them with a rifle while sitting & that they have not unfrequently risen & flown an hundred or more yards after the ball had passed entirely through the body. It is a particularly laborious sort of shooting as there are no hunting dogs here & at the first rise the game often flies for a mile or more before it rests, so that it is generally impossible to mark it down with any accuracy & the walking on the plains covered with wormwood is difficult & tiresome. With a brace of good pointers the difficulty would be very much lessened & the sport would be excellent, but it would always be necessary to have a game carrier in consequence of the great weight of the birds. Three of those that I procured to day are young & moulting, but the fourth is an adult male, the plumage very nearly perfect, a most beautiful fellow & a far better specimen than I expected to get at this season. He measures 32 inches in length & weighs 6lbs . *[JKT footnote: *This bird is very poor; when fat they often exceed 8lbs.] There is a very small species of Hare inhabits the prairies in the neighbourhood which I have sometimes seen in my excursions, but it is not plentiful & very shy, so that I have not been able to get it* [**Mountain Cottontail** and/or **Pygmy Rabbit**]*. The large Hare (Lepus variabilis?)* [**White-tailed Jackrabbit**] *is very abundant* (MS Journal).

13th. ... This morning an indian brought me a living Badger which I killed & skinned. I value the present specimen[1] more than those I have heretofore had, on acct. of the skull being perfect. Those killed by the indians invariably have the crania broken to fragments. They are generally killed with the handle of their riding whips. The animal is caught at a distance from his burrow, & as he does not run rapidly is easily overtaken & knocked on the head (MS Journal).]

AMERICAN BADGER

[16th. I crossed the Columbia this morning with an indian & spent most of the day shooting the large Grous [**Greater Sage-Grouse**]. I saw 5 or 6 packs, in all perhaps 30 or 40 birds & bagged eleven. This, although it does not sound large in the way of slaughtering game, is yet, everything considered, doing tolerably well. The birds almost invariably fly so far at the first rise that it is impossible to mark them, & alighting on the broad & monotonous plain, one might search the whole day, (without a dog,) & not fall upon them. In several instances I killed two at the first flushing, & once or twice a single bird that happened to be separated from his companions. My poor indian lad no doubt thought that I had enough game as he staggered under his load, & before we reached home was almost worn out. The bird is very little inferior in size to the domestic Turkey, & 10 or 12 form as much load as a man wishes to carry in this warm weather*. (As a footnote: *These dry & uninteresting details will probably only weary my beloved sister & perhaps will be considered very stupid & tiresome by all who happen to read them; my only excuse is that at the time they were written I had no occupation more interesting than shooting birds & then scribbling about them) (MS Journal).

20th. I collected a party of 12 indians this morning & crossed the [Walla Walla?] river to hunt the Small Hares. The Indians had no other weapons but their small bows & arrows but with these they exhibited great dexterity. Immediately as we landed, several of the animals were seen & almost as soon as seen, were killed. We recrossed the river about noon with 9 small & one large Hare most of them captured by the indians. During the remainder of the day & evening I was busied in preparing my specimens (MS Journal).]

American Badger, plate XLVII (detail). Although Audubon and Bachman say they examined the American Badger skulls and skins collected by Townsend, Audubon does not actually say if he used them in the preparation of this plate.

Burrowing Owls, plate CCCCXXXIII (detail). The owls used as models for the plate may or may not have been from Townsend, Audubon does not say. Townsend remarks that "This species inhabits the plains near the Columbia River and the whole extent of the Rocky Mountains, residing in the forsaken burrows of the Marmots [= prairie dogs and ground squirrels] and American Badgers ... Nothing can be more unpleasant than the bagging of this species, on account of the fleas with which their plumage swarms, and which in all probability have been left in the burrow by the Badger or Marmot, at the time it was abandoned by these animals. I know of no other bird infested by that kind of vermin. The species suddenly disappears in the early part of the month of August, and the Indians assert with great confidence that it retires into its burrow, and spends the winter there in a torpid state" (Audubon 1840-44, 1: 119).

MOUNTAIN COTTONTAIL

Under the name "Wormwood Hare" Townsend gave a detailed account of cottontail hunting: "This small hare inhabits the wormwood plains near the banks of the streams in the neighborhood of Fort Walla-walla. I cannot define its range with any degree of certainty, but I have reason to believe that it is very contracted, never having met with it many miles from this locality. It is here abundant, but very shy and retired, keeping constantly in the densest wormwood bushes, and leaping with singular speed from one to another when pursued. I have never seen it dart away and run to a great distance like other hares. I found it very difficult to shoot this animal, for the reasons stated. I had been residing at Fort Walla-walla for two weeks [in July and August 1836], and had procured only two, when, at the suggestion of Mr. Pambrun, I collected a party of a dozen Indians, armed with bows and arrows, and sallied forth. We hunted through the wormwood within about a mile of the fort, and in a few hours returned, bringing eleven hares. The keen eyes of the Indians discovered the little creatures squatting under the bushes, where, to a white man, they would have been totally invisible. This hare, when wounded and taken, screams like our common species" (Townsend 1839a: 328-329, and quoted in Bachman 1839c: 94-96).

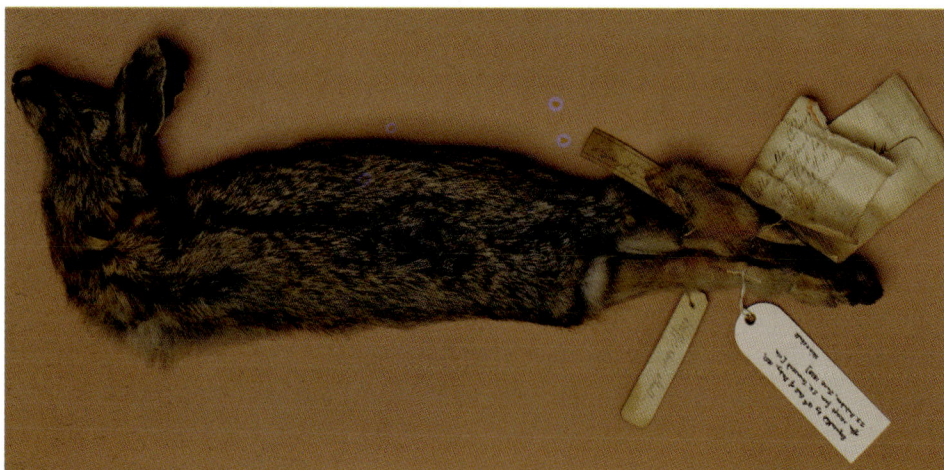

WHITE-CROWNED SPARROW

Townsend collected a single male White-crowned Sparrow "towards the close of August [1836?], in the willow thickets bordering the Columbia, near Fort Wallah-Wallah". This was named as a new species, Gambel's Finch, by Nuttall (1840: 556-557), but is now the *gambelii* race of White-crowned Sparrow. The type specimen has been lost and the exact date of its collection is not known. Nuttall was rather vague about dates so it is possible that it was collected in early September 1834, on their first arrival at Fort Walla Walla and the Columbia River.

Sept. 1st.– Mr. John M'Leod, a chief trader of the Hudson's Bay Company, arrived this morning from the rendezvous, with a small trading party. I had been anxiously expecting this gentleman for several weeks, as I intended to return with him to [Fort] Vancouver. He is accompanied by several Presbyterian missionaries, the Rev. Mr. Spalding and Dr. Whitman, with their wives, and Mr. Gray, teacher. Doctor Whitman presented me with a large pacquet of letters from my beloved friends at home. I need not speak of the emotions excited by their reception, nor of the trembling anxiety with which I tore open the envelope and devoured the contents. This is the first intelligence which I have received from them since I left the state of Missouri, and was as unexpected as it was delightful. [*The pacquet contained about a dozen letters, the most of them from our dear family, & also remembrances from my friends Doct. Morton, B. Leedom, J.C. Allen, & N.P. Gibbons. My beloved sister H. – the excellent girl – has furnished me with a complete history of the chief events which have transpired in our little circle ever since I left it* (MS Journal).]

Mr. M'Leod informed me of the murder of Antoine Goddin, the half-breed trapper, by the Blackfeet Indians, at Fort Hall … We also hear, that three of Captain Wyeth's men who lately visited us, had been assaulted on their way to Fort Hall, by a band of Walla-walla Indians, who, after beating them severely, took from them all their horses, traps, ammunition, and clothing …

I have had this evening, some interesting conversation with our guests, the missionaries. They appear admirably qualified for the arduous duty to which they have devoted themselves, their minds being fully alive to the mortifications and trials incident to a residence among wild Indians; but they do not shrink from the task, believing it to be their religious duty to engage in this work. The ladies have borne the journey astonishingly; they look robust and healthy. [*Mrs. Whitman has borne the journey astonishingly. She is singularly robust & healthy looking. Her person is above the middle height, commanding & noble, & she has the soul of a giantess. I had a long, & to me a most interesting conversation with her last evening in private. The poor girl described the heart-rending scene of her*

Narcissa Whitman and her husband, Marcus, arrived at Fort Walla Walla in 1836 and established a Protestant mission among the nearby Cayuse Indians. Their daughter was the first white child born in the northwest. In 1847 an epidemic of measles (brought by overland emigrants from the east) wiped out half the tribe, and convinced that they were being deliberately poisoned, Cayuse warriors attacked the mission killing the Whitmans and eleven others.

parting with a large & interesting family & circle of friends, knowing too that the farewell was to be eternal. Oh! how inexpressibly trying that parting must have been. I am deeply interested in this lady. I never met with one who at first sight so completely enlisted my feelings, & if perfect strangers can so love her, what must she have been to her family, father, mother & eight brothers & sisters (MS Journal).]

On the same day, Narcissa Whitman wrote: "Soon the Fort appeared in sight and when it was announced that we were near Mr. McLeod, Mr. Pambrun, the gentleman of the house, and Mr. Townsend (a traveling naturalist) sallied forth to meet us. After usual introduction and salutation we entered the Fort and were comfortably seated in cushioned armed chairs. They were just eating breakfast as we rode up and soon we were seated at the table and treated to fresh salmon, potatoes, tea, bread and butter. What a variety, thought I" (Whitman journal, 1st September 1836).

Downriver to Fort Vancouver

3d.– Mr. M'Leod and myself embarked in a large bateau, with six men, and bidding farewell to Mr. Pambrun and the missionaries, were soon gliding down the river. We ran, to-day, several rapids, and in the evening encamped about fifteen miles below the mouth of the Utalla river …

We are enjoying a most magnificent sight at our camp this evening. On the opposite side of the river, the Indians have fired the prairie, and the whole country for miles around is most brilliantly illuminated. Here am I sitting cross-legged on the ground, scribbling by the light of the vast conflagration with as much ease as if I had a ton of oil burning by my side; but my eyes are every moment involuntarily wandering from the paper before me, to contemplate and admire the grandeur of the distant scene. The very heavens themselves appear ignited, and the fragments of ashes and burning grass-blades, ascending and careering about through the glowing firmament, look like brilliant and glorious birds let loose to roam and revel amid this splendid scene. It is past midnight: every one in the camp is asleep, and I am this moment visited by half a dozen Indian fishermen, who are peering over my shoulders, and soliciting a smoke, so that I shall have to stop, and fill my calamet.

*[5th.– Indians are numerous along the river. They are still engaged in fishing. As we pass along we frequently see them posted upon the rocks overhanging the water intently surveying the boiling & roaring flood below for the passing Salmon. In most instances an indian is seen entirely alone in these situations, often standing for ½ an hour perfectly still, his eyes rivetted upon the torrent & his long fish spear poised above his head. The appearance of a solitary savage thus perched like an eagle upon a cliff is sometimes, when taken in connexion with the wild & rugged river scenery, very picturesque. The spear is a pole about 12 feet long at the end of which a long fork is made fast, & between the tines is fixed a barbed iron point. They also in some situations use a hand scoop net & stand upon scaffolding ingeniously constructed over the raging torrents. Their winter store of dried fish is stored away in little huts of mats & branches closely interlaced, & also in caches under ground. It is often amusing to see the hungry ravens [**Common Raven**] teasing & tugging at the strong twigs of the houses in a vain attempt to reach the savory food within (MS Journal).]* … In the afternoon, we passed John Day's river, and encamped about sunset at the "shoots." Here is a very large village of Indians, (the same that I noticed in my journal, on the passage down,) and we are this evening surrounded by some scores of them.

6th.– We made the portage of the shoots this morning by carrying our boat and baggage across the land, and in half an hour, arrived at one of the upper dalles. Here Mr. M'Leod and myself debarked, and the men ran the dall … we soon embarked again, and proceded to the lower *dalles*. Here it is utterly impossible, in the present state of the water, to pass, so that the boat and baggage had to be carried across the whole portage. This occupied the remainder of the day, and we encamped in the evening at a short distance from the lower villages. The Indians told us with sorrowful faces of the recent death of their principal chief, Tilki. Well, thought I, the white man has lost a friend, and long will it be before we see his like again …

We see great numbers of seals [**Harbor Seal**] as we pass along. Immediately below the Dalles they are particularly abundant, being attracted thither by the vast shoals of salmon which seek the turbulent water of the river. We occasionally shoot one of them as he raises his dog-like head above the surface, but we make no use of them; they are only valuable for the large quantity of oil which they yield …

At 11 o'clock next day [7th] we arrived at the cascades, where we made the long portage, and at nine in the evening encamped in an ash grove, six miles above *Prairie de Thé*.

On the 8th, reached [Fort] Vancouver, where we found two vessels which had just arrived from England.

Narcissa Whitman followed a few days later and gave some details not mentioned by Townsend: "We are now in [Fort] Vancouver, the New York of the Pacific Ocean. Our first sight, as we approached the fort, was two ships lying in the harbor, one of which, the Neriade, Captain Royal, had just arrived from London. The Columbia, Captain Dandy, came last May, and has since been to the Sandwich Islands, and returned. On landing we first met Mr. Townsend, whom we saw at Walla Walla. He is from Philadelphia, and has been in the mountains two years. He is sent here by a society to collect the different species of bipeds, and quadrupeds, peculiar to this country. We brought a parcel of letters to him, and the first he had received since he had left home. Mr. Townsend led us into the fort. But before we reached the home of the chief Factor, Dr. McLoughlin, we were met by several gentlemen, who came to give us a welcome, Mr. [James] Douglas, Dr. Tolmie and Dr. McLoughlin, of the Hudson's Bay Company, who invited us in and seated us on the sofa. Soon we were introduced to Mrs. McLoughlin and Mrs. Douglas, both natives of the country – half breeds. After chatting a little we were invited to walk in the garden.

What a delightful place this is … Here we find fruit of every description …" (Whitman journal, 12th September 1836).

SPECIMENS AND OBSERVATIONS FROM FORT VANCOUVER, 8th – 24th SEPTEMBER 1836

Townsend was at the fort from 8th to 24th September. There are only two records of specimens from this period. On 22nd September 1836 he collected a **Golden-crowned Sparrow** (ANSP 24067, see Stone 1899). Although not a new species this specimen was the type of Nuttall's "Yellow-crowned Finch *Fringilla aurocapilla*" (Nuttall 1840: 555-556). He also secured a **Gray Wolf** "captured by me in the month of September, 1836, on the plains of the Columbia river, one and a half or two miles west of Fort Vancouver" (Townsend 1850). There are four wolf specimens at ANSP but no data to link any of them with this particular wolf (Koopman 1976: 21).

Townsend makes a trip to Fort George, Young's Bay and Cape Disappointment

On the 24th, I embarked in a canoe with Indians for Fort George, and arrived in two days. Here I was kindly received by the superintendent, Mr. James Birnie, and promised every assistance in forwarding my views.

[*27th. This morning I walked with Mr. B. along the beach & found a number of fossil shells of several species, Pecten [Pectin], Nautilus, Territella [Turritella] &c., imbedded in the stones. These shells are evidently very ancient; none of the species are at present found there in a recent state. Birds appear to be rather scarce in this neighbourhood: there are none except a few Pelicans* [**Brown Pelican**], *Shags* [**cormorant species**], *Gulls &c* (MS Journal).]

Brown Pelicans on rocks,
Cape Disappointment.

30th.– I visited to-day some cemeteries in the neighborhood of the fort, and obtained the skulls of four Indians. Some of the bodies were simply deposited in canoes, raised five or six feet from the ground, either in the forks of trees, or supported on stakes driven into the earth. In these instances it was not difficult to procure the skulls … There are but few Indians here, and I do not therefore incur much risk; were it otherwise, there would be no little danger in these aggressions … [*The corpses of several different tribes which are buried here are known by the difference in the structure of the canoes, & the <u>sarcophagi</u> of the chiefs from those of the common people, by the greater care which is manifested in the arrangement of the tomb. Among the skulls I procured to day is one a <u>Killemook</u> [Tillamook] chief, (a most gigantic fellow he must have been from the enormous size of his head,) one of the <u>Clatsap</u> [Clatsop], & one of a <u>Chinook</u>[5]* (MS Journal).]

Cape Disappointment

SPECIMENS AND OBSERVATIONS FROM NEAR THE MOUTH OF THE COLUMBIA RIVER, 24TH SEPTEMBER – 5TH NOVEMBER 1836.

30th September 1836, **Mountain Quail**, NYHS 1863.17.423 (Not personally collected by JKT).

6th October 1836, **California Gull**, USNM 2772, Cape Disappointment.

6th October 1836, **Western Gull**, USNM 2769, card catalogue, Cape Disappointment.

7th October 1836, **Western Gull**, USNM 2767, card catalogue, Cape Disappointment.

8th October 1836, **Brandt's Cormorant**, USNM 2742, Cape Disappointment.

11th October 1836, **Gray Jay**, ANSP 162471, ex Harris collection (Street 1948: 182.)

21st October 1836, **Black-bellied Plover**, USNM 2775.

22nd October 1836, **Ring-billed Gull**, USNM 2771.

29th October 1836, **Bushytail Woodrat**, LM D353a (see page 149).

-- Oct 1836, **Pelagic Cormorant**, USNM 2004. This must be the specimen painted by Audubon (Plate CCCCXII) because Townsend says he only collected one of these cormorants.

2nd November 1836, **Oregon Vole**, the type specimen described by Bachman (1839b: 60-61) is now missing (Koopman 1976).

"November 1836", **Surfbird**, AMNH 156652, collected "on the base of the rocky cape at the entrance of the Columbia river", i.e. Cape Disappointment (Townsend 1839: 349-350).

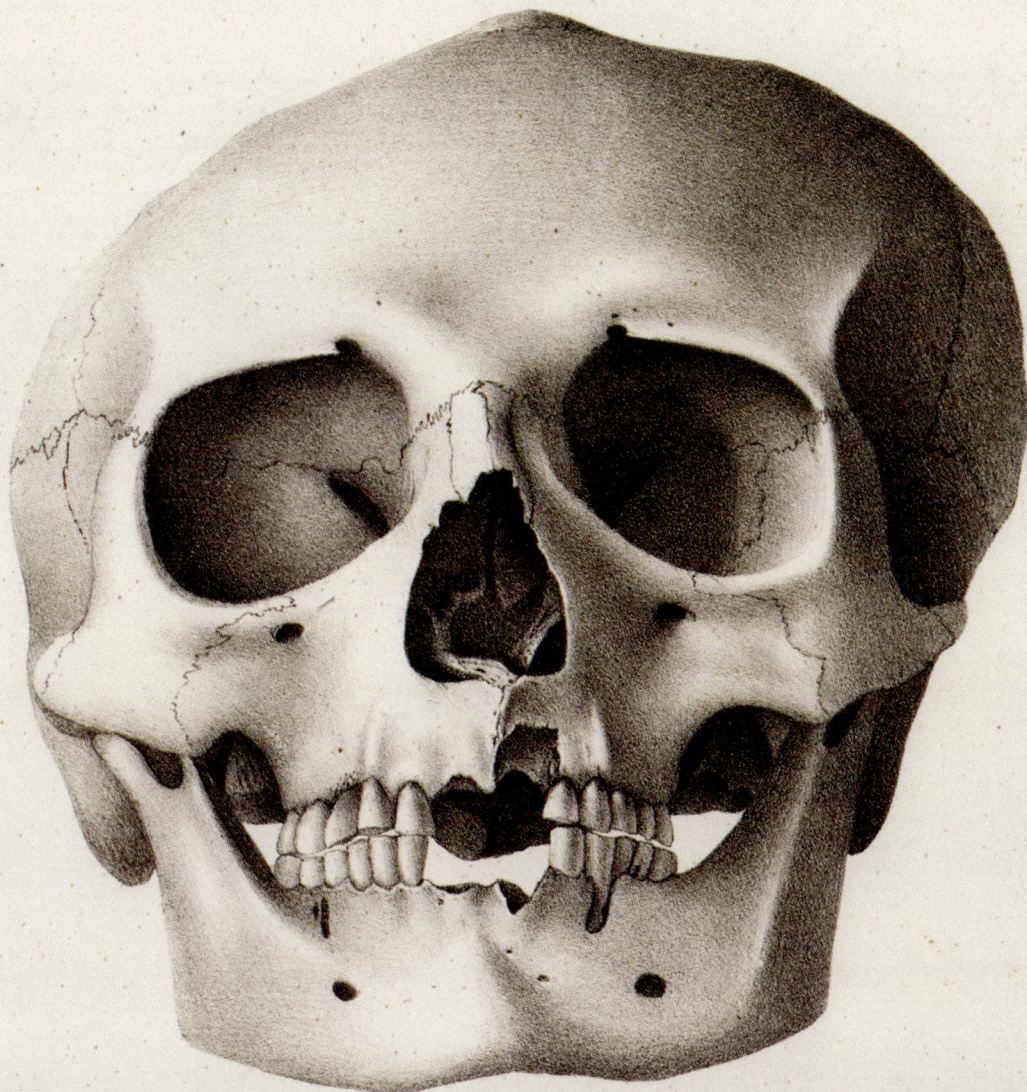

CLATSAP

FROM COLUMBIA RIVER

Drawn from Nature and on Stone by J. Collins

Head flattening of infants was practised by many of the tribes on the lower Columbia River. This misshapen skull of an adult Clatsop Indian was collected by Townsend and presented to his Philadelphia friend Dr Samuel Morton who included it in his *Crania Americana* (1839) as plate 46.

MOUNTAIN QUAIL

Only the female specimen was supplied by Townsend and there is some doubt about whether he collected it himself or not. In his text Audubon says "The following notice by Mr. TOWNSEND shews that it is entitled to a place in our Fauna. "This bird inhabits the dense woods along the tributary streams of the Columbia river, and is said to extend south into California [Its range extends from Washington to Baja California]. It is at all times a very scarce species, going in coveys of from six to ten, and is rarely seen away from its favourite places of resort. In all my rambles through the Oregon country I was never so fortunate as to meet with this pretty bird, the three specimens which I have received having been procured for me by others."

One of these specimens has been forwarded to me by Mr. TOWNSEND, and as it proved a female, I made a drawing of the male from a superb specimen now in the Museum of the Zoological Society of London" (Audubon 1840-44, 5: 69).

Both Nuttall and Audubon contradict Townsend's statement: Nuttall (1840: 791) says "Mr. Townsend met with small coveys of this fine species of 10 to 15 individuals each, in the woods of the Wahlamet, not far from the Columbia. It appears here to be a very scarce species." And Audubon's original watercolour of the Mountain Quail (NYHS 1863.17.423) is inscribed " … (female) / was shot 9th Mo 30th 1836 / by J.K. Townsend." On that date Townsend was at Fort George (Astoria) near the mouth of the Columbia. The only known Mountain Quail specimen that can be linked to Townsend is at Washington (USNM 2831, no locality or date with specimen).

BRANDT'S CORMORANT AND PELAGIC CORMORANT

Brandt's Cormorant (at right). Unknown to Audubon, who wanted to name this cormorant after Townsend, the species had already been described and named by Johann Friedrich von Brandt, in 1838, from a specimen in the zoology museum at the Academy of Sciences at St Petersburg. It had been collected by a Russian visitor to the Pacific Northwest (Mearns and Mearns 1992: 109-113). Audubon named it "TOWNSEND'S CORMORANT" adding that "Two specimens of the Cormorant here represented were sent to me by Mr. TOWNSEND, who procured them at Cape Disappointment in the beginning of October, 1836. They are both marked as males …" (Audubon 1840-44, 6: 438-439). Townsend observed that "It is seldom seen near the sea, but is mostly observed high upon the river. It is, like most species of its genus, particularly gregarious, and is fond of resting in company … It is very shy and cautious, and is seldom killed even by the Indians, who are fond of its flesh" (Townsend 1839a: 351-352). There is a specimen at Washington dated 8th October 1836 (USNM 2742).

Pelagic Cormorant (at left). Audubon thought that this was a new species but it later proved to have been described by P.S. Pallas in 1811 from specimens from the Aleutian Islands. Audubon named it "VIOLET-GREEN CORMORANT" noting in his text that "This Cormorant, the most beautiful hitherto found within the limits of the United States, was obtained by Mr. TOWNSEND[6] at Cape Disappointment, near the entrance of the Columbia river. The specimen from which the figure in the plate was taken, was transmitted to me by that zealous student of Nature. The beautiful gloss of its silky plumage suggested the specific name which I have given to it" (Audubon 1840-44, 6: 440). Townsend agreed with Audubon, remarking that "This most splendid of all the species of cormorants yet discovered, inhabits in considerable numbers the Rocky Cape at the entrance of the Columbia river … The procuring of the only specimen which I was ever enabled to kill, almost cost the lives of myself and eight men. Our boat was carried with frightful velocity into the furious breakers, and a full hour was consumed in unremitting efforts to escape the danger towards which the swift current was hurrying us. (Townsend 1839a: 350-351).

Townsend's only specimen, which must therefore be the one represented above, is at Washington (USNM 2004).

MARBLED MURRELET

Audubon gave this bird the name "SLENDER-BILLED GUILLEMOT URIA TOWNSENDII" adding that "I have received not less than four specimens of this small Guillemot from Mr. TOWNSEND, who procured them on the north-west coast of America, not very far from the mouth of the Columbia river. The changes of colour in birds of this genus are well known to be considerable; and I have represented two individuals, supposing one to be an adult, and the other a young bird in its first plumage" (Audubon 1840-44, 7: 278). Townsend merely observes that it "Inhabits the bays of the N.W. Coast of America, in latitude 38° to 40°. The specimens were shot and presented to me by Captain W. Brotchie, to whom I am under very great obligations for the addition to my collection of several fine species" (Townsend 1839a: 352).

Western Gull.

WESTERN GULL

Audubon was the first to describe the Western Gull, though he gave no reason for not painting the specimens and including them in *The Birds of America*. The second specimen that Audubon refers to below (unless there were more gulls collected on 6th October 1836) has subsequently been identified as a California Gull (a species not officially described until 1854, by George N. Lawrence, from a specimen from Stockton, California):

"WESTERN GULL. LARUS OCCIDENTALIS, *Aud*. (NOT FIGURED.) Two specimens of this Gull have been sent to me by Mr. Townsend. One of them, an adult, is marked, "Male, Cape Disappointment, October 7th, 1836," the other, a young bird, "Young, Male, Cape Disappointment, October 6th, 1836." The iris of both is stated to have been light hazel. This species, which I presume to be undescribed, as I have not met with any account of it, is about equal to *Larus marinus* [Great Black-backed Gull *L. marinus*] in size, and resembles *L. argentatus* [Herring Gull *L. argentatus*] in colour, but differs from both in many respects, as will be seen from the annexed description …" (Audubon 1840-44, 7: 161).

October 14th. – I walked to-day around the beach to the foot of Young's bay, a distance of about ten miles, to see the remains of the house [Fort Clatsop] in which Lewis and Clark's party resided during the winter which they spent here. The logs of which it is composed, are still perfect, but the roof of bark has disappeared, and the whole vicinity is overgrown with thorn and wild currant bushes.

One of Mr. Birnie's children found, a few days since, a large silver medal, which had been brought here by Lewis and Clark, and had probably been presented to some chief, who lost it. On one side was a head, with the name "Th. Jefferson, President of the United States, 1801." On the other, two hands interlocked, surmounted by a pipe and tomahawk; and above the words, "Peace and Friendship."

15th.– This afternoon I embarked in a canoe with *Chinamus*, and went with him to his residence at Chinook [on the north bank of the Columbia River, perhaps a little to the east of today's Chinook.] The chief welcomed me to his house in a style which would do no discredit to a more civilized person. His two wives were ordered to make a bed for me, which they did by piling up about a dozen of their soft mats, and placing my blankets upon them, and a better bed I could not wish for. I was regaled, before I retired, with sturgeon, salmon, wappatoos, cranberries, and every thing else that the mansion afforded, and was requested to ask for any thing I wanted, and it should be furnished me. Whatever may be said derogatory to these people, I can testify that inhospitality is not among the number of their failings. I never went into the house of an Indian in my life, in any part of the country, without being most cordially received and welcomed ...

I remained here several days, making excursions through the neighborhood, and each time when I returned to the lodge, the dogs growled and darted at me. I had no notion of being bitten, so I gave the Indians warning, that unless the snarling beasts were tied up when I came near, I would shoot every one of them. The threat had the effect desired, and after this, whenever I approached the lodges, there was a universal stir among the people, and the words, "*iskam kahmooks, iskam kamooks, kalak'alah tie chahko*," (take up your dogs, take up your dogs, the *bird chief* is coming,) echoed through the village, and was followed by the yelping and snarling of dozens of wolf-dogs, and "curs of low degree," all of which were gathered in haste to the cover and protection of one of the houses.

[*16th. I visited the Cape [Disappointment] to day, but had very little success* (MS Journal, entire entry for this day).]

An excursion to Willapa Bay

October 17th.– I left Chinook this morning in a canoe with Chinamus, his two wives, and a slave, to procure shell-fish, which are said to be found in great abundance towards the north. We passed through a number of narrow *slues* which connect the numerous bays in this part of the country, and at noon debarked, left our canoe, took our blankets on our shoulders, and struck through the midst of a deep pine forest. After walking about two miles, we came to another branch, where we found a canoe which had been left there for us yesterday, and embarking in this, we arrived in the evening at an Indian house, near the sea side, where we spent the night [at the south end of Willapa Bay, still renowned for its oysters, clams, crabs, salmon and steelheads].

In our passage through some of the narrow channels to-day, we saw vast shoals of salmon, which were leaping and curvetting about in every direction, and not unfrequently dashing their noses against our canoe, in their headlong course. We met here a number of Indians engaged in fishing. Their mode of taking the salmon is a very simple one. The whole of the tackle consists of a pole about twelve feet long, with a large iron hook attached to the end. This machine they keep constantly trailing in the water, and when the fish approaches the surface, by a quick and dexterous jerk, they fasten the iron into his side, and shake him off into the canoe. They say they take so many fish that it is necessary for them to land about three times a day to deposit them ...

The original Fort Clatsop where Lewis and Clark spent the winter of 1805-06 has long since disintegrated. These logs are part of a reconstructed fort that now stands on the same site.

Willapa Bay.

Sand Dollar *Dendraster excentricus* (top) and Pacific Razor Clam *Siliqua patula* (below).

18th.– Last night the wind rose to a gale, and this morning it is blowing most furiously, making the usually calm water of these bays so turbulent as to be dangerous for our light craft. Notwithstanding this disadvantage, the Indians were in favor of starting for the sea, which we accordingly did at an early hour. Soon after we left, in crossing one of the bays, about three-quarters of a mile in width, the water suddenly became so agitated as at first nearly to upset our canoe. A perfect hurricane was blowing right ahead, cold as ice, and the water was dashing over us, and into our little bark, in a manner to frighten even the experienced chief who was acting as helmsman. In a few minutes we were sitting nearly up to our waistbands in water, although one of the women and myself were constantly bailing it out, employing for the purpose the only two hats belonging to the party, my own and that of the chief. We arrived at the shore at length in safety, although there was scarcely a dry thread on us, and built a tremendous fire with the drift-wood which we found on the beach. We then dried our clothes and blankets as well as we could, cooked some ducks that we killed yesterday, and made a hearty breakfast. My stock of bread, sugar, and tea, is completely spoiled by the salt water, so that until I return to Fort George, I must live simply; but I think this no hardship: what has been done once can be done again.

In the afternoon the women collected for me a considerable number of shells, several species of *Cardium* [**cockle spp**.], *Citherea* [**Cytheria spp**.], *Ostrea* [**California Oyster** *Ostrea lurida*], &c., all edible, and the last very good, though small.

The common pintail duck, (*Anas acuta*,) [**Northern Pintail**] is found here in vast flocks. The chief and myself killed *twenty-six* to-day, by a simultaneous discharge of our guns. They are exceedingly fat and most excellent eating; indeed all the game of this lower country is far superior to that found in the neighborhood of [Fort] Vancouver. The ducks feed upon a small submerged vegetable [Eelgrass, *Zostera sp.*] which grows in great abundance upon the reedy islands in the vicinity.

The next day [19th] we embarked early, to return to Chinook. The wind was still blowing a gale, but by running along close to the shore of the stormy bay, we were enabled, by adding greatly to our distance, to escape the difficulties against which we contended yesterday, and regained the slues with tolerably dry garments. [*This morning I observed some immense swarms of a bird which I have never before seen in this country: A Godwit* [**Marbled Godwit**], *of a species with which I am not unacquainted [having seen them on the east coast]. The great flocks of this bird reminded me forcibly of the flights of the Reed Buntings [Bobolink] which I have seen on the Delaware & Schuylkill. I do not wish to exaggerate, but in many*

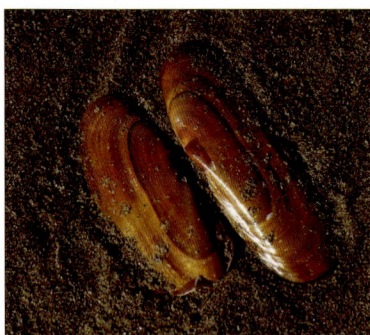

instances they actually darkened the air & as soon as our canoe passed the sands on which they were resting, the whole body would rise <u>en masse</u> with a roar like the dying cadences of heavy artillery. At the time when the great body of these birds was visible the wind was so high that I was not willing to risk deviation from our course to procure any, but in the afternoon I killed several stragglers that passed us (MS Journal).]

At about 10 o'clock, we arrived at the portage, and struck into the wood, shouldering our baggage as before. We soon came to a beautiful little stream of fresh water, where we halted, and prepared our breakfast. In this stream, (not exceeding nine feet at the widest part,) I was surprised to observe a great number of large salmon. Beautiful fellows, of from fifteen to twenty-five pounds weight, darting and playing about in the crystal water, and often exposing three-fourths of their bodies in making their way through the shallows. I had before no idea that these noble fish were ever found in such insignificant streams, but the Indians say that they always come into the rivulets at this season, and return to the sea on the approach of winter. Our slave killed seven of these beautiful fish, while we made our hasty breakfast, his only weapon being a light cedar paddle.

Chinook River.

We reached Chinook in the evening, and as we sat around the fires in the lodge, I was amused by the vivid description given to the attentive inhabitants by Chinamus and his wives, of the perils of our passage across the stormy bay. They all spoke at once, and described most minutely every circumstance that occurred, the auditors continually evincing their attention to the relation by a pithy and sympathising *hugh*. They often appealed to me for the truth of what they were saying, and, as in duty bound, I gave an assenting nod, although at times I fancied they were yielding to a propensity, not uncommon among those of Christian lands, and which is known by the phrase, "drawing a long bow."

[*21st. The wind yesterday was so high that I did not consider it safe to attempt the passage [across the Columbia River] to Fort George. This morning it was rather more calm & we started in a large canoe at sunrise. When we arrived near the middle of Young's bay the wind again rose & the water was dashing over us in fine style, so that we were compelled to go in shore & wait until it subsided. We lay by about an hour & then put off again to tolerably smooth water & reached Fort George at 11 o'clock* (MS Journal).]

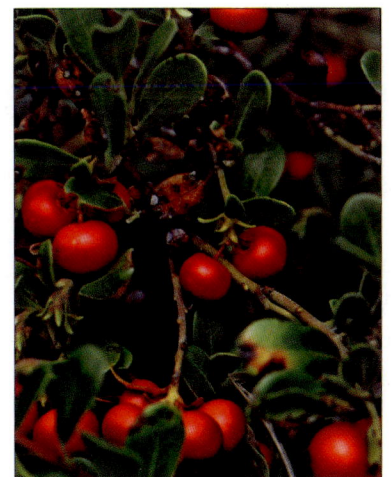

Cranberries *Oxycoccus macrocarpus*.

SURFBIRD

"TOWNSEND'S SURF-BIRD ... In order to exhibit its characters to the best advantage, I have figured it flying in two different aspects. The following note accompanied the specimen sent to me by Mr. TOWNSEND:– "I shot this bird, the only one I have ever seen, on Cape Disappointment, at the entrance of the Columbia river [in November 1836]. ... It was sitting on the edge of the steep rocks, and the heavy surf frequently dashed its spray over it as it foraged among the retreating waves. When it started, it flew with a quick, jerking motion of its wings, and alighted again at a short distance. It was a female. The stomach was remarkably strong and muscular, and contained fragments of a small black shell-fish which adheres to the rocks in this neighbourhood"" (Audubon 1840-44, 5: 228; based on Townsend 1839a: 349-350).

Although Audubon and Townsend both thought that the Surfbird was a new species it had already been discovered in Prince William Sound, Alaska, during Captain Cook's third great voyage, and a description had been published as far back as 1789.

TOWNSEND'S SURFBIRD SPECIMEN AND A HOMOSEXUAL MURDER IN NEW YORK

Mary LeCroy of the Department of Ornithology at the American Museum of Natural History tells the intriguing story:

"Of the Audubon types we have the holotype of *Aphriza townsendi*, which Audubon says Townsend sent him! This was a mounted specimen. It has had a checkered history and is in terrible condition. Apparently at some point it was not housed with the types and was stolen, then later recovered. I was the person who identified it as an AMNH specimen for the police and brought it back. I was dumbfounded when I looked it up in the catalogue with the idea of discarding it and found that it was an Audubon TYPE! It is now in the collection in pieces.

The AMNH number is 156652. It is catalogued along with the other supposed types that were sent us on deposit by Vassar [College] and then later given to us. It has no original label, only an AMNH label with the number and a female symbol. This is all the information that is in the catalogue. It was formerly mounted and almost certainly was part of the Jacob Giraud collection. Professor Orton (1871, *American Naturalist* 4: 716) listed it as a type. If it had other labels, presumably the catalogue would have noted it. I added a type label when we got the specimen back.

I will tell you the bizarre story.

I received a call one day [sometime after 1978] from the head of Security that a former museum guard had been murdered (by his male lover) and that the police had found items in his apartment stolen from AMNH. Would I come and have a look at some bird specimens? A number of us, from various departments represented by the items found with AMNH tags, went up to the Bronx to the bloody murder scene. There were three birds, and I said I wanted them, despite their awful appearance, because they had AMNH numbers and I wanted to check them against the catalogue before destroying them. The NYC police assured us that everything would be carefully guarded and returned to us when they completed their investigation.

There was no word from the police for a couple of years and we had all given up ever hearing again. Then one day a call came to come pick up our things. The NYC police had inexplicably put them, along with personal belongings, in large boxes in the garbage room of the former guard's apartment house. Of course, the tenants in the building thought the boxes contained items being thrown away. Most of the equipment and books were gone, but by digging in the boxes, I found the three pitiful bird specimens. I nearly fainted when I checked the catalogue and realized what the Surfbird was!" (Mary LeCroy, Department of Ornithology, AMNH, *pers comm.*, reproduced by permission from emails dated 18th January, 6th and 12th April 2003).

This Surfbird specimen (AMNH 156652) was collected by Townsend at Cape Disappointment. It was painted by Audubon in two different poses to create plate CCCCXXVII of *The Birds of America* (see opposite page). In the 1970s it was stolen and turned up at a murder scene in New York.

233

OREGON VOLE

This species was first described by Rev John Bachman who gave it the name "OREGON MEADOW MOUSE" adding that "This diminutive species is another of the discoveries of Mr. Townsend … The above was an old male, captured at the Columbia river, November 2d, 1836" (Bachman 1839b: 60-61). On that date Townsend was on the south bank of the Columbia River at Astoria, Oregon. The specimen cannot now be traced.

Did Townsend collect the Northwestern Crow?

Townsend and other early naturalists in the Pacific Northwest refered to the Fish Crow *Corvus ossifragus* even though it is a bird that resides exclusively on the east coast of the United States, For example, Audubon states that "The Fish-Crow is … plentiful on the Columbia river, according to Mr. TOWNSEND, who brought specimens from that country" (Audubon 1840-44, 4: 98). The Fish Crow is also listed in Townsend 1839a and 1839f. Nuttall considered the Fish Crow to be "common on the banks of the Oregon [i.e. Columbia River], where they were nesting in the month of April" (Quoted in Jobanek 1994: 13). These naturalists did not know of the existence of the Northwestern Crow because it was not described until 1858 (by S. F. Baird, from specimens from Fort Steilacoom, Washington).

The Northwestern Crow *C. caurinus* does not normally occur on the Columbia River since the southern part of its range does not extend much beyond Puget's Sound. However, it is just possible that American *and* Northwestern Crows were seen by Townsend at the mouth of the Columbia or more likely in Willapa Bay. Like the Fish Crow, these two west coast crow species can spend a lot of time scavenging on the shore. However, Jobanek (1994:13-14) considered all sightings of Fish Crows by Townsend (and Nuttall, Titian Peale and Newberry) to refer to the small Oregon race of the American Crow *C. brachyrhynchos hesperis*.

Stone (1891: 446) does not refer specifically to any Townsend crow specimens and recent searches suggest that there are no surviving crow specimens collected by Townsend to clarify the matter.

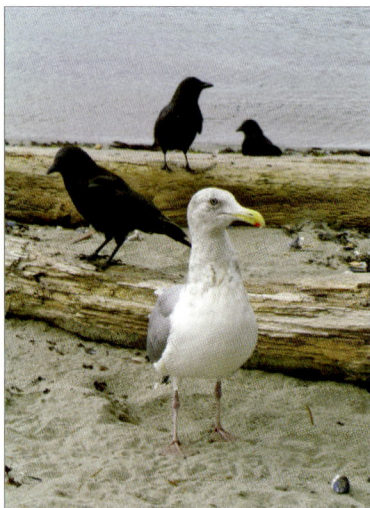

Northwestern Crows and Western Gull.

AMERICAN BLACK OYSTERCATCHER

Townsend brought back at least two oystercatcher specimens[a] and Audubon decided that there were two different species. Audubon named one of them (at left) after the Rev John Bachman and the other (at right) after Townsend. Townsend's text and Audubon's text for the two birds were as follows (though Townsend only included one species in his appendix to his *Narrative* having realized Audubon's mistake):

"WHITE-LEGGED OYSTER CATCHER *Haematopus Bachmani* ... This fine species was shot near Puget's sound, by my friend William Fraser Tolmie, Esq., surgeon of the Honorable Hudson's Bay Company, by whom it was presented to me. I was anxious to give it the name of its discoverer, but I have been overruled by Mr. Audubon, who has probably had good reason for rejecting my proposed specific appellation, *Tolmiei*" (Townsend 1839a: 348-349).

"BACHMAN'S OYSTER-CATCHER HAEMATOPUS BACHMANI ... According to my friend Mr. TOWNSEND, this species is abundant along the whole of the north-west coast of America, as well as in Regent's Sound [? Pugets Sound], but is rarely seen on the shores within Cape Disappointment. The specimen sent to me by him is ticketed as a male, shot in June 1836 ... The discovery of two new species is very remarkable ..." (Audubon 1831-39, 5: 245, 1840-44, 5: 243).

"TOWNSEND'S OYSTER-CATCHER HAEMATOPUS TOWNSENDI ... A specimen of this species, which very closely resembles the last, but is much larger, and differs in its proportions, was also forwarded to me by Mr. TOWNSEND, but without any notice respecting its habits or distribution" (Audubon 1840-44, 5: 245).

a A Black Oystercatcher labelled "Sitka – N.W. Coast America June 1836 J.K. Townsend" may be the type of *Haematopus bachmani* (AMNH 45404). Stone (1899) says both types were lost but an unlabelled Black Oystercatcher in the Edward Harris collection (now missing again, perhaps AMNH 45404) was probably from Townsend (Street 1948).

Townsend prepares to leave the Columbia River

On the 5th of November, I returned to [Fort] Vancouver, and immediately commenced packing my baggage, collection, &c., for a passage to the Sandwich Islands, in the barque Columbia, which is now preparing to sail for England. This is a fine vessel, of three hundred tons, commanded by Captain Royal; we shall have eight passengers in the cabin; Captain Darby, formerly of this vessel, R. Cowie, chief trader, and others.

[Also on board was a consignment of specimens from Dr W.F. Tolmie including plants from Fort Walla Walla collected by Townsend and a second box containing seventeen species of land birds (see Tolmie list, appendix 3) all collected by Townsend (W.F. Tolmie to W.J. Hooker, 15th November 1836).]

On the 21st, we dropped down the river, and in two days anchored off the cape. We have but little prospect of being able to cross the bar; the sea breaks over the channel with a roar like thunder, and the surf dashes and frets against the rocky cape and drives its foam far up into the bay.

I long to see blue water again. I am fond of the sea; it suits both my disposition and constitution; and then the reflection, that now every foot I advance will carry me nearer to my beloved home, is in itself a most powerful inducement to urge me on. But much as I desire again to see home, much as I long to embrace those to whom I am attached by the strongest ties, I have nevertheless felt something very like regret at leaving [Fort] Vancouver and its kind and agreeable residents. I took leave of Doctor McLoughlin with feelings akin to those with which I should bid adieu to an affectionate parent; and to his fervent, "God bless you, sir, and may you have a happy meeting with your friends," I could only reply by a look of sincerest gratitude. Words are inadequate to express my deep sense of the obligations which I feel under to this truly generous and excellent man, and I fear I can only repay them by the sincerity with which I shall always cherish the recollection of his kindness, and the ardent prayers I shall breathe for his prosperity and happiness.

[*28th. … the bar is very rough, the Sea constantly breaking over it. We have made one attempt to get out, but failed & were obliged to stand in again. It is becoming rather dull here; we go ashore generally several times a day but there are very few birds or any thing else that is interesting"* (MS Journal).] Nevertheless, Townsend perhaps collected a second specimen of the **Black Oystercatcher** at this time. His only **Surfbird** specimen was collected "on the base of the rocky cape at the entrance of the Columbia River, in November 1836." This was a very exciting find for him but was probably collected earlier in the month or he would have been a bit more positive about this trip ashore.]

30th.– At daylight this morning, the wind being fair, and the bar more smooth, we weighed anchor and stood out. At about 9 o'clock we crossed the bar, and in a few minutes were hurrying along on the open sea before a six-knot breeze. We are now out, and so good bye to Cape Disappointment and the Columbia, and now for *home*, dear home again!

15th November 1836. Back in Philadelphia on this date, a paper entitled "Description of Twelve New Species of Birds, chiefly from the vicinity of the Columbia river" was read under Townsend's name at a meeting of the Academy of Natural Sciences. It was published in the Academy's *Journal* under Townsend's name (Townsend 1837); Nuttall was amongst those who had actually written it.

1. The books would have been Nuttall's *A Manual of the Ornithology of the United States and Canada: The Land Birds* (1832) and the companion volume on *The Water Birds* (1834) – if the latter had been published in time for the naturalists to take a copy with them. They could have had one or more of Charles Bonaparte's publications, perhaps his four-volume continuation of Alexander Wilson's *American Ornithology* (1825-33). Although it seems highly probable that Townsend and Nuttall would have taken some reference texts with them, one or both of the books could have come from McLoughlin or Tolmie (see Jobanek 1999: 271-72, note 15).

2. The Dalles is well out of range of Richardson's Ground Squirrel. Townsend's MS Journal for this day says only two were shot but four of his specimens were used by Audubon, one or more of them collected earlier, on 22nd June 1835, see original watercolour of "Douglas' Marmot" inscribed with these details (Tyler 2000: 127).

3. The area of the Blue Mountains visited by Townsend, in or near today's North Fork Umatilla Wilderness, rises to just above 5000 feet. They are heavily forested throughout the higher elevations and except for some additional woodpecker species, it is unlikely that he would have found much that was new to him even if he had arrived earlier in the season.

4. There are three Badger specimens at ANSP but no data to link this Badger with any particular specimen.

5. These three skulls are illustrated by Morton (1839: 202-216) in plates XLV (Killemook = Tillamook), XLVI (Clatsap = Clatsop), XLIII (Chinook).

6. Townsend claimed to have seen, but not collected, two additional new cormorant species. He did not list them in his appendix to the *Narrative* (Townsend 1839a) but he did list them in the *Journal of the Academy of Natural Sciences of Philadelphia* (Townsend 1839f), referring to Audubon's names for the supposed two species that had appeared in the final volume of the first edition of the *Ornithological Biography* (but not in the 1840-44 octavo edition):

"Species seen within the limits of the United States, but not characterized ...
WHITE-TAILED AND WHITE-RUMPED CORMORANTS.
PHALACROCORAX LEUCURUS. PHALACROCROCORAX LEUCONOTUS.
    "At Cape Disappointment," says Dr TOWNSEND in his notes transmitted to me, "there are two Cormorants, at least in spring, about the size of *Phalacrocrax resplendens* [Pelagic Cormorant], one with a white tail, the other with a white rump'" (Audubon 1831-39, 5: 334, 336).

On this stretch of the Pacific coast the only cormorants that have white near their rumps are adult breeding Pelagic Cormorants. No Pacific coast cormorants have white tails (though he may possibly have seen birds with tails that had become soiled by white droppings).

The extinct Oʻahu ʻOʻo. Townsend and Deppe both collected specimens of this remarkable bird, indeed they were the last naturalists to see it alive as it has not been reported since 1837. For this painting of the Oʻahu ʻOʻo (from *Aves Hawaiiensis* by S.B. Wilson & A.H. Evans, 1890-99) the artist F.W. Frohawk used a specimen at the British Museum (Natural History) collected when the *Queen Charlotte* called at the Hawaiian Islands in the 1780s.

Second visit to the Hawaiian Islands, 1837

"*Several days ago Mr. Deppe and myself visited Nuano [Nuʻuanu] valley [on Oʻahu], where we hired a native house, in which we are now living. Our object has been to procure birds, plants &c. and we have so far been very successful. I have already prepared about eighty birds which I procured here.*"

Townsend's *Narrative*, 15th January 1837.

Voyage to the Hawaiian Islands[1]

[*13th [December 1836]. Since we left the [Columbia] river, until yesterday we have had such unpleasant weather, & such a heavy rolling sea that I have not been able to write with any degree of comfort, & so have neglected my journal. For the space of ten days the weather was constantly squally with heavy S.E. gales, accompanied with rain, hail, & lightning. Every thing in the vessel was turned topsy turvy: it was not possible to take the least exercise on deck … About a week ago as we were sitting in the cabin in the evening we were startled by the dreadful cry of "a man overboard." We all rushed to the deck instantly & there we saw one of our poor fellows buffeting the heavy sea at the distance of at least 1/4 of a mile astern. The vessel had been going about 6 knots, but she was instantly put aback, the jolly boat was cut away & lowered in a moment, Capt. Darby himself & 4 men sprang into her & in a few minutes the poor tar was aboard the ship again. He was very sick in consequence of the large quantity of sea water that he had swallowed, & completely exhausted by his efforts to keep himself afloat … On Friday last while the ship was lying to in a squall, we were surrounded by great numbers of a small species of Petrel [**Leach's** and **Fork-tailed Storm-petrel** are the most common in this area] & getting out our hooks & lines, we took, in the course of the day, upwards of twenty of them, besides one of the brown Albatrosses of these seas [**Black-footed Albatross**, see appendix 10 for a discussion concerning Townsend's seabirds]. Our observed latitude yesterday was 30° 45′ N. Longitude 131° 37′ W.*

14th. This afternoon the cry of "Sail ho" called us from the cabin & upon ascending to the deck we saw a fine ship of about 300 tons bearing right for us at the distance of a league on our weather quarter. We immediately bore up for her, hoisted our ensign, & in a few minutes we saw the colors of Great Britain floating from her gaff. It has been truly remarked that a ship under full sail is one of the most beautiful objects in the world … "What ship's that"? sang out our commander. "The Sir George Cockburn of London," was answered in return. "Whither bound"? "Cruising for whales" was the reply. "Have you seen any whales" asked the Captain of the strange craft. "No." "We are thirty days from Oahu" said the Captain of the whale ship. A few questions & answers relative to our vessel closed the short conference, & each again proceeded on his way … (MS Journal).]

December 16th.– We are now in the delightful tropics, and more lovely weather I never saw – clear, warm and balmy, but not in the slightest degree debilitating – and a fine trade wind, before which we are going eight and a half knots. This morning we saw a number of beautiful tropic birds [**Red-tailed Tropicbirds**] flying around the vessel. This is one of the loveliest birds in the world. With a plumage of the most unsullied white, a form which is grace itself, and with long red tail-feathers streaming in the wind, it looks like a beautiful sylph sporting over the desolate ocean.

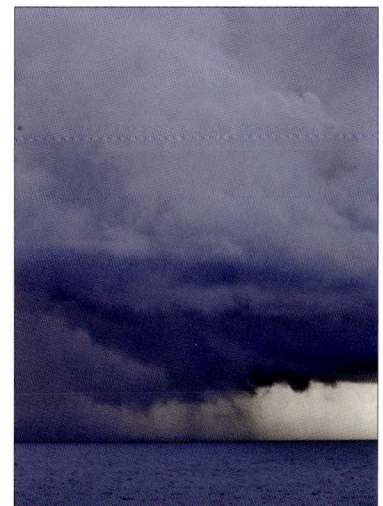

Storm clouds off the Oregon coast.

On the 22d, we made the island of Maui, distant about twenty-five miles. This evening is a most delightful one, as indeed are all the evenings in this latitude. The moon is shining most brilliantly. The atmosphere is deliciously warm, and we are sailing over a sea as smooth as a lake, with the island of Moroka'i [Molokai] about ten miles on our weather beam.

Arrival at O'ahu

On the morning of the 23d, we made Oahu, and as we rounded Diamond Hill, Adams, the pilot, boarded us, and brought us close outside the harbor, where we anchored for want of wind. The captain, Mr. Cowie, and myself, went ashore in the pilot boat, and paid our respects to a number of old friends who were assembled on the wharf to meet and welcome us.

January 1st, 1837. – Since we arrived, we have been so constantly engaged in visiting, receiving visits, and performing the usual *penance* imposed upon strangers visiting this island, that I have not had the opportunity of continuing my notes. I am now so much in arrears that I scarcely know where to begin, and many little circumstances, in themselves, perhaps, trifling enough, but which at the time of occurrence interested me, must of necessity have escaped my recollection.

On my arrival, Mr. George Pelly, agent of the Honorable Hudson's Bay Company, kindly invited me to his house, where I remained three days, and at the end of that time, Mr. Jones procured for me a neat and very comfortable grass cottage [*with an excellent bed, good furniture &c. &c. I am under the protection of Mr. Mills, (sometimes called Deacon Mills,) the high constable, who lives next door to me & from whom my domicile is rented. I have kind and obliging neighbours, & a good active servant who waits upon me with the greatest assiduity, so that my bachelor's hall is no such contemptible establishment. Breakfast I generally take with Mr. Cowie who has another bachelor's hall in the neighbourhood, & as to dining & teaing I always have so many invitations that I can never be at a loss. Indeed they are so exceedingly hospitable to strangers here that sometimes it even becomes irksome* (MS Journal).]

The society of the town has been considerably augmented and improved since my last visit, by the importation, from the United States, of some four or five young ladies, and they have routes and balls, and lu au parties in abundance, [*far too much so indeed for one of my sober turn, & I have therefore come to a determination to cut the fashionables & their dissipation even at the expense of losing a popularity which perhaps I never possessed. The natives are fast becoming acquainted with me & my pursuits & are bringing me birds, shells, fishes &c., in considerable numbers. The birds they take by liming the branches of the trees with a sort of tenacious gum, & they are invariably brought alive & mostly uninjured. I have met here a Prussian gentleman, – a naturalist, – named Deppe, who arrived a few days since from California where he had been making collections. He states that he has been very successful, particularly in ornithology, but most unfortunately his collections were all shipped from Monterey, so I have not had the pleasure of seeing his specimens* (MS Journal).]

3d.– This morning, Mr. Cowie and myself called at the palace to see the king [Kamehameha III] …He gave us a glass of excellent Madeira, and a cigar, and we smoked and chatted with him very pleasantly for half an hour. He does not look so well as when I last saw him; is even more careless in his person, and he never was remarkable for neatness or particularity in his attire. Some allowance should, however, be made for him now, as he is suffering great distress of mind on account of the extreme illness of his favorite and only sister, the princess *Harieta Nahienaena* [Harriet Nahi'ena'ena by Keopuolani (1815-1837)]… Should Harieta die, the royal Hawaiian line will be broken forever, the insignia of Sandwich Island rank will be buried in her tomb, and the children of her reigning brother will not inherit their father's rank … She has just given birth to a child, which died a few hours later, and she is now suffering from severe puerperal fever. Dr. Rooke, who attends her, feels very slight hopes of her recovery.

4th.– This afternoon Mr. Josiah Thompson, Mr. Cowie, and myself, rode down to Waititi [Waikiki], and to Diamond Hill. The day has been a most delightful one, our horses were excellent, and we enjoyed the ride highly. In the upper part of the Waititi district, we passed through an extensive and beautiful cocoanut grove, probably a mile in length, by half a mile in width.

In the midst of this grove, we came to a small group of native houses, the inhabitants of which came running around our horses, the younger branches, in a state of perfect nudity, and capering about like so many little imps of darkness. I told them I should like to have a couple of nuts from one of the trees. I had hardly spoken before two of them ran to the nearest, and commenced mounting, one each side, and then, best fellow reached the top soonest. They climbed just like monkeys, placing their arms half way round the tree, and their feet flat against it, and then actually *jumping* up the perpendicular trunk. The exhibition was so exceedingly ludicrous, that I was indulging myself in a hearty laughing fit, when my cachination was suddenly interrupted by two nuts falling so near me that I felt in some danger of having my brains knocked out. I suspended my mirth, to shake my fist threateningly at the young urchins, who immediately slid to the ground, and with the greatest good humor, held out their hands to receive a *rial* for their trouble.

The milk of the young cocoanut, when fresh from the tree, is peculiarly delicious and refreshing; no idea can be formed of its excellence by those who have only tasted the stale fruit at home.

6th.– Yesterday the princess Harieta died. Scarcely was the circumstance known in the town, when it was announced to all by the most terrific and distressing crying and wailing amongst all ranks and classes of people. The natives, particularly the women, walked the streets, weeping bitterly and loudly, and real briny tears were falling from their eyes in plenteous showers. This most lugubrious exhibition is common on the occasion of the death of any of the high chiefs; but in the present instance there is evidently evinced much real feeling. In the afternoon, Mr. Deppe, (a Prussian gentleman,) and myself, walked to the king's palace to see the mourners who were collected there. We found the large enclosed space surrounding the house, filled with natives of both sexes, to the number of perhaps a thousand, all of whom were weeping in their loudest key …

15th.– Several days ago Mr. Deppe and myself visited Nuano [Nuʻuanu] valley, where we hired a native house, in which we are now living[2]. Our object has been to procure birds, plants &c. and we have so far been very successful. I have already prepared about eighty birds which I procured here. We have a very good and comfortable cottage, and a more delightful country residence I certainly never saw. The valley here is narrow, only about a quarter of a mile across, and the mountains on either side, at least two thousand feet in height, are clothed with the most beautiful verdure. Within gunshot of our dwelling, there is a cascade of delightfully cold mountain water, which falls perhaps thirty feet; the basin below gives us an excellent bath, and we can take a shower when we wish it, by standing under one of the jets. As our cottage is situated upon elevated ground, we have a fine view of the town of Honoruru [Honolulu], five miles from us, as well as the lovely harbor and the shipping. I am so pleased with this residence, on every account, that I shall be loath to leave it; I have escaped from the bustle, and confusion, and dissipation of the town, from the, at times, almost insufferable heat which prevails there, and am living exactly as I wish, in a retired and quiet manner. We never suffer from heat here, and although this is the rainy season, we have had, so far, fine, clear weather.

Coconuts on Oʻahu.

241

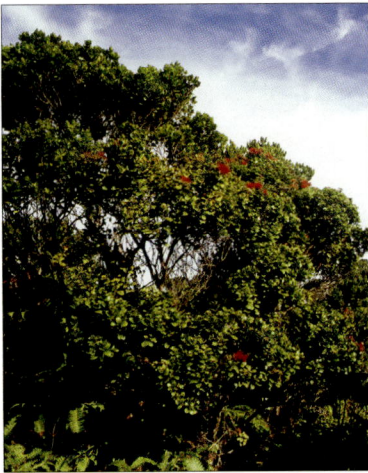
O'ahu forest, not far from Honolulu.

Apapane on Ohia blossom, a favourite source of nectar for many of the Hawaiian honeycreepers.

FERDINAND DEPPE (1794-1860)

Ferdinand Deppe was a natural history collector with several years' experience in Mexico and California. He was the first to provide specimens of such North American species as Franklin's Gull, Brown Jay, Bronzed Cowbird, Greater Pewee and Common Black Hawk, though most of his discoveries were written up by museum curators rather than by himself. He was also one of the few early naturalists (in addition to Townsend) to acquire a California Condor. He was an artist of some talent and painted the earliest known scene of a Californian mission station.

Once a gardener from the Royal Gardens in Berlin, Deppe was recommended to the wealthy sponsor Count von Sack by Professor Hinrich Lichtenstein of the Berlin Zoological Museum for the task of bringing back exotic specimens from Mexico. Deppe gave up his job in 1821, trained himself to skin birds, to preserve plants, to speak Spanish and English, and set out in 1824 on an expedition that was to last until April 1827. Deppe and William Bullock junior (whom he met near Temascaltepec) were the first men to carry out scientific collecting in Mexico, their birds being described by William Swainson, Johann Wagler, Charles Bonaparte, John Gould, Hermann Schlegel and others.

Deppe returned to Mexico in 1828 with the botanist Wilhelm Schiede, hoping to make a living by collecting, and selling their discoveries to Europe, but had to give it up as a failure in 1830, Schiede dying soon afterwards. Deppe became an agent for a merchant company based in Acapulco and Monterey. He suspended his collecting activities and travelled throughout northern Mexico purely on business, though David Douglas met him in California and said that he was only 'partly engaged in mercantile pursuits'. In 1836 he appears to have been tricked out of his earnings and he decided to return to Germany. Consequently he started to build up a collection of birds from near Monterey so that he could sell them on his return, and he continued to collect when he stopped briefly at Oahu (Harris 1941, Stresemann 1954). Deppe's specimens from Hawaii were better studied than Townsend's, many of them being worked up at Berlin by Lichtenstein and later by Jean Cabanis. There is no known portrait of Deppe.

242

TOWNSEND'S BIRD COLLECTIONS FROM O'AHU

The earliest know bird collecting by Europeans on O'ahu took place during May-June 1786 when *HMS Queen Charlotte* called at Honolulu. In May 1825 *HMS Blonde* arrived and a few birds were taken by the Reverend Andrew Bloxam. In September 1834 the Italian collector Botta preserved some birds (Olson and James 1994). Townsend was next in line, so he is amongst the earliest to have worked there. He was on O'ahu during the following periods (but not always in the field): 5th January – 10th February 1835 (with Thomas Nuttall); 16th – 26th March 1835 (with Thomas Nuttall); 23rd December 1836 – 9th February 1837 (with Ferdinand Deppe for part of this time); 27th February – 18th March 1837. His most productive collecting locality was in the Nu'uano valley, just inland from Honolulu. On his second visit there in 1837 he afterwards reported that he had collected over 80 specimens. Specimens from Townsend's 1835 visit are not now separable from those collected in 1837.

Audubon sold these four Hawaiian birds to Lord Derby in 1838, on Townsend's behalf. In 1851 the Derby Museum was transferred to Liverpool where early curators seem to have struggled with the source of the specimens (see comments on labels above and on p. 249). Clockwise from top left: Lesser 'Akialoa (LM D511c), 'Apapane (LM D5359c), 'Akepa (LM D4872), O'ahu 'Amakihi (LM D5361).

243

Two illustrations by F.W. Frohawk from *Aves Hawaiiensis* (Wilson, S.B. & Evans, A.H., 1890-99).

The Nukupuʻu (at left) was probably painted from a Townsend specimen that was bought in Scotland by Sir William Jardine. On Jardine's death the specimen was acquired by the Museum of the University of Cambridge where Evans, and presumably Frohawk, examined it. The Greater ʻAkialoa (at right) has no known connection with Townsend.

27th.– I went, this morning, again to the palace of the king, accompanied by Captain Charlton, the British Consul, to view the remains of the princess … The coffin rests upon a tressel in the centre of the large house, and underneath it is a native mat of the finest and most delicate workmanship. This mat is considered a great curiosity. It was made in the time of Tamehameha [Kamehameha], and was presented to his queen. The grass of which it is composed is about the thickness of a horse hair, and the fabric is soft and pliant as a silken cloth. The coffin is covered with a large cloak, made of the splendid yellow feathers of one of the native birds [from **mamo** and **oʻo** species], and is surrounded by about a dozen of the magnificent insignia of royalty, called *kahiles*. These kahiles are made of the feathers of different birds, and some of the tail plumes of the common dung-hill cock, fastened together with light pieces of bamboo, and arranged cylindrically on a long pole. Many of the kahiles are as large in circumference as a hogshead, and some few not thicker than a man's leg. Including the handle, they are, most of them, from eighteen to twenty-five feet in length. The handle is composed of alternate rings of tortoise shell and fine ivory, so accurately fitted, and beautifully polished, as to appear at a short distance like one piece. I observed that one of these handles was tipped with the bone of a human leg, and upon inquiry learned that it had belonged to one of the ancient kings. This was also finely polished, and looked like ivory, but the joint by which it was terminated had a rather grim and ghastly appearance. One of these kahiles is of so ancient a date, that the natives have no tradition respecting its fabrication. It is indeed a most antiquated looking affair, composed of a wiry sort of white feather from a bird which is now either entirely extinct, or which had been brought from a distance[3]. All these splendid and costly ornaments will be buried in the tomb of Harieta, where they will always remain.

February 3d.– This was the day appointed for the ceremony of carrying the body of the princess to the church … [a three-page description of the event follows].

A cruise among the islands

9th.– Mr. French kindly offered me a passage in his brig Diana, Captain Hinckley, to make a short tour of the islands. The object of the trip is to carry lumber to several of the ports, to trade with the natives, and to bring to Oahu a cargo of live stock, cattle, &c. The time allowed will be so short that I shall probably not be able to do much in my vocation, but I shall at least be furnished with an opportunity of visiting several islands, and as we have pleasant companions as passengers, besides our agreeable and accomplished captain, we anticipate a delightful trip.

We stood out of the harbor in the afternoon, and the next evening made Maui, but as we came under the land it fell calm, with a heavy ground swell, and we were tossing about most uncomfortably the whole night.

244

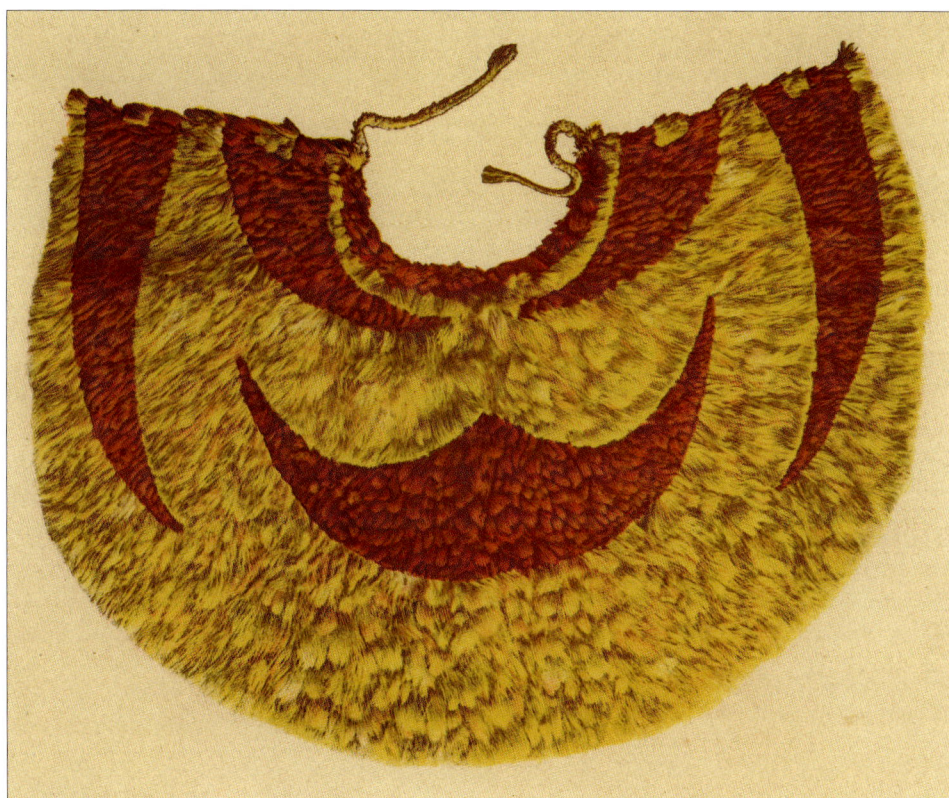

Feather cape at the Bishop Museum, Honolulu, from W.T. Bingham's *Hawaiian Feather Work* (1899). The tinted photograph was coloured incorrectly – the central crescent is made of black feathers.

Maui

14th.– Yesterday we made Maui again, after having been cruising around the islands at the mercy of contrary winds, since Friday. [*On Sunday we were running along Hawaii within 8 miles of the land & we could easily have brought in to the bay of Karakakua [Kealakekua], but that we have lumber on board which must first be landed at Maui* (MS Journal).] Several of our passengers have been constantly sea-sick, and our anticipated pleasure has been thus very much lessened.

When I arose this morning, we were off the pretty village of Lahaina, and in about two hours after, we dropped our anchor within half a mile.

While the ship's people were engaged in discharging cargo, Mr. Paty, (one of our passengers,) and myself went ashore to see the town. The village is one of the prettiest I have seen: many of the houses are built of stone, handsomely whitewashed, and, as at Honoruru, a very picturesque looking fort frowns upon you as you approach the anchorage. These forts, although they add greatly to the appearance of the harbors in which they are situated, yet appear to me better calculated for show than service, as in case of an attack from the sea, they could not act efficiently, not being provided with bastions …

The houses, composing the village of Lahaina, are, many of them, so obscured by cocoanut and kou trees, (*Cordia sebestena*,)[4] that you cannot see the whole of the town from any single point of view, even from the offing. On a high hill, two miles back of the town, stands another village, called *Lahainaluna*, (or upper Lahaina,) composed entirely of white stone houses. It is here that the missionaries chiefly reside. The high school here is a large building of stone, thatched with grass, and stands on an elevated piece of ground, so as to be distinctly seen some miles out at sea. I called, with Mr. Paty, upon Mr. Andrews, to whom I had a letter of introduction from the Rev. Mr. Dieill [Diell], and here I met several other missionaries, Mr. Baldwin, Mr. Rogers, and Mr. Dibble. These gentlemen are all more or less concerned in the management of the high school, but Mr. Andrews is the principal. It was commenced by him in the year 1831 … [*Mr. Paty & myself had a very pleasant visit at his house; and dined with him & his family, & took tea in the evening at Mr. Baldwin's in Lahaina* (MS Journal).]

245

Hawai'i

[15th. This evening we weighed our anchor & stood out of the harbour of Lahaina, but the wind being light we made very little progress during the night. The next morning a heavy gale commenced, which compelled us to reduce our canvass to a double reefed maintopsail & foresail, the sea was tremendously agitated & our brig danced like a bubble on the waves, but although this was somewhat uncomfortable, we could endure it cheerfully as we were proceeding on our course at the rate of 5 or 6 knots. At 3 in the afternoon we made Hawaii distant 45 miles. Shortly after this the wind fell, & during the night it was nearly calm with a smooth sea (MS Journal).]

On the morning of the 17th, we made the island of Hawaii, and, approaching with a free wind, soon [*at 2 o'clock*] let go our anchor in the bay of *Karakakua* [Kealakekua]. The land here is composed almost entirely of rough and irregular masses of lava, but towards the summit of the hills, as in Oahu, vegetation is abundant. The shore, for miles, in both directions, is sprinkled with the little sylvan looking hamlets of this country, and they are sometimes so thickly grouped together, as to form the most picturesque and beautiful villages. On the hill fronting the bay is one of these, at which the missionary, Mr. Forbes, resides, and about eighteen miles from this, there is a considerable town called *Kairua* [Kailua], the residence of the chief, John Adams, governor of Hawaii. In the afternoon Mr. Paty and myself went on shore, chiefly for the purpose of seeing the spot on which Captain Cook was killed, in the year 1779.

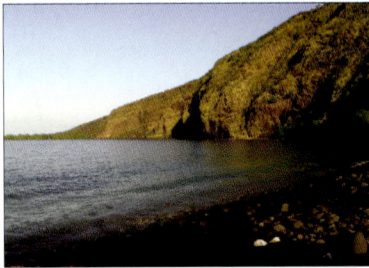

Kealakekua Bay by John Webber (from the Atlas to Captain Cook's Third Voyage of 1776-1779). In 1874 a white obelisk was erected close to the spot were Cook was killed.

When we made this inquiry after we landed, a number of natives ran to the beach, and pointed out to us the exact spot where the gallant mariner received his death blow. I need not attempt to describe, for my sisters can in a measure understand the emotions with which I viewed the rock on which this brave and excellent man offered up his life in the service of his country. I had read the voyages of Cook, with great interest, when I was a child; I had pondered over his dangers, his magnificent discoveries, the intense excitement of his life, and his premature and violent death, but if at that time any one had told me that I should ever visit the scene of his discoveries, and stand upon the identical rock which was pressed by his bleeding bosom, I should have smiled at it as too chimerical for belief; here I am, however, although at times I can scarcely realise the possibility of it.

The rock is somewhat isolated, and at high tide the water breaks over its summit. It is said to be at present not one-fourth its original size, as almost every visitor, for a number of years, has been in the habit of carrying away a fragment of it as a relic. A French man-of-war [named as *La Bonité* in MS Journal], which was lately here, is said to have taken off about *a ton* of it; and some Spaniards, who visited the island several years since, not only took specimens of the rock, but the whole ship's company knelt upon it, and offered up a prayer for the repose of the hero's soul.

There is perhaps no one unfortunate circumstance connected with foreigners, that has ever occurred here, which the natives of these islands so deeply regret, as the death of Cook. They all speak of it as a lamentable event, and some of the elder of them are said even to shed tears when the subject is mentioned. They have canonized him, and he is universally known by the title of "*Olono*," a particular deity.

18th.– This morning I met Mr. Forbes[5], the missionary of this station, at the lower village, and after delivering to him a letter from Mr. Dieill [Diell], accompanied him to his house on the hill, a distance of three miles. At about one mile from the shore on the hill is a monument[6], erected in 1825 by Lord Byron, Captain of his Britannic majesty's frigate "Blond," [*H.M.S. Blonde*] to the memory of Captain Cook. It consists of a simple wall of lava about five feet high, embracing a square of twenty feet, in the centre of which is a cedar post, twelve feet in height, and near the top a copper plate, with this inscription:

"In memory
of
Captain James Cook, R.N.,
Who discovered these Islands,
in the year of our Lord,
1778.
This humble monument is erected
by his fellow countrymen,
in the year of our Lord,
1825."

This post is completely covered with the initials of persons who have from time to time visited the spot, chiefly the masters, officers, and crews of vessels, and among them I noticed the well known name of "*Coffin, Nantucket.*"

[I dined at Mr. Forbes' & afterwards walked through the woods in the vicinity for several miles. Saw some birds that I have never noticed on the other islands, & regretted the absence of my gun. I purchased from one of the men here, three of the beautiful native Geese [**Nene / Hawaiian Goose**] *for a dollar each. This is a sort of Brant, very similar in many respects to our common one at home* [Brant *Branta bernicla*]. *The membranes separating the toes are much more deeply scalloped, & on the neck the feathers are formed in longitudinal curving lines which give the bird a very singular & beautiful appearance. These Geese are said to possess an uncommon degree of sagacity, very soon becoming acquainted with their master, & even following him about like a dog. They are also very good watchers at night, giving alarm by a loud <u>cackling</u> on the approach of a stranger. I shall use my utmost endeavours to carry my pretty pets home. After tea I bade adieu to my kind entertainers & rode down to the shore on a horse furnished me by Mr. Forbes, & went off to the ship in a canoe.*

Rev Cochrane Forbes. He offered hospitality to Townsend (and later to Titian Peale of the US Exploring Expedition). A few years later Forbes sent Townsend the type specimens of the Hawaiian Crow and Hawaiian Hawk.

Nene or Hawaiian Goose.

Koa *Acacia koa*. Townsend failed to get into the higher altitude Koa forests where he could have discovered such birds as the Kona Grosbeak, Greater Koa Finch, Lesser Koa Finch and Palila.

19th. Last evening at 11 o'clock we weighed anchor & went out of the bay with the land breeze, but this soon failing us, & the current carrying us rapidly to leeward, we were compelled to bring in to Karakakua [Kealakekua] again, where we again dropped our anchor at 1 o'clock. We shall now remain here until Monday night as Capt. Hinckley is desirous of completing his cargo of wood; & for myself I am very much pleased by the detention, as I shall be furnished with an opportunity of visiting the shore again, & probably of procuring some birds. For this purpose I have made arrangements for going on the hill at an early hour tomorrow morning.

20th. At 4 o'clock this morning Mr. Paty & myself walked up to Mr. Forbes' & after an early breakfast, procured two native men, & spent the day in the neighbouring woods looking for birds. One, a Cassicus with yellow axillary feathers, (which are used by the natives in making the ornament called "Re", & worn on the head,) was pretty plentiful but shy [**Hawai'i 'O'o**]: *other birds were rather scarce so that I procured but little to compensate me for my trouble* (MS Journal).]

Mr. Paty and myself spent the day in traversing the extensive forests of this island, in search of birds, but met with very little success. The walking was extremely difficult, and sometimes dangerous, in consequence of a thick undergrowth of bushes, intermixed with large masses of rough, porous lava. There is here a small species of crow [**Alala / Hawaiian Crow**], said to be numerous at times, but we did not see any, as, in consequence of a long drought, they, as well as most other birds, have retired back into the mountains to procure water. We returned to Mr. Forbes' house late in the afternoon, and found him preparing his baggage, &c., for a passage to Oahu in our brig. He takes his wife and two children with him. [*At 7 in the evening they came on board, & at 9 we got under weigh with the periodical land breeze* (MS Journal).

[*21st. When I rose this morning I was surprised to find that we were lying nearly stationary almost opposite Karakakua [Kealakekua] which we left last evening. Although during the night we had a three knot breeze, yet the current was so rapid as to overcome this, & cause us to go astern. At 9 o'clock the wind freshened & we made some headway. The remainder of the day the wind continued free, & the next morning at 9 we anchored off Kawaihae* (MS Journal).]

22d.– …This is a barren and most unattractive looking place, a rambling sort of village, containing about fifty houses, but no vegetation except a few scattered cocoanuts, and an occasional kou [Kou *Cordia subcordata*], and tutui tree [Kuku'i *Aleurites moluccana*]. The soil is composed entirely of volcanic earth, or the pulverization of lava and basalt. I observed none of the handsome taro patches here that form such a relief to the eye when scanning this rugged country in other places. From our anchorage we have a view of several of the colossal mountains and peaks of this island, among which the majestic point of *Mauna kea* stands pre-eminent. I have not yet seen *Mauna roa* [Mauna Loa], except from a considerable distance at sea, and I suppose that now the gratification of a nearer view will not be afforded me. I cannot too much regret that I have had no opportunity of visiting this celebrated and stupendous volcano.

23d.– Yesterday morning I went on shore with Captain Hinckley and others, and called upon Mrs. Young, widow of the late John Young, the oldest foreign resident of the Sandwich Islands. He came hither in the year 1789, remaining until his death, which took place about a year ago, in his 90th year. Mrs. Young, is a sister [niece] of old king Tamehameha [Kamehameha I], and is now probably sixty-years of age, a very pleasant and lady-like old woman.

In the afternoon we visited a large *heiau*, or temple, in the neighborhood [Pu'ukohola Heiau]. This temple, (which of course has not been used as such since the abolition of idolatry) was built in the early part of the reign of Tamehameha; in it were deposited the gods of wood and stone, which the natives worshipped, and at regular intervals, a human victim was offered as a sacrifice to their imaginary deity …

The heiau is built of stones laid together, enclosing a square of about two hundred feet. The walls are thirty feet high, and about sixteen feet thick at the base, from which they gradually taper to the top, where they are about four feet across. In the centre, is a platform of smooth stones, carefully laid together, but without any previous preparation, raised to within ten feet of the top of the wall. It was on this platform that the victims were sacrificed, the gods standing around outside in niches made for their accommodation.

Kou *Corda subcordata*.

249

Pu'ukohola Heiau at Kawaihae.

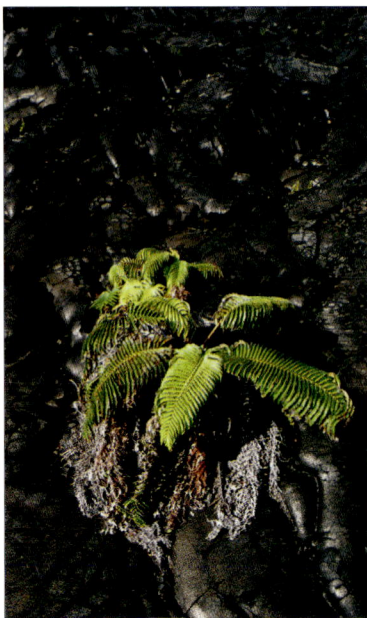

Young tree fern *Sadleria cyatheoides* on fresh lava, Hawai'i.

There is, near the heiau, another very similar, though smaller edifice of stone, called a *morai* [Mailekini Heiau]. This was used for nearly similar purposes, and, in addition, it was the place to which the bodies of the dead chiefs were carried, previous to interment. After lying here in state for a longer or shorter time, according to the grade of rank held by the deceased, the flesh was stripped from the bones, and buried in the sea; the bones were then taken and deposited in caves, or subterranean vaults, which concluded the ceremony …

24th.– The ship's people have been engaged the whole day in taking cattle on board, and we are now deep in the water, having upwards of one hundred and twenty head stowed under the hatches. These cattle are procured wild, on the island, by Spaniards, who live here for the purpose. They take them by means of lassos, and display great dexterity in the business. This operation has been so often described, that I need not repeat it here, suffice it to say, that all the bullocks on board have been taken expressly for us, by three Spaniards, since our arrival here on Wednesday. [*At ½ past 8 in the evening a most brilliant, and very unusually large meteor was seen to start from a point near the zenith, & take its course to the N.E. describing the arc of a large circle. It then scintillated, & did not entirely disappear until it had very nearly reached the horizon. Its size appeared about the head of a hogshead, & the light emitted was so glaring & intense that all on deck were compelled to veil their eyes. Although the night was unusually dark, yet all the spars & rigging of the vessel were for a moment most distinctly seen. The natives, of whom we have a number with us, have never before seen one which approached it in size* (MS Journal).]

Return to O'ahu

25th.– We were under way at daylight this morning, bound for Oahu. We passed, in the course of the day, the islands of *Maui, Kaawalawi* [Kaho'olawe], *Ranai* [Lana'i], *Morokai* [Moloka'i], and *Morokini* [Molokini]. The weather was rough, and the sea high, and as usual, most of our passengers have been suffering from sea sickness, and at times the scene on the quarter deck is quite distressing. [*It is the more so as this troublesome malady admits of no remedy, & I believe it is generally conceded that it is one of the most unendurable of temporary diseases with which poor weak stomachs are assailed. I have often thought I could not feel too thankful that I have never suffered in the slightest degree from this disagreeable complaint. I have seen some pretty rough weather too, but, when all around me have been retching & vomiting & groaning, I have never experienced the slightest qualm* (MS Journal).]

Our vessel is now literally stowed full, so much so as to be somewhat unpleasant for passengers. All forward of the mainmast, both above and below, is crowded with cattle; the 'tween-decks are stowed with hides, and the quarter deck with passengers of all colors, from the fair skinned European, to the deep copper-colored native, not

Mauna Loa, seen from near Kawaihae.

omitting the intermediate grade of half-castes. Men, women, and children, of various families, are all huddled together in a mass, lolling about, talking and smoking during the day, and sleeping and grunting like swine at night. The effluvia arising from the mass of native bodies, during a still, warm evening, is not comparable to otto of roses, and I have often been compelled to forego the pleasure of a nocturnal lounge on deck, and dive to the cabin for purer air. This effluvia is owing to a common habit among these people, and particularly of the women, of anointing the hair and body with cocoanut oil. The oil, in a recent state, possesses an aromatic, and rather agreeable odor, but when allowed to become rancid, it is most insufferably rank and disgusting. When in the rancid state, its cosmetic properties are supposed to be improved, and it is then applied in large quantities to the whole person. Were it not for this disagreeable and unsavory practice, the women here would be well calculated to please the taste of a stranger, as many of them are truly handsome, and remarkably graceful in their deportment. I believe that most of those who are married to the foreigners have given up this disgusting practice.

[*26th. During last night the weather was nearly calm, but this morning at 9 o'clock a light breeze sprang up which soon carried us in sight of Oahu, distant about 30 miles. We had expected to have arrived early this morning & no doubt should have done so if the weather had favored us. The cattle on board are suffering extremely from confinement & want of air: several have already died & we shall probably lose a number more if we should not reach the harbour to day. My pretty native Geese* [**Nene / Hawaiian Geese**] *are doing well – they have become quite domestic & appear to know me. If I should succeed in carrying these beautiful birds home they will be very valuable & a great curiosity*[7].

27th. Last night at 9 o'clock the wind again freshened, (the weather having been calm during the greater part of the day) & continued free during the whole night, so that when we rose this morning we found our vessel off Cocoa-head [Koko Head]. We soon after rounded Diamond Hill, took the pilot on board, & at 9 o'clock anchored in the harbour & drew up close to the wharf to discharge our cargo of cattle (MS Journal).]

On the 27th we anchored in the harbor of Oahu, and from this time, until the 16th of March, I was busily engaged in packing my multifarious collections, making calls upon my friends, &c., preparatory to embarking for Valparaiso, via Tahiti, in the ship Europa, Captain Shaw, of this port. I have now been here nearly three months; much longer than I expected to have been detained. My time has been employed chiefly in pursuing my scientific avocations, collecting specimens, &c., in which I have been as successful as I anticipated. In this pursuit I have received much and very steady assistance from many of the resident foreigners, and, as a parting word, I wish them to accept my most unfeigned thanks, both for this and for the uniform hospitality and kindness with which they have treated me. To J.C. Jones, Esq., – the American consul,

Nene or Hawaiian Geese by F.W. Frohawk (from *Aves Hawaiiensis* Wilson, S.B. & Evans, A.H. 1890-99) painted from British specimens.

251

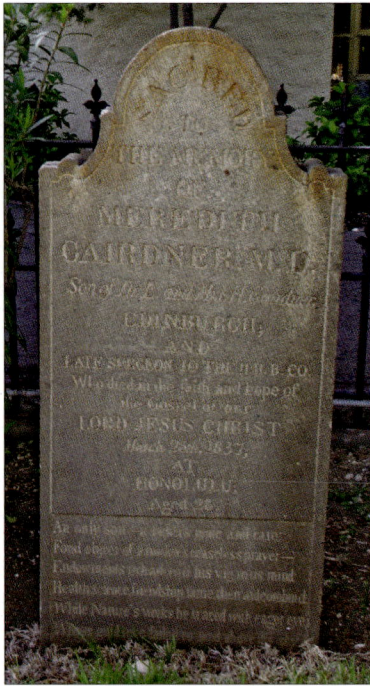

Dr Meredith Gairdner's gravestone at Kawaiahao Church, Honolulu. A former surgeon at Fort Vancouver, he was temporarily replaced there by Townsend when ill health forced him out to the Hawaiian Islands. Gairdner is commemorated in the scientific name of the Rainbow Trout *Salmo gairdneri*. He died ten days after Townsend sailed to Tahiti.

– my acknowledgements are particularly due. I shall always remember, with gratitude, the many favors he has conferred on me.

[*March 16th … I anticipate a pleasant passage, both from the circumstance of performing it with one whom I do highly esteem [the owner of the ship, Henry Skinner of London], & from the kind & obliging manner of our Captain. The ship is a very excellent one[8], a fast sailer & comfortable sea boat. We have also two other passengers, Capt. Winkworth, formerly of this vessel, for Tahiti, & W. Murphy, - an irish gentleman – for Valparaiso* (MS Journal).]

1. There is no entry in the published *Narrative* until 16th December 1836.

2. Unfortunately there is no corresponding account in Deppe's letters and journals because this period is missing from his papers in the Humboldt-Universität Museum für Naturkunde, at Berlin (Dr Hannelore Landsberg *pers. comm.*, 18th November 2002).

3. From an unidentified species, perhaps White-tailed Tropicbird *Phaethon lepturus*.

4. *Cordia sebestena* is the West Indian Geiger tree. It has been widely planted in the Hawaiian Islands where it is known as 'kou haole' (foreign kou). Townsend is more likely to be referring to the similar Kou *Cordia subcordata*, introduced by the first Hawaiians.

5. Rev Cochran Forbes was born at Goshen, Pennsylvania, in 1805 and died at Philadelphia, in 1880. On Hawai'i he lived at Kuapehu about two miles above the landing at Kealakekua Bay. In his journal there is an annoying gap of several months that includes the time when Townsend visited (Forbes 1984). Forbes later sent specimens to Townsend that Titian Peale used when making the formal description of Hawaiian Crow and Hawaiian Hawk as new species. Judging by the content of the journals it seems a little unlikely that Forbes would have spared the time to collect these birds himself.

6. In November 2005 we found lava walls and a wooden post fitting this description, just a few yards off the old trail leading inland from the long-abandoned coastal village of Ka'awaloa. The post is very weathered and leaning at about 45°. It may or may not be the original post but certainly the copper plaque is long gone. The site is little visited despite its proximity to the popular route down the hill from Captain Cook village to the large Cook monument erected on the shore of Kealakekua Bay in November 1874.

7. Hawaiian Geese were first seen by Europeans travelling with Captain Cook in 1778 (Ellis 1782, quoted in Medway 1981) and subsequent voyagers attempted to introduce the species alive into Europe. In 1832, for example, geese were sent to England by David Douglas and the first scientific description of the species was made by N.A. Vigors from these birds that had been exhibited alive in the gardens of the Zoological Society of London. In 1834 there were a pair of live geese at Lord Derby's menagerie at Knowsley, near Liverpool (Olson 1989). Three years later Townsend attempted to bring this little known species back alive to the eastern United States, as testified by his MS Journal entries for 18th February, 26th February, and 1st June 1837 (see main text).

There is no further mention of the geese in Townsend's journal or in any other of his surviving papers. The birds may well have died while Townsend was ill in Valparaiso or he may have given them to friends there who had helped him. If they died on board ship, as at least one of them did (see 1st June 1837), he would surely have preserved them. The Academy of Natural Sciences of Philadelphia has two Hawaiian Goose specimens: one from the Rivoli collection that has no possible connection with Townsend: the other (ANSP 6001) has no data with it so may or may not be one of Townsend's geese.

8. Townsend later changed his mind about this voyage and always regretted the day he boarded the *Europa*.

Homeward bound

"It was a most unfortunate & ill-starred voyage for me, as it utterly incapacitated me from making collections in Chili – a country exceedingly rich in Natural history – & has left me with deep seated complaints of which I fear I shall never perfectly recover."

J.K. Townsend, 25th September 1837 (MS Journal).

First stop: Tahiti

18th [March1837] .– We cleared Oahu yesterday, and this evening, are sailing along delightfully before an eight knot breeze. I think that of all enjoyments I have ever experienced since I became a dweller in distant lands, there is none that has ever excited in me such a thrill of delight and pleasure, as an evening sail upon a moon-lit sea. I can hang for hours over the gunwale, as the ship ploughs the deep blue waters; I gaze upon the lovely moon, and turn my face towards my father-land, and then, oh then, do I fancy I can see my quiet, peaceful home, and commune with the loved objects there! All, all rise before me with a distinctness at times almost startling. I see my excellent and affectionate father, my beloved and tender mother, my dear sisters, brothers, all whom I love, and I think I can see them beckoning to the wanderer, and entreating him to turn his footsteps homeward. These images have risen before me, this evening, with uncommon vividness. It is now eight bells in the middle watch; the officer is pacing the quarter deck, muffled in his large pea jacket, the helmsman stands by the wheel, the drowsy watch are lolling on the forecastle, and all else are asleep. But I cannot sleep, nor would I if I could, on such a glorious night as this. [*Let me doze away the day but on such a night as this it would be sacrilege to sleep* (MS Journal).]

April 1st.– Nothing important has occurred to vary the monotony of a sea voyage. We have generally been favored with good breezes, though the sea has been mostly rough. On Thursday last, we crossed the [equatorial] line, and our latitude is now 3° 52´ south.

8th.– Yesterday morning at 10 o'clock, "*land, ho!*" was sung out by a man at the mast head, and we ascertained it to be Dean's island, distant about fifteen miles. We had a fine seven knot breeze, and we rapidly approached, and soon passed it within five miles. This is a very long, low island, profusely covered with vegetation, very undulating, and with a fine sand beach surrounding it, upon which the surf breaks furiously. It is said to be sparsely inhabited by people of a very wild and unsocial nature. Ships rarely, if ever, touch here, as the island produces nothing to tempt the cupidity of our mariners. In the evening we had a heavy squall, with rain, and incessant and very vivid lightning. We shortened sail immediately, and lay to, under a double reefed maintopsail and reefed foresail, for about an hour, when the gale subsided, and a dead calm of about the same duration succeeded. During the storm, we observed a little speck of brilliant light, like a star, resting upon the *main truck* or top of the mainmast. In a few minutes after, a similar light appeared upon the summits of both the other masts, and continued visible for about an hour. This is what sailors call a "*complaisant*," and is of course occasioned by an excess of electricity in the atmosphere.

In the afternoon we made Tahiti, (or Otaheite,) [Society Islands] and the next morning [9th] approached to within two miles of it, brought our vessel to, in a fine breeze, and hoisted our signal for a pilot. After waiting about two hours, a native, who

Coconut palms.

spoke English well, boarded us in a whale boat, and announced himself as authorized pilot of the port. The charge of the vessel was of course given into his hands, and in another hour we were riding at anchor in a beautiful, and very safe harbor. Tahiti, like most islands in these seas, is nearly surrounded by a coral reef, a narrow passage only being found for entrance, but the native pilot appears to be skilful, and I am told that no accident has ever happened here.

The outline of this island is exceedingly uneven and rugged, being formed of high hills and valleys alternately, but the whole of the land is profusely covered with vegetation. The bay in which we are anchored, (Papeeté,) is one of the most beautiful I have seen; the water in the harbor is at all times so smooth and placid that no motion whatever can be felt on board a vessel riding at anchor, and the shore, fringed with cocoanut, bread-fruit and banana trees, with the neat white-washed cottages sprinkled amongst them, forms a view at once striking and lovely. There are about eight whale ships now in the port, and several of the masters of these, as well as some resident gentlemen from the shore, visited us shortly after we came to anchor. Among the latter were the missionary of this station, the Rev. Mr. Pritchard, Doctor Vaughan, Mr. William Henry and others. Soon after, Mr. Skinner, the supercargo, and myself, went on shore, and called upon Mr. Moerenhaut, the U.S. consul, to whom I had a letter of introduction from Mr. Jones of Oahu. He received us kindly, and we spent an hour with him very pleasantly. We partook of a good dinner at the house of Mr. Henry, after which Mr. Skinner and several other gentlemen with myself, took a stroll back of the village. If I was pleased with the appearance of the harbor from the anchorage, how much more was I delighted with the opportunity of rambling in the interior. Soon after we left the house, we entered upon an excellent turnpike road made by natives, chiefly convicts, and extending nearly the whole circuit of the island. This, as is almost every part of this lovely isle, is a complete orchard of the most delicious of the tropical fruits; vast groves of oranges, lemons, guavas, &c. &c., growing wild, and in the most prodigal profusion, patches of pine apples, interminable forests of bananas, cocoanuts, and Vi's [a forest tree with a pear-like fruit], and all without an owner. Well may it be said, this is a highly favored, and most fruitful land. The natives do not require to cultivate the earth; it teems with every luxury that their unsophisticated palates crave. For a meal, they have but to enter the forest, and gather a mess of bread-fruit, bananas, and guavas, and kill a pig from the large droves which are constantly roaming the country, in a half wild state, and fattening to obesity on the ripe and luscious fruit which every where strews the ground. [*I have remarked a number of beautiful birds this evening such as I have never before seen. One in particular is very numerous, – a Thrush* [**Tahiti Reed Warbler**], *which sings most delightfully* (MS Journal).]

10th.– I strolled, during the whole of this day, through the woods, and procured a number of very pretty birds, all new to me. [*Among the most remarkable of them, are the fine Thrush mentioned before, a beautiful & very small parroquet, blue, with white throat* [**Blue Lorikeet**]*; a lovely Dove, green, with vinaceous breast & azure crown* [**Gray-green Fruit Dove**]*, several Flycatchers* [**Tahiti Monarch**, adults and immatures] *&c., &c. As we shall probably sail in a very few days, it is necessary for me to be busy in order to procure specimens of the whole* (MS Journal).] In this expedition I was accompanied by a stout boy, a Sandwich Islander, whom I have engaged as my servant while I remain. This is a convenience, inasmuch as I am not acquainted with the language of the Tahitians, but am sufficiently familiar with that of the Sandwich Islanders, to ask for whatever I want, and understand ordinary conversation. In my ramble through the forest to-day, I was surprised to hear a stave of the old familiar song, *Jim Crow*, sung by a little puling voice, but with singular fidelity of tone and time, and after a short search, I perceived a little naked native girl, of not more than four years of age, washing her only calico garment in a creek which flowed by, and amusing herself at her work, by singing "wheel about, and turn about, and do just so." The child attempted to escape when she found she was observed, but I caught her, and by dint of persuasion, and the offer of a *rial*, induced her to sing several verses to me.

12th.– I went, with the consul, to the palace of the queen … [a description of the queen and her husband follows].

15th.– This day, although with us, in our ship account, Saturday the 15th, is Sunday the 16th, at Tahiti. This is accounted for by the fact of the early missionaries having made the passage around the Cape of Good Hope instead of Cape Horn, and making no allowance for easting, consequently gained nearly a day in their reckoning. This mistake has never since been corrected, and at the present time it would perhaps not be advisable to do so.

I attended with most of the gentlemen of the place, the native church, at 9 o'clock in the morning. Mr. Pritchard performed the service, and I was pleased, not only with the order and regularity of the exercises, but with the strict and decorous deportment of the audience. The hymns were sung with much taste and skill, and many of the voices, particularly of the females, were sweet, and well trained.

The chapel is a very neat and pretty piece of workmanship, somewhat in the style of those at the Sandwich Islands, but more tasteful and lighter. The roof, instead of a thatch of grass, is neatly covered with the large leaves of a species of *Pandanus*, handsomely and ingeniously worked on light reeds, and the beams are wrapped, for about one-fourth of their length, with alternate strips of fine sinnit and mats of different colors, and adds very much to the general appearance of the building … [*At 11 o'clock we attended the English Chapel, where Mr. Pritchard also officiated.*

16th, This evening Mr. Skinner & myself paid a visit to the mission family. Mr. Pritchard is a man of considerable intelligence & appeared well qualified for the station he holds. His wife is in every respect a lovely woman.

20th. We expected ere this to have been again on the wide ocean & sailing towards Valparaiso, but Mr. Skinner has found profitable business here which has obliged him to detain the vessel. We shall remain here several days longer. This detention has been to me a very welcome one, as it has enabled me to explore the country much more extensively than I expected, & to collect a considerable number of birds (MS Journal).]

Information about Townsend's collecting activities on Tahiti is mostly derived from his MS Journal and from the labels on his surviving specimens. From these and other sources, it is known that he collected about 110 specimens of 16 species. Only about 16% of these specimens can now be traced (see appendix 12).

These specimens are all at the Academy of Natural Sciences of Philadelphia. At left, from top: Gray-green Fruit Dove (ANSP 13139), Venerated Kingfisher (ANSP 21456, 21455), Blue Lorikeet (ANSP 22118). At right, from top: Tahiti Monarch (ANSP 1103, 1102, 1101, 1100), Tahiti Reed Warbler (ANSP 17347, 18189).

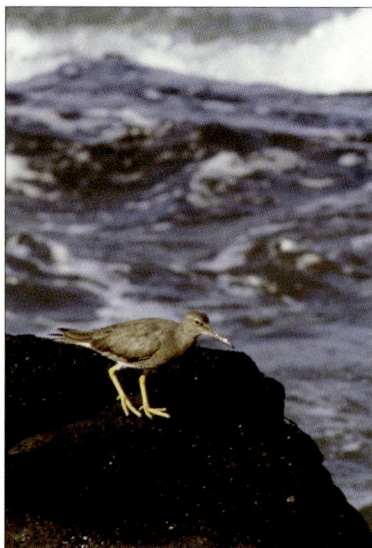

Wandering Tattler. There is a Townsend specimen of this species at Liverpool (LM D3096). For a summary of the birds he collected in Tahiti see appendix 12.

Red Junglefowl, native to south east Asia, brought as domestic birds by ancient Micronesian and Polynesian peoples to most of the inhabited Pacific islands.

[Although not mentioned in the printed or MS version of his *Narrative*, Townsend also passed the time by going out alone in a canoe beyond the reef to look for hours at the corals and brilliantly coloured fish. On one occasion he quickly hired six natives to row him out to see a Sperm Whale hunt that was just beginning beyond the harbour, and was able to witness at close hand the chase, harpooning, death, and subsequent rendering process (Townsend 1847: 163-166).]

May 2d.– We are now quite ready for sea, and are only waiting a breeze to go out. All my little matters are packed & sent on board the ship, & I am now as anxious as any one to make a start. I have done all that is to be done here in the ornithological way, & there are no plants in flower so that I have had no botanizing. I have prepared 16 species of birds, (all the varieties I could find,) most of them I think peculiar to this island. Number of specimens about 110. Within a few days past I have killed three specimens of a very beautiful & large Cuckoo (Cuculus) remarkable for the great length of its tail [**Long-tailed Cuckoo**]. *I have every reason to believe this fine bird a new species1 – if it should prove so it will surely be a noble addition to the oceanic fauna. It is at all times rare & very difficult to be procured, & I am told by the residents here that it has never been obtained by any collectors who have formerly visited this island* (MS Journal).

May 2d.– ...The common dunghill fowl is found wild in the forests here. Some of the residents think that it is a jungle fowl, peculiar to the country, but, upon examination, I have not been able to perceive any material difference between it and the domesticated bird, and therefore incline to the belief that it is the common species returned to its original habits [**Red Junglefowl**]. In my excursions, I have killed about a dozen of them. Their plumage is generally more rich and brilliant than that of the domesticated bird, and there is not so much variety in the color of different individuals. Their flesh is exquisite. They are very shy, running away with singular rapidity, and concealing themselves on the approach of the sportsman. When flushed, they fly with great vigor and swiftness, and where the trees and bushes are not too dense, afford a very good mark.

4th.– This morning, the wind being fair, we took the pilot on board, and at 8 o'clock, stood out. While in the middle of the passage, the breeze fell very light, and our vessel began to swing towards the high and dangerous reef which was just beside us. For myself, I gave our good ship up for lost, and was waiting to see her dashed upon the rocks, which I thought was inevitable. This was evidently the opinion of our captain also … At the instant when I fancied, (and I believe correctly,) that we were in the most imminent peril, a light breeze struck our sails, which were soon filled, and the ship made some headway; then followed a strong puff, and in about five minutes more, we were past all danger. The captain sprang down from the rail, ejaculating, "thank God, thank God!" and he had reason. A fine ship, a valuable cargo, and many lives still more valuable, in all probability, depended on that single puff of wind.

After congratulating ourselves upon our escape, we all turned anxiously to look at the situation of a whale ship, which attempted the passage a few minutes after us … He appeared to have more wind than ourselves, and was coming out beautifully, when suddenly, from a cause to us unknown, he sheered towards the reef, and the next moment, to our consternation and horror, the fine ship struck, hung by her keel, and leaned over 'till her yards were in the water … Captain Shaw, Mr. Skinner, and myself went off to him in our boat … Nothing more could be done for her safety, and all the efforts of the crew were directed to getting out the cargo of oil. Meanwhile, we who could render no service, concluded to go ashore …

Onward to South America

[*5th. This afternoon Mr. Skinner & myself went off to our ship in a boat, (Capt. Shaw went yesterday), set sail & went out to sea immediately. The wreck of the whale ship is still in statue quo & has not yet gone to pieces which is surprising. All hands are busy in getting out her cargo, but it is probable they will not save more than ¹/₃ᵈ. of it. Many of the casks are already stove, & it is probable more will be lost in the attempt to get it ashore. The name of the vessel is the "Oregon of Fairhaven" – Capt. Harding – cargo 2200 barrels of oil – worth perhaps 80.000 dollars! We sailed this afternoon along the N. side of Eimeo [Moorea], a beautiful island only 20 miles from Tahiti, & the next morning, (having had a good wind during the night,) the loom only of the land was seen astern. We are now again upon the wide ocean & I confess I feel more happy than I was when ashore. I love the deep blue sea & I am hastening to my home, that home which whether sleeping or waking is never absent from my thoughts.*

18th. Nothing has lately occurred worth recording. We have had pleasant weather, but the winds have mostly been contrary; we have however an excellent ship, good accommodation & good company, so that we pass our time pleasantly enough. We read, converse & walk the decks during the day, and smoke a segar [sic] & have a rubber of whist in the evening.

25th. We have now a fine fair wind & we are proceeding on our way most gaily, but it has become so extremely cold that we have been compelled to don all our warm clothing & most of us have been sick in consequence of the sudden change of climate. It is now the beginning of winter in this latitude. How often have I compared the weather here with that of our beloved country at this season. We shall however be in Valparaiso at the pleasantest time of year. It is said to be excessively hot there in the summer & very pleasant in winter. **Albatrosses, Cape Pigeons & several species of Petrel** *have been abundant around our ship for some days.*

28th. Our latitude today is 34° 24′ South. Long. 113° 41′ West. & we are making about an average of 3° of longitude per diem. At this rate we hope to reach Vaparaiso in 12 days.

June 1st. A few days ago I had the misfortune to lose one of my pretty Sandwich Island Geese [**Hawaiian Geese/Nene**]. *It had been unwell for a week previously with dyspepsia & died in spite of my utmost exertions for its restoration. The three² that remain are perfectly well, the cold weather appears to have no effect upon them. We are proceeding on our course most beautifully, making 3½ & sometimes 4 degrees of longitude pr. day. Our latitude yesterday was 34° South, Longitude 103°. 20′ West.*

June 9th. 1837. For several days past we have had head winds & light, so that we have made but little progress towards our destination. Our obsᵈ. latitude to day was 34° 17′ S. Long. 81° 7′ W. At 10 o'clock in the morning we made the island of Massafeura [MasAfuera or Isla Alejandro Selkirk] distant about 60 miles (MS Journal).]

June 10th. – When I rose this morning, the island of Juan Fernandez was in sight, distant about thirty miles. The outline is very uneven and rugged, being composed of alternate rough peaks and vallies. We soon approached so near that I distinctly saw, with the glass, a herd of goats bounding over the rocky heights. I felt anxious to set my foot on the shore, hallowed by the romantic narrative of De Foe, but this was impracticable under the circumstances, and I was compelled to abandon it. This is the Botany Bay of Chili; the number of convicts at present is about two hundred.

Arrival at Valparaiso

12th.– We have had fine breezes since Saturday, and this morning, at 8 o'clock, we made the coast of Chili [Chile], distant about fifty miles. The day has been a lovely, clear one, so that we had a fine view of the land until evening closed. We were then within about eight miles of Valparaiso point, and as it was deemed unsafe to attempt to run in during the dusk, we were compelled to lie to all night. The coast here appears exceedingly bold, with a very small portion of level beach. In the back ground, hills rise upon hills to the far distance, where their summits are crowned by the snow-capt Andes. As the sun sank this evening, and gilded with his departing rays the frozen peaks of these lofty mountains, the effect was truly magnificent.

The hills in the vicinity of the shore appear to be totally devoid of vegetation, nor can the eye discern a single shrub in the whole of the vast space comprehended within the range of vision. From our present station, we can see two flag-staffs erected in different situations, upon the tops of two of the highest hills, intended, doubtless, as guide for mariners; and in one of the little valleys, we observe a small, but neat looking village of white houses. The harbor of Valparaiso is deeply embayed within the hills, so that we have not yet had a sight of it.

13th.– Early this morning we passed the point, and came immediately in view of the town of Valparaiso. The houses appeared thickly grouped together, but without any attention to order or regularity, and between these groups, there often intervened large uninhabited spaces, producing the appearance, from the harbor, of several towns. Immediately in front of the bay; and for the space of a quarter of a mile east and west of it, is the principal part of the city, the place of commercial business and fashion. Back and westward of this, are three large groups of houses, occupying the summits and sides of three hills, commonly known to foreigners by the nautical names, *fore-top, main-top, and mizzen-top* ...

Immediately after we dropped our anchor, the captain of the port came on board for the purpose of examining the ship's papers. Then followed the custom-house officers, who also made the requisite investigations, and in about an hour we were allowed to go on shore. We landed accordingly on a large mole in front of the custom house, and Mr. Skinner and myself called upon Mr. Chauncey, of the house of Alsop & Co., to whom we had letters from Oahu, and by whom we were politely received. After sitting about an hour, we strolled out to look at the town. Every thing here is quite new to me; the style of building, the manner in which the streets are laid out, the customs, and even the language of the inhabitants. It is now more than three years since I saw a town which had any pretensions to civilization, and though so far inferior in every respect to our cities at home, yet from my first landing, I have enjoyed the opportunity of seeing an approximation to polite society, generally diffused. I do not mean that I have seen no polite society since I left home; far from it, but the little which I have seen has been so surrounded by baser material, that here where civilization predominates, I am more deeply impressed with the contrast.

Townsend severely ill

August 12th.– Here a considerable *hiatus* occurs in my journal, occasioned by a severe fit of illness which confined me for several weeks to my bed, and from which I did not wholly recover during my residence of two months in Chili [see appendix for surviving specimens from Chile]. I was so fortunate as to meet here a gentleman from Philadelphia, Doctor Thomas S. Page, by whom I was assiduously attended, and to whose skilful and judicious treatment I consider myself indebted for my recovery. I also received much kindness from Captain E.L. Scott and his estimable lady, as well as from a number of the foreign residents and British naval officers in the port.

The political affairs of the country, and the events to which certain important and recently adopted measures have given rise, are worthy a slight notice. [Townsend then gives an eleven-page account of a recent insurrection and the public executions that he witnessed.]

[*At Valparaiso, Chili. August 12th. 1837. I have fallen my dear Sisters, into my usual <u>non-scribbling</u> habits. I cannot exactly say why it is so … But it is perhaps due to myself to say, in excuse for my negligence on the present occasion, that ever since my arrival here I have been very unwell in consequence of a heavy cold contracted at sea, & it has had the effect, not only to produce a great repugnance to writing, but to cause me to view with indifference & sometimes even contempt many things which under other circumstances would perhaps have been gratifying to me. For more than three weeks I was confined closely to my bed, & this induced another disease called I believe by the learned ones, <u>Nosalgia</u> [sic], but which is now commonly known by the title <u>home sickness</u>. I suffered with this complaint most severely, & why? Because I never before had an opportunity of knowing, from actual experience, what I had lost in being separated from all those to whom I am united by indissoluble ties. I tossed upon my bed of pain & suffering, but no one was beside me to pour in the balm of consolation, or to lighten my load by friendly solicitude. Oh did I not wish my excellent father, my mother & my sisters then! I thought of them almost every minute of every day. I had no other occupation than thinking, for during most of the time, my eyes were severely inflamed; it was necessary therefore to keep my chamber constantly darkened, & the least movement of the organs occasioned excessive pain. Are you surprised then that I became home-sick? Do you think I would not almost have given one of my limbs to have been at home? Yes, & I determined to embrace the very first opportunity that offered of whatever kind, & in three months, three little months, I would join you all again. Returning health has not induced me to abandon my resolution, & I have taken passage in a fine brig which sails in a few days direct for Philad[a]* (MS Journal).]

TOWNSEND'S SPECIMENS AND OBSERVATIONS FROM VALPARAISO, JUNE - AUGUST 1837

Although Townsend complained that his illness had "utterly incapacitated me from making collections in Chili" he did manage to gather together over 50 bird specimens of at least 18 species (see appendix 13). Most of them are common and widespread species that he could easily have collected within a few days, and some could have been bought in the local markets. The following extract from Townsend's POPULAR MONOGRAPH OF THE ACCIPITRINE BIRDS OF N.A. – No.1. confirms that at one point, at least, he was well enough to explore the surrounding country:

"THE CONDOR OF THE ANDES …

During my residence in Chili some years since, I had an excellent opportunity of observing this fine bird. Scarcely a day passed in which I did not see dozens of them sailing over the elevated plains back of the town of Valparaiso. Their flight is exceedingly graceful, and is performed by slow flappings of their long falcate wings; sometimes sailing, without any apparent motion of their pinions, the long wings set wide, and the heavy bird passing, while in this position, over a great extent of the country, and at a very rapid rate.

The two sexes are very nearly alike in plumage. Their color is a dark brown or slate, with occasional reflections of bronze or copper. The male is distinguished, both by his greater size, and by having a singular fleshy caruncle on the ridge of the upper mandible. Both sexes, when adult, have a beautiful ruff around the neck, white, and resembling swan's-down in softness and fineness of texture. I killed, while in Chili, a male which measured across the extended wings, twelve feet; and a female whose extent was nine feet. I skinned and prepared this pair of birds, and deposited them, after my return home, in the Academy of Natural Science[a], of this city [Philadelphia].

It is not uncommon, in the town of Valparaiso, to see these birds in confinement; and after having been kept some time they appear to know their master, and to appreciate

An early illustration of the way Andean Condors were captured and killed, just as Townsend described.

kindness at his hands. They are frequently taken about the streets by the Chilian boys, and offered for sale to strangers. The birds being too heavy to be carried, they are dragged along by a cord fastened through their nostrils, which are so large as readily to admit a man's finger being passed through them.

The manner in which these birds are captured is rather singular, and I hope will prove interesting. A large *pen* of from twelve to fifteen feet square is constructed of rails, and raised to the height of four or five feet, but without any covering. In this pen a dead animal, such as a sheep or goat, is placed, and this contrivance forms the whole of the trap, in which the condor is to be taken. The owner then conceals himself, and in a very short time the great vulture is seen sailing with extended wings over the plain in search of food. His quick and piercing eye detects the tempting bait at an immense distance, and with the most beautiful sweepings and gyrations, he rapidly descends, and comes down with a surge directly upon the body of his prey. The man may then approach, if he chooses, even to the enclosure itself, and if the Condor be hungry, – as is almost always the case, for he is one of the greatest *gourmands* in existence, though you will agree with me, not much of an epicure, – he will not leave the repast until he is gorged so perfectly that he can scarcely move … The owner of the pen … not unfrequently secures in one pen several condors, and to remove them he has but to throw his *lasso*, (or noosed cord,) over their heads and drag them out …" (Townsend 1848: 249-251).

a. Only Townsend's male Andean Condor specimen is currently listed there (ANSP 37).

260

AUDUBON'S INFAMOUS MORTON'S FINCH

Audubon erroneously included the Rufous-collared Sparrow *Zonotrichia capensis* in his *Ornithological Biography*, amongst the *Species found in North America but not figured in the "Birds of America"* under the name "Morton's Finch". It was subsequently illustrated in the octavo editions of *The Birds of America*. In his original text he says "A single specimen of this pretty little bird, apparently an adult male, has been sent to me by Dr TOWNSEND, who procured it in Upper California. Supposing it to be undescribed, I have named it after my excellent and much esteemed friend Dr MORTON of Philadelphia, Corresponding Secretary of the Academy of Natural Sciences of that city" (Audubon 1839, 5: 312-313). Audubon has managed to make three mistakes in this short extract: Townsend never went to Upper California, the species does not occur there, and it had been described long ago.

Rufous-collared Sparrow. Audubon erroneously included this South American species in the octavo edition of *The Birds of America*. It appeared in plate 190 as Morton's Finch, a supposed new species discovered by Townsend.

The Rufous-collared Sparrow is a widespread and common species in much of Central and South America, occurring in open country, brush, gardens, etc., and there is little doubt that Townsend brought this species back from Valparaiso, Chile. Townsend was blamed by Stone (1899) for mislabelling the specimen but he was later of the opinion that Audubon "was the one at fault" (Stone 1934). The type specimen shown above (ANSP 10614) has lost Townsend's label but has one that says, in part, "original specimen from Dr. J.K. Townsend Collection. Chili. Pres.[ented] by Dr. Woodhouse" (see also Grinnell 1932: 323).

For some further Central and South American land birds incorrectly included in *The Birds of America* see appendix 15.

The final leg

On the 22d of August, I embarked on board the brig B. Mezick, Captain Martin, bound for Philadelphia, and in the evening, sailed out of the harbor of Valparaiso.

[At Sea – 22ᵈ. August. We sailed yesterday from Valparaiso & so good bye to Chili & now for home. I never loved Chili & need not say that I love home dearly, never so dearly as now when I am flying to it on the wings of the wind. I have long been progressing towards home: Keeping that as the goal, & pursuing my way steadily towards it, but I have met with so many detentions of one sort or other, & been compelled to spend so much time at various places on the route, that until now I have seemed almost as far from home as when I started. Now however I have embarked on board a Brig, (& a noble Brig she is, of 360 tons) which goes right to Philadelphia & does not stop any where by the way, so that at least there seems a fair prospect of my getting home within a reasonable time. If we can only go 'round Cape Horn with fair winds & tolerable weather – a fig for all the rest of it. Nice Captain, excellent Mate, splendid state room, first rate living, & a fast ship – are not these enough to make a man comfortable, – or a woman either? – So as I said before, a fig for all care & hurra for blue water, Cape Horn & home within three months! –

29ᵗʰ. We have now been a week at sea but have yet made but little progress in consequence of light & baffling winds. The weather is dull & gloomy, & occasionally rainy & cold. The dead lights are in our cabin & skylight windows, caulked & sealed, & every thing on board made <u>snug</u> for bad weather. Cape Horn is yet 12 degrees from us, but the breezes which we expect in these latitudes ought to carry us 'round in 8 or 10 days. We have not taken them yet, but are in daily expectation of them. I had no idea that I shd. Have felt the cold so severely, but my peregrinations in warm countries, under the bright sun of the tropics have so enervated my poor frame that I am constantly shivering as though I had an ague. While our people are lolling about in their shirt sleeves & with open bosoms, I am running up & down the quarter deck enveloped in a prodigious pea jacket & trying to get warm. Its very odd, but I supposed I shall soon become accustomed to the change, & shall then probably be better in health than before. I have never been quite well since my arrival in S. America. Birds are very numerous around our vessel. **Albatrosses, Cape Pigeons,** *&* <u>Mother Cary's Chickens</u> [here, predominantly **Wilson's Storm-Petrels**]. *We occasionally catch the larger of them with fish hooks.*

Sept. 1ˢᵗ. Yesterday the expected breeze came & still continues. We are going 10 & 10½ knots by the log. Weather unusually pleasant for these latitudes, but the sea tremendously high. The water is flying & dashing over us from stem to stern, the companion way is closed, & the cabin is so dark & gloomy that I can scarcely see to write. I am braced hard against the table, holding on to the legs of it with my feet & swaying from side to side with the merciless rolling of the vessel. I cannot use an inkstand as it would not stop an instant in my neighbourhood unless it were lashed fast – this operation I am too lazy to perform, so must scribble away with my pencil.* (*JKT footnote: *It is scarcely necessary to remind you that this is a copy of the original.*) *How lucky it is that none of us are sea-sick men. What a sorry pickle we should be in. The sun appeared a little while this morning & we had a glimpse of his pretty face at noon, so we got an observation. Find our latitude to be 46° 53´ South. Longitude 82° West. We have made over 5 degrees of latitude in two days. Excellent – if we go on in this way we shall be 'round the Cape in a jiffy* (MS Journal).]

["About five days after leaving the port of Valparaiso, a single **Albatross** made his appearance, the first we had seen – I happened, at the time, to be practicing with a pair of large horseman's pistols; my target being a porter bottle suspended from the foreyard. As the bird hove in sight, our Captain seized one of my pistols and fired. The ball passed through one of the wings, breaking a long feather, but doing the bird no further injury. Strange to say, this bird kept with us, being easily recognised by the broken and dangling feather. It became a habit with me each morning, to look for the Albatross, and I never was kept long waiting. Indeed the huge bird seemed to have taken a fancy to our ship, (although it must be acknowledged he had not been treated very kindly by us,) and night or day he appeared never to leave us. Whenever the moon gave light, our *consort*, (as we were wont to call him,) was always near us, and for the space of more than three weeks, during which time we voyaged about two thousand five hundred miles, we never knew him to alight upon the surface of the sea. Finding it impossible to procure specimens of these and other marine birds by the use of the gun, I was compelled to resort to the common, though more cruel mode of taking them – baiting a hook and hauling them in by a line like a fish. The **Cape Pigeon** (*Procellaria Capensis*) and many other of the small sea birds are very readily secured in this way; but with the large Albatross the case is widely different. When hooked in the bill it resists with all its might, spreading its long and powerful wings over the surface of the sea, and catching every wave as it is drawn towards the ship. Sometimes the hook is torn out, and then, the evident suffering endured by the poor bird is so painful to behold, that even the callous and unsympathizing naturalist is ready to desist.

It is a curious fact, that neither the Albatross, nor any other of the large sea birds, is capable of rising from the deck of a ship when once landed upon it. They require a yielding surface, such as the element upon which they live, to enable them to commence their flight.

Cape Pigeons.

The little "Mother Carey's Chicken," or "Stormy Petrel," as it is often called, (Thallassidroma Wilsonii,) [**Wilson's Storm-Petrel**] is, I believe, never seen to alight upon the water, it picks up its food, – which consists chiefly of small sea-nettles [jellyfish], and any fatty matter, floating upon the sea, – while on wing, pattering constantly, with its little delicate feet upon the surface. From this well known habit, it originally acquired its name, – Petrel, – from its walking upon the water, like Peter attempted to do, when he would have met his Divine Master upon the sea of Galilee … (Townsend 1847: 91-92).]

September 7th.– During the past week we have had some Cape Horn weather – rain, snow, and hail, but happily, no ice. The sea has been tremendously high, and still continues so, with the weather excessively cold. We may, however, consider ourselves peculiarly favored, as not a day has passed, in which we did not see the sun and ascertain our longitude. Probably the greatest difficulty and danger of this vicinity is the constant darkness and gloom which is its usual characteristic. You are in consequence, unable

Light-mantled Sooty-Albatross, plate CCCCVII. Audubon used a Townsend skin in the preparation of this plate. Audubon thought that the specimen had been collected off the Oregon coast but it is far more likely to have come from the southern oceans (see appendix 10). Audubon's text for this species is as follows: "DUSKY ALBATROSS DIOMEDEA FUSCA … The skin from which I made my drawing of this species was prepared by Mr. Townsend, who procured the bird near the mouth of the Columbia river. Of its habits or distribution I am entirely ignorant. Having failed in finding any figure or description of an Albatross agreeing entirely with it, I have been induced to consider it as new." (Audubon 1840-44, 7: 200). The only surviving Townsend specimen of this albatross lacks details about the collecting date and locality (USNM 2718). In his *Ornithological Biography*, Audubon also described the "Yellow-nosed Albatross" [Gray-headed Albatross *Thalassarche chrysostoma*], "Slender-billed Fulmar" [Southern Fulmar] and "Gigantic Fulmar" [Northern Giant Petrel *Macronectes halli* or Southern Giant Petrel *Macronectes giganteus*]. All his descriptions were based on Townsend specimens that were supposed to have been obtained off the Oregon coast, yet Audubon did not depict any of these three species in *The Birds of America*. Once again they were almost certainly all collected in the southern oceans (see appendix 10).

to ascertain your true position by observation, and dead reckoning furnishes but an insecure guide when powerful currents are impelling you to leeward, and drifting your vessel towards the most frightful of all dangers, a rocky lee-coast. We have now doubled the cape, and are steering N.E., the island of Diego Ramirez bearing W. 130 miles. We have therefore left the Pacific, and our now in the South Atlantic ocean.

[*25ᵗʰ. My sisters will no doubt say that I neglect my journal overmuch, that I ought to write a little every day, & if nothing happens worth recording, to describe my thoughts, my feelings &c. Some persons might do this, & perhaps I might myself, did I ever feel in the vein, but I know not why it is, the moment I get on ship board I lose all inclination for scribbling unless some event occurs which I consider worth mentioning. As events are very rare on board ship, & nothing at all happens to vary the dull monotony, I lose my itch for writing, & spend the time in reading, in conversation with the Captain, in working the time, & and in lolling about the deck. Besides I am not quite well, never having entirely recovered from the heavy cold taken on board the ship from the Sand. Islands. By the way – I only lately discovered the real cause of my taking this heavy cold, & if I had known of it before I embarked, the Sand. Islands might still hold me before I would have encountered such a risk. The reason was, that the place called the run, which is a space directly under the cabin, was filled with salt as cargo. This salt was wet, & the consequence was that the cabin was always damp. So long as we were in warm weather no bad consequences resulted from it, but the moment we entered a*]

Northern Giant Petrel.

colder climate, its effects were visible on all the passengers, not one of whom escaped. It was a most unfortunate & ill-starred voyage for me, as it utterly incapacitated me from making collections in Chili – a country exceedingly rich in Natural history - & has left me with deep seated complaints of which I fear I shall never perfectly recover. If I ever again see the Capt. of the Europa, I shall certainly expatiate warmly upon the impropriety of carrying salt in the run of his vessel. Well as I was saying, my being a little sick is another reason why I do not write more in my journal, but enough of this – The weather has now become so warm that we have been compelled to throw off most of our winter clothing – the sea is smooth & every thing about us is delightful. Why? We are approaching the lovely tropics & are now within about 10 days sail of the line … I look at the chart about 20 times every day to ascertain the progress we are making, & as often make <u>accurate</u> calculations about the time of arrival. My impatience is becoming very great, & you know I never was remarkable for a superabundance of Job's virtue … We are now sailing along the east coast of Brazil, Rio Janeiro bearing west, about 400 miles (MS Journal).]

October 8th.– We are within about two degrees of the tropical line, and, with good breezes, only about twenty-five days sail from the capes of Delaware. Oh, who can describe the anxious longings of him who is approaching his beloved home, after having been long separated from it, or depict his feelings, his ardent soul absorbing feelings, in the prospect of soon holding to his bosom the dear beings who are twined around every fibre of his heart!

[*Oct. 8th. …Yesterday we passed the latitude of Pernambuco, and are now steering N.W. along the northern coast of Brazil … This neighbourhood is a common piratical cruising ground. It is infested by the buccaniers of the western coast & is moreover in the very track of vessels bound to the opposite shore of Africa for cargoes of slaves. These slavers are said to be dangerous people to meet on the open sea at any time but particularly on their return passage when they [are] often disappointed in their plans relative to their execrable traffic, & are then willing & glad to supply the deficiency by plundering the rich merchantmen who may happen to fall in their way. The worst of it is, that they are not generally satisfied with plundering simply, but murder the captives & scuttle the ship to prevent exposure. Now this is a sort of end that I very much wish to escape, & methinks it would seem rather hard to be made food for fishes here within a few days of home, after having – but I say no more. I do not wish to magnify any danger or difficulty that I may have escaped during some three & ½ years of wandering by flood & field … I have no ambition to be a hero, nor the least particle of a "lion", & if when I arrive at home, people expect me to talk & endeavour to entertain them, they will find them selves much mistaken, that's all – for I shall be as "silent as an oyster" – I have become very taciturn, very – I can't talk at all except to my dear Sisters & to them I shall pour out my heart …*

The birds, (Albatrosses, Cape Pigeons &c) all disappeared from our track about a week ago, & in their place we have now the little fairy Petrels called <u>Mother Cary's Chickens</u> [here, predominantly **Wilson's Storm-petrel**]*, an occasional Tropic Bird* [**Red-billed** and/or **White-tailed Tropicbird**]*, myriads of bright flying fish, & often a splendid Dolphin is seen glancing like silver & gold amidst the flashing waters, & sometimes following for miles in our wake. We caught one of these beauties with a hook a few days ago, a fine fellow of about 12Hs. weight, & I had a good opportunity of observing his surpassing & rapidly varying lustre as he fluttered & flounced in his dying struggles upon the quarter deck. I think I have described this before, so shall not repeat it now, suffice it to say that it is <u>wonderfully</u> beautiful. The little <u>Mother Cary's Chickens</u> are very pretty & curious birds which you have no doubt often heard of. They appear never to rest on the water like most other sea birds, but when engaged in feeding on the surface, keep their little webbed feet constantly pattering upon it, & their wings elevated & slightly in motion & at such times are nearly stationary. For many days I exerted myself almost incessantly to catch one of them with a fish hook as most sea birds are taken. I prepared a light line & fastened to it a number of bent pins which I baited with little pieces of pork, but the small creatures were too cunning for me. They would always stop & take the fragments of grease that I scattered on the sea as a lure, but had no notion of touching those which were attached to the hooks. Observing that they sometimes heedlessly struck the line in their flight, an idea of another sort was suggested to me. I attached a number of long lines of fine thread to the stern & suffered them to veer out in the wake of the vessel without any baits. Then I threw from time to time* *a quantity of grease on the water. The little hungry birds darted at it immediately & catch it who can was the order of the day, but the catcher was soon caught. A poor little [word missing] flew against one of the lines, it twisted around his wings & I brought him struggling & fluttering 'till within reach of my hand. Then my murderous finger & thumb were insinuated into his pretty plumage, & my old method of <u>compression</u> terminated his captivity & his life. In a few minutes afterwards I took another in the same manner & the next day prepared them both. I have described at length my mode of taking these little birds because I believe it is generally supposed impracticable to catch them at Sea, but I am often inclined to believe that a person <u>determined</u> to succeed, seldom fails in any undertaking how difficult soever it may at first appear. And now farewell for the present – for it is too warm here in the cabin to write any longer, & I must go on deck ... (MS Journal).]*

On the 13th of November, we made Cape Henlopen [Delaware], and took a pilot on board, at the distance of four miles from land. The next day we ran in, and anchored within view of the light-house, during a heavy N.E. gale. In the night we were so unfortunate as to lose successively both our bower anchors, and were compelled to run out to sea again. The day following, however, was clear; we procured another anchor at the breakwater, and had a fine run of forty-eight hours to the city. I again trod the shore of my native land, after an absence of three years and eight months. I met again the dear relatives and friends, from whom I had been so long separated, and who had been spared in mercy to welcome the wanderer to a participation in the inestimable blessings of Home[3].

1. The Long-tailed Cuckoo was first described by Anders Sparrman, a naturalist on Cook's second Pacific voyage (1772-75). It breeds in New Zealand, but winters throughout the islands of central and southern tropical Pacific.

2. Townsend says he bought three geese on 18th February 1837 (MS Journal). He seems to have acquired a fourth goose because after one dies he talks about the three that remain (MS Journal 1st June 1837).

3. In the original edition of the *Narrative* the concluding page is followed by a scientific appendix detailing Townsend's principal mammal and bird discoveries.

TOWNSEND'S BIRD SPECIMENS FROM BUENOS AIRES, ARGENTINA

In the current USNM database Townsend is given as the source of at least twenty bird skins from Buenos Aires, of at least eleven different species. It has been stated that one of these, *Myospiza humeralis* USNM 25261 was collected by Townsend, and by implication also the remaining specimens (Deignan, 1961, p. 638). There is no evidence that Townsend collected these birds himself. In Townsend's MS Journal he says that he boarded a ship at Valparaiso bound directly for Philadelphia and indeed no stops are mentioned. Moreover, there are no Buenos Aires birds mentioned in Lord Derby's list of duplicates. These Buenos Aires specimens must have been obtained from a correspondent after Townsend's voyage was completed, or obtained indirectly by purchase or exchange while back in the United States. It may be significant that a James Baird of Buenos Aires was supplying birds to Sir William Jardine in 1829 (Jackson and Davis 2001, pp. 213, 217).

PLATE CCCLIV.

A classic Audubon painting from *The Birds of America*, perhaps the very best of all those featuring songbirds (Plate CCCLIV). Western Tanagers (above) painted from Townsend's specimens and Scarlet Tanagers (below).

Townsend's later life

Audubon paints Townsend's birds

"I have purchased Ninety Three Bird Skins! Yes 93 Bird Skins! – Well what are they? Why nought less than 93 Bird Skins sent from the Rocky Mountains and the Columbia River by Nuttall and Townsend! – Cheap as Dirt, too …"

Audubon to Rev John Bachman, 23rd October 1836.

Back home at last

Townsend sailed up the Delaware River to Philadelphia in mid November 1837 after an absence of three years and eight months. It was now only a short journey from the wharf into the city to 138 South 10th Street[1] where at last he was re-united with his parents and those brothers and sisters that still lived there; later he would see the rest of the family, his uncles, aunts and cousins, as well as his friends and neighbours.

He attended the next meeting of the Academy of Natural Sciences which happened to be a few days later, on 21st November, and heard the news concerning his collections. His first small consignment of birds, sent back from Independence at the end of April 1834, had been virtually ignored. He had left instructions for a description of a new species (later to become known as Harris's Sparrow) to be published by the Academy but there had been no such outcome (see pp. 47-50). Only within the last few days would he have learnt that the much larger and more precious specimen collection, sent from the Columbia River a year and a half earlier, had arrived intact. With any such shipment there was always the worry of attack by insects, damage by salt water, theft or loss, and the not too unlikely possibility that the ship would sink on the passage around Cape Horn. It must have been a huge relief to find that all had arrived safely, but there must have been little else to please him. Some of his new species had been described in his absence and Audubon had made off with a lot of the birds, thereby denying Townsend the right to describe all his discoveries in the way he wanted, with the names he wanted, in the publication he wanted. On top of this, one of his sponsors started to demand a share of the collections.

A view of Philadelphia, from J. Cundee's *The Stranger in America* (1807).

Townsend versus the American Philosophical Society

Before setting out on his expedition across the Rocky Mountains, Townsend had appealed for financial support from the Academy of Natural Sciences of Philadelphia and the American Philosophical Society. Indeed there was a written agreement with the treasurer of the Society to the effect that the sum received, $125, would go "towards his expenses on a Journey &c. for making collections of Natural History for the Cabinets of the American Philosophical Society and the Academy of Natural Sciences, who subscribe a like sum. The two societies to be put on equal footing ... After supplying both Societies, the duplicates to be the property of Mr. Townsend." This was signed by Townsend and witnessed by Charles Pickering. The underlining appeared when this agreement was quoted back to the Academy by the Philosophical Society because Townsend refused to hand over the Society's share of the birds and mammals, and because the Society wanted a meeting with officials at the Academy to discuss the problem (APS to Morton, 21st December 1837. APS).

Dr Samuel Morton, at that time Corresponding Secretary of the Academy, had already declined to form a committee to discuss the issue on the grounds that the Academy "as a body" had not subscribed to Townsend's expenses, the contribution having come instead from individual members. Morton also indicated that Townsend would repay the full amount (Morton to G. Ord, 19th December 1837. APS).

This did not please the Society, who quoted back extracts from an earlier letter from the Academy (of 21st July 1836) that announced the arrival of the first specimens from the Columbia River. It was pointed out that Morton had said that "it was proposed as most in accordance with his (Mr. Townsend's) wishes, that the collections should remain undivided until his return". It is added – "Of the Shells, however, a set has been taken out, for the Academy; and if desired by the Phil. Socy., a similar set shall be at once sent to that Institution." (APS to Morton, 21st December 1837. APS). In other words, it had suited the Academy to act as a body in securing a share of the shells.

The dispute dragged on for months, and though Townsend met the Society's committee either as a group or individually, all they got from Townsend was "a small collection of shells, a few ordinary specimens of rocks, and some Indian Vocabularies: the whole of trifling value." The committee was increasingly frustrated that "Mr. Townsend positively refused to give to the Society any of the Birds and Quadrupeds &c of which he has a large and valuable collection; although by his agreement with the Treasurer, Mr. Vaughan, he came under contract to do so" (APS [to ANSP], 30th April 1838. APS).

The American Philosophical Society never did get their share of specimens from Townsend, and nearly three years later their Historical Committee was still discussing what to do with Townsend's Indian vocabularies (APS Minutes, 10th July 1840 and 8th January 1841). Quite why Townsend was so reluctant to give up his specimens was never made clear, though he obviously believed that $125 was insufficient claim on the specimens that were worth far more. There may also have been some personal animosity towards George Ord who is known to have been a difficult character and was very much a champion of the late Alexander Wilson rather than Audubon (see e.g. Peck 1999). The situation was probably complicated by the fact that many of the members of the Philosophical Society were also members of the Academy. Indeed, Townsend later found himself working under Charles Pickering and Titian Peale, both of whom had been on the Philosophical Society Committee.

In order to continue our story, we must now go back a few years to the time when the first fruits of Townsend's labours in the west started to arrive at Philadelphia.

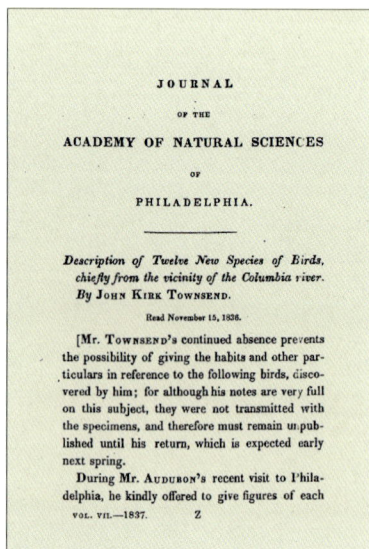

Some of Townsend's discoveries were described while he was still away, on the Columbia River.

In Townsend's absence

In September 1835, when Nuttall left the Columbia River on board the *Ganymede*, he took part of Townsend's main collection with him as well as his own. He spent about four months in the Hawaiian Islands before heading home via California and Cape Horn, but Townsend's consignment either continued in the *Ganymede* or was transferred to another ship, arriving on the east coast in the following year. On 12th July 1836 *Waldie's Circulating Library* announced that "Last week the Academy of Natural Sciences of Philadelphia received safely from them [Nuttall and Townsend] many large boxes; among Mr. Townsend's collection alone are three hundred birds and fifty quadrupeds, many of which are unknown to naturalists." Nuttall did not arrive at Boston until 20th September 1836, on the *Alert*, so he could not have travelled all the way back with the specimens as has sometimes been suggested (e.g. Herrick 1938, Simpson 1999). Rumours about the number of new species obtained were rife. Audubon was particularly aroused by news from Richard Harlan who said "Nuttall & Townsend had forwarded about 100 New Species of Birds from the Pacific Side of the Rocky Mountains!" (Audubon to Harris, 7th September 1836, in Corning 1930: 127). And while Audubon was in New York he heard news so tantalising that he wrote at once to Bachman:

> "I had a letter from Ed*d* Harris yesterday, in which he enumerates the New Species of N.A. Bird collected by Nuttall & Townsend as follows. – "Several Hawks, Clarkes Crow, a Jay resembling the Floridian [Scrub-Jay], Lewis Woodpecker and several New others, Evening Grosbeaks, Purple Grosbeaks [Lazuli Bunting?], several New Fringillas [finches], several thrushes, one Swallow, a black & white Sylvia [Black-throated Gray Warbler], a Towee Bunting [Spotted Towhee] (Pipilio arctica a good species [now separated as Eastern Towhee] which I found in Labrador also), Arctic blue Bird [Mountain Bluebird], Red winged Woodpecker [Red-shafted Northern Flicker] and several others, a minute Titmouse [Bush-tit], several regulus [kinglets] & Wrens, Two or Three Tringas [sandpipers], a New Snipe, a vireo, several Fly catchers, Band Tailed Pigeon, A New Shrieke [shrike], and Many others. Harlan writes me that this collection belongs to the A.N.S. of Phil*a* but that Townsend has Duplicates of all of them: Now I am anxious to Pourtray all those, and to publish them in an appendix to My present Work; but I have some doubts whether these Gents will allow me to do so?" (Audubon to Bachman, 10th September 1836, in Corning 1930: 130-131).

In reply to Harris's letter, mentioned above, he wrote on 12th September 1836, from New York:

> "Whilst running over the interesting list of the Species of Birds procured by Nuttall & Townsend in the Rocky Mountains, and the shores of the Pacific, I became so completely wrapped up with the desire to see these as soon as possible that I have concluded to go to Philadelphia tomorrow by the 10 o'clock boat … You well know how anxious I am to make my work on the Birds of our Country as compleat as possible within my power: you know that to reach this end I have spared neither time, labours or money … Well the desiderata has come to Philadelphia at least in part, and if I could be allowed to pourtray the new species now there as an appendix to the Birds of America, I should be proud and happy to do so, but do you think that the Academy is likely to indulge me in this my wish?" (in Herrick 1938, 2: 147-148).

Audubon's question was soon answered: the Academy would not allow him to examine the birds until Nuttall himself was present. Completely thwarted, Audubon returned to New York and on 20th September he travelled by steamer and train up

Edward Harris (1799-1863), a wealthy landowner who befriended and assisted Audubon. Harris and Audubon travelled together on the Gulf of Mexico and Missouri River expeditions.

to Boston, in pursuit of subscribers. There he met Nuttall. A few days later, on 25th September 1836, he wrote to Harris to let him know that: "Nuttall has arrived – he breakfasted with me the other day – gave me 6 new species of Birds [see appendix 14] and tells me that he will urge both Townsend and the Society at Philadelphia to allow me to portray all the species which they have procured within the limits of our Territories" (in Herrick 1938, 2: 151). But how was Townsend to agree to this? He was at the mouth of the Columbia River and only contactable by a lengthy and unreliable process. It should also be remembered that the birds found in the vicinity of the Columbia River were not from "within the limits of our Territories" i.e. the United States, but were from British territory.

Audubon desperately wanted to include all the new birds in the last part of *The Birds of America* that he was trying to bring to a conclusion. It was easy to persuade Nuttall to part with his few bird skins because he had more than enough to do with the hundreds of plant specimens he had to work through, besides which his reputation as a naturalist was well-established, he cared little for more credit, honour or recognition and was just happy to help out. Audubon wrote a rather garbled message to Bachman, already beginning to revise the early estimate of the total number of new species: "I found Nuttall when he arrived at Boston from California he has given me 6 new species of Birds found *within the limits* of our Territories, and there most positively fully 40 more new ones procured by himself and Townsend on both sides the Rocky Mountains. Nay Good Friend he assures me that they *saw* many more, among which were 3 species of Jays! But Wonders will never end – no, not even in Ornithology." (Audubon to Bachman, 2nd October 1836, in Corning 1930: 133-134).

On 15th October Audubon set off from New York for Philadelphia where he stayed with Richard Harlan. It may have been easy to secure Nuttall's birds, but the skins held at the Academy were proving much more difficult to get hold of, because the skins really belonged to the Academy where members and officers, particularly George Ord, were not inclined to favour Audubon – a rival to the sacred memory of Alexander Wilson. Audubon was regarded by Ord as a "contemptible impostor", so when the artist arrived at the Academy he still had some tough negotiating to do:

"Dr Townsend's collection was at Philadelphia; my anxiety to examine his specimens was extreme; and I therefore, bidding farewell to my Boston friends, hurried off to New York, where, in a week, I added names to my list of subscribers … Once more my son and I reached Philadelphia, where at once we placed ourselves under the roof of my ever staunch and true friend Richard Harlan, Esq., M.D., with whom we remained several weeks … Having obtained access to the collection sent by Dr Townsend, I turned over and over the new and rare species; but he was absent at Fort Vancouver, on the shores of the Columbia River; Thomas Nuttall had not yet come from Boston, and loud murmurs were uttered by the *soi-disant* friends of science, who objected to my seeing, much less portraying and describing those valuable relics of birds, many of which had not yet been introduced into our Fauna. The traveller's appetite is much increased by the knowledge of the distance which he has to tramp before he can obtain a meal; and with me the desire of obtaining the specimens in question increased in proportion to the difficulties that presented themselves. Having ascertained the names of the persons best able or most willing to assist me on this occasion, and aided by Thomas Nuttall, who had now arrived, Drs Pickering, Harlan, S.G. Morton, Secretary to the Academy of Natural Sciences, M'Murtrie, Trudeau, and above all by my friend Edward Harris, who offered to pay for them with a view of presenting them to me, I at length succeeded. It was agreed that I might *purchase duplicates, provided* the specific names agreed upon by Mr Nuttall and myself were published in Dr Townsend's name. This latter part of the affair was perfectly congenial to my feelings, as I have seldom cared much about priority in the naming of species. I therefore paid for the skins which I received, and have now published such as proved to be new, according to my promise. But, let me assure you, Reader, that seldom, if ever in my life, have I felt more disgusted with the conduct of any opponents of mine, than I was with the unfriendly boasters of their zeal for the advancement of ornithological science, who at that time existed in the fair city of Philadelphia" (Audubon 1831-39, 4: xi).

Audubon wrote at once to Bachman, rather giving himself away as he gloated over his success:

"Now Good Friend, open your Eyes! Aye, open them tight!! Nay place specks on your probosis if you chuse! Read aloud!! Quite aloud!!! I have purchased *Ninety Three Bird Skins*! Yes 93 Bird Skins! – Well what are they? Why nought less than 93 Bird Skins sent from the Rocky Mountains and the Columbia River by Nuttall and Townsend! – Cheap as Dirt, too – only one hundred and Eighty-Four Dollars for the whole of these, and hang me if you do not echo my saying so when *you see them*!! Such beauties! Such rarities! Such Novelties! Ah my Worthy Friend, how we will laugh and talk over them!" (Audubon to Bachman, 23rd October 1836, in Corning 1930: 135-136).

Well, dear Reader, what do you make of that? It seems to us that Audubon was not so much thinking about the advancement of science, nor the absent Townsend from whom he so unashamedly stole all the glory, but was rather more thinking of himself. For years he had belittled those who sought to win priority in the naming of new species but now he was sucked into doing so himself, conscious too that the more new species in *The Birds of America*, the easier it could be to persuade new people to subscribe, though he knew too that additional plates brought extra labour upon himself.

Audubon soon set to work on the skins, by 6th November sending his painting of the Mountain Plover to the plate makers in London (Corning 1930: 137). He then went from Baltimore down to Charleston to stay with the Bachmans for the winter, where "I opened the box containing Dr. Townsend's precious series of birds ... drew upwards of seventy figures of the species which I had procured in Philadelphia, assisted in the finishing of the plants, branches of trees, and flowers, which accompany those figures, by my friend's sister-in-law Miss M Martin ..." (Audubon 1831-39, 4: xiv). More paintings were then shipped to London, via his friends the Rathbones in Liverpool. On 13th February, feeling over-tired and morose, he wrote to his family outlining what should happen if he died before *The Birds of America* was completed: his son Victor, and William MacGillivray, were to continue with the *Ornithological Biography* using Audubon's journals and notes, the notes of Bachman and Nuttall, possibly with "the assistance of Young Townsend should he return from the Pacific". In case the last two batches of drawings be lost at sea, he had taken the precaution of making outline copies with the suggestion that the birds could be filled in by his sons John and Victor, or MacGillivray (Corning 1930: 142-143).

	Double-Elephant Folio	Royal Octavo Edition
	435 plates	500 plates
	1065 figures	970 figures
	489 named species	500 named species
Minimum number of plates with Townsend specimens	31 (of 41 species)	45 (of 45 species)
Minimum % of plates with Townsend specimens	7%	9%

Townsend's contribution to *The Birds of America*.

Fries (1973: 99) says the Nuttall-Townsend collection provided about 70 figures for more than a tenth of the 435 plates in the Double-Elephant Folio. Nuttall's few birds provided only about seven figures so the percentage of plates with Townsend specimens may be as high as 10%.

The number of species recognised today is slightly lower than the number claimed by Audubon.

All the birds in plate CCCXCIII of *The Birds of America* were painted by Audubon from Townsend's specimens. Townsend's Warbler (male, top left); Mountain Bluebird (male and female, below); and Western Bluebird (male and female, at bottom).

JOHN JAMES AUDUBON AND *THE BIRDS OF AMERICA* (1827-38)

The life of Audubon (1785-1851) is too well documented to bear much repetition here (see e.g. Herrick 1938, Ford 1964, Chalmers 2003). Suffice to say that he was born in Haiti, the illegitimate son of a French sea captain and his mistress who died shortly after the baby's birth. When John was six years old his father took him to France to be raised by his wife. Audubon emigrated to the United States in 1803, married Lucy Bakewell in 1808 and after a series of failed business ventures hit upon the idea of illustrating all the birds of North America, representing each bird as life size.

With a wife and two growing sons to support, Audubon bravely embarked upon a business venture of huge proportions and extraordinary vision. The project consumed his life for more than twenty years, involving travels throughout the eastern United States in search of subscribers and specimens. Having failed to win support for his work in Philadelphia he was advised to seek out an engraver in Europe, so he sailed three times to Britain to examine birds in museums, oversee the production of the coloured plates and to liaise with his co-author on the text.

The first ten plates were produced in Edinburgh by William Lizars but owing to a disagreement with the colourists Audubon took his custom to London where Robert Havell produced all the remaining plates. Over the years Audubon provided 433 paintings for the Double-Elephant Folio edition, mostly watercolours, and these were turned into 435 plates. For the first 352 plates there was one species per plate (except CXLI which had two). After that, as more species came to light and as subscribers began to grow weary of the expense, he began to produce plates with several different species in order to speed up the process. It is unlikely that more than 200 complete sets of plates were produced.

The text to accompany the plates was issued as the five-volume *Ornithological Biography, or An account of the habits of the birds of the United States of America* (1831-39). It was written by Audubon, with the assistance of the Scottish naturalist William MacGillivray who polished up Audubon's English, added scientific descriptions and provided some anatomical drawings. *The Birds of America* was far too expensive for most people so a smaller, seven-volume, octavo edition (1840-44) was produced in which each species had its own text and plate.

The subjects for Audubon's paintings came from a range of sources. Some of the birds he collected himself, of course, others were given to him by friends or bought in the markets. When he was in Britain he visited the museums of Edinburgh and London that contained birds brought back from the Canadian fur countries or from the high arctic by British naval expeditions in search of the Northwest Passage. Just when Audubon thought his project was tending towards completion, Townsend returned from the West with more birds. Some of them were completely new to science, some were very rare in collections and were new to Audubon, while other specimens were simply better than he had seen before. Audubon acquired more than 134 of these skins[a], of an unknown number of species, so Townsend probably provided more skins for *The Birds of America* than any other individual (with the exception of Audubon himself). But it is difficult now to be sure just how many Townsend specimens Audubon painted because he did not always specify who provided his subjects. Sometimes he refers to Townsend specimens in the text, but does not actually say if they were used in the preparation of the plate[b]. At other times he was more specific, or we know that Townsend skins were painted because there was no other choice available at that time.

a. Audubon is known to have used: 12 bird species examined from Townsend's first batch sent from the Columbia River (Townsend 1837); and 93 bird skins (Audubon to Bachman, 23rd October 1836); and 29 skins in the McEuen list of 25th October 1839; plus an unknown number of additional specimens.

b. Not all the Townsend specimens mentioned in Audubon's text were painted, e.g. the California Condor juvenile and some of the seabirds. Sometimes single specimens were painted from two or more different angles, e.g. Townsend's only Surfbird specimen was painted from two different viewpoints, and the original watercolour of Townsend's Bunting portrays the unique specimen in three ways.

A Townsend discovery finds its way into *The Birds of America* and the *Ornithological Biography*

Mountain Plover shot and preserved by Townsend near South Pass, Rocky Mountains, June 1834.

↓

Specimen carried overland by Townsend to the Columbia River. In September 1835, sent by ship to Philadelphia, via the Hawaiian Islands and Cape Horn

↓

Specimen arrives at Academy of Natural Sciences of Philadelphia, July 1836.

The "Rocky Mountain Plover" is described by Nuttall and others, during Townsend's absence. Description appears in the *Journal ANSP* in 1837 under the scientific name *Charadrius montanus*.

↓

Specimen purchased by Edward Harris for Audubon, October 1836.

Audubon paints the Mountain Plover watercolour by 6th November 1836. The painting is sent by ship to England.

↓

In London, Robert Havell copies the watercolour, engraving it onto copper.

↓

The Mountain Plover appears in *The Birds of America* as plate CCCL, dated 1836.

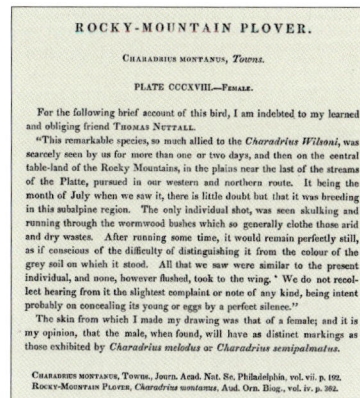

Audubon writes the draft text on the Mountain Plover from Townsend and Nuttall's notes and the specimen label.

↓

Audubon sends his text to Edinburgh where William MacGillivray revises and corrects Audubon's English.

↓

ROCKY-MOUNTAIN PLOVER.

CHARADRIUS MONTANUS, *Towns.*

PLATE CCCXVIII.—FEMALE.

For the following brief account of this bird, I am indebted to my learned and obliging friend THOMAS NUTTALL.

"This remarkable species, so much allied to the *Charadrius Wilsoni*, was scarcely seen by us for more than one or two days, and then on the central table-land of the Rocky Mountains, in the plains near the last of the streams of the Platte, pursued in our western and northern route. It being the month of July when we saw it, there is little doubt but that it was breeding in this subalpine region. The only individual shot, was seen skulking and running through the wormwood bushes which so generally clothe those arid and dry wastes. After running some time, it would remain perfectly still, as if conscious of the difficulty of distinguishing it from the colour of the grey soil on which it stood. All that we saw were similar to the present individual, and none, however flushed, took to the wing. ' We do not recollect hearing from it the slightest complaint or note of any kind, being intent probably on concealing its young or eggs by a perfect silence."

The skin from which I made my drawing was that of a female; and it is my opinion, that the male, when found, will have as distinct markings as those exhibited by *Charadrius melodus* or *Charadrius semipalmatus.*

CHARADRIUS MONTANUS, Towns., Journ. Acad. Nat. Sc, Philadelphia, vol. vii. p. 192.
ROCKY-MOUNTAIN PLOVER, Charadrius montanus, Aud. Orn. Biog., vol. iv. p. 362.

Mountain Plover text appears in 1838 in the *Ornithological Biography*, volume IV, p. 362; and again in 1842 in the octavo edition, volume 5, page 213 (above).

Rev Dr John Bachman (1790-1874), Lutheran pastor and naturalist. He described the new mammals collected by Townsend and wrote the text for Audubon's *Viviparous Quadrupeds* (1845-54).

Bachman's house in Charleston, South Carolina, where Audubon painted many of Townsend's birds (from F.H. Herrick's *Audubon the Naturalist* (1938)).

THE REV DR JOHN BACHMAN (1790-1874)

John Bachman was a lifelong naturalist, born of Swiss-German descent at Rhinebeck in New York State. His interest in wildlife was originally stimulated by one of his father's slaves, who taught him how to trap birds and mammals in the nearby woods. While a young student, and afterwards for a couple of years while teaching Latin, German and French at Milestown School, he enjoyed being part of the dynamic group of naturalists centred on Philadelphia. He improved his plantsmanship at Bartram's Garden, hunted birds in the company of Alexander Wilson and began corresponding with Alexander von Humboldt after meeting the German scientist when he visited the city. From an early age Bachman recognised the importance of basing his zoological research on the careful observations of wild creatures. He loved to ride far afield, was a superb shot and had an exceptionally good ear for songs and calls, a talent that helped to make him a skilful nest-finder.

In 1815 he became pastor of St John's Lutheran Church in Charleston, South Carolina where he ministered for the next fifty-five years. He married Harriet Martin who bore him fourteen babies (five died in infancy) and they shared their home with her mother and Harriet's petite younger sister, Maria.

Bachman kept rodents and a large aviary. A pet Ruddy Turnstone was allowed to peck crumbs off the dining room table; a Brown Thrasher roosted on his writing chair and Bobwhites, hatched under his bantams, often wandered around beside him in the study as he wrote his sermons. It was no wonder that Audubon felt completely at home there during his first visit to Charleston in October 1831. This was the beginning of a stimulating, joyful, turbulent and hugely productive association between the two families. The pastor contributed bird skins, observations and ideas to Audubon's great ornithological enterprise, and was rewarded by having three birds named after him: Bachman's Warbler, Bachman's Sparrow and the Black Oystercatcher *Haematopus bachmani*. Bachman's Warbler and Swainson's Warbler were unknown to science until the pastor collected them in the countryside near his home.

Both of Audubon's sons fell in love with Bachman girls: John married Maria in 1837 and Victor married Mary Eliza in 1839, but the women, like many others in their family, were afflicted with tuberculosis and the men were quickly widowed, John being left with two little daughters.

In 1839 Bachman described nine new mammals collected in the west by Townsend (see appendix 2). During the next thirteen years he devoted almost all his spare time to providing the text for *The Viviparous Quadrupeds of North America* (1845-54). He warned the Audubons at the outset that it would be a much harder task than writing about the North American birds, since so little was known about the mammals, especially the nocturnal species. He lamented, "The books we have are really worse than nothing" and insisted, "the Book must be original & credible — no compilation & no humbug". His professional responsibilities prevented him from visiting distant libraries and museums as much as he would have liked, so he depended heavily on the Audubons to send him skins, notes and papers, but the usually mild-mannered and patient pastor was frequently angered and frustrated by their lack of support. Fortunately, he already had a large network of more reliable naturalists, on both sides of the Atlantic, willing to provide him with specimens and vital information. In all, he described just over twenty new North American mammals, seven of them ostensibly with Audubon as co-author.

Two years after the death of his first wife, Bachman married Maria Martin, who, having helped to illustrate *The Birds of America*, now devoted herself to helping her husband with the mammals, reading to him when his eyes were sore, taking dictation, transcribing notes from borrowed books and helping to edit text. John James Audubon died in 1851 and during his closing years, as he became increasingly demented, his sons had to take over the burden of all the artwork and business responsibilities. Finally, on 26th March 1852, Bachman wrote to a friend, "The last mammal on hand was described five minutes ago. I threw my hat to the ceiling, kicked books, papers, rabbit and squirrel skins and bats about the room, and felt that the nightmare of some years was off my breast" (Bachman to J.P. Kirtland, quoted in Shuler 1995: 210). The awesomely ambitious undertaking had damaged his health and strained his relationship with all the Audubons, but it assured his lasting reputation as one of the great naturalists of his generation.

Bachman and his beloved flock suffered greatly during the Civil War, but he survived all the dangers and outlived both wives, his sons-in-law and most of his friends. He died on 24th February 1874, following a stroke, at the age of eighty-four.

In the spring, Audubon went on a collecting trip along the Texas coast as far as Galveston Bay and Houston, with his son John and Edward Harris, but the Audubons were back on the east coast by early June. He was not becoming paranoid about the safety of the new bird skins, but was just being sensible, when he wrote from Edward Harris's home at Moorestown, New Jersey, on 2nd July 1837, asking the Bachmans to "lose not a moment in dividing up My Collection of Rocky Mountain Bird Skins as equally as you can and ship the better half by [the *Superb*] to Liverpool care of Messrs Rathbone Brothers and Co and the other half by the Vessel that takes the Rummed Birds, the Blue Grosbeak, the large case of Books &c and the Flying Squirrels which we altogether forgot to mention – the latter Vessel goes to London" (in Corning 1938: 161-162).

Audubon sails to Britain

The reason for all this activity, of course, was that Audubon was off to Britain again to oversee the production of the plates for *The Birds of America* by Robert Havell in London, and to visit MacGillivray who was working away in Edinburgh on the text for the *Ornithological Biography*.

He also wanted to do some comparative work in the British museums and was able to report from London on 4th October 1837 that he had proved that Clark's Crow was a true nutcracker – he said it was the third species known to exist, but only two nutcrackers are now recognised. He found that Catesby had long ago published the "Rocky Mountain Blue bird of Nuttall", and that the Least Flycatcher given to him by Nuttall was the *Muscicapa nigriscens* of Swainson [Black Phoebe] – "so that after all poor Townsend did only obtain 8 species that were yet unknown" – that is, of the twelve first published, not the whole collection because Townsend had not yet arrived home. But Audubon was wrong on both counts when he said that "the Rocky Mountain Garrot of Swainson, (Fuligula Borrowi [sic]) [Barrow's Goldeneye] is only the Summer plumage of Clangula vulgaris [Common Goldeneye], and next that his Artic [sic] three toed woodpecker [Black-backed Woodpecker], is only the young of the one which I have figured" (4th October and 20th December 1837, in Corning 1930: 184, 192).

While still in London, Audubon heard from Harris that Townsend had at last reached home. He replied to Harris on 26th October 1837 (in Rhoads 1903): "The return of Dr. Townsend to our happy land has filled me with joy, and trebly so when you tell me that he is as friendly disposed to me as I ever have been towards him. I congratulate you, my dear friend, in the step which you have so kindly taken in my favour, by first selecting all such Birdskins as you or Townsend have considered as new, and also in having given freedom to Dr. Morton [who was acting as Audubon's agent] to pay Dr. Townsend Fifty Dollars for the skins selected by you". Audubon went on to say: " I wish you to urge him in forwarding me either his own manuscripts or a copy of all such parts as appertain to Birds, as soon as possible, knowing (I think) that he will not undertake to publish them himself under his present (I am sorry to

William MacGillivray (1796-1852) Scottish naturalist and friend of Audubon. He co-wrote Audubon's *Ornithological Biography* (1831-39) the text that accompanied the plates in *The Birds of America* (1827-38).

say) embarrassed pecuniary circumstances." He also wanted as much information from Townsend as he could supply and wanted him to go through all the plates for *The Birds of America* at the Academy and note those that were seen in the Rocky Mountains or the Columbia River, or at sea off the Pacific coast. He asked Harris to urge Townsend to be careful in naming birds as new, offering to name all the birds in London where he had access to more published works. "Do not take this as egotism far from it, it is in friendship and for his sake that I venture on undertaking such an arduous task." He further reduced his estimate of Townsend's first batch of new birds that had been described in the Academy's *Journal*, from twelve novelties to only six – and complained that it would cause him problems when it came to the text because the names would be different from the plates. He also urged Harris to get eggs from Townsend and made a particular request for goldeneye skins.

Referring again to the new birds from Townsend and Nuttall, Audubon said that they were causing difficulties with his subscribers. Now that he had the specimens, they were proving to be a bit of a curse. He had to draw them, he had to include them in *The Birds of America*, he had to get information about them and he also had to find even more subscribers to make the extra labour worthwhile. There happened to be a financial crisis in many quarters and even the very wealthy were pulling out of their obligation (that over ten years had cost them up to $100 a year). At first Audubon had insisted on one bird per plate, then one species per plate (e.g. male, female and juvenile together) but now in the latter stages there were many composite plates with two to six species depicted: Townsend's smaller birds, especially, suffered from this treatment. Further cutbacks were also necessary as many patrons pulled out or refused to take any additional plates. The pictures of birds' eggs had to be omitted for lack of space, and then Townsend had inconveniently turned up with yet more new species. As a final drastic compromise, a full description of some of Townsend's birds appeared in the *Ornithological Biography* but with the words NOT FIGURED preceding the text (e.g. Western Gull, Black-footed Albatross, Gray-headed Albatross, Giant Petrel, "Pacific Fulmar", "Slender-billed Fulmar") or, were not mentioned at all, as in the case of Vaux's Swift.

In the meantime, Audubon wrote from London to Samuel Morton, thanking him for making the $50 available for Harris so that a portion of Townsend's birds could be acquired. Though it was never fulfilled, he also noted his good intention to work closely with Townsend: "Would that I could spend some three weeks with Townsend, & indulge in the hope that by then, he would have confided to my care, the whole of his <u>notes</u> upon the <u>Birds</u> which he has observed during his several tedious journeys, or sojourn at [Fort] Vancouver." He could not resist adding a request as a postscript: "P.S. Please to tell Townsend that [if] he has brought any Eggs and Nests of Birds to lend them to me for a while. – especially those of the Carthartes Californianus [California Condor], which are <u>said</u> to be black??? And all <u>possible</u> information which he may please put at my disposal for publication as <u>coming from him</u>!" (Audubon to Morton, 26th December 1837. APS). In the same letter, Audubon showed some sympathy for the young collector when he ruefully reflected upon his humble homecoming: "To me it seems that Townsend on his return from the shores of the Pacific, laden with knowledge unknown, would have been hailed in his native city … That some one hundred … very wealthy Gents, would at once have proffered not only their hand to him, but … in an instant, made a purse of at least one poor thousand dollars for him but alas! It seems that this was not, or has not yet been the fact …"

Two days later Audubon also added a postscript to a letter he had drafted to Bachman, full of expectation:

"28th December 1837. – My Dear Bachman – Since writing the *few* lines therein, I have received a letter from Friend Ed*d* Harris, announcing the return to Phil*a* of Doc*r* Townsend with an increase of new ornithological species amounting if all correctly told to upwards of Twenty species, some of which I expect are very curious, and others beautiful. – Harris has bought specimens of each for me, and I suppose that at this moment they are under way from New York to London as they were to have been shipped on the 20th Instant. Townsend also is sending me a parcel of them for sale on his account – I also expect from him some information respecting Habits &c I am glad that he has sent you his new Hare and other quadrupeds[2]. – Should the Birds come in time I will of course publish the whole of the new species, and my Work will be *the* Work indeed!" (in Corning 1930: 195).

Here was important news. Good news in that Townsend was home with yet more birds and mammals, but Audubon and the skins were an ocean apart and he had to decide how to present extra bird plates to an increasingly restless list of subscribers. In January Audubon wrote to Morton, sending him "5 More numbers of my Work, in which you will find all the New Species of Townsend purchased by me, when at Phil*a* last. I shall think that [when] his Enthusiastic eye glances over them, he would be pleased; at least I hope that he may be so! I regret that Townsend's whole collection was not at once entrusted to me; for between you and I, and the post, methinks I might have done more for him [in London] than will be effected in America?" (Audubon to Morton, 19th January 1838. APS).

On 6th February 1838 Audubon wrote to Harris expressing his bitter disappointment that he had not shipped out the Townsend skins at once so that they would have arrived six weeks before. It was frustrating, but there were difficulties in Philadelphia where Townsend was still in dispute with the American Philosophical Society over their share of specimens. "My Publication is almost at a stand," complained Audubon, "for as I am forced to finish my Work in as few number of Plates as possible (not to Lose any more subscribers in this country) I am forced to introduce as many new species of the same Genera in the same plate, as circumstances will afford. – And Now I am at a stand-still for the Birds which you have mentioned in your interesting lists ... If I do not figure the new species of Townsend, no one else will, for it would not pay any one but me, and the Skins would be suffered to rot in the Public Institutions of Phila ... (in Corning 1930: 196-198). Audubon urged Harris to stir up Townsend to send to London a complete list of all the birds seen in the West, where they were seen and when, their relative abundance, whether they were migratory or resident, their measurements, habits and plumage changes and so on. "If he has brought *any Eggs*, ask him to *lend* these to me. – My Plates of Eggs will be the last to come out." Audubon had heard that Trudeau was thinking about purchasing some of Townsend's birds and he knew already that Lord Derby was also keen. He even thought that Townsend's fishes would sell well in London. But mostly Audubon railed against those who were causing him so much trouble: "I am sadly afraid that all my Sayings are useless and that the Philadelphia *Sharks* have already preyed upon them all. – Wonderful Friends of Sciences the Messrs Ord & Co! – *If I cannot have the Skins*, pray send me *Townsend's close descriptions* of all those which you both consider New, that I may at least publish this in my last Vol. of letter press" (in Corning 1930: 196-198).

Audubon spent a total of three years in Edinburgh between the years 1826-39. (Canongate Tolbooth, Edinburgh, from T.H. Shepherd's *Modern Athens* (1829)).

The artist wrote again to Harris on 4th April 1838: "I have made bold enough to name a New Woodpecker after you, it is another species of Hairy Woodpecker from the Columbia River sent to me by Townsend, and I think you will be quite astonished to see that at this moment no less than 19 species of this interesting Tribe are in my published Plates – Townsend has sent Two New Oyster Catchers [both were Black Oystercatchers], Two New Cormorants [Pelagic and Brandt's Cormorants], a new something [Surfbird] which I intend to name in the way of a subjenus [sic] allied to Tringa, a New Burrowing Owl or one described by Temminck and a New species of Bombicilla [waxwing] … I have drawn One hundred Birds; indeed I feel quite fagged, and now think that I never laboured harder than I have done within the last Two Months" (in Corning 1930: 202). Though all the birds mentioned were rare in collections at that time only the Black Oystercatcher was undescribed (though it had been painted in Russia (Mearns and Mearns 1992: 113)).

If Harris and Bachman had passed on Audubon's request for measurements and other information, Townsend seems not to have responded[3]. Audubon was now asking Townsend directly for measurements as well as breeding biology and distribution data. "Harris tells me that you are sending your skins &c over for me to sell on your account. I shall do as well as I can and send you returns immediately, had you sent 20 Cocks of the Plains [Greater Sage-Grouse] they would have brought 100 pounds! And your insects would sell well also. At the British Museum they are anxious to have these if in good Condition, but send them at once and whatever else you choose, I will try my best for you." He went on to say that his work in England would be finished in 21 days, that plates up to 425 had been delivered, and only ten more were due, mostly already etched on to copper. He then listed the species and plate numbers from 381 to 434, many of which had been supplied by Townsend, and asked him to send him notes on their distribution in the west. He was still pestering Townsend about sightings without specimens and asking for endless details. "What sort of a <u>Pheasant</u> with a long tail did you see on the mountains? What teal with red shoulders? … Let me know in what situations the Californian vulture breeds, on the ground or on trees? How many eggs? Colour of those? And the colour of the eyes of this bird? … Do they attack living animals. Do they feed on reptiles? Fish? Or what. – are they gregarious or not, where do they roost?" On and on he went, scribbling several more lines of questions, before finishing with "my Dear Townsend God bless you" (Audubon to JKT, from London, 31st May 1838, in McDaniel 1963: 208-210). It would have been better if he had simply sat down with Townsend in the same room for a week or two – and paid him for the information.

The batch of birds that Audubon was to sell on behalf of Townsend evidently arrived very soon after this letter was written, because only a fortnight later, on 15th June 1838, Lord Derby copied out a list of the birds and made a selection, perhaps at Derby's London house rather than at Knowsley Hall, his Liverpool home. There were 278 bird specimens on the list and 4 mammal specimens. The birds were from the Hawaiian Islands (121 specimens), Chile (56), Tahiti (31), Columbia River (42), North West Coast (2), California (1), Atlantic Ocean (2) and the Pacific Ocean (3). The mammals were from the Columbia River. Lord Derby selected not more than twenty specimens – a disappointing total (see appendix 4).

On 20th June 1838 Audubon saw the final folio plate of *The Birds of America* completed by Robert Havell. It showed a pair of American Dippers (a single Townsend specimen drawn in different positions) beside a waterfall, supposedly near the Columbia River. But Audubon still needed to complete the *Ornithological Biography* so he travelled up to Edinburgh towards the end of June to confer with MacGillivray. While he was there he off-loaded the rest of Townsend's skins to Macduff Carfrae, a natural history dealer. Surely Audubon could have done better than that? Any wealthy collector in Edinburgh[4] or London would have given him a better price. (For an account of these Townsend birds and their subsequent fate, see pp. 319, 321, 323).

Robert Havell junior (1793-1878), the London engraver for almost all the plates for *The Birds of America* (1827-38). The first ten plates were engraved in Edinburgh by W.H. Lizars, and Havell's father had a hand in some of the other early plates.

Still short of information for some North American species, Audubon wrote home on 1st July 1838 to his son John, asking him to look around the house and find all the birds' nests brought back from the Columbia River by Townsend and Nuttall, including one of a Bewick's Wren, and to pack them all up and send them across to Britain, along with a few other things, mostly notes and journals. A few days later John Bachman arrived in Edinburgh to join Audubon for a couple of weeks, before going back down to London and on to Germany to speak to an international gathering of naturalists at Freyburg. Bachman had with him several of Townsend's mammals, because only in Europe could he find suitable reference collections where he could compare skins and skulls and come to satisfactory conclusions about the identity of each specimen. Fortunately, even while Bachman was in Edinburgh, a long letter from Townsend caught up with him. It was full of notes about the various mammals collected in the West and was of course referred to when Bachman read a paper on North American tree squirrels to the London Zoological Society on 14th August 1838 (Bachman 1839a); and he quoted from it extensively in other papers describing the majority of Townsend's new mammals in the *Journal of the Academy of Natural Sciences of Philadelphia* (Bachman 1839b, 1839c. JKT letter untraced, presumed now lost or destroyed).

Audubon wrote to Bachman on 27th October 1838 and seemed decidedly pleased that Nuttall and Townsend had supposedly missed a spectacular bird during their travels[5], reinforcing his opinion that they had not done as much field work as he himself would have done: "The Columbia Jay is found there [on the Columbia River and in California] but has escaped Nuttall & Townsend!" (in Corning 1930: 206). But this bird, the colourful Magpie-Jay, is restricted to Central America, an area never visited by either of the two naturalists. Although Audubon reported that it was given to him by "a friend who had received it from the Columbia river" it may have come from William Swainson, who acquired a number of Mexican species from William Bullock. The mistake was Audubon's or, to be generous, perhaps it was the fault of MacGillivray or the anonymous friend.

Still belittling Townsend's efforts, Audubon wrote to Bachman with typical arrogance, on 27th October 1838: "I have this day received a very long letter from Townsend, but alas containing so very little, that I cannot understand how he spent his whole days and years at Fort Vancouver – Only think of his not having taken the measurement length and breadth of above 20 species during the whole time. – That he saw many new birds which he has not procured I am now certain but good God, what a Pity that you or I, or Harris had not been there, oh what might we have done, that he has left undone. – Pheasants with long Tails, almost black, several remarkable Woodpeckers, one as large as the Ivory billed or Larger [see p. 110], &c &c &c but who Knows? I may yet ransack that country!!" (in Corning 1930: 207-208).

The way in which Townsend spent his time at Fort Vancouver is obvious: besides working as a surgeon for the Hudson's Bay Company for nearly six months, he hunted for birds and mammals, laboriously prepared skins and skulls, gathered insects, shells, reptiles and a few plants (see appendix 1). When Audubon, the great naturalist, eventually *did* go West, up the Missouri in 1843 with a few companions, he spent so much of his time Buffalo hunting that he neglected one of the primary aims of the trip – catching small mammals – and on his return earned a scathing rebuke from Bachman: "I am afraid the broad shadows of the elk, buffalo and big-horn hid all the little marmot squirrels, jumping mice, rats and shrews. Why man, what poor trappers you proved yourselves to be" (in Ford 1965: 411). Eighty-five mammal specimens were prepared but they found not one new mammal[6]. Audubon described twelve new birds from this expedition but not one of them seems to have been actually collected by Audubon himself and the number of true species still credited to him from this expedition is only six. Whereas most of Townsend's specimens had labels with collecting dates and broad localities, Bachman complained that Audubon and his friends had not taken good care of the specimens, had not made good notes and there was "as usual nothing labelled" (see Tyler 2000:162).

1. While Townsend was away, although his family had long ceased to live at his childhood home at 105 Chestnut Street, it had been sold by his father and completely demolished to make way for three "modern stores". The sale of the three houses for $17,000 had allowed his father to pay off all his debts leaving "a clever balance besides" (H. Townsend to JKT, 29th November 1835. Family papers).

2. Townsend evidently lost little time in sending some mammal specimens to the Rev Bachman since Bachman referred to them as early as 12th December 1837, saying: "I have before me a box full of Rocky Mountain Quadrupeds. When I shall have time to describe them, I know not, but think of going to work in a day or two." (J. Bachman to Edward Harris, 12th December 1837, in C.L. Bachman 1888: 159).

3. Audubon's published version of the events in the *Ornithological Biography* (1831-39: 4: xxi-xxii) is slightly at odds with the difficulties he experienced in extracting specimens and information from the collector: "Dr. Townsend was extremely desirous that everything new or rare belonging to our Fauna, should be given to me, I received only a few [,] weeks before closing the engraving of my plates. A few others did not reach me until several days after ... All Dr. Townsend's species as well as some received through different channels have been published." By this Audubon meant that descriptions of all the birds were included in the text, although, as we have already mentioned, several species were never painted by him and some were not included.

4. One notable exception was Sir William Jardine, a wealthy landowner near Lockerbie in the south of Scotland, who was especially keen on foreign birds. Unfortunately for Townsend, Audubon and Jardine were no longer the friends they had been in 1826. At Edinburgh, they had seen each other often and Audubon had more than once demonstrated his painting techniques to Jardine who became an eager early subscriber to *The Birds of America*. Their relationship had cooled when Jardine's brother-in-law, W.H. Lizars, had abruptly ceased work on *The Birds of America* after engraving only ten plates. The rift was completed when Jardine, without permission or acknowledgement, copied plate VI, depicting a Wild Turkey hen and chicks, to use in volume 20 of his own *Naturalist's Library* (Chalmers 2003: 59, 159-161, 164). Jardine never completed his subscription and Audubon criticised the *Naturalist's Library* as "trash" – because many of the volumes were compilations lacking original work (Fries 1973: 10, 65, 88). Audubon would not have wanted to resume contact with Jardine just for the sake of a few Townsend specimens.

5. Townsend and Nuttall did miss a jay, but not the Magpie-Jay. While travelling rapidly between Laramie's Fork and Fort Hall, they passed through the breeding range of the Pinyon Jay *Gymnorhinus cyanocephalus* but did not collect or even report seeing one. It was discovered in 1833 by Prince Maximilian of Wied, between the Marias and Yellowstone Rivers though Wied did not publish a description until 1841.

Townsend did not miss many other birds during his time on the Columbia River: White-tailed Ptarmigan and Mountain Chickadee because he did not go to high elevations; Black-backed Woodpecker; Barrow's Goldeneye perhaps; a few shorebirds such as Black Turnstone; Western (and Clark's) Grebes he saw but could not shoot. Others such as California Gull were collected but not recognised as new species at the time. Even Cassin's Vireo was collected, although he did not differentiate it from the Solitary Vireo. It is true that Nuttall missed quite a lot in California but he did not travel extensively there, nor was he making bird collecting a priority. In due course, Nuttall's protégé William Gambel found most of the western birds remaining to be discovered: Nuttall's Woodpecker, Mountain Chickadee, Wrentit, Gambel's Quail, California Thrasher and Plain Titmouse [now split into Oak Titmouse and Juniper Titmouse].

6. The Black-footed Ferret *Mustela nigripes* was described by Audubon and Bachman from a specimen sent from Fort Union by Alexander Culbertson of the American Fur Company – after Audubon and his companions had returned home.

Townsend in Philadelphia

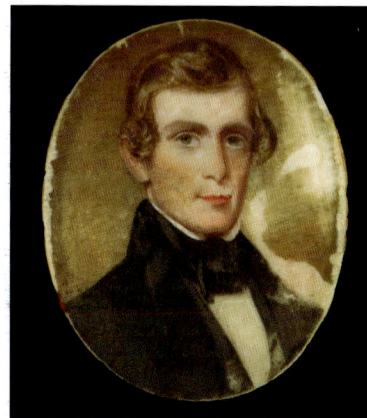

John Kirk Townsend, watercolour on card, presented in 1904 to the Academy of Natural Sciences of Philadelphia by Townsend's son, Frank Holmes Townsend. It may have been painted by Townsend's wife, Charlotte, a portrait painter.

"You have acted wisely in sending your birds to Audubon – his pencil will immortalize them – depend on it."

Rev John Bachman to J.K. Townsend, 7th February 1838.

Now that he was back in Philadelphia, Townsend's most important task was to earn a living. The financial contributions from the American Philosophical Society and the Academy had barely sustained him during his time in the West and he had supplemented his capital only when working for the Hudson's Bay Company as a doctor at Fort Vancouver. He had a slight income from the sale of some specimens but, as we have seen, he often received much less than he should have done. Townsend therefore decided to take advantage of his newly found modicum of fame and set to work on an account of his experiences.

During the first part of 1838 he visited family and friends and once again became a regular attendant at the Academy meetings. At the end of January he began his long and happy correspondence with the Rev John Bachman who was already examining some of Townsend's mammal skins at his home in Charleston, South Carolina. In time, Bachman would describe and name several of Townsend's mammals, but he wrote back asserting that he was reluctant to describe anything as new: "I believe this may be traced to the fact of many of my brother naturalists being such a set of jealous quarrelsome gentry and the fear that I may unintentionally be found preaching on the ground preoccupied by someone else." He went on to offer Townsend advice on his forthcoming book, suggesting that everything that Townsend knew about the birds and mammals should be placed in an appendix to the main narrative. Those mammals named after Townsend would need to be given under Bachman's name but the remainder could be named jointly or by Townsend on his own. "You have worked for these specimens, and yours is all the credit, and that which will induce most to your interest and fame will please me best," was Bachman's generous response (J. Bachman to JKT, 7th February 1838. PUL). He concluded by saying, "You have acted wisely in sending your birds to Audubon – his pencil will immortalize them – depend on it. Audubon's Book is destined to become a standard work on Ornithology. It will I hope however come to us in a small size and cheaper form."

Bachman wrote again a few months later, thanking Townsend for another batch of mammal skins, making detailed comments and queries about each of them, and summarising those that he believed to be new to science. He confessed that his health was poor and that his doctor had advised him to take a long holiday. He was therefore planning to go to Europe soon, but hoped to complete his current researches on the Townsend material before leaving (J. Bachman to JKT, 24th May 1838. PUL). Townsend replied quickly, sending a letter in June about the mammals collected in the West, reserving the longest passage for the White-tailed Jackrabbits that he had often hunted near Fort Walla Walla. But Bachman had already left, sailing for Liverpool at the beginning of the month, and the letter did not catch up with him until he was in Edinburgh with Audubon.

Dr Ezra Michener (1794 - 1887), a Pennsylvania naturalist who compiled a list of birds of Chester County (1863). Townsend stayed with Michener while writing his *Narrative of a Journey across the Rocky Mountains* (1839).

Deciding eventually that there were too many distractions in Philadelphia, Townsend moved out of the city to New Garden to stay with his friend Dr Ezra Michener. In rural isolation he was able to make much better progress and it was here that he completed the bulk of his famous *Narrative*. In early autumn, Townsend wrote to his friend Dr Morton saying:

"I am snugly ensconced here at Michener's & engaged in writing – I make so much better progress here where I can be quiet & retired, that I am in great hopes of soon being able to finish. Michener is very kind & obliging, & we get on very smoothly. As a recreation I occasionally mount a dried skin for him ..." (JKT to S. Morton, 6th September 1838. APS).

In his *Autobiographical Notes*, Michener confirms that the bulk of Townsend's book was written at his home about thirty miles to the west of the city:

"He was a correct ornithologist, an expert taxidermist, and a good shot. His leisure and inclination led him often to my place, where he spent days and even weeks at a time; indeed after his return, the larger part of the Narrative of his journey to the Columbia River and Sandwich Islands was written in my studio" (Michener 1893: 48-49).

Townsend's book was published by Henry Perkins of Philadelphia, with the long title of *Narrative of a Journey Across the Rocky Mountains, to the Columbia River, and a Visit to the Sandwich Islands, Chili, &c.*, and was dated 1839. It is said to have sold out quickly, but there was no second edition in the United States. Some issues of the Philadelphia *Saturday Museum* were issued with chapters from the *Narrative*, and later, without Townsend's approval, some chapters were offered in newspaper form to new subscribers. There were hopes to issue "an octavo, pamphlet edition [of the whole *Narrative*] which can be afforded at 25 cents to 37½ cents" – but nothing seems to have materialised (JKT to H. Townsend, 6th June [1843]. Family papers).

After 1838 had come to a close, Townsend was one day walking around Philadelphia when he was somewhat surprised to recognise a twenty-year old full-blooded Chinook whom he had known on the Columbia River, now in the care of Dr William Blanding. After three years with western missionaries the Chinook spoke good English, had adopted the name William Brooks, and happily allowed Dr Morton to take his skull measurements for Morton's forthcoming *Crania Americana* (1839). Townsend must have noted with satisfaction his own contribution to Morton's book: the provision of skulls from various tribes on the lower Columbia River, and a wooden cradle used in the head flattening process. Some of the skulls were hideously distorted and most of them had been filched from their resting places at no little risk to his safety. Many times had Townsend attended to the sick but his care and concern for the physical well-being of those living along the river seems never to have extended to those that had gone to the grave.

During 1839 he began to catch up on ornithological matters, and brought out a series of papers in the *Journal of the Academy of Natural Sciences*. The first two were rather unnecessary as his own descriptions of the same species appeared at an earlier date in the scientific appendix to the *Narrative*, but perhaps he had submitted entries to both to make sure that whichever appeared first would carry the descriptions, the earliest publication date being critical in matters of nomenclature. The first paper was a "Description of a New Species of Cypcelus, from the Columbia River" read on 5th March (Vol VIII, p. 148). At first he had thought that the swifts seen nesting in hollow trees on the Columbia were Chimney Swifts (Jobanek and Marshall 1992: 8, 11) but comparison with other skins had now convinced him that it was a new species, and he named it "Cypcelus *Vauxii*" dedicating it to his friend "Wm. S. Vaux, Esq., of Philadelphia". It subsequently acquired the English name Vaux's Swift (for Vaux, see Mearns and Mearns 1992: 467-469).

VAUX'S SWIFT

The type specimen of Vaux's Swift, collected by Townsend in the forests bordering the Columbia River (ANSP 24169). Townsend named it in honour of his friend William Sansom Vaux (1811-1882), an eminent mineralogist and patron of the Academy of Natural Sciences of Philadelphia.

Townsend's descriptions of Vaux's Swift in the *Journal of the Academy of Natural Sciences* and in the scientific appendix to the *Narrative* were very similar, except that in one he gave the bird an English name and in the other he only gave it a scientific name:

"Vaux's Chimney-Swallow …is common on the Columbia river; breeds in hollow trees, forming its nest in the same manner as the *pelasgius* [Chimney Swift *Chaetura pelagica*], and lays four white eggs" (Townsend 1839a: 348).

"Cypcelus *Vauxii* … It differs from the C. [Cypcelus] *pelasgius* [Chimney Swift *Chaetura pelagica*], with which it has been confounded, in several very striking particulars. It is one inch shorter, and two inches less in extent; the body is proportionably smaller in every aspect, and the color much lighter.

This species (which I dedicate to my friend, Wm. S. Vaux, Esq., of Philadelphia) is common on the Columbia river; breeds in hollow trees, forming its nest in the same manner as the *pelasgius*, and lays four white eggs" (Townsend 1839d: 148).

There is no mention of Vaux's Swift in Audubon's *Ornithological Biography* and no plate in *The Birds of America*, even though this species is listed as No. 117 among a batch of birds lent to Audubon by the Academy of Natural Sciences of Philadelphia sometime prior to 25th October 1839 (see appendix 5). Field numbers in Townsend's catalogue suggest that he first took this species in spring 1835 (see appendix 6).

Townsend's second paper, "Description of a New Species of Sylvia, from the Columbia River," was read on 2nd April, and is a straightforward description of "Sylvia *Tolmoei*, Tolmie's Warbler" (now MacGillivray's Warbler *Oporornis tolmiei*) that he named for his friend "W.F. Tolmie, Esq., of Fort Vancouver" (for Tolmie, see Mearns and Mearns 1992: 436-440).

The year of 1839 also saw the publication of John Bachman's papers dealing with Townsend's mammals. As we have seen, one of these was first read at the London Zoological Society in August of the preceding year, and it concentrated on tree squirrels collected by David Douglas, Townsend and others; the Townsend specimens described were Downy Squirrel [Red Squirrel], Columbia Pine Squirrel [Red Squirrel] and Douglas's Squirrel [Chickaree]. The other papers were read in Philadelphia in Bachman's absence on 7th August 1838, and appeared in print in the 1839 *Journal* of the Philadelphia Academy under various titles. They included the original descriptions of most of the new mammals collected by Townsend, namely Townsend's Mole, White-tailed Jackrabbit, Least Chipmunk, Townsend's Ground Squirrel, Townsend's Vole, Oregon Vole and Townsend's Pocket Gopher.

For recreation Townsend re-visited his old haunts on the New Jersey coast, in July collecting such birds as Northern Mockingbird, Eastern Towhee and Red-winged Blackbird (that still survive at Oxford University Museum of Natural History, see appendix 7A). There were probably other attractions there besides birds, as it may be assumed that he had by now made the acquaintance of his future wife, Charlotte Holmes[1]. It is thought unlikely that he knew her well in the months before he went West, because she would have been less than 16 years old and when he wrote home there was no mention of her in his letters to his sisters. So it may have been around this time that he first began to be a regular visitor to the Holmes' farm[2] just north of Cape May Court House. Charlotte was the third of five children of Robert Holmes and Mary Leaming, and although Robert Holmes was to die in 1840 he had provided a basic education for all his children. Charlotte might also have received some artistic training because she was already known locally as a portrait painter. The extensive property ran along Holmes Creek to Holmes Cove, forming the southwest corner of Great Sound and included woodland, farmland and tidal saltmarsh, an ideal mix of habitats in which Townsend could go a-birding, perhaps with one or other of Charlotte's three brothers.

For some time Townsend had thought that there could be a market for a small book on the birds of North America, something more up-to-date and wider in geographical scope than Wilson's *American Ornithology*, something cheaper and more accessible to the public than Audubon's vastly expensive *The Birds of America* and the *Ornithological Biography*, and yet better illustrated than Nuttall's *Manual of the Ornithology of the United States and of Canada*. From Cape May, he wrote to Samuel Morton informing him that he had teamed up with publisher John B. Chevalier and that they had "concluded to procede with the "Ornithology" so long under Consideration". He wanted Morton to provide a certificate signed by himself and the vice presidents of the Academy confirming that "the enterprise will be eminently deserving the approbation & patronage of the American public". Townsend felt that "As our frd. Mr. Ord is out of the way, I fancy there will not be much difficulty in the matter". He also reported that he had been confined indoors for two weeks "by a severe fit of my "old enemy" (JKT to Morton, 4th August [1839]. APS). This is not necessarily a reference to the illness he had endured in Valparaiso because in a later letter to his sister he says "I have been quite unwell with a complaint which has caused me much Suffering at intervals for years, & with which our dear Mother has been frequently afflicted" (JKT to H. Townsend, 6th June 1843. Family papers).

Unfortunately for Townsend, his bird book was doomed to failure. Audubon had just returned to the United States and wrote to Morton (9th September 1839. APS) from New York outlining plans for his own forthcoming books:

"Having determined to publish the Quadrupeds of North America in our Country, of a handsome folio size, all the objects drawn on stone by myself, I hope to be able to present that work to the public at a price which ought to enable me to meet the expenses of such a publication without difficulty? I am also going to Publish immediately a New Edition of the Birds of America (Royal octavo) the figures of which will all be reduced from the plates or original Drawings of my Large Work. These two publications will be issued in mo^{thly} numbers, along with the matter appertaining to each of the subjects &c. &c. &c."

Audubon finished the letter by asking: "How is Townsend, what is he doing, and has he kept any of his specimens of the quadrupeds which he procured during his sojourn on or about the Rocky Mountains? Is he still engaged with M [= Monsieur]. Chevalier and has the latter continued his publication of Birds?" At that time it seems as though Townsend's book was still progressing but news that the long-rumoured octavo edition of the birds was definitely going to be published must have been a bitter blow. Bachman had warned him of the possibility back in 1838 (J. Bachman to JKT, 7th February 1838. PUL), but perhaps Townsend hoped that Audubon's mammal project would distract him[3].

At the meeting of the Academy on 10th September 1839 Townsend (1839e) read a note that was a thinly veiled criticism of the way Audubon had dealt with "Tolmie's Warbler". Townsend had always believed that it was a new species and yet Audubon had assumed it to be the Mourning Warbler, a well-known eastern species. When Audubon saw Townsend's description of Tolmie's Warbler in the scientific appendix he realised his mistake and in volume five of his *Ornithological Biography*, he called it *Sylvia Macgillivrayi* – this name giving rise to the common name in use today: MacGillivray's Warbler. Townsend commented that "If I had been aware, before the publication of my appendix, of Mr. Audubon's wish to name this bird, I should have adopted his appellation with cheerfulness; but as his intention was never communicated to me, the name of S. *Tolmoei* [sic], which I have given it, having priority, must of course be retained." It has been retained (but with his earlier spelling *tolmiei*) thus explaining how this bird came to commemorate two different naturalists.

For the 10th September meeting Townsend also compiled a *List of the Birds Inhabiting the Region of the Rocky Mountains, the Territory of the Oregon, and the North West Coast of America*. Regrettably, it was just a list of birds seen, without embellishment, and of so little use with its all-embracing geographical span that it caused confusion among later ornithologists, some of whom claimed species for Oregon State on the strength of this list even though certain species were only ever seen by Townsend on the plains a little to the west of Independence, Missouri (see also Jobanek 1994).

On 25th October Townsend got together with Thomas McEuen at the Academy and together they wrote out a list of birds that had been lent to Audubon, presumably so that the specimens could be retrieved from him at some future date. These were not the duplicates Audubon had bought but the more precious specimens, several of them unavailable at that time in any other museum in the world, e.g. Mountain Plover, Sage Thrasher and Chestnut-collared Longspur (see appendix 5). In December Townsend became one of the Academy curators and continued in that position for twelve months.

In 1840 Audubon started work on the paintings for *The Viviparous Quadrupeds* having already secured the help of the Reverend Bachman who assumed responsibility for the text. He wrote to Bachman saying that "I have seen a good deal of Townsend of late, and am sorry to say have lost much towards him, he has become or perhaps always was Lazy and careless in the extreme and hardly speaks of those *who have befriended him when in need* in sufficient words of gratitude" (Audubon to Bachman, 15th February 1840, in Corning 1930: 232). This oft-quoted passage casts Townsend in a very bad light (see e.g. Ford 1967: 364), yet it can be examined in another way and the question that needs to be asked, is how exactly had Audubon befriended and helped Townsend? Audubon had bought specimens from Townsend and painted them. He had published Townsend's observations and made Townsend's name more widely known. But these actions were motivated by self-interest, to improve the scope and quality of his own projects, so that he himself could prosper. Although he may have performed some unrecorded and truly charitable act towards Townsend there is no evidence that he gave him any generous help at all. As has been so perceptively and succinctly observed, "Audubon had, unquestionably, the gift of being able to extract from every sympathetic soul the last ounce of service to his interests. Wherever his scientific pursuits led him … there seems to have appeared, in city and in wilderness, a man Friday to minister to his needs. Bachman, Harlan, and Townsend, all, were angels of grace in this respect" (W.B. McDaniel 1963). In Townsend's case that last ounce of goodwill had become pretty well exhausted.

The reality was that Audubon had sucked Townsend completely dry. He had acquired Townsend's bird specimens cheaply, indeed he had been pumping Townsend for specimens and data ever since Townsend had returned from the West, first for birds, nests and eggs and now for his mammals, always expecting help for nothing. On top

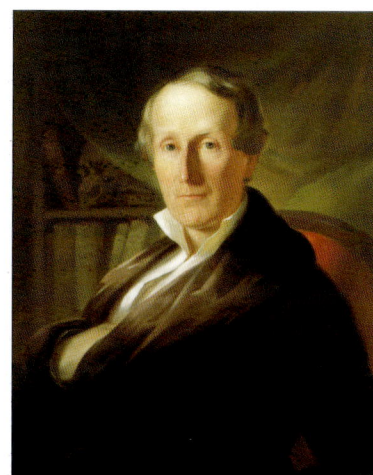

Samuel George Morton (1799-1851). A long serving member of the Academy of Natural Sciences of Philadelphia, becoming vice-president in 1840 and president in 1849. His *Crania Americana* (1839) examined the diverse origins of the races of man. He spent over $10,000 on his collection of skulls, though most of them were gifted by friends, including Townsend.

Townsend's ill-fated *Ornithology of the United States of North America* (1839)

Townsend's short introduction paid tribute to preceding ornithologists and outlined his reasons for "obtruding on public notice another work on the Ornithology of the United States" – portability, affordability and completeness. It was Townsend's opportunity to present his new discoveries:

"ALEXANDER WILSON was the great pioneer in this branch of American science; and who that appreciates his chaste and eloquent style, his accurate and happy dilineation of a class of the most lovely objects in nature, can fail to experience the greatest delight in reviewing the pages of the "American Ornithology." Next among the labourers in this field, followed the amiable and enthusiastic AUDUBON, a man of whom America may well be proud, and who has given to the world a work of magnificence which will forever remain a monument of his surpassing genius, perseverance and research. Next in order was that eminent naturalist, THOMAS NUTTALL, and to him are we indebted for a Manual of American Ornithology, highly valuable as a text book, which, like all that has proceeded from his pen, exhibits in an uncommon degree the qualities of patient investigation and sound judgement, with great beauty and eloquence of description.

The present publication is not expected to rival in their appropriate sphere those which have preceded it, but it is considered desirable to offer the public a work of portable dimensions and generally accessible form ...with all the newly discovered species, and a faithful and accurate figure of each" (Townsend 1839g: iii - iv).

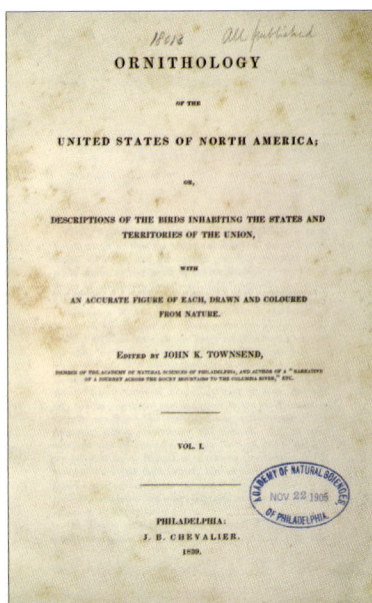

The title page of Townsend's failed enterprise the *Ornithology of the United States of North America*. Species accounts and colour plates were produced for only four of the birds of prey before the whole project was abandoned, making it one of the very rarest items in ornithological literature. The species completed were California Condor (text and plate reproduced on pages 191-192), Black Vulture, Turkey Vulture and Crested Caracara (plate IV at right).

Caracara Eagle. (1/3 nat size.)

of this, Audubon had set about issuing an octavo edition of *The Birds of America*. It is still rather surprising that Townsend ever contemplated producing his own bird book, knowing that he would be competing with Audubon whose first miniature volume appeared in 1840, and would inevitably be a success. Townsend's planned multi-volume *Ornithology of the United States of North America* never really got underway; only four plates and four species accounts were ever published, dated 1839 but said not to have appeared until 1840 (see box). Once again his hopes and expectations had been dashed, so, after all this aggravation, is it any wonder that Townsend was becoming somewhat reticent towards Audubon, being simply tired of furthering Audubon's career for scant reward? Poor Townsend was even put in the position of having to recommend his own publisher, Chevalier, for Audubon's octavo edition.

The charge of being careless could well be turned around towards Audubon himself because there are plenty of mistakes in the *Ornithological Biography*. George Ord, in a letter to Charles Bonaparte, a celebrated taxonomist of the times, described the book as "the biggest hodgepodge of stupidity and lies that I have ever seen." (quoted in Peck 1999). This was rather unfair of Ord who now seems best remembered for his animosity towards Audubon but there was some truth in his criticism. Whitmer Stone (1906) lists Audubon's errors of reference and errors of quotation without comment, presumably because he thought such discrepancies were inevitable in a work of such huge scope. Some other mistakes are inexcusable, particularly where Audubon and MacGillivray have no clear idea of the basic itineraries of either Townsend or Nuttall (see appendix 15). While it would certainly be ridiculous to call Audubon lazy, considering the Herculean effort expended in the compilation of his books, there were some species included in the *Ornithological Biography* that he did not paint for *The Birds of America* even though he had perfectly good specimens. Nor did he ever update the text to include new discoveries by other naturalists, e.g. Pinyon Jay described by Prince Maximilian in 1841, and numerous new species from California described by William Gambel in 1843.

After the excitement of the preceding years, 1840 was a more relaxing time for Townsend. One would have thought that his spare time would have been taken up with courting but on 14th February he set out for Arkansas, arriving at Little Rock a month later in order to visit the nearby hot springs. This seems to have been intended for the sake of his companion's health rather than his own, but they spent little time there and afterwards planned to go northwards to Missouri. The two men must have been away for at least eight weeks and Townsend missed the first members' meeting in the new premises of the Academy that was held on 18th February.

Washita Springs, Arkansas

In early March 1840 Townsend and John Jackson[4] made a trip to Arkansas, principally to visit Washita Springs for the sake of Jackson's health (JKT to C. Townsend, 4th & 14th March 1840. Family papers). Even though this is now designated as Hot Springs National Park, Townsend was not particularly impressed:

"We went to the house of a man named Sabin, who keeps a log cabin for the accommodation of visitors, & after seeing our horses properly stabled, visited the celebrated springs which constitute the attraction of this quarter. There are some four or five of these springs of various degrees of temperature. The hottest, is at the fountain, about 170 of Fahr., & in this they cook an egg in about 20 minutes, or scald a pig in 35. Near the springs they have little bath houses erected with wooden baths into which the water is conveyed by troughs, and in the various houses you may take a bath at almost any temperature, from tepid to scalding hot. The country in the vicinity is very hilly and rocky, abounding in quartz & limestone, & beautiful rock-chrystals are often found. And now I have said all that is to be said in favour of the Washita Springs. When you have seen the hot water, you have seen every thing that is interesting in this region. The people living in the neighbourhood are rough, & not very polite; the accommodations for strangers are of a very inferior quality, there is but little game, & no scarce birds at all" (JKT to C. Townsend, 14th March 1840. Family papers).

After sharing a room with another traveller who "had the <u>itch</u> awfully" they were pleased to return to the house six miles from the springs where they had spent the previous night in the company of a refined young widow, her father and daughter. The house was clean, had plenty of books and the food was good, all in sharp contrast to the primitive places at which they had often been forced to stay. It turned out that the widow knew of Townsend because she was a subscriber to *Waldie's Circulating Library* that had contained extracts from his journal. She had often wondered if he had completed the book that the editor had hinted at and when Townsend confessed that he had a copy of the *Narrative* amongst additional baggage stored some fifteen miles back along the road he was more or less obliged to retrieve it! When he returned they read his book eagerly, asked him lots of questions, and he was pleased to regale them with some of his western experiences. His friend John was far from happy because while he was away the young girl had teased him mercilessly – and Townsend was beginning to support the girl.

The subsequent plan was for the two men to descend the Arkansas River, take a boat up the Mississippi to St Louis, walk overland to Independence and go west to the first Indian villages, in effect repeating some of the early part of Townsend's journey of 1834. His friend wanted to camp out, shoot game, see the prairies and meet the Indians but he had proved to be rather fastidious, was constantly complaining, and Townsend found him to be "the most complete hypochondriac I ever heard of". He had already refused Townsend's advice to remove his gun cover while riding and had missed the chance of shooting a deer with his shiny new weapon. He had also failed to fire at some Wild Turkeys until they were out of range: "Bang went the gun, but he might as well have fired at the moon," chortled Townsend. Whether they tolerated each other enough to get as far as Independence is unclear. There are no Townsend bird specimens from as far west as that, dated 1840, to suggest that they did.

The Carolina Parakeet once ranged over much of the eastern United States from the Great Lakes southwards. It is now extinct, slaughtered by the thousand because of its destructive habits in field and orchard. Townsend collected several but only one can now be traced (USNM 12272).

The few known Townsend specimens from this trip confirm only that Townsend passed through Cairo, Illinois, at the junction of the Ohio River and the Mississippi, and that he may have gone 300 miles northwards to Peoria. Details on USNM specimens are: Kentucky Warbler USNM 12197, Illinois, no date; Red-eyed Vireo USNM 25238, Pe—a [illegible = Peoria?], Illinois, May 1840; Red-eyed Vireo USNM 25239, Pe—a [illegible = Peoria?], Illinois, May 1840; Warbling Vireo USNM 25240, Cairo, Illinois, April 1840. A Carolina Parakeet USNM 12272, Illinois, is dated 1834 on the back of a label with a reference to *Cassinia* (1903: 2) in which Stone refers to "Carolina Paraquets" seen by Townsend in 1834. Since Townsend's 1834 Carolina Parakeet specimens went to the Academy of Natural Sciences of Philadelphia it seems more likely that USNM 12272 was collected in 1840. This may be corroborated by the fact that on 13th September 1841, among a list of other birds, Townsend is recorded as having donated "1 Psittacus carolinensis, *Illinois*" to the National Institution (later the National Institute) – the precursor for the USNM (*Proceedings of the National Institution* 1841: 103).

Another edition of the *Narrative*

The year 1840 also saw the publication of the London edition of Townsend's *Narrative*, re-titled, in the hope of improving sales, as *Sporting Excursions in the Rocky Mountains, including a Journey to the Columbia River, and a Visit to the Sandwich Islands, Chili, &c.* (1840). Perhaps encouraged by this he wrote to Victor Audubon at New York and asked him to put in a good word for him with the Harpers in the hope that they would be interested in publishing a new American edition of the *Narrative*. But the Harpers already had seventy or eighty works to be completed that year so they were not at all interested (V. Audubon to JKT, 2nd June 1840. Houghton Library, Harvard University).

The only surviving Townsend correspondence for the latter part of the year is a surprisingly perky one-page letter to Spencer Baird at Carlisle, Pennsylvania, requesting specimens of Rough-winged Swallow and Seaside Sparrow, and keeping Baird up to date with the results of a recent shooting spree:

> "You requested that I should tell you about Rail shooting. We went, but had not very good success, indeed much less than Common – of five boats, your humble servant came in first, with 45 birds, [George] Leib got only 26. & three other boats averaged but 7 each. So you see I was cock of the walk for once, but I bore my triumph with most becoming decorum & only smiled knowingly at the excuses to which unsuccessful sportsmen always resort, of course attributing their failure to any cause rather than to a want of skill on their part. Yesterday Leib & I had an afternoon's partridge shooting. Fifteen birds were killed, my friend being ahead, one. Next week we take two days shooting at Attleboro, & I may perhaps let you know the result" (JKT to S.F. Baird, 10th October 1840. SI Archives).

1. Charlotte's Protestant forebears emigrated from Northern Ireland to Philadelphia in about 1700. The farm near Cape May Court House was purchased in 1775 by John Holmes; flax being the most important crop in the early years. Her father was Robert Morris Holmes, the inventor of the Holmes lifeboat. Her mother, Mary Leaming, was a native of Reading, Pennsylvania. Charlotte was born on 29th May 1818 and had three brothers and one sister, Harriet, who was born on 8th May 1822.

2. The house remained in the family until 1935. It had two new owners before being purchased by the Cape May County Historical and Genealogical Society in 1976. It is now open to the public as the John Holmes House museum. Amongst the varied collection of historical items that it contains are three portraits by Charlotte Holmes: G.H. Hand at age 14, painted in 1838; Mary Leaming Holmes (her mother) at age 66 or 67, painted in 1855; and John Hand Holmes (her nephew) at age 16, painted in 1858.

3. It is hard to agree with Porter (1979) who argues that Townsend's *Ornithology* was suppressed because travelling field naturalists were no longer respected by academicians or regarded as their peers.

4. This was probably the John Jackson who was a neighbour of Townsend's friend Dr Ezra Michener at New Garden, near Philadelphia, where Townsend stayed while writing his *Narrative*. John Jackson is described as an enthusiastic florist who taught botany to Michener (*Dictionary of American Biography* (1828-36), vol. XII: 596-597).

Thomas McEuen's list of Townsend specimens lent to Audubon in October 1839. For a transcription of this document see appendix 5.

At the National Institute

"Townsend can skin, stuff and sew up a bird, so as to make it look far superior to any I have ever seen, in five minutes."

William Baird to Spencer F. Baird, 1st February 1842.

In 1841 Townsend moved to Washington, D.C. to begin work as a taxidermist at the National Institution for the Promotion of Science, which the following year changed its name to the National Institute. Initially it was housed in the basement of the Patent Office but later moved upstairs to the large gallery. His principal duties were the preparation of birds and mammals, but as time went on he became more generally involved, preparing other kinds of specimens, encouraging friends and organisations to contribute, and became Recording Secretary, responsible for cataloguing acquisitions and thanking donors. He occasionally went to the coast for fresh specimens for the museum.

The collections were rather overwhelming for Townsend and the others involved there, because over twenty tons of material collected by the US Exploring Expedition had arrived. Cases and cases were waiting to be sorted through and most of the specimens needed attention if they were to be conserved properly. William Baird visited the Patent Office in July 1841 and commented to his brother and fellow bird collector:

"The collection of specimens which has been sent home is enormous, the shelves of one room are piled with bird skins; very few have been mounted in the cases; of the birds sent home by the Exploring expedition probably not more than 150, principally parrots & pigeons. Although three or four persons are engaged in stuffing and mounting birds and arranging shells minerals &c, not more than one or two birds are finished in a day. The skins in the state sent by the expedition look amazingly rough, but when mounted present a well finished appearance. I should like very much to tumble over and examine the rough skins, but nothing is allowed to be touched, and the room in which they are deposited (separate from the exhibition room) is kept locked and persons admitted but for a few minutes. Dr Leib's birds are beautifully preserved and mounted, most of them looking, as to position, eyes &c perfectly lifelike & natural. It is not so complete a collection of the birds of Pennsylvania as ours ..." (W. Baird to S.F. Baird, 2nd July 1841. SI Archives).

Townsend wrote to Thomas Nuttall asking if he would be prepared to arrange the plants in the new herbarium. The good botanist cautiously agreed (Nuttall to JKT, 7th April 1841. Gratz collection at HSP), knowing that neither he nor the Institute had a general botanical library and knowing that there were no plants in Washington with which he could make comparisons, so he would have to keep referring back to the Philadelphia Academy. Nuttall began at Washington in early May but by 13th July he was at Philadelphia and had begun to wind down all his North American commitments in order to settle in England on the estate that he had inherited from his uncle. He did not leave until the end of December, so he had ample time to personally convey his best wishes to Townsend for his forthcoming marriage.

Townsend surprised everyone by falling for a lady who was not a Quaker, opting for a civil ceremony in order to marry Charlotte (Lottie) Holmes, a Baptist from Cape May Court House. The mayor of Philadelphia, John Swift, conducted the wedding on 12th

The National Gallery at the Patent Office in Washington, D.C., where Townsend worked as a taxidermist.

Dr Charles Pickering (1805-1878), an early member of the Academy of Natural Sciences and American Philosophical Society in Philadelphia. From 1838-42 he served as a naturalist with the US Exploring Expedition and for a time was a curator of the expedition collections at the Patent Office in Washington, D.C.

October 1841 and afterwards they lived in Washington. In 1847, Lottie's sister Harriet married William Baird, brother of Spencer Fullerton Baird who later became Assistant Secretary of the Smithsonian Institution and one of the most influential of all North American ornithologists. Although William went on to choose a law career, in the early days he was just as knowledgeable about birds as his brother and was a regular visitor to the National Institute in order to keep up to date with the birds that were arriving there and to chat with Townsend. As we shall see, William's letters are a frequent source of comment upon the Institute's general progress and development.

Townsend felt very settled at the National Institute and was confident that it would become one of the leading museums of its day, a world-class collection that would rival or surpass those at Philadelphia. This confidence is evident in the *Proceedings of the National Institution* for September 1841, where Townsend is recorded as the donor of "One hundred and twenty-one specimens of Birds, from North and South America, East Indies, and the Sandwich Islands; four specimens of Quadrupeds and Reptiles; twenty-two specimens of Shells, Unios [?], &c., from the American waters; twenty-nine specimens of Shells from the Sandwich Islands, Columbia River, &c." The list of birds and mammals that follows is a miscellaneous assortment, some of which he could have had for years, or bought or exchanged recently; mostly they were from places he had not visited such as Brazil, Argentina and New Caledonia, together with a few that he had obviously collected himself. The latter included "Nectarinia rubra", "Nectarinia viridis", "Fringilla cinnamones", "Pyrrhula psittacina" all from the Hawaiian Islands, "Columba taitensis" from Tahiti, one "Sorex (new)" – a shrew from the Columbia River – as well as a few birds from his recent trip to Arkansas and Illinois.

The main drawback was the startling lack of funds. The Institute was hoping for a large appropriation from Congress but in the meantime Townsend reported that "we who are working for the government are receiving no pay; "working for nothing, & finding ourselves" … I really have no money to pay my board for the present month." He was so desperate that he wrote to a man in Philadelphia to recover about $250 belonging to his wife (JKT to H. Townsend, 29th December 1841. Family papers).

The US Exploring Expedition and the National Institute

Charles Wilkes led the first US government-sponsored circumnavigation of the world (1838-42), which has been described as the most ill-prepared, the most controversial and probably the most unhappy expedition that ever sailed. Twenty-three men died, 127 deserted and of the original five ships only the *Vincennes* and *Porpoise* returned. The *Sea Gull* sank off Cape Horn and the *Peacock* was wrecked at the mouth of the Columbia River and others became unfit for service. Wilkes made a disputed claim to be the first to discover the Antarctic continent and was criticised for his ruthless leadership. Landfalls included Australia and New Zealand, Argentina, the Philippines, the Hawaiian Islands and the Oregon coast. The natural history material gathered included 50,000 plant specimens of about 10,000 species, 1100 bird specimens of over 500 species and 400 other zoological items about half of which were new.

If the expedition was controversial, so too was the fate of the collections, and some of the subsequent scientific reports. When the 150 crates of specimens were transported from Philadelphia to Washington, space was found for them in the Patent Office, part of it already occupied by the National Institute. Nearly a fourth of the crates had been marked as private even though the contents had been gathered on a government expedition – and the chief culprit was thought to be Titian Peale, the expedition ornithologist and mammalogist. This was rectified but the specimens began to suffer not only from the damp basement but also from pillaging by scientists and souvenir hunters, most of whom should have known better. Rather notoriously, some molluscs in glass jars lost all their scientific importance when the numbered labels were removed because the tin foil was discolouring the specimens

Peale had a very turbulent time of it. He claimed that 180 of his birds were missing and that one pair of birds had been cobbled together by the taxidermists to form one specimen. He later complained about the difficulties that beset the report preparation and bemoaned the fact that more than $1 million had been spent on the expedition, but only 100 copies of the "Account of the Discoveries" were to be printed for the 18,000,000 people of the United States who had paid for the expedition. Even then a considerable number of the completed books were sent abroad to foreign governments. Peale's own report on the birds and mammals was condemned by Wilkes who suppressed it and commissioned John Cassin to rewrite the text – the new version appeared sixteen years after the end of the voyage, in 1858, with Peale's original plates (Porter 1979).

The National Institute had begun in May 1840 when the Secretary of War, Joel Poinsett, had formed an organization originally called the National Institution for the Promotion of Science. Poinsett suggested that his organisation was the ideal one to care for the collections of the US Exploring Expedition and he was backed by cabinet members and congressmen, high-ranking government administrators such as Colonel Abert of the Topographical Engineers and Francis Markoe of the State Department. The new Patent Office building had plenty of space because all the old patent models had been destroyed by fire in 1836 and had not yet been replaced or superseded. The collection was soon moved to the upstairs gallery, an immense hall measuring 273 feet in length. This did not please the Patent Office, one officer complaining that he had no idea that "birds & beasts were to be stuffed & prepared in the very Hall itself & still less so that living animals such as rattlesnakes, foxes, etc., would be kept there." (Viola and Margolis 1985: 230).

Nor did it please the Expedition's scientists who saw that there was a definite conflict of interests: one of the Institute's main aims was to prepare specimens for display for the stream of visitors that came to see them (100,000 in 1845); but the scientists needed to be able to study the specimens to prepare their reports. In 1842, however, when Charles Pickering replaced Dr King as curator, the Institute's control of the Expedition collections came to an end. After Wilkes himself took over from Pickering and rearranged the displays the new layout included only a small number of National Institute exhibits.

Meanwhile, James Smithson, a wealthy Englishmen, had left a small fortune to the United States "for an Establishment for the increase and diffusion of knowledge among men." After years of debate a new building began to arise on the Mall, almost opposite the White House: it was the Smithsonian Institution. The Patent Office immediately saw this as a chance to clear out the collections of both the National Institute and the US Exploring Expedition but had to wait until 1858 before they could be transferred. Fortunately, Spencer Baird was on hand at the Smithsonian Institution to receive them.

At this time Dr King was in charge of the National Institute, an unfortunate appointment that hindered its early progress. William Baird thought he seemed "a very poor sort of a character, and to know very little about the business," adding after another visit that King was "a humbug and I don't care about knowing him." Even so, the collection continued to expand – "in the most astonishing manner" according to William Baird, once again our chief source of information:

"The collection of reptiles is immense, shells ditto, minerals very large. Birds ditto, both of the South Sea [US Exploring] expedition of which some are mounted and set up daily, and also of our own birds which they buy in the market whenever an opportunity offers, and set up in the most splendid manner, besides having large numbers presented. They will no doubt get a pretty good appropriation from Congress, if this is accomplished, they will no doubt go on even more rapidly. I did not attend the meeting of the Society in December,

Titian Ramsay Peale (1799-1885), artist and naturalist. He took part in Major Long's expedition to the foothills of the Rocky Mountains, 1819-20, and the US Exploring Expedition, 1838-42. Townsend knew him at Philadelphia and at Washington, D.C.

having a very bad toothache, but will go to the next; the meetings are said to be very pleasant. Townsend is very clever, if I had time I could soon learn from him his manner of mounting as it appears very easy" (W. Baird to S.F. Baird, 7th January 1842. SI Archives).

Later in the month Townsend's close friend George Leib, one of the curators at the Philadelphia Academy, sent him shells, bird skins and birds' eggs, and teased him about the Institute's *Proceedings* that were so slow in appearing; he said it was important for Townsend to contribute to show that he was "alive and kicking". He urged Townsend to present his wolf skins from the far West to the Academy to ensure their preservation. Leib said that the Academy could never afford to buy them, and warned him that the Exploring Expedition wolf skins could deprive him of that which money could not bestow – the honour of discovery (G. Leib to JKT, 23rd January 1842. HSP) – if indeed his giant wolf was a new species.

William Baird visited again a couple of weeks later and sent off another report to his brother:

"The National Institute collection is increasing in the most wonderful manner; donations of every description are daily pouring in. Townsend buys up everything worth having in the market, and mounts them in the most splendid manner. There are about 450 species of birds mounted (besides those of the Exploring expedition, which have not been set up, to the number of some hundreds probably) of these in the neighborhood of 180 are North American. Quadrupeds are about fifty in number ... The collection of crustacea is said to be the largest in the United States. The number of shrimps, crabs, lobsters, &c., is immense. The collection of reptiles and fish, which are principally from the Exploring expedition is also large, but they are so mixed up together in the bottles that there is very little satisfaction in looking at them. The cabinet of animals is very fine and there is a good collection of geological specimens. Besides these are many things too numerous to mention: skulls, mummies, coins, &c. (W. Baird to S.F. Baird, 1st February 1842. SI Archives).

With all this practice Townsend was able to hone his skills as a taxidermist to a fine degree. Some of his Hawaiian specimens prepared in the 1830s had been sufficiently well prepared for a leading ornithologist to pass comment on their excellent state of preservation – some fifty years later. Yet, every time William Baird wrote to his brother he seemed to remark on Townsend's further progress in the art:

"I have arrived at the same conclusion as yourself, that arsenical soap is not the best thing for preserving skins. Those put up with the powder look a great deal better and are much less trouble, even if the soap were as efficacious. At the end of the letter I will give you Townsend's receipt for preparing the powder, and which he always uses. It is safe, while there is a danger of pure arsenic injuring the lungs. I have noticed a great many little points in Townsend's manner of stuffing, which will be of great assistance to me when I prepare any skins myself. I looked at his tools and will try to get some like them. The value of proper instruments is very great in the saving of time as well as the appearance of the skins. Townsend can skin, stuff, and sew up a bird, so as to make it look far superior to any I have ever seen, in five minutes ...

<u>Townsend's Preserving Powder</u>

Tanner's Bark finely powdered			1 lb
Arsenic	"	"	1 lb
Burnt Alum	"	"	2 oz.
Gum Camphor	"	"	4 oz.

Powder the Camphor with a small quantity of alcohol in a stone or Wedgewood mortar, and mix in the other ingredients gradually.

NB Keep in boxes or jars securely closed.

———————

P.S. The pure arsenic if you prepare many birds will not only injure the lungs, but also gets under the finger nails and forms abcesses which take months frequently to heal. The above preparation never produces these effects.

The Alum is placed in lumps on a shovel over the fire so as to completely drive off all the water of crystallization, when sufficiently burnt it becomes snow white and crumbles" (W. Baird to S.F. Baird, 1st February 1842. SI Archives)

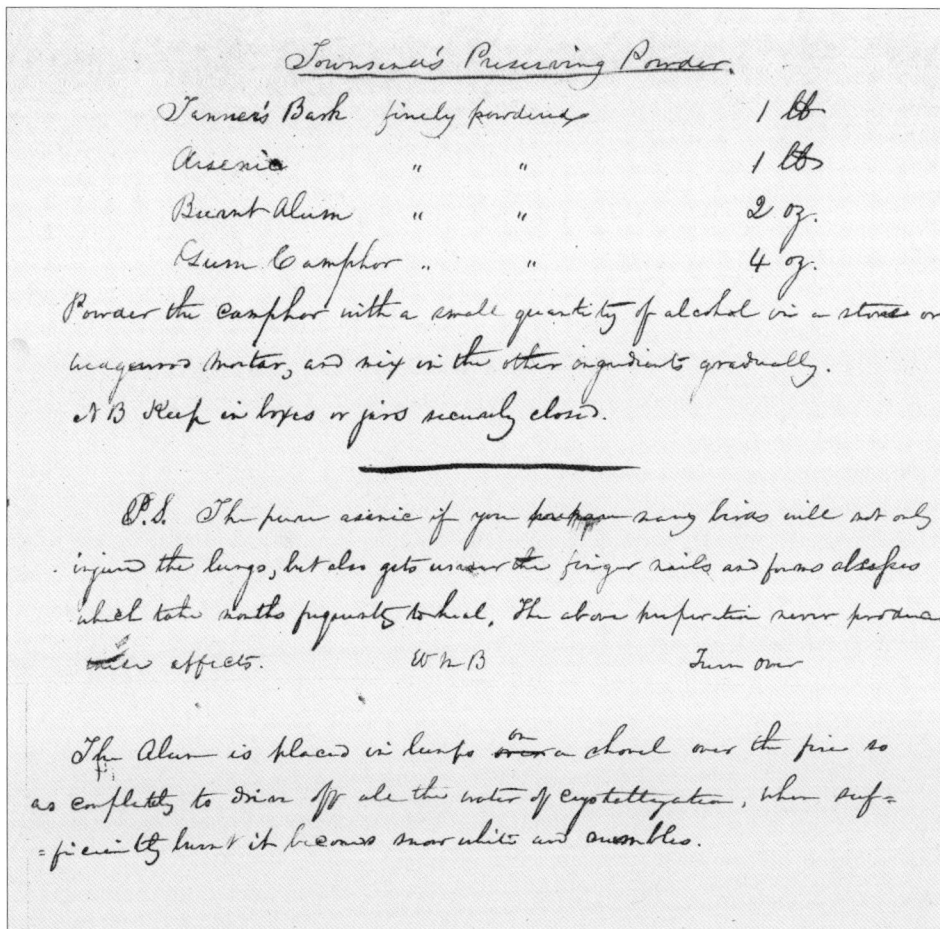

The ingredients of Townsend's preserving powder (from a letter sent by William Baird to his brother S.F. Baird, 1st February 1842).

Another visitor to the Institute was Mahlon Kirk who had married one of Townsend's sisters, Elizabeth. He wasn't so sure about the safety of Townsend's concoction and years later recalled that he "often saw John when employed by the government to mount specimens in Washington, bending over a big tray of arsenic in what was then called the patent office, now the department of the interior, enveloped in a cloud of the dust, which being a cumulative poison, destroyed his health and ended in his premature death" (in Stone 1903: 54).

William Baird (1817 - 1872) (above) and Spencer Fullerton Baird (1823-1887) (below). At one time the two brothers were equally knowledgeable about birds, but while William concentrated on a career as a lawyer, Spencer became Second Secretary at the Smithsonian Institution and directed the course of American ornithology by placing naturalists on government surveys in the West, and through his influential publications.

In February William Baird reported that "Townsend proposes, if he can get enough subscribers to make him safe, to publish another edition of his journal across the Rocky Mountains to the Columbia River & at the Sandwich Islands; with descriptions of the new birds which he procured &c. The first edition was exhausted in about three weeks after publication. The price will be $1.50 a copy ... It will (if there are subscribers enough) be out some time in March. As Townsend is a clever fellow I put down my name for two copies ..." (W. Baird to S.F. Baird, 1st February 1842. SI Archives). This edition was never published, probably because Townsend lacked the time to drum up support.

By 14th February 1842 Townsend had made another donation of specimens, much smaller than the previous one, consisting of twenty-eight birds of sixteen species, and four mammals of two species. The birds included six Seaside Sparrows and two Sharp-tailed Sparrows, a Canada Warbler and Northern Parula from "Pennsylvania". Only "Lepus Nuttalli (unique specimen,) Rocky Mountains" was from his long absence in the West. A footnote in the *Bulletin of the Proceedings of the National Institution* (1842: 147) says that this was the specimen shot by Townsend "on the north branch of the Platte river" and described as new by Bachman (1837: 345) in the *Journal of the Academy of Natural Sciences of Philadelphia* after failing to find another like it in the museums of the United States and Europe.

Around this time William Baird had a minor disagreement with Townsend, no more than a difference of opinion, over the identity of some *Empidonax* flycatcher skins, notoriously difficult to differentiate even now. The Baird brothers had found some flycatchers near their home town of Carlisle, not far from Philadelphia, and just could not match them up with any existing descriptions. Townsend pronounced one of them to be the *Tyrannula pusilla* of Swainson but the Bairds, quite remarkably, by careful study of flycatcher specimens and descriptions, had found two new species right on their doorstep: Yellow-bellied Flycatcher *E. flaviventris* and Least Flycatcher *E. minimus*, that were published as new in 1843. William Baird gently commented that "Townsend ...is very practical, and is a very good ornithologist, quite well acquainted with our North American Birds and tolerably with the foreign ones. He however makes mistakes occasionally as for instance he pronounced our quondam pusilla, at once, a true pusilla, in which he is evidently wrong. Your friend [James] Trudeau however, makes more mistakes than that ..." (W. Baird to S. F. Baird, 28th March 1842, SI Archives). At that date, even William Baird was unconvinced that the Acadian and Traill's Flycatchers were different species.

It so happened that there were still several more *Empidonax* flycatchers waiting to be discovered in North America but Townsend had little time for detailed bird study. In any case, he was becoming more and more interested in insects. He had been appointed to a sub-committee to examine and report back on the vast collection of Count Castlenau, the US Consul at Lima. He found that it consisted of four large wooden crates, each of which contained a number of smaller boxes, all containing, he estimated, upwards of sixty thousand insect specimens. Many were in very poor condition, and yet more were suffering from mould so it was decided to employ someone to examine and arrange the items and to ensure their preservation. It was a huge task and not until July 1844 did the Institute receive a report on the collection, by the Reverend John G. Morris, a reputable entomologist from Baltimore, outlining what he and Townsend had done to preserve the Castlenau collection.

Townsend's fieldwork still continued from time to time, though to what extent this was funded as part of his official duties seems unclear. In September 1842 he was reported to have "collected for the National Institute" some 237 birds, along with an assortment of shells, small mammals, crustaceans, fishes, reptiles, birds' eggs and skulls, insects, and even a few plants. The majority of the birds were from Cape May, as good a place to hunt for a variety of birds as can be found on the entire east coast of North

America, and conveniently situated not far from the home of his wife's family. The only specimen of note was an Eskimo Curlew, though this could have been a misidentified Whimbrel that is now among the only surviving specimens from this period at the US National Museum: Dunlin, Red Knot, Whimbrel, Wilson's Snipe (2), Willet (3), Northern Flicker and Common Nighthawk (all collected by Townsend from Cape May, dated either April or May 1842[1]).

By mid September 1842 Dr King had been ousted and was replaced by Dr Charles Pickering, a frustrating blow to Townsend who had hoped for this appointment. Pickering and Townsend already knew each other, indeed Pickering was one of the members of the Philadelphia Academy who had proposed Townsend for membership back in September 1833. He had joined Lieutenant Wilkes in 1838 as one of the naturalists of the US Exploring Expedition and therefore had a personal interest in the fate of the collections, doing his utmost to ensure that the specimens were properly preserved and displayed. William Baird noted an immediate difference: "Dr Pickering, is a different sort of curator from Dr King, he keeps the fellows busy every moment of their time, having that sort of Yankee energy, besides understanding the business, which the other did not" (W. Baird to S.F. Baird, 1st November 1842. SI Archives). Pickering realised at once that things were not all they should have been. The interior arrangement of the main hall was not how he would have liked it but he was willing to conform to the original plan. What worried him most was that there was no provision for any further increase in any of the departments, no room for a geological series of specimens from the United States, no room for a library, nor for a gallery of fine arts. He regretted that there were no artefacts or written material concerning the native tribes of eastern North America, except a few hatchets and arrowheads. For the western tribes there was much more already at the Institute, including an extensive series of portraits from the war department that together he considered to be a "National Monument". He was satisfied with the existing staff and did not think (Pickering MS, 22nd November 1842. AMNH) that any increase was needed in addition to:

> "Mr. Varden, having the immediate supervision of the Hall & fixtures.
>
> Mr. Dana, having charge of Mineralogy & Geology, and also of Corals & Crustacea.
>
> Mr. Brackenbridge, having charge of the Greenhouse, & all Botanical Collections.
>
> Messrs. Townsend & Pollard, Taxidermists, also having charge of the Ornithological Department.
>
> Mr. Falconer, Carpenter, constantly occupied.
>
> Mr. Campbell, Messenger, & General Assistant."

Despite this supposed sufficiency of staff, William Baird reported in November that an Englishman by the name of Naylor was engaged at the Institute making glass eyes and for the first time it seems Townsend had been able to place coloured eyes into those birds that needed them. But there was no requirement for coloured eyes for two dark-eyed Hawaiian species that William Baird was now shown: "Townsend has just received from the Sandwich Islands, a box containing two specimens male and female of a crow [Hawaiian Crow/Alala], and a specimen of a hawk [Hawaiian Hawk/Io], both of which he thinks new, and which he will describe in the bulletin of the National Institution (W. Baird to S.F. Baird, 23rd November 1842. SI Archives). They were indeed new species but they were not described until 1848 when Titian Peale included them in the *Mammalia and Ornithology* of the US Exploring Expedition.

HAWAIIAN CROW/ALALA AND HAWAIIAN HAWK/IO

The type specimens of these two species were sent to Townsend by the Rev Cochran Forbes, an American missionary who lived on the island of Hawaii at the village of Kuapehu a mile or so above the shores of Kealakekua Bay, where Capt Cook was killed. Townsend had met the missionary there in February 1837, and it so happened that Titian Peale and some of the other scientists from the US Exploring Expedition had stayed with Forbes in November 1840 (Forbes 1984). At that time the crow and hawk were conspicuous and familiar birds on Hawaii but they had not been made known to science. Townsend did not see the birds because he was there so briefly and did not travel far inland. Peale managed to get specimens of both species but they were lost in the wreck of the *Peacock* at the mouth of the Columbia River.

Townsend never published anything about the crow or hawk but generously loaned the skins to Peale, knowing that Peale would paint them and describe them in his report for the US Exploring Expedition, not expecting that it would be more than five years before the publication of Peale's description nor that the paintings would not be printed until after Townsend's death. Perhaps Townsend also felt that Peale had a greater right to describe the birds because he had seen them on Hawaii and unluckily lost his own specimens.

The plates of Hawaiian Crow and Hawaiian Hawk in J. Cassin's *Mammology and Ornithology* (1858) were painted by Titian Peale from the type specimens below. The skins of Hawaiian Crow (ANSP 2830, 2831) and Hawaiian Hawk (ANSP 2304) were sent to Townsend from Hawaii by Rev Forbes.

Peale's *Mammalia and Ornithology* (1848) was suppressed by Wilkes who was unhappy with his standard of nomenclature and classification. He sacked Peale and replaced him with John Cassin, a more experienced taxonomist, whose new report appeared in 1858, together with a lavish folio atlas of 53 plates, 32 of them by Peale. When Cassin first saw Peale's painting of the Hawaiian Hawk he thought that the honeycreeper held in its talons was inappropriate. He ignored Peale's first-hand field observations and instructed the engraver to replace the honeycreeper with a fish, in the mistaken belief that the hawk was closely related to the fish-eating Osprey!

Some months earlier, Townsend had requested certain birds from Rev Bachman but in November he finally received word that all attempts to trap Common Ground Doves on islands off the South Carolina coast had been unsuccessful, and that the Painted Buntings that Bachman had caught had lost much of their colour in captivity. The good news was that Bachman hoped to be in Washington the following May (J. Bachman to JKT, 2nd November 1842. PUL).

Around this time the Institute received eighty-seven birds from the Jardin des Plantes at Paris, originating from various parts of Africa, India, Indonesia and New Guinea. And Leib sent thirty-four species of birds, mostly from Pennsylvania, but also some from Michigan including an Eskimo Curlew, plus a few foreign birds. His single Rufous Hummingbird from Oregon probably originated from Townsend. William Baird took in some birds to swap but found that there was no regular exchange system in operation, so Townsend agreed to give him "quietly, without letting it be known" some duplicates from Cape May – which suggests that his collecting trip there was official business after all. An uncle of the Bairds had also had a quiet word with Colonel John Abert, an influential government administrator, to see if perhaps a position could be found for Spencer Baird at the Institute, but there was no possibility because all the arrangements had already been made with the removal of Dr King and the appointment of Pickering in September. Indeed, because of the generally poor economic state of the country, and the Institute in particular, poor Townsend found that his salary had been cut down to $1090 a year instead of the $1200 he had been led to expect (W. Baird to S.F. Baird, 14th October & 23rd November 1842. SI Archives). So now there was a financial setback on top of his career disappointment of not succeeding Dr King, all at a time when Lottie was heavily pregnant.

Nevertheless, in the spring the Townsends were rejoicing in the birth of their son Francis (Frank) Holmes Townsend – their only child. William Baird broke the news to his brother in bachelor ornithologist style[2]:

"The only item of news that occurs to me is that Townsend has the proud gratification of becoming a father a day or two ago, (it will not be as fine when it gets the cholic, and he has to get up and warm the catnip). The baby is a boy, and the mother better "than could be expected'. Besides the boy he thinks he has got a new mocking bird" (W. Baird to S.F. Baird, 10th April 1843. SI Archives).

Two months later, the proud father wrote to his sister Hannah in Philadelphia with the standard comments on the oppressive summer heat in Washington, an unusual irruption of myriads of small beetles, a long discourse on the topic of original sin prompted by his helpless new offspring, and a progress report on the baby: "Although still small, he has grown finely, is very healthy, scarcely ever has any cholic, & is an exceedingly good baby. I really think, & many others agree with me, that he is without exception, the prettiest baby I ever saw of his age. He has a fine clear complexion, bright, uncommonly bright blue eyes, inclining to hazel, like his mother's (about the best feature in her face,) a beautiful molded chin, rather pointed like our father's, and the prettiest little mouth I ever saw …Oh! How I love this dear helpless little creature! Really it seems to soften the rough heart of a man to love a creature so perfectly pure & innocent, so nearly allied to the angels of heaven. I feel a new tie binding me to existence; a new motive is felt for worldly exertion, & an additional impulse given to all my desires & feelings. I feel also that another sweet tie binds me to my excellent wife, & I love her better, if possible than ever before. With what delightful emotions have I watched Lottie while engaged in her various maternal tasks; She is So attentive, So patient, So self-sacrificing, but she must be almost less than woman to be otherwise with so sweet a piece of mortality dependant for its very life upon her" (JKT to H. Townsend, 5th June 1843. Family papers). They had no intention of having Frank

Rev Dr John Bachman, from C.L. Bachman's biography of Bachman (1888).

christened, both parents agreeing on "the uselessness of the ceremony". The father because he was a Quaker; and the mother, as a Baptist, would have been opposed to infant baptism but in favour of believer's baptism, when a person, either as an adult or a child could decide for his or her self.

Lottie's mother and sister Harriet came to stay during or soon after the birth, and afterwards the couple were inundated with more family, including country cousins, coming to see the baby. Then there was a surprise:

" … we had a visitor with us recently[3] with whom I was more delighted than all the rest. This was no other than my old friend and correspondent, D^r. Bachman of Charleston. Thee knows we have corresponded with considerable regularity for five or six years, but never have met until this time. He called on me at the office & introduced himself, & I know thee would have been gratified to have seen our meeting. He spent but three days in the city, only about half of which time I was permitted to enjoy his company alone, & although we made good use of our time, we did not discuss half the subjects which were on my mind. He is one of the most delightful men in every respect that I ever met; his mind is most richly stored with every sort of lore, & in general society he shines as much as when in company with his brother naturalists. He said he had a conversation in Georgia with Joel R. Poinsett, President of the Institute, in which he (D^r. B) recommended me as curator of the Society. He told me that Poinsett expressed a desire that I should receive the appointment, & said he [had] always expected that on the resignation of D^r. King, I was to take his place. He promised the D^r. to use all his influence, (which you know is great,) to advance me in the Institute. Whether any thing will ever come of it, I am unable to say, but will live in hope at least for the present" (JKT to H. Townsend, 6th June 1843. Family papers).

One of the subjects discussed by Townsend and Bachman would have been Audubon's current expedition up the Missouri River. Unknown to them both, Audubon, Edward Harris, John Bell and a few other companions were just short of Fort Union, at the junction of the Yellowstone River, and would arrive there a few days later on 12th June, basing themselves at the fort for several weeks before returning. Townsend had been considered for this expedition, being mentioned by Audubon when he tried to recruit young Thomas Brewer: "How would you like to trip it over the Rocky Mountains next spring in company with Ed. Harris, Townsend and about forty others … Harris tells me that such an expedition is now *on talk*" (in Tyler 2000: 126). Whether Townsend was actually asked, or whether he had fallen out of favour by 1843, is unrecorded but Townsend no doubt felt fairly settled at the National Institute and it would also have been difficult to leave when Lottie was expecting. Bachman, Brewer, Spencer Baird and Victor Audubon were amongst those who had also had to forego the exciting trip up the Missouri, the main purpose of which was to study and collect western mammals for *The Viviparous Quadrupeds*. The reason for Bachman's visit to Townsend was much the same: to discuss the habits and habitats of western mammals and perhaps to secure a few additional specimens.

Audubon and *The Viviparous Quadrupeds* (1845-58)

Planning for a grand work on the North American mammals began around 1839, before Audubon had finished the *Ornithological Biography* and *The Birds of America*, octavo edition. He naively expected to take a couple of years to complete the mammal artwork and to wait another year for his co-author at Charleston, the Reverend John Bachman, to complete the text. Yet Bachman had warned him that such a book would not be "child's play", that it would be far more difficult than the birds because: "Books cannot aid you much. Long journeys will have to be undertaken ... the ever varying Squirrels seem sent by Satan himself, to puzzle the Naturalists" (Bachman to Audubon, 13th September 1839). Bachman knew that procuring skins and illustrating them would be easy compared to the problem of winning accurate information about the behaviour and distribution of the many secretive and nocturnal small mammals. A few plates appeared quickly in 1842 (including "Townsend's Hare" as plate III) but these were early samples to encourage subscribers and the last plates were not completed until 1848, with the final volume of text not issued until 1854, which extended the project to a full fifteen years. The original estimate of 100 species of mammals (from north of the Tropic of Cancer through Mexico northwards) proved to be far too low and by completion of the 1854 quarto edition the plates depicted no less than 167 named species with a further 27 mentioned in an addendum. Even then, all the bats, seals and whales had been omitted.

Audubon painted some of the mammals while still in Britain, overseeing the production of *The Birds of America*. On his return he set about contacting his network of friends and acquaintances, asking them for specimens, notes on breeding biology and habits, and the names of potential subscribers. He visited museums and private collections, he hunted for mammals around Charleston with Bachman, he sent his son, John Woodhouse Audubon, to Texas to collect mammals and then had him sail to Europe where he spent eleven months painting specimens in the museums of London. In the end John Woodhouse was responsible for at least 72 of the 150 plates in the imperial folio edition, as well as five extra plates in a later supplement. Audubon's other son, Victor Gifford Audubon, helped by painting in backgrounds as well as by performing many of the routine business tasks that were required.

Townsend possessed excellent mammal specimens, including several new species that Bachman described and named in the *Journal of the Academy of Natural Sciences of Philadelphia* (1837, 1839), but he also had specimens of other species that he was increasingly reluctant to hand over. Nevertheless, after repeated entreaties from Bachman and Audubon, he gradually gave them up. Most of them had labels with

▶

Detail of plate III of Audubon's *Viviparous Quadrupeds* (1845-58), showing Townsend's Hare – the White-tailed Jackrabbit.

John Woodhouse Audubon (at left) and Victor Gifford Audubon (at right), from J.H. Herrick's *Audubon the Naturalist* (1938).

301

collecting dates and localities, a fact confirmed by Audubon's original watercolours that are sometimes inscribed with the details, though this information is often now missing from the surviving specimens because the labels have vanished. At least eighteen of the plates (12%), perhaps as many as twenty-two (14%), were drawn from Townsend's specimens. Plate CXLVII contains two species collected by Townsend.

In 1843, feeling the need for more specimens from the West, Audubon and his friend Edward Harris went on an expedition up the Missouri River as far as Fort Union, near the mouth of the Yellowstone River, where they stayed from mid-June to mid-August (Boehme 2000). Eighty-five mammals, of twenty-seven species, were collected but Audubon, his friends and helpers took their eye off the real target by spending too much time hunting buffalo, with the result that no new mammal species were collected and precious little information gathered. On his return, Audubon felt guilty and embarrassed, so much so that he refused to let Bachman read his journal until late in 1845, and then only because Bachman threatened to pull out of the project. Bachman was outraged when he realised how Audubon had squandered his time. He curtly told him that "your expenses were greater than the knowledge was worth" (Bachman to Audubon, 6th March 1846, in Peck 2000: 86) and reckoned that Lewis and Clark had found out more when they passed through Missouri some forty years earlier.

As Audubon declined mentally and physically, the two sons took on more and more of the work, yet failed to send Bachman the specimens and journal extracts that he requested for finishing the accounts of certain species: "They have not sent me one single book out of a list of a hundred I gave them, and only six lines copied from a book after my having written them for four years." Once more Bachman threatened to abandon the project, forcing the Audubons to be more attentive. Even so, Victor managed to upset him again by his heavy handed and careless editing of the first volume of text. John infuriated Bachman by not even having the sense to copy the information from the labels of the museum specimens that he painted (Shuler 1995: 204).

The Audubons had their own difficulties, often with tardy or impatient subscribers. Fortunately for the Audubons they had simplified the task they had set themselves by insisting that *The Viviparous Quadrupeds* was produced entirely in the United States. The lithographer J.T. Bowen, who had produced the octavo edition of the birds, also agreed to produce the imperial and octavo editions on the mammals, volume one of the first 50 plates appearing in 1845, volume two with the second fifty plates in the following year, and the final volume with the remainder of the plates in 1848; the text volumes were issued in octavo size in 1846, 1851 and 1854. A supplement to the last volume included five extra plates, depicting eight species, including Harris's Antelope Squirrel that was said to have been from a Townsend specimen (see pp. 328 and 353). An octavo or miniature edition that dropped '*Viviparous*' from the title was published 1851-54 in three volumes

Though far less known and admired than *The Birds of America*, *The Viviparous Quadrupeds* had a much broader scope than John Richardson's 1829 volume on mammals in the *Fauna Boreali-Americana* that pertained mainly to northern species. Though the *Quadrupeds* was obviously a team effort, its scientific success was largely due to Bachman who had persisted despite failing eyesight, the loss of his first wife, the deaths of two daughters, and countless other hindrances. This success was somewhat temporary as the *Quadrupeds* was quickly superseded by Spencer Baird's *Mammals of North America* (1859) that contained the early results of the first railroad surveys across the American West.

Townsend's role in the *Quadrupeds* was the same as in *The Birds of America*: he helped to make the work more complete by providing new or rare western species, as well as notes for the text. Financially he fared no better than for the birds, his main reward was another batch of North American species named after him, this time: Townsend's Mole, Townsend's Ground Squirrel, Townsend's Chipmunk, Townsend's Vole, Townsend's Pocket Gopher and *Lepus townsendi*, the White-tailed Jackrabbit.

J.J. Audubon in about 1848, from J.H. Herrick's *Audubon the Naturalist* (1938).

Although the talk was mostly of mammals, Townsend was able to show Bachman his first ever Mourning Warbler, a specimen killed locally by William Baird, that Townsend had stuffed a few days previously. Indeed Townsend was preparing quite a few birds for the Bairds (a Least Bittern, Prairie Warbler, White-eyed Vireo and Ruby-throated Hummingbirds are mentioned). In fact Townsend was specialising in hummingbirds and had set up a system whereby he could slaughter them with a long glass tube and putty balls, a method that scarcely damaged their plumage. He even used his blow-pipe on hummingbirds that entered the hall of the Patent Office and at one time he had more than sixty specimens that he was using for exchanges. It was evidently an old trick of his, as he said that he used to go around the gardens in Philadelphia where he could kill birds as large as a sparrow at a range of twenty yards, and up to ten warblers in an hour.

The Fourth of July was eagerly awaited as a date when Townsend and William Baird could take to the field outside Washington, to Beltsville, where they could spend the whole day collecting. They did not get much, a few Henslow's Sparrows being the only species of interest to them. It was around this time that Townsend began to arrange for the Bairds to contact his old friend Ezra Michener so that they could obtain a few duplicates from him (W. Baird to S. F. Baird, letters May-July 1843, SI Archives). In July, William Baird took the opportunity to go to Cape May Island, staying at Schellinger's Landing for a week or so. Few people lived there then, and the maze of creeks and tidal saltmarshes provided excellent opportunities to hunt for shorebirds, gulls and terns. While he was there one of Townsend's brothers-in-law came to see him and invited him to stay at the Holmes farm about ten miles northwards near Cape May Court House. Dr George Leib had been at the Holmes farm "all summer" making a collection of birds and William was happy to go up there too for some more shooting (W. Baird to S.F. Baird, 16th July 1843. SI Archives). This may or may not have been when William first met Townsend's sister-in-law, Harriet Holmes, his future bride.

Meanwhile, at the Institute, the first part of 1843 continued much as before with a steady influx of more natural history items, periodicals and books. In May, as Recording Secretary, Townsend described the contents of two boxes of birds and insects from the US Consul in Brazil, and was responsible for thanking John Cassin for a donation of African specimens from the Philadelphia Academy. Later in the year the financial crisis worsened. In June Townsend reported that "In a few months however our last appropriation will be exhausted, & then what we shall do I cannot say. I hope that in that case the members of the Institute will continue me, at my present salary at least (JKT to H. Townsend, 6th June 1843). But Congress failed to support the Institute and it began to go into decline. In July Pickering resigned to concentrate on his scientific report and was replaced by Captain Wilkes himself. Wilkes immediately claimed the main hall for the expedition collections and set about re-arranging the displays. It was rumoured that Townsend had lost his job (S.F. Baird to W. Baird, 30th October 1843, in Dall 1915: 104), and this was later confirmed:

"Townsend's case is this. Soon after Wilkes was appointed to superintend the operations of the men employed in the preparation of the articles brought home by the Exploring Expedition he began to show great hostility to the National Institute and the men employed by it. He often talked about making the Institute take away their collections, and I was told has threatened to throw them into the streets. After he had been there three months, about a month since, he wrote to Townsend without any previous intimation and told him that his services would no longer be required, giving as a reason that the appropriation was nearly exhausted. I don't know whether this is a fact, but if it is Wilkes has wasted the money shamefully by his changes of the [display] cases, cutting up the old cases, making new ones, altering the arrangements &c. Most persons here believe he has no authority for what he has done, and the matter has been before the cabinet where it created quite an excitement. The Secretary of State is now investigating the business. Wilkes' plan was to go on with the birds, by retaining Pollard (who gets $600) with Titian Peale to direct him, but I have heard since that Pollard has also been discharged[4]. Peale is no more fit to superintend the ornithological Department than the poorest ornithologist in Philadelphia. Indeed I don't think he knows half as much about birds as our Tom [their younger brother]. Mr Spencer & Judge Upshur who are great friends of the Institute were very angry at Wilkes and have promised whenever they can to provide for Townsend by giving him a clerkship, in which he can remain until Congress does something for the Institute, in which event T will be the curator. It is abominable that Wilkes should go on in this way, never paying the least attention to the Institute collections, which are fifty times as valuable as those of the Exploring Expedition" (W. Baird to S.F. Baird, 1st November 1843. SI Archives).

This was a significant blow because had Townsend continued with Wilkes he might well have been transferred onto the staff of the forthcoming Smithsonian Institution. Instead, he had to work as a copy clerk in the State Department in order to survive, but without relinquishing his attachment to the Institute and his various time-consuming administrative duties connected with it (JKT to H. Townsend, 16th April 1844. Family papers). If he was downhearted about the way life was treating him he was tactful enough not to show it when writing to his bed-ridden sister: "We are, as I said comfortably fixed, & as happy as we can be. We enjoy excellent health, all of us – little Frank is as hearty as a buck, altho' just now a little fretful from recent vaccination & teething. We have enough to live on comfortably, with proper economy, & I will venture to say, a happier family is not to be found" (JKT to H. Townsend, 12th April 1844. Family papers).

Townsend continued to attend the meetings of the members of the National Institute as before, even if the number of attendees began to slowly dwindle. In February 1844 he wrote to an old friend, Samuel Haldeman, a wealthy and respected scientist, imploring him to come to an important April meeting to read a paper on any subject he wanted, preferably including in it something favourable to the grand designs of the Institute, underlining "don't fail to come" as if desperate to demonstrate the scientific credibility of the Institute; and he offered him hospitality at his home on New York Avenue between 9th and 10th Streets. Instead, Haldeman sent a lecture that was read in his absence. It was a further setback when Congress adjourned without making any provision for the Institute and began instead to turn its attention towards supporting the Smithsonian Institution.

Ironically, the National Institute's conference was a great success. "It lasted 8 days, & was altogether highly interesting. Many very interesting papers were read on various scientific & popular subjects, & the meetings were fully attended, not only by the scientific & literary portion of the community, but by the elite of the metropolis, and a fair sprinkling of ladies. I attended every meeting, & it being fortunately, just at the time when Calhoun assumed the direction of our Department, & before he had got it perfectly organized, my absence was not noticed ... my name was frequently mentioned creditably, & I several times thought that if my sister H. had been present, she would have enjoyed hearing the praises of her unworthy brother trumpeted forth to the world. Senator Walker, who delivered the Opening Address, was the first who named me. In his enumeration of the American citizens who had exerted themselves to illustrate the Natural history of our country, he alluded to me as the "excellent and enthusiastic Townsend – and then stated that I had traversed sea & land; crossed the wide & inhospitable rocky barriers ...&c. &c. The Rev. D[r]. Morris of Baltimore also mentioned me, in a paper which he read "on the past & present state of Entomological Science in the U. States." D[r]. Morris & I have become firm friends" (JKT to H. Townsend, 16th April 1844. Family papers).

In May, having recovered from helping to organise the conference, he wrote to his old friend, Samuel Morton, congratulating him on an honour that had been bestowed on him by the Royal Academy in Paris, using this as an excuse to ask Morton to chase up William Cooper for the return of some bat specimens (see appendix 17). Morton knew Cooper and had already asked for the return of the specimens, and in due course a package had been returned from New York. This had been forwarded to Townsend who was not at all pleased with what he found within – and said so to Morton, becoming uncharacteristically belligerent as he went on:

> "I surely have a right to ask what has become of them. Will you therefore do me the favor to write to Cooper on the subject, & make enquiries about the missing specimens. I do not know Cooper & have never seen him, or corresponded with him, & as you sent these specimens to him for description, I trust you will not object to write to him on the subject. If I can recover the two specimens of Plecotus, I shall be satisfied, though I might very properly inquire why the third – the one sent me – was destroyed.
>
> Does it not seem at least, that I have been unfairly treated in regard to several portions of my Oregon collections? In addition to the present matter, I fear that all the beautiful & rare insects collected by me in that region have been either destroyed, or surreptitiously appropriated. I have written to Dr. Harris on this Subject, & shall leave no stone unturned to recover at least any portion of this fine collection which may not have been destroyed, lost, or taken" (JKT to S. Morton, 8th May 1844. Library Company MSS at HSP).

His general mood seems not to have improved in June when he wrote a friendly letter to Spencer Baird but included a swipe at those who had been picking away at his hummingbird collection, particularly John Varden who had gone off with six males after promising to collect insects for him. "But to this day I have never seen one insect of his capturing, & he informed me a few days since that he had not had time to look for them – so it goes. Of 50 persons promising to collect for me, about one individual fulfils his pledge." As for personally collecting birds this had almost ceased, much to his regret: "My time has been so entirely occupied by my office & Institute duties as to allow me scarcely a moment of leisure for several months, which I could devote to the shooting of Birds. My clerical duties occupy me, as you know, during the greater part of every day, and in the short intervals, – which I in fact require for relaxation from constant & excessive labour, – I am compelled to give my Services to the Institute ..." (JKT to S.F. Baird, 7th June 1844. SI Archives).

So, although Townsend was still working as a clerk in the State Department, he was keeping up his interest in the Institute, still persistent in the hope that a paid position would be found for him there. He confided these hopes to his sister, urging her to keep secret the news that he been told by an unnamed friend: "that it has been proposed ... in order that the mass of valuable materials now in possession of the Institute may not altogether be lost to the country, that a small appropriation be voted for the purpose of paying a Curator, whose duty it shall be to take proper care of the collections, & attend to their preservation & arrangement. Now if a Curator is to be appointed, who but thy brother ought to be a candidate. My friend thinks that I shall certainly be the person, & I have lost no opportunity of talking with my influential friends about it, all of whom promise to assert themselves in my favor. This would suit me exactly, & surely, if any one deserves the appointment, I am the man, for I have been connected with the Institute from its foundation, & have since served it gratuitously for months together (JKT to H. Townsend, 16th April 1844. Family papers).

William Cooper (c.1798-1864), one of the founders of the New York Lyceum of Natural History. In 1837 Cooper described and named Townsend's Big-eared Bat in the *Annals of the Lyceum*, from three specimens brought from the Columbia River by Townsend, illustrating only the head of the bat, in plate 3, figure 6 (bottom left).

In late January 1845 the first volume of imperial folio plates of *The Viviparous Quadrupeds* was issued, and Townsend would surely have been pleased that plate number three was a well executed rendition of "Lepus Townsendii Townsend's Rocky Mountain Hare" and that many more of his specimens had been painted. On 29th January Townsend was elected Recording Secretary again. Joel Poinsett, the originator of the notion of the National Institute, had declined to be re-elected as president, the society's publications lost their scientific content and were more or less discontinued, and the impressive list of 350 resident members and 1250 corresponding members gradually decreased. In July Townsend contacted Edward Harris at Moorestown, asking him if he could supply live Wood Ducks for a friend, hoping his new gun dog puppy was developing well, and quizzing him about the status of the Cliff Swallow on his patch and to the eastward. Townsend wanted to know about these swallows for a paper (never published) that he was preparing for the *Literary Record and Journal of the Linnaean Association of Pennsylvania College* (JKT to E. Harris, 4th July 1845. ADAH). In August he was on the New Jersey coast where he was spotted by John Cassin: "Harris & I were at Cape May, & saw Townsend, – had a days shooting on Saturday but got nothing of consequence – lots of Godwits, Martins, Terns, Gulls & tilt tails of one sort & another – a vile place that seashore – I have been sick ever since my return ..." (J. Cassin to S.F. Baird, 21st August 1845. SI Archives).

Townsend attended all of the Institute meetings of 1845. He was there for the meeting of 8th September though there was no quorum in June, July, or August. It may have been that the members had simply moved out of town as usual to escape the oppressive summer heat, as Townsend had done, but the reality was that the Institute was doomed. In the following year, the fourth *Bulletin of the Proceedings of the National Institute* carried a report, dated 25th November 1846, that Congress had again failed to assist them: "more than a thousand boxes, barrels, trunks &c., &c., embracing collections of value, variety, and rarity in literature, in the arts, and in natural history, remain on hand unopened ... For the preservation, reception, and display of these the Institute has neither funds, nor a suitable depository. Usual meetings of members have been suspended for a considerable period." The only good news was that the bulk of this material and the Exploring Expedition collections were later transferred to the newly formed Smithsonian Institution and much of it still survives there (or in other museums).

No longer able to justify remaining in Washington, Townsend abandoned the city before the end of 1845 and took his family to Philadelphia. By December he was once again one of the curators at the Philadelphia Academy of Natural Sciences, though this was surely an unpaid position.

1. The Museum of Vertebrate Zoology at Berkeley, California, has a Wilson's Snipe, collected by J.K. Townsend in April 1842 (MVZ 100853).

2. William Baird later wrote to his brother (on 10th February 1848) to tell him that he "was very much pleased last night to learn by a letter from Mother that a specimen of Homo sapiens – female young – locality Carlisle Penna. had been added to your collection". He was referring to the birth of Lucy Hunter Baird (1848-1913). Lucy's Warbler *Vermivora luciae* was named after her in 1861 by J.G. Cooper.

3. William Baird says that "Dr. Bachman was here for a day or two last week, I did not get to see him, not having heard of his arrival until after he had gone. He was with Townsend most of the time (W. Baird to S.F. Baird, 31st May 1843. SI Archives).

4. This put a stop to Pollard who was slipping a few U.S. Exploring Expedition specimens to William Baird and probably to other collectors too. "He says he got them from one of the Exploring Expedition men, but inter nos [between ourselves] I think he <u>pooked</u> [pinched] them himself. However I am not bound to know this, and I will get what good birds I can in this way" (W. Baird to S.F. Baird, 5th July 1848. SI Archives).

Philadelphia once more

"Can you tell me whether John Townsend & his family are in Phila[delphia] ... I want [my wife] Mary to know the cleverest people of my acquaintance."

S.F. Baird to W. Baird, 4th August 1846.

At the start of 1846 John Cassin was optimistic about bird research at the Academy of Natural Sciences of Philadelphia because several fieldworkers had recently returned from their travels. In a letter to Spencer Baird, after listing a few of these naturalists, he noted that "We have Townsend also, in the city this winter, & he expects to stay here permanently – he is studying Dentistry ..." This, if you need reminding, was the profession of his three brothers: Elisha, the eldest, being particularly eminent as co-founder and Dean of the Philadelphia School of Dental Surgery. But it was not only Townsend that needed qualifications and livelihood, as Cassin soon realized: "Our ornithological corps is doing nothing – Heerman[n] is in Baltimore studying Medicine – Gambel & Woodhouse here, doing the same thing – Townsend has set up his pole as a dentist – Harris lives at home like a gentleman, as he is, & your humble servant (in his lucid intervals) tries to mind his business, with more or less success" (J. Cassin to S. F. Baird, 13th January and 20th February 1846. SI Archives).

In some ways Townsend was probably pleased to be based in Philadelphia, living at 9th and Cherry Streets, much closer to his family, friends and old acquaintances in the city and nearer to his favourite haunts on the New Jersey coast. He had often visited the city while living in Washington but now he had more time for socialising. His main problem once again was a shortage of money and he certainly could not afford the second volume of *The Viviparous Quadrupeds* when it came out in March.

Following the sad demise of the National Institute, he now rarely saw his old friend William Baird. But William had not lost touch and asked his brother Spencer to spy upon the Townsends when he visited Philadelphia, or to be more precise, to spy on Lottie Townsend's sister Harriet, who was staying with them. William and Harriet had met several times, but he wanted his brother's opinion as to whether she would make a suitable wife (W. Baird to S.F. Baird, 14th March 1846. SI Archives).

In June, Spencer Baird wrote to his brother asking him to quiz Townsend about the fate of the birds brought back from California by Colonel Frémont (S.F. Baird to W. Baird, 15th June 1846. SI Archives). This was evidently a reference to a small collection gathered by Edward Kern, a Philadelphia artist and member of the Academy who had travelled with Frémont's third expedition. Kern had deposited his specimens at the Academy and Townsend must surely have admired the two California Condors contained amongst them (Harris 1941: 29) and perhaps wondered why it was Kern and not himself who had acquired the Academy's first specimens of Pinyon Jay. Townsend had quickly passed through the jay's breeding range (while traversing Wyoming) so it seems he was just unlucky not to come across it. But it was Cassin who was supposed to be working through the Kern collection, though as late as 1850 the over-worked taxonomist confessed that it was still "positively untouched (J. Cassin to S.F. Baird, 14th February 1850).

John Cassin (1813-1869), a leading 19th century taxonomist who studied the huge collection of American and foreign birds at the Academy of Natural Sciences of Philadelphia.

Thomas Nuttall, who crossed the Rocky Mountains with Townsend in 1834, returned to the United States for a few months in 1847.

Around the end of July, Townsend had a well-earned break, taking his family to Lottie's old home at Cape May Court House. Spencer Baird got married in August and at last planned to carry out his proposed visit to Philadelphia. He wanted his bride to meet Townsend, and he had William's instructions to scrutinize Townsend's sister-in-law. Spencer must have been favourably impressed because late in the following year, on 2nd December 1847, William Baird married Harriet Holmes in Philadelphia. The Townsends and Bairds were thus brought closer together, the men sharing a mutual interest in birds and now also related by marriage. It is amusing to note the topping and tailing of three letters that Townsend wrote to Spencer Baird. "My dear Sir ...very truly J.K. Townsend" (10th October 1840); Dear Baird ...Your friend Sincerely John K. Townsend (7th June 1844); and "Dear Spencer ...I am, dear Spencer, your assured friend, John K. Townsend" (13th January 1849) – the last being about as informal as it was possible to be in those days.

More important to the Academy and its curators than a few birds gathered by the Frémont expedition, was the news that Dr Thomas B. Wilson had purchased for the Academy the famous collection of Victor Masséna, the Duke of Rivoli – some 12,500 specimens from around the world, and had paid for an extension to the new building in order to house them. Wilson also bought the Bourcier collection of parrots and tanagers, small collections from the Verreaux brothers, Captain Boys's Indian birds and John Gould's Australian birds. Over the next few years, birds collected by William Gambel, Dr Heermann, Dr Woodhouse and many others were added. In time, Wilson presented his entire collection of 26,000 birds to the Academy – and it proved to be a great public attraction. Could Townsend prosper from all these developments? The answer was: hardly at all. In 1847 Joseph Leidy was elected Curator in preference to Gambel. John Cassin was already well ensconced in the museum as the chief ornithologist, and a very prolific and diligent one, even though he only worked at the Academy in his spare time. He was somewhat jealous of his unofficial position as ornithological curator and tended to resent intrusions by others (being particularly critical of young Gambel). He devoted nearly two years to arranging the Wilson acquisitions, and spent much of the rest of his life handling these skins, and other collections, including those of Townsend, as he compiled a series of important ornithological papers, scientific reports for government surveys, and his well-regarded *Illustrations of the Birds of California, Texas, Oregon, British and Russian America* (1856).

Thomas Nuttall paid a brief visit to the United States and was at the December meeting of the Philadelphia Academy. He remained in the city until March 1848 and presumably met up with Townsend for a spell. The third and final volume of Audubon and Bachman's *Viviparous Quadrupeds* appeared in that year. Edward Harris donated 118 bird skins to the Academy in October – these were some of Townsend's western specimens that Harris and Audubon now no longer required, as well as the fruits of Harris's 1843 trip up the Missouri with Audubon.

Perhaps Townsend no longer wanted a position at the Academy but it seems more likely that he simply could not afford to be there. He still maintained an interest, being listed as one of the curators for 1846, but as time went on he was not as actively involved as he had been in former times, neither here nor at the National Institute. In the latter part of the year he is recorded as exhibiting a live House Mouse "which possessed the remarkable peculiarity of uttering notes resembling those of a singing bird" (Townsend 1847b), hardly an item of great significance. An 1848 list of the most active members working in the Academy includes Samuel Morton, John Cassin, Joseph Leidy, William Vaux, William Gambel and S.S. Haldeman – but there is no mention of Townsend (Nolan 1909: 13-14).

John Kirk Townsend and his wife, Charlotte Holmes Townsend. Watercolour, possibly by Charlotte who was a portrait painter [1].

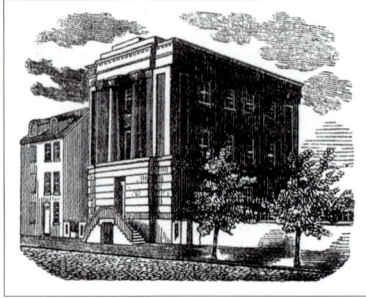

Pennsylvania College, Philadelphia, in 1847.

The main reason for this was that Townsend had begun to attend Pennsylvania College, a medical school in Philadelphia. Founded in 1839, it offered courses in the theory and practice of medicine, obstetrics, materia medica and therapeutics, anatomy and physiology, the principles and practice of surgery and medical chemistry. Townsend enrolled for the session that began in early November 1847, attending lectures and instruction at the college in Filbert Street, with two mornings of clinical practice each week a few blocks away at the Pennsylvania Hospital in Pine Street (where his father had attended to the clocks). The course finished in March and he graduated M.D. sometime later in 1848, having presented his thesis entitled "Intermittent Fever of Oregon Territory". Townsend's connection with the college seems to have been for much longer than his six-month course because, as we have already noted, he was gathering data as early as 1845 on Cliff Swallows for a paper for the *Literary Record and Journal of the Proceedings of the Linnaean Association of Pennsylvania College*. He may have already begun to fulfil some of the college's requisites for graduation that, besides stipulating that candidates must be over twenty-one years old and of good moral character, demanded that they apply themselves to the study of medicine for three years and become a pupil for at least two years with a respectable practitioner of medicine. Townsend seems to have fulfilled some of these obligations long before, but further study may nevertheless have been required of him, particularly in connection with dentistry.

Although his paper about the swallows was never published, he contributed a series of accounts to the *Literary Record*, the first of which harked back to his time in the Pacific and was entitled "Sketches of a Voyage" (1847a), and the second was his "A Popular Monograph of the Accipitrine Birds of N.A." (1848). The latter contains his only full account of his killing of a young California Condor on the Willamette River (see pp. 192 - 193) as well as observations on raptors in South America and the eastern United States close to Philadelphia.

Early in the following year, 1849, Townsend wanted to raise money from the sale of his western mammals, amounting to nearly sixty specimens, but the Academy could not afford to buy them. Instead, he thought he would first try Hermann Schlegel at the natural history museum at Leiden and then J.E. Gray, Keeper of Zoology at the British Museum (Natural History). To help things along, he wrote to Spencer Baird asking for two testimonials from him, the same one for each letter, even including a draft of what Baird should write:

"I have frequently seen and examined the skins of Mammalia brought from Oregon by Mr. Townsend and can safely say that it would almost be impossible to find anywhere the same numbers of specimens in equally good condition. They are of the very first order, perfect in every part. It is unnecessary for me to dwell on the extreme rarity of many of them, several being even unique specimens, very few of them are contained in any collections in the United States" (JKT to S.F. Baird, 13th January 1849. SI Archives).

Townsend may never have written to the Netherlands but he certainly wrote to London (JKT to J.E. Gray, 13th January, 21st March and 15th June 1849. BMNH). In the first letter (transcribed in appendix 18), he expressed his hope that his mammals would "command a higher price in England than in this Country, because you are more capable of appreciating the value of such rarities than our more plodding & money-loving people". Gray wrote back asking Townsend to say how much he wanted for the skins and Townsend replied in March saying that he was hoping for $300, the sum he thought he could have got from the Academy – if they had had any money. He said he would take $250 if the money were sent soon because "I am now very much in want of such a sum." He even wanted the money before Gray had seen the skins, suggesting that Gray should arrange for someone in the States to look at them for him. Gray does not seem to have replied to this impertinence, so Townsend wrote a third time. The price was still $250 "although, considering the rarity of the specimens, the distances from which they had been brought & the labor & risk of procuring them, I consider the sum inadequate … If I do not hear from you in a reasonable time, I propose to put the specimens in our Acad. Nat. Sc., agreeing to receive for them whatever the Institution can afford to pay." Gray declined because of lack of funds and because he said that London had three quarters of the specimens already. This was no doubt close to the truth because the Hudson's Bay Company had been sending furs from North America to London for about 180 years.

Another way Townsend could have raised (or lost) money was to have gone to California to join the scramble for gold. He confided to Spencer Baird that, "If I were a single man it would not take me long to make up my mind about going, but my "duck" objects so strongly that I dare not even think about it." But he evidently *had* thought about it because he suggested that Baird's brother William, who had been ill, should go out to the "diggings" for the sake of his health and because "he might return, after a few months absence with the means of living comfortably, without toiling, for the remainder of his life …" (JKT to S.F. Baird, 13th January 1849. SI Archives).

ANOTHER TOWNSEND SPECIMEN AND ANOTHER MURDER

Towards the end of 1849 Townsend was shocked to hear of the murder and dismemberment of Dr George Parkman at Harvard. One of Townsend's House Wren specimens from the Columbia River was displayed prominently in Parkman's home – a gift from Audubon who had named it Parkman's Wren *Troglodytes Parkmanii* in 1839 and later had it mounted for him. Townsend, and the rest of the country, followed the sensational eleven-day trial of Professor John Webster in the daily newspapers, a story that ended with Webster's conviction and execution (Thayer 1916, Palmer 1928: 291).

Audubon thought that Parkman's Wren was a new species but it is now recognised as the western form of the House Wren. In his *Ornithological Biography* he had this to say:

"PARKMAN'S WREN … A single specimen of this Wren which differs considerably from *Troglodytes hyemalis* [Winter Wren] and *T. europaeus* [also Winter Wren], has been sent to me by Mr. TOWNSEND, who procured it on the Columbia river, along with several others, all exactly similar … Feeling perfectly confident that this species is distinct from any other, and not finding it anywhere described, I have named it after my most kind, generous, and highly talented friend GEORGE PARKMAN, Esq., M.D., of Boston, as an indication of the esteem in which I hold him, and of the gratitude which I ever cherish towards him" (Audubon 1840-44, 2: 133-135).

In 1916 John E. Thayer reported that he had acquired the original specimen of Parkman's Wren "mounted on a twig, in a paper box with a glass front … the bird is in excellent condition." In addition to four original watercolour paintings by Audubon there were two letters containing references to the bird.

The first letter, by Audubon to Parkman, is headed "New York, June 20th 1841". Audubon regrets not having time to send the specimen: "My Dear Friend, – I intended having written to you yesterday … but I was deeply engaged on a drawing of Rocky Mountain Flying Squirrels … The "Parkman Wren" well mounted will soon be on your chimney mantle!"

The second much shorter letter, by Audubon to Parkman, is headed "New York, August 13th 1841" and continues: "My Dear Friend, – By Mr. Legaré who revisits your City, I have the pleasure of sending to you, the "Parkman's Wren" and I hope you will receive it in good order. We found it necessary to recast the position of this little fellow on account of the many shots that passed through its neck when killed …"

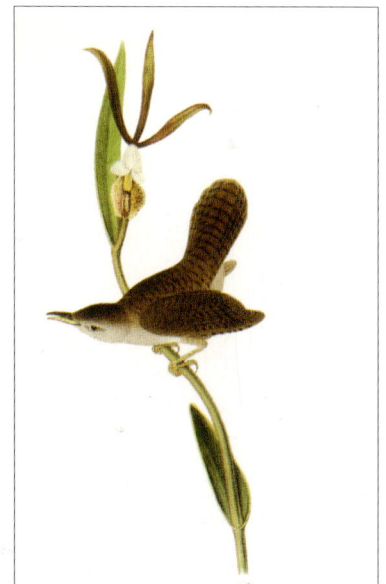

"Parkman's Wren" did not appear in the folio edition of *The Birds of America* but was included in the later octavo edition, plate 122.

In 1931 Thayer donated his massive collection of skins, nests and eggs to Harvard University. After his death in 1933 Harvard also received his collection of 3500 mounted birds. The Museum of Comparative Zoology continues to house Townsend's specimen of Parkman's Wren (MCZ 140279).

Townsend's Giant Wolf *Lupus gigas* (at front) and Peale's *Lupus occidentalis* (at rear) as depicted by Peale in Cassin's *Mammology and Ornithology* (1858) of the US Exploring Expedition.

Gray Wolf skull (ANSP 2266) described by Townsend as belonging to the Giant Wolf *Lupus gigas*.

TOWNSEND'S "GIANT WOLF"

"I travelled from St. Louis to Fort Vancouver, gun in hand, and shot, not only the Canis occidentalis of Richardson [Gray Wolf], but also the C. latrans of Say repeatedly [Coyote], and twice the C. gigas" (Townsend, 1850: 78).

Towards the end of his life Townsend decided to write up his wolf discoveries, strongly believing in the existence of an extra-large wolf species. He submitted a four and a half page article to the *Journal of the Academy of Natural Sciences of Philadelphia* (1850) entitled "On the Giant Wolf of North America. – LUPUS GIGAS". In it he gave a full description of a skin and skull from a wolf "captured by me in the month of September, 1836, on the plains of the Columbia river, one and a half or two miles west of Fort Vancouver." He does not say where his second wolf was shot, but in his *Narrative* (1839a) he does mention another skin that he purchased from a trapper "who took an enormous wolf in a beaver trap" on 19th December 1835 near Fort Vancouver. In the appendix to his *Narrative* he lists a "Dusky Wolf, *Canus [sic] nubilus*?" with a footnote: "This is probably a new species. It is much larger than *nubilus* [Western Gray Wolf *Canis lupus nubilus*], as described [by Thomas Say], and differs much in its habits."

Lewis and Clark referred to a large brown wolf of the wooded regions of the Columbia, and Titian Peale described a large wolf collected by the United States Exploring Expedition on the west coast, though it was not quite as large as Townsend's specimens (Cassin 1858; 16-18). Indeed it was Peale's naming of the wolf as *Lupus occidentalis*, wrongly linking it to a wolf of that name described by Dr John Richardson, that prompted Townsend's attempt to clarify the situation.

There are currently three Townsend specimens of "Lupus gigas" in the Academy collection. Two skins are marked in the catalogue "Rocky Mountains, 1834," and one skull has "Canis gigas Dr. J.K. Townsend" written in ink upon it, and it is noted in the catalogue as being from the "Rocky Mountains" (without a collection date). There is nothing to suggest that the skull belongs to either of the two skins. The discrepancy in dates and location is of little consequence as these probably just refer in a general way to Townsend's 1834 trip across the Rocky Mountains.

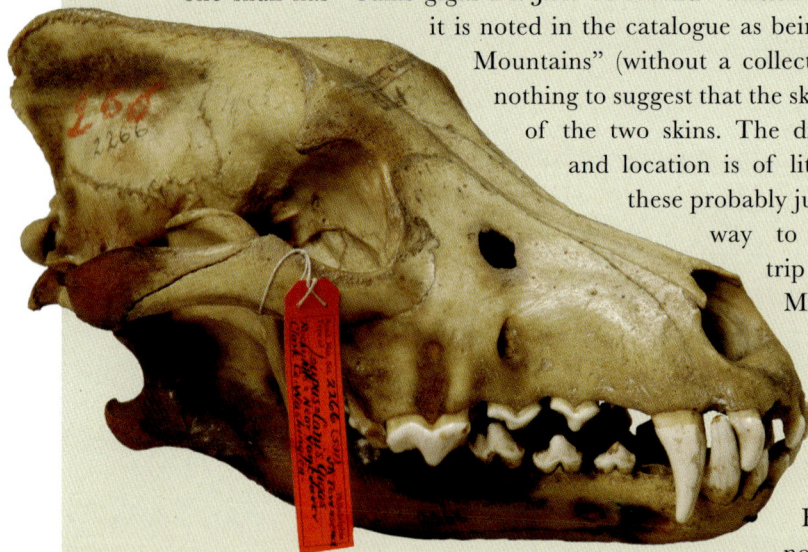

There was never any doubt about the large size of the Townsend specimens. However, they are not now considered to be of a new species, but are referable to the race *Canis lupus fuscus* of Richardson. Indeed, all wolves from the northern part of North America are classified as the Gray Wolf, even though they may be black, grey, brownish, almost white, or a mixture of colours and pattern. The largest Gray Wolves are found in Alaska and the Mackenzie River region of Canada.

In 1850 Townsend's hard-won mammal skins went to the Academy after all and he at last published his report on his Giant Wolf in the *Journal* of the Academy. This was also the year that he was elected a Life Member. In July, Spencer Baird was appointed Second Secretary to the Smithsonian Institution, an appointment that allowed him to direct the course of North American ornithology for decades. Baird received information and specimens by careful placement of naturalists on government expeditions that were then surveying all parts of the States in preparation for the western expansion of the railroads and in order just to see what was out there, even if mainly for future economic development. Ironically, William Baird had once written to Spencer telling him that he must choose a career and advising him that "no means of livelyhood however is to be obtained in America from Ornithology" (W. Baird to Spencer Baird, 23rd November 1842. SI Archives).

Whilst Townsend was no doubt pleased for his friend, he must have wondered why he had never been fortunate enough to carve out a proper career for himself in natural history. Instead, he had somewhat reluctantly set himself up as a dentist in Philadelphia, though he cannot have practiced for long. An 1851 city directory lists him as a doctor and dentist at 360 Chestnut Street but the directory may have been compiled in the preceding year, and was probably out of date by the time it was printed because by then Townsend had other plans. He was often ill, so much so that in the summer it seemed that he would not survive. Although he apparently made a full recovery, these periodic bouts of illness were nevertheless a great worry to his wife and family. His mother confided in her diary: "I have felt for him what a Mother's feeling only knows – I am sometimes doubtful of his ever being better – having been with him and his dear Lottie, (in mind) much of to day I have desired to be resigned to the event – knowing all is in the hands of a kind Providence who doeth all things right" (P. Townsend, diary No. 5. Family papers).

It was particularly difficult for Lottie, struggling to cope with her ailing husband and concerned too for her sister Harriet, who was poorly following the birth of her second child (the first having been stillborn); the new baby proved to be constantly sickly. In June 1850 Lottie and her mother went up to Reading to stay with William and Harriet Baird for a short spell. Early in 1851, on 18th January, John, Lottie and Frank left Philadelphia to visit the Bairds but it cannot have been a very happy time as Harriet and her baby, and then young Frank, all went down with chicken pox. They returned on the 21st, John reportedly looking much better than he had for some time, and he was soon off to Washington, leaving Lottie behind with his parents. At Washington, Townsend was, of all things, planning to sign up for a government cruise to Africa under Commodore John H. Aulick. But he fell ill again.

Supposing that he would soon recover, three days passed before Lottie was sent for and by then it was almost too late. His mother recorded: "She found him still suffering, but less oppressed and rejoiced to see her, and conversed with her til the moment almost of his departure which was two hours after her arrival. – He told her he thought he should not live, and when the D^r said he thought there was no immediate danger, he looked up and said to him "Doctor I am not afraid to die" and in a few moments breathed his last – Disease of the heart no doubt – He was an affectionate Son, brother, Husband and Father – but in this great bereavement we have much to console and reconcile us, first and greatest, the sweet and undeniable assurance that he is at rest, in a state of acceptance. Next that he was with those who loved him next to his own family, that every comfort that could alleviate his suffering was bountifully bestowed, the best Medical skill was procured – and his dear Lottie saw him alive and sensible and we have the testimony of our dear Rosanna, who was with him day and night, that he was "The sweetest, most patient sufferer she ever saw" – Had he been at sea amongst strangers and his remains consigned to the bottomless deeps – how widely different would have been our reflections – I feel that I cannot be thankful enough, for in all I

John Kirk Townsend and his mother Priscilla Kirk Townsend (Family papers). The portrait at top is probably the source of the well-known reproduction that accompanied Witmer Stone's biographical sketch of Townsend in *Cassinia* (1903).

Elizabeth Townsend (1825-1869). By the end of 1851, Elisha, Edward, Charles and Elizabeth were the only four of Charles and Priscilla Townsend's twelve children that were still alive.

behold the hand of a kind Providence, and I feel now that I can say in great sincerity "Thy will be done" (P. Townsend, diary No. 5. Family papers). Five months later, his mother had much to reflect upon, as Townsend's beloved sisters, Hannah and Mary, both passed away in July. Of her twelve children, eight had now died, leaving only Elisha, Edward, Charles and Elizabeth. She faced her sorrows bravely, with a certain degree of resignation: "We do <u>not murmur</u> but rather feel that we can say "He gave and he hath taken away, blessed be his name – Now let our concern be to follow them as they followed Christ" and then we shall be at liberty to enjoy the blessings we have left us … of dear affectionate children who are anticipating our every wish – and of every earthly comfort too." (P. Townsend, diary No. 6. Family papers).

John Townsend's death was on 6th February 1851 and two days later he was buried in Washington, at the Congressional Cemetery, just off Pennsylvania Avenue, beside the Anacostia River. Many of the family, including his mother, were unable to attend the service but a letter from a friend (that Priscilla quotes in her diary No. 5) reported that: "he looked as calm – composed and tranquil as though he had "Wrapped the drapery of his couch about him And laid down to pleasant dreams". This has been the testimony of others who saw him there, confirming our feeling that all is well with him."

Townsend's early death has long been attributed to the effects of arsenic poisoning, due to a lifetime of preserving specimens, perhaps casually as a boy, then more frequently during his travels and later it was day after day when he became involved in museum taxidermy. We have already read his brother-in-law's report that Townsend was sometimes enveloped in clouds of arsenic powder (as reported in Stone 1903) and we have modern evidence from radiographs of some of Townsend's Hawaiian birds that he tended to use more arsenic per specimen than some other collectors (Olson 1996: 20). John Cassin was frequently ill from arsenic poisoning and it is well accepted that this was due to prolonged handling of specimens from the Duke of Rivoli collection at the Academy. When describing his own symptoms, at various times between 1848 and 1850, Cassin said that he had "*congestion of the lungs,* and most violent headache and fever … pain in my breast … a most abominable ulceration of my tongue", the latter being so bad that for three months he had no proper meals, only mush (Peck 1991, I-24-25, I-33). Although Cassin died in 1869 from "remittent fever" his failing health due to exposure to arsenic must surely have been a contributory factor. It is curious that Townsend seems not to have mentioned the same symptoms experienced by Cassin. If he had had similar ailments it would have only been natural for the two of them to compare notes and remedies. Instead, when Cassin received the news of Townsend's death he said, "I heard of it with great regret and some surprise having seen him and had considerable conversation with him in my room at the Academy but a short time since … What was the matter with Townsend?" (J. Cassin to S.F. Baird, 12th and 16th February 1851).

The truth is, we do not really know the cause of Townsend's death. His mysterious illness at Valparaiso may have helped to shorten his life by some means. Arsenic poisoning may have weakened his health but is unlikely to have been the immediate cause of death because he had reduced his involvement with museum work for the previous four years or so. His mother believed that he had died of heart failure and several years previously Townsend himself had confessed to a sister that he had a troublesome medical condition that he shared with their mother (JKT to H. Townsend, 6th June 1843. Family papers). And yet his mother lived to within a few weeks of her seventy-seventh birthday.

The deaths of Townsend and Audubon were just ten days apart. After a long decline in mental and physical health, Audubon died at Minnie's Land, his home near New York, on 27th January 1851. One wonders how Townsend felt about that? Perhaps he never even heard the news – when the artist's death was announced in the papers his name was given as Anderson. Audubon lived to be sixty-five while Nuttall, another key character in the Townsend saga, pottered about his country estate in England until 1859, dying at the age of seventy-three. Townsend had died aged only forty-one.

> "… with Townsend there prematurely perished
> one of the humblest, gentlest, and therefore
> truly greatest of Nature's noblemen"
> (Rhoads 1903: 377).

EPILOGUE

What became of Townsend's wife and their small boy?

Until recently, little was known about Charlotte except that she was buried at Cape May Court House in 1876. Likewise, there was only scanty information about Frank, gleaned from a relative by Witmer Stone while preparing a short biography of John Kirk Townsend that appeared in *Cassinia. The Proceedings of the Delaware Valley Ornithological Club* (1903). Stone had managed to get in touch with Mahlon Kirk, Townsend's brother-in-law, and from him began to piece together the latter years of Townsend's life. Kirk still had some of Townsend's letters and copies of journals and after discussion with his relatives decided that "John's correspondence and M.M.S. [sic] relating to his journey across the Rocky Mountains should be in the archives of the Academy of Nat. Sciences … [Those letters to] his Sisters, his expressions of affection to his Parents, and his family, reveal a side of his character in harmony with his loving nature, often wanting in scientific men, who, engrossed with their pursuits, often neglect or ignore, these finer feelings of our nature which go to make up the companionable man. I consign the whole to you in the belief you will find in them, much to interest, much to admire, and much – if you think so with me – to preserve, though written half a century ago" (M. Kirk to W. Stone, 27th February 1904. ANSP).

In his search for information, Witmer Stone traced Townsend's son, who in 1904 was just over sixty. Frank had served in the navy during the American Civil War, resigning in the midst of it, in 1864, for some unknown reason. He spent many years as a bookkeeper for the Reading Rail Road and seems never to have prospered. When Stone heard about him he passed on Frank's address to Kirk, who was in the process of sending someone to find out more about him, when he wrote again to Stone:

> " Frank grew up to be a nice boy. We got him a position in the U.S. Navy as engineer where he did well for two years. On his Grandfather's death [in 1859] he [later] came in possession of a legacy of several thousand dollars [in 1865]. I was very desirous, and urged him to let me invest it for him. But some young naval friend, persuaded him to join him, and they put their money in some patent payement [?], with the result, that in six months their money was all lost. His Mother died about this time [1876], and he drifted further and further away from our influence, and from our sight. We then heard he had married some Irish girl far beneath his social station. For twenty-five [years] I have heard nothing concerning him until you gave me what had come to your knowledge.

Thus matters rest. Of course I must know something of his surroundings before I can do anything for him. But it is all very sad to me ...

Should I hear anything favourable of Frank I will let you know, for his father's sake. If on the contrary, he has drifted to a level below recognition, and is alive I see nothing we can do, but regret that another young man – and there are many – has taken a downward path that leads, inevitably to degradation and obscurity – tho he is heir to a honourable name" (M. Kirk to W. Stone, 15th September 1905. ANSP).

Kirk painted a rather depressing picture of Frank and should perhaps have waited until he met him again before passing judgement. There is no surviving correspondence from Kirk and Stone to explain the follow up to all this but it would have been a pity to let the matter rest there.

New research by Donald Townsend Little offers a much less pessimistic picture of Frank. He certainly seems to have led a fairly unremarkable life (without developing his father's interest in natural history) and although things were undoubtedly tough at times for both himself and his mother, there is nothing to suggest that he was a reckless spendthrift nor that he had sunk as low as Kirk was keen to suggest. Frank may have been quite content, or even very happy, for most of his life, we simply do not know. Indeed, there are still many gaps in the record as well as key questions that may forever remain un-answered. The following brief account is based entirely on Donald Little's discoveries.

After ten years of marriage Charlotte was suddenly left a widow with a child who was still less than eight years old. She had no income but was blessed with kind relatives who never deserted her and were always willing for her to stay with them for long periods, being welcomed particularly by her late husband's parents and his youngest brother, Charles. Nevertheless, Charlotte spent the next twenty-five years roaming around various boarding houses in Philadelphia and its vicinity. She may have gone to Cape May Court House from time to time, but to eke out a living as a portrait painter she really needed to be in or near the city. It cannot have been easy for her, as this was a time when photography was in the ascendancy, when everyone seemed to be heading for the photographic studios for individual and family portraits.

Shortly after being widowed, Charlotte "went West" – which could have been anywhere, but was probably only as far as Indiana where her young married brother John was living. In April 1855 she was back in Philadelphia staying with her husband's parents again: " Lotte and Frank are here, but Lotte is not at-all well, had gone to lie down, dear child, she is very excitive in trying to keep up, but she had taken cold and I saw bed was the best place for her ...we are very glad to have them, Frank is a sweet good boy – his Uncle Charles says of him he is the very best child he ever saw, he is a smart little fellow and very affectionate. He went to Market with his Grand Father this morning and carried one basket, he is ready and willing to do any-thing, and seems to enjoy a book as much as any-thing else ..." (P. Townsend, 28th April 1845. Family papers).

In 1855 Charlotte no doubt spent some time at Cape May Court House because during the year she painted a portrait of her mother, and was probably back there in 1858 when she painted her nephew, John Hand Holmes. The portraits are both in the Cape May County Historical and Genealogical Society collection in the John Holmes House and although they show considerable artistic ability Charlotte has managed to escape the National Portrait Gallery on-line index and appears in the New-York Historical Society Dictionary of Artists in America only because she is mentioned as an artist in the 1860 census. Charlotte and Frank were then staying in a Philadelphia boarding house with another 19 occupants. Three artists were listed there, Anni Marsh aged 26, Charlotte Townsend aged 36 (actually 42), and John Neagle, one of Philadelphia's best portrait painters, aged 66. Frank was listed as a machinist, aged 17.

As we have heard, Frank joined the US Navy as an engineer during the American Civil War. He resigned in 1864, perhaps because he knew he was due to inherit about $4200 from his late grandfather. By 1870 he had married Mary, probably an Irish Catholic, a course of action that may have set him against Mahlon Kirk, particularly as Frank had also chosen to disregard his investment advice. In the 1870 census Frank is noted as being in "Bolt Manufacture" and had assets of $5000, so if there was a financial setback it must have happened later.

With Frank now married, Charlotte was on her own, though at the time of the 1870 census she was staying with her sister Harriet and brother-in-law William Baird at Reading, Pennsylvania, still listed as a portrait painter. How long she stayed there is not known. She died six years later, on 25th October 1876, and was buried in the Holmes plot of the Baptist cemetery at Cape May Court House.

By then Frank had become a clerk and remained in this relatively humble position for much of the rest of his life, never owning property but living in a series of rented accommodation in Philadelphia. He had lost his wife Mary after just a few years, indeed, by about 1874 had already remarried. His second wife was a girl from Virginia, by the name of Armenia, who was with him for the next 26 years until she died, apparently without bearing any children. Frank may have lost touch with Mahlon Kirk, but he kept up with his uncles and cousins because in 1904 he spent the summer with them at Cape May Court House. In October of that year (perhaps prompted by his contact with Witmer Stone) he donated a portrait of his father to the Academy of Natural Sciences. It is tempting to think that Charlotte had painted it, but this seems unlikely as judging by the naturalist's apparent age it would have had to be painted when she was fifteen years old or less.

Frank's latter years were spent at a home for war veterans in Tennessee. He died at the age of sixty-seven on 12th February 1911, while on a visit to Philadelphia, and it must have been his cousins who ensured that he was buried beside his mother at Cape May Court House.

1. Donald Townsend Little found this unlabelled watercolour amongst family papers in 2005 and believes that it shows John Kirk Townsend, aged 32, and Charlotte Holmes Townsend, aged 23, about the time of their marriage.

Little has made the following observations. The style of the clothing is common to Quaker folk of Pennsylvania in the 1830s and a few decades thereafter. The facial features and ears of the man are those of the Townsend family (cf. portraits of his father and brothers, and confirmed portraits of J.K. Townsend, especially the miniature donated to the Academy of Natural Sciences of Philadelphia by John and Charlotte's son, Frank Holmes Townsend). Charlotte earned a living as a portrait painter and Little believes that she not only painted this picture of the two people but also the miniature.

In addition, if some speculation is allowed, we can ask – is the book in the painting simply a conventional prop or is it Townsend's *Narrative of a Journey across the Rocky Mountains*? This was a book of which the holder, the artist and the rest of the family would have been justly proud. For earlier generations of the Townsend family, perhaps it was so obviously J.K. Townsend and his *Narrative* that there was no need to label the picture.

John Kirk Townsend's grave stone in the Congressional Cemetery at Washington, DC.

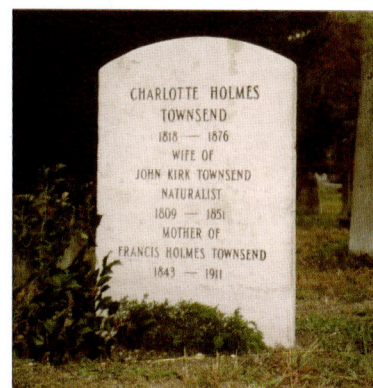

Charlotte Holmes Townsend's grave stone in the Baptist cemetery at Cape May Court House, New Jersey. It was originally inscribed only with the words " In / memory of / Charlotte H. Townsend / Died / Oct 25 1876 / Aged 58 Years." In 2004 the family added the above inscription to the other side.

Part of a small collection of Townsend's specimens at Oxford University Museum that still have their original labels. Clockwise from top left: Lewis's Woodpecker (OUM 3523), Northern Flicker (OUM 3299) and Downy Woodpecker (OUM 3658).

APPENDIX 1. TOWNSEND'S NATURAL HISTORY COLLECTIONS

General content and subsequent dispersal

Townsend collected specimens all his life, starting while just a schoolboy and continuing to his last days. It was not a long career because his life was cut short, but his collections are significant because he secured a relatively high proportion of North American type specimens, of both birds and mammals. Indeed, Townsend was among the last of the early naturalists in North America to discover a substantial number of new species.

Birds were his main passion, but being a competent taxidermist, he was also able to preserve mammals, and at certain times he turned his attention to other orders that were comparatively easy to collect and preserve such as marine and freshwater mollusc shells and beetles.

The specimens now of most interest are those that were found during his journey across the Rocky Mountains, his time on the Columbia River, in the Hawaiian Islands, Tahiti and Chile. An additional aspect is that many of the specimens were painted by Audubon for *The Birds of America* and the *Viviparous Quadrupeds of North America*.

The indications are that Townsend carefully labelled almost all his birds and mammals, including duplicates. If specimens do not still have their original labels then they were either removed by Audubon (because they were a nuisance when he was posing the specimens so that he could paint them) or the labels were removed and replaced by museum curators (as often happened, unfortunately) or they came off accidentally. Each Townsend label had the exact date, or sometimes only the month and year. The places where specimens were collected were not given accurately, tending to be rather vague, as was then customary, often with only the words "Rocky Mountains", "Columbia River", "Oregon" or "Sandwich Islands". He noted the sex, for that would have been determined during preparation, and if the eyes were not dark brown he noted the colour of the eyes, saying "irids yellow" or whatever colour they were. He was also in the habit of giving specimens a field number, so that all specimens of a particular species had the same number. Additional notes were probably kept in his field catalogue (see appendix 6).

An assessment from surviving documents and specimens indicates that during his long journey he prepared and safely deposited at Philadelphia in excess of 1000 bird skins[1], 100 mammals, 10 human skulls, 300 insects, more than 30 crustaceans and a substantial number of shells. The birds and mammals have a chequered history, analogous in some ways to the works of some famous artists, being bought and sold by private collectors and dealers and ultimately ending up in the care of public institutions, with some items along the way being lost, destroyed or stolen. The wide dispersal of Townsend's collections is perhaps little different from many others of the period, though the link with Audubon adds a certain piquancy to the story.

Tracing the current location of specimens has not been easy. Not all museum collections are computerised, and some that have been, are not searchable by the name of the collector. Instead, some of the larger museums have thousands of index cards that are searchable by species name only, and it was just not feasible to examine them without some definite target. In addition, some specimens are currently hard to access because of staff shortages and storage difficulties. Nevertheless, we believe that we have traced the majority of the surviving specimens.

The irony of it all is that almost all the birds and mammals that Townsend collected and wrote about on his journey to the Columbia River are still considered fairly common and these specimens that are scattered around the museums of North America and Europe are of minor significance for study (except for the type specimens). His Hawaiian birds, on the other hand, are now significantly more valuable, because a tragic number of Hawaiian species and races have since become extinct, with little or nothing known about their habits or behaviour. Unfortunately, Townsend considered Hawaiian birds to be too exotic (coming from outside of North America) and too difficult to study (with comparative material only available in Europe). Although not without its difficulties, he could have compiled a small book or even a short paper on the birds of each of the Hawaiian Islands that he visited and significantly added to his reputation.

In the following assessment of Townsend's specimens, almost all the birds and mammals he collected on the east coast of the United States are ignored, though surviving specimens can be found at ANSP, USNM and several other museums.

Birds in North America (see fig 1)

The bulk of Townsend's bird skins and skulls are at the Academy of Natural Sciences of Philadelphia (ANSP). Most of them were deposited there directly, in a large consignment sent back while Townsend was still on the Columbia River, and in a second batch that he brought with him when he returned home. Townsend presented or exchanged a few specimens with friends and members of the Academy, particularly Samuel Morton who received skulls, but the majority of such specimens in due course found their way to the Academy. Other sources include more recent acquisitions that came to the Academy via J.F. and P.B. Street in the late 1930s and probably also from Westtown School at an undetermined date.

The second largest holding of Townsend specimens is at the United States National Museum of Natural History (USNM) in Washington, D.C. These specimens have a more complex history, their exact movements not always being completely traceable. The majority seem to have been used by Audubon at some time, either purchased for the artist by Edward Harris, sold to Audubon directly by the ANSP, or received by Audubon from Townsend himself (an exception is Townsend's Bunting received via the Ezra Michener collection). Audubon had a smaller bird collection of his own than might be expected as he often had scant regard for specimens once he had finished painting them. He passed many to Harris and Spencer Fullerton Baird who, when he became Assistant Secretary of the Smithsonian Institution, donated his large personal bird collection to the Smithsonian – in effect the USNM.

More complex still are the specimens that Harris retained. On his death they passed to his son who, not realising their historical importance and being somewhat short of money, used the specimens to pay off a plumbing bill. Some of them were eventually rescued by J.F. and P.B. Street (Street 1948) and subsequently dispersed to ANSP, Princeton University Zoology Museum (who passed them to the Field Museum, Chicago) or to Newark Museum, Delaware (see appendix 16).

A few specimens at the Museum of Comparative Zoology (MCZ) at Harvard University can be traced back to Brown University – but no further. Of particular historical interest is the single specimen of Parkman's Wren that Audubon presented to Francis Parkman a few years before Parkman was murdered; it was later acquired by J.E. Thayer and went to MCZ with the rest of the Thayer collection.

Another intriguing batch of specimens is at the American Museum of Natural History (AMNH) at New York: a Violet-green Swallow "Presented by J.W. Audubon – from the collection of J.J. Audubon. Columbia River 14th July 1838 [= 1835] J.K. Townsend"; a Mountain Bluebird "Presented by V.G. Audubon from J.J. Audubon's collection"; a Black Oystercatcher that is possibly the type specimen; and the Surfbird painted by Audubon that was involved in a *second* bizarre murder case.

Charleston Museum was long known to have housed several Audubon specimens that were deposited there at the close of 1850. Only a few years later most of the museum records, and no doubt most of the specimens, were lost during the Civil War (Sass 1908). The only Townsend specimen known to have survived the war was a Black-headed Grosbeak, but even that seems now to have vanished. Bachman himself wrote that "my whole library and all my collections in Natural History, the accumulation of the labors of a long life, were burnt by Sherman's vandal army" (Herrick 1938, 2: 285).

At one time it was said that James Trudeau was keen to purchase some of Townsend's birds but we have been unable to trace the current whereabouts of any specimens from that collection. More skins probably exist in other North American museums, some still with original labels and others, perhaps, with labels missing and no demonstrable link with Townsend.

Birds in Europe (see fig 2)

Almost all the Townsend specimens in Europe are derived from a batch of duplicate birds that Townsend sent to Audubon while the artist was in Britain in 1838, with the understanding that they were to be sold to raise as much cash as possible. The consignment contained 278 specimens (121 from Hawaiian Islands, 32 from Tahiti, 56 from Chile and the remainder from the Columbia River area). Lord Derby was apparently the first to look through them and on a list he marked 17 specimens that he wished to have. In fact he seems to have acquired a few more than that, since the specimens he actually selected do not exactly match the list (see appendix 4). Derby's collections went directly to Liverpool Museum in 1851.

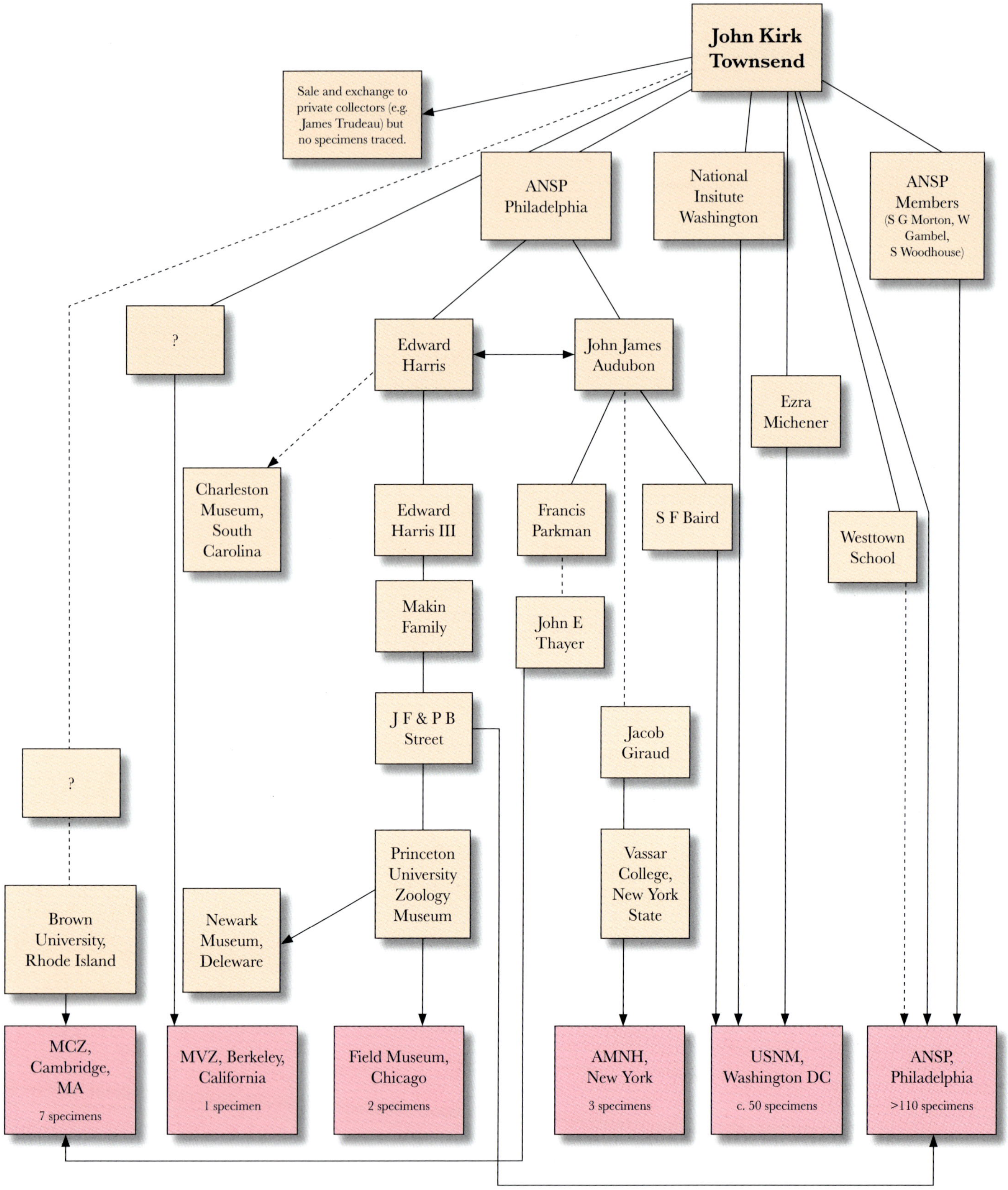

Fig. 1. The dispersal of Townsend's bird collection in the United States. Institutions able to confirm the possession of Townsend specimens in 2003 are in pink boxes. Dotted lines indicate probable but unconfirmed links.

320

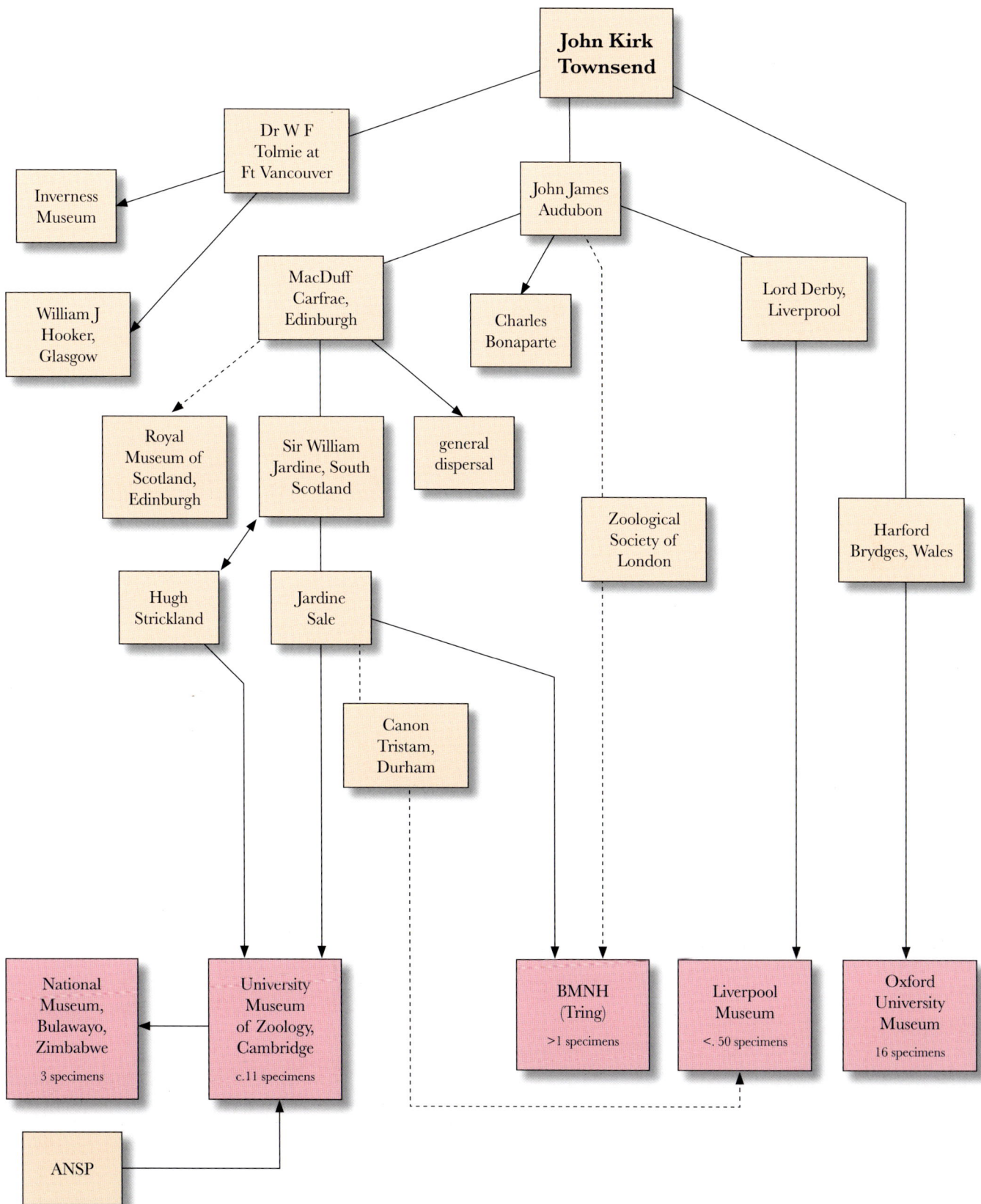

Fig. 2. The dispersal of Townsend's bird collection in Europe. Institutions able to confirm the possession of Townsend specimens in 2003 are in pink boxes. Dotted lines indicate probable but unconfirmed links.

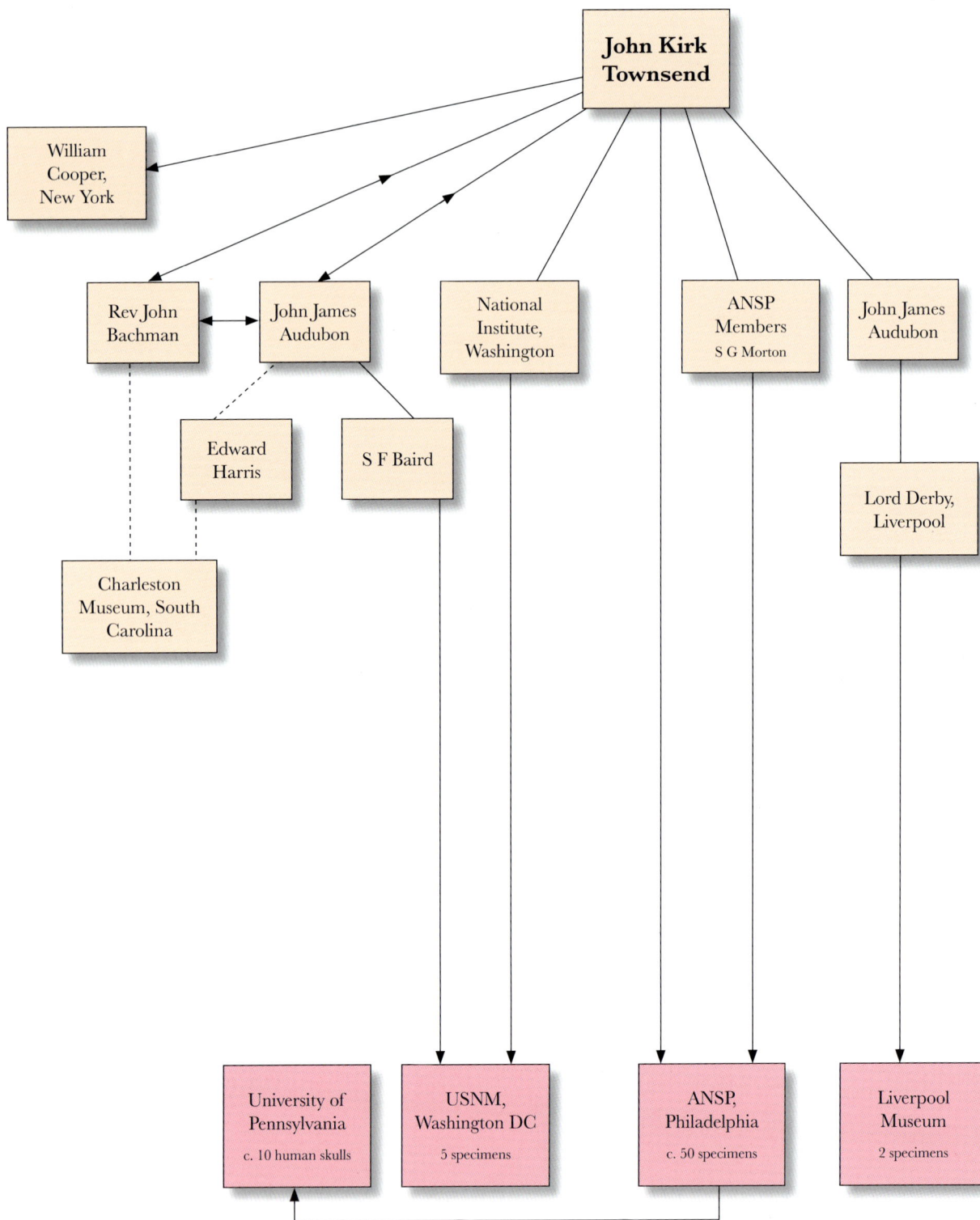

Fig. 3. The dispersal of Townsend's mammal and human skull collections in the United States and Europe. Institutions able to confirm possession of Townsend specimens are in pink boxes. Dotted lines indicate probable but unconfirmed links.

Audubon took the remainder of Townsend's birds up to Edinburgh where he sold them all to a dealer by the name of MacDuff Carfrae. Sir William Jardine, of Dumfriesshire, purchased some of the Hawaiian specimens from Carfrae and perhaps others too, but the majority seem to have been generally dispersed to small private cabinets and taxidermists who mounted the skins for resale as household ornaments. A few may have found their way to Edinburgh Museum but the collections there are currently difficult to access. The Hugh Strickland collection (now at Cambridge, UK) has at least one Townsend specimen from Hawaii derived from Carfrae but since Strickland was Jardine's son-in-law it seems more likely that he got the bird directly from Jardine.

When Jardine's collection was sold by auction in 1886 all, or almost all, the Hawaiian specimens seem to have been bought for the Museum of Zoology at Cambridge University. Nevertheless, there is a suspicion that Canon H.B. Tristram acquired a Townsend Hawaiian specimen at the sale; now at Liverpool Museum, it is listed as from the Jardine sale but there is no proven link to Townsend (M. Largen in litt.). The British Museum (Natural History) also bought 516 specimens from the Jardine sale and at least one of them, from Chile, was collected by Townsend; perhaps there are more but the museum collection is not currently searchable by collector, only by species. In more recent times, in 1975, three Hawaiian duplicates at Cambridge University that were collected by Townsend went to Zimbabwe by exchange.

Although it is not clear when it happened, Audubon says that he sold a specimen of the Willow Flycatcher *Empidonax traillii* to Lord Derby and two others to Charles Bonaparte, and yet another perhaps to William Swainson (see *Ornithological Biography* text for "*Muscicapa Traillii*" and Herrick 1938, 2: 176). These flycatchers were almost certainly collected by Townsend; Derby's specimen cannot be traced with certainty (it is possibly LM D4473a) and though Bonaparte's collection went to the Museum National d'Histoire Naturelle at Paris the museum at one time had a policy of dispersing specimens to assist provincial and foreign museums – with the result that the Bonaparte collection has virtually disappeared (C. Jouanin in litt.).

Other specimens that came to Europe include a few that Townsend exchanged with Dr. W.F. Tolmie while at Fort Vancouver. Tolmie is known to have sent plants and birds gathered by Townsend to William Jackson Hooker, Professor of Botany at Glasgow University; it seems that the 17 bird species in the only known consignment were intended as a gift for his son, William Dawson Hooker, and have now vanished. Tolmie also donated birds to Inverness and Glasgow museums but not one of Tolmie's birds appears to have survived (J. Watt and R. Sutcliffe in litt.) so it is impossible to know if there were any Townsend skins among them.

Reserved until last is one of the most interesting clusters of Townsend specimens, now at Oxford University Museum (housed in the very chamber where Bishop Wilberforce and Thomas Huxley had a famous dispute about the theory of evolution at an 1860 meeting of the British Association for the Advancement of Science). Although numbering no more than sixteen specimens, almost all of them still have their original labels and three of them also have field numbers. Thirteen of the birds are from the Columbia River but they are unlikely to have come to Britain via Audubon because the other three specimens are from the New Jersey coast and are dated after the time that Audubon sold birds to Lord Derby and Carfrae. They seem to have been bought directly by Sir Harford James Jones Brydges (son of the famous orientalist Sir Harford Jones Brydges), a collector who resided at Boultibrooke, near Presteigne in Wales. This assumption is supported by a reference to a Mr. Brydges in a letter that John Torrey wrote to Townsend: it seems that Townsend was planning another trip across the Rocky Mountains and was looking for support from Brydges, though Brydges wanted Townsend to go with him to Hudson Bay, ideas that never came to fruition (J. Torrey to JKT, 2nd February 1841. APS).

North American Mammals (see fig 3)

The dispersal of Townsend's mammal specimens is a much simpler story. The majority are at the Academy of Natural Sciences of Philadelphia, deposited there by Townsend on his return from the West in 1837; much later, shortly before he died, a second batch was added. He had wanted to sell these latter specimens to the Academy but because there was no money available he attempted to sell them in Europe (see appendix 18). Since this was unsuccessful he held onto them and eventually donated them to the Academy. The Academy also received at least 25 mammal skulls and a few human skulls from S.G. Morton, originally given by Townsend to Morton (see below).

While Townsend was working at the National Institute he added some of his Columbia River mammal specimens to the museum and these were later transferred to the Smithsonian Institution (USNM). We also know that Townsend lent some bat specimens to William Cooper at New York but the bats in best condition were retained by Cooper and have not been traced.

From time to time, Townsend's mammal specimens were examined by both Bachman and Audubon. Bachman described some of them as new species in various journals and even took a few of them to Europe to compare with other museum skins and skulls. They were handled again by Bachman while compiling the text for *The Viviparous Quadrupeds of North America* and the types and other specimens were used by Audubon (or his son, John) as models for some of the book's colour plates, probably far more than can now be positively determined. The majority of specimens they looked at seem to have been returned to Townsend but others could have gone via Bachman to the Charleston Museum (though there are none now traceable there) and some certainly went to the Smithsonian Institution via S.F. Baird.

Only four mammal specimens have been traced to Europe, all sent by Townsend to Audubon while the artist was in Britain. Two were sold to Lord Derby and are now in the Liverpool Museum (see appendix 4); the remaining two are unaccounted for.

Human skulls

Townsend made a special effort to collect examples of skulls of all the Indian tribes on the lower Columbia River. He did this partly out of his own scientific curiosity and more particularly because of his friendship with Samuel Morton who was making a study of human skulls. He knew that Morton would be especially pleased to receive any that were misshapen by the practice of head-flattening carried out by some of the tribes.

At no small risk to himself, Townsend raided the burial places of the Indians for skulls, on one occasion carrying off a whole mummified body (that he was afterwards forced to replace, see Narrative entry for 3rd February 1836). He also brought Morton one of the wooden cradles that had been used for head-flattening and another is in the possession of Townsend's relatives.

Eight skulls acquired by Townsend were illustrated by Morton in a special chapter on "The Flat-Head Tribes of Columbia River" in his *Crania Americana. Or, a comparative View of the Skulls of Various Aboriginal Nations of North and South America* (1839). Using Morton's spelling, the skulls were from the Chinouk, Klatstoni, Killemook, Clatsap, Kalapooyah, Clickitat and Cowalitsk peoples, one from the latter being strikingly distorted. In Morton's skull catalogue (1840) there are an additional two Townsend items, one of which was of a Kanaka (a Hawaiian Islander) though the skull could actually have come from the Columbia River where Hawaiian Islanders were employed. The other skull was that of a four-year-old Chinouk child "much flattened by art".

Existing skull specimens have been transferred from ANSP to the University of Pennsylvania, in Philadelphia.

Reptiles

All the reptiles that Townsend had gathered on the journey across the Rocky Mountains and while on the Columbia River were destroyed when one of Captain Wyeth's men drank all the alcohol that was preserving them. This happened while Townsend was away on his first visit to the Willamette Falls in September 1834 (see *Narrative* for 4th July 1835). The two-gallon jar was almost full of specimens and would surely have been of great interest to herpetologists. It is unclear if Townsend began another reptile collection or not. He may have had difficulty obtaining more alcohol or he may have considered that it was not worth the risk of losing a second batch of reptiles in the same way. After Townsend's return to the east coast, Bachman once asked him about his reptile collection and suggested that he send them to John Edwards Holbrook (J. Bachman to JKT, 7th February 1838). This may have been an assumption that Townsend had reptiles or Townsend may indeed have brought home a few specimens. Hanley (1977, p. 218) says that reptiles collected by Townsend *were* sent to Holbrook who refused to include them in his *North American Herpetology* because of the lack of accompanying data and because alcohol had destroyed the natural colours. Hanley does not give any references for this statement (nor, as it happens, for anything else in his 339 page book, which even lacks an index).

Fish

The ANSP database has three specimens said to have been collected by J.K. Townsend. Two have localities (Costa Rica and Virgin Islands) that suggest that there may be confusion with C.H. Townsend. The other specimen has a note saying that it cannot be from the Hawaiian Islands, as catalogued, and must be from the Western Atlantic.

Molluscs

Nuttall was especially interested in marine and freshwater molluscs and spent hours searching for shells near the mouth of the Columbia River, in the Hawaiian Islands and on the coast of California. We know that Townsend occasionally helped Nuttall to look for shells, indeed while stranded on Kaua'i without ammunition or preserving materials there was little else for the two of them to collect. Nuttall's shells were examined and described by T.A. Conrad in 1837 in the *Journal of the Academy of Natural Sciences of Philadelphia* in a paper entitled "Descriptions of new Marine Shells from Upper California. Collected by Thomas Nuttall, Esq." In fact, the 143 species were from scattered localities as far apart as the Azores and the Hawaiian Islands, with the majority from California, (and a few of them were not even from Nuttall). There is no mention of Townsend but there are several species from the mouth of the Columbia River and from the Hawaiian Islands, some of which were from Kaua'i "on rocks bare at low water". Some of the shells may have been gathered by Townsend and this may also apply to the freshwater and land snails collected by Nuttall while on the Columbia River. These molluscs were worked through by Isaac Lea and described in "New Freshwater and Land Shells" in the *Transactions of the American Philosophical Society* (1839-46).

Townsend evidently had a shell collection of his own, as it is referred to in American Philosophical Society correspondence of 21st December 1837: "Of the Shells, however, a set has been taken out for the Academy; and if desired by the Phil. Socy., a similar set shall be at once sent to that Institution." In the ensuing correspondence the American Philosophical Society complained that it was "a small collection of shells" (see p. 268).

Crustaceans

The shrimps, lobsters, crabs and hermit crabs collected by Nuttall and Townsend were first examined by J.W. Randall in 1839 for a paper in the Journal of the Academy of Natural Sciences of Philadelphia entitled "Catalogue of the Crustacea brought by Thomas Nuttall and J.K. Townsend, from the West Coast of North America and the Sandwich Islands". The paper contained descriptions of about 30 species that were found by Townsend in the Hawaiian Islands (as well as others found in the islands and in California by Nuttall, and a few from other localities from other collectors). Most of the species were already known to European naturalists.

Insects

The first beetles to be collected in Washington and Oregon were taken by Townsend between 1834 and 1836 (Hatch 1953). Nevertheless, only a relatively small proportion of Townsend's insects were new discoveries, because he had been preceded in the Pacific coastal areas by Frederick Eschscholtz who had collected hundreds of insects between 1815 and 1818 while travelling with the Russian explorer Captain Kotzebue.

Most of Townsend's insects were probably gathered while he was based at Fort Vancouver or moored next to Sauvie Island because he would then have had the means to dry and pack them up ready for shipping to Philadelphia. On his return he passed the insects to Thaddeus William Harris, the Librarian at Harvard University and one of the most eminent North American entomologists of the day. Harris eagerly examined them and began a paper in 1838 entitled *Descriptions of the Insects collected by Mr. John Kirk Townsend in Oregon, near the mouth of the Columbia river*. In the introduction he acknowledged the care and hard work that had gone into making the collection: "Mr. John Kirk Townsend, in his recent travels

to Oregon, collected about one hundred and thirty species of insects, belonging chiefly to the order Coleoptera [beetles], which, notwithstanding the perils of a long transportation, were received in Philadelphia in pretty good condition. When the small size, the comparative insignificance, & the fragile nature of insects are considered, we must admit, that no small credit is due to Mr. Townsend for his exertions in the cause of entomological science." For some reason the paper was never published but a thirty-page manuscript survives with Harris's notes on the first forty species (Harris MS, 5th April 1838, MCZ 6Mn.1308.41.4). For about half of the species Harris gives the number of specimens that were collected, sometimes as many as seven or ten, but with an average of about 2.5 specimens per species; this suggests that the whole collection would have exceeded 300 items. Harris later published a paper in the *Boston Journal of Natural History* (1839) on the genus *Cychrus* in which three new snail-eating ground beetles collected by Townsend were named: *C. tuberculatus, C. angulatus* and *C. cristatus*.

Other entomologists got hold of some of the Townsend specimens and described additional new species, in European journals: in 1838 Louis Reiche named two wingless tiger beetles as Omus dejeani and O. audouini; in 1843 E.F. Germar worked through some click beetles naming *Ctenicera bombycina, C. glauca, C. leucaspa, Ampedus phoenicopterus* and *A. pullus*; and, again in 1843, W.F. Erichson named *Amphicyrta chrysomelina*. In 1857, John Lawrence LeConte prepared a paper upon insects for *Reports of explorations and surveys to ascertain the most practicable and economical route for a railroad from the Mississippi River to the Pacific Ocean*; in it LeConte listed 233 species, about 50 of them collected by Townsend (Hatch 1953).

Townsend was not at all pleased with the way that Harris had treated his insect collection, rather suggesting that the insects that went to Europe had been sent there without his knowledge or had been purloined from Harris by visiting naturalists[2]. Townsend was so incensed that he wrote a letter to Samuel Morton, dated 8th May 1844, complaining bitterly about what had befallen: "I fear that all the beautiful & rare insects collected by me in that region have been either destroyed, or surreptitiously appropriated. I have written to Dr. Harris on this Subject, & shall leave no stone unturned to recover at least any portion of this fine collection which may not have been destroyed, lost, or taken."

Some of Townsend's insect specimens may still survive but we have made no serious attempt to locate them.[3]

Plants

Townsend must surely have helped to build up Nuttall's western herbarium from time to time, bringing back plants for him on those occasions when they were foraging separately. We know that Townsend was not averse to collecting botanical items because he exchanged plants (and birds) with Dr. W.F. Tolmie while they were both at Fort Vancouver. Tolmie sent a package of plants from Fort Vancouver to William Jackson Hooker, Professor of Botany at Glasgow University, and acknowledged that eleven species and some others "in [a] packet marked "Wallawalla" were procured from an American ornithologist, who is here at present, in exchange for duplicates." This can only have been Townsend, as the rest of the letter makes plain (Tolmie to W.J. Hooker, 15th November 1836).

1. Townsend's first main consignment of specimens, sent back by ship, was said to number "three hundred birds and fifty quadrupeds" (*Waldie's Select Circulating Library*, 12th July 1836). His second batch, which he brought back with him, is believed to have amounted to considerably more. After depositing bird specimens at the Academy of Natural Sciences of Philadelphia, selling some to Audubon and other American collectors, he could still send 278 birds to Europe to be sold to collectors there (see appendix 4). It therefore seems reasonable that the total number of bird specimens was slightly more than a thousand.

2. Harris himself was not happy about Europeans describing insects that he had already described and in 1829 had proposed a rule to give priority in scientific nomenclature to the first describer (Sorenson 1995, p. 13).

3. MCZ has one of the most important collections of North American beetles, renowned because of the extensive number of type specimens. No Townsend specimens were located on the MCZ website but it is not searchable by collector.

APPENDIX 2. NEW BIRDS AND MAMMALS IN THE TOWNSEND COLLECTION

BIRDS

* Denotes a species published under Townsend's name while he was absent in the west. Nuttall seems to have written the paper using Townsend's MS names from the specimen labels, with the exception of Townsend's Warbler that Nuttall is credited with naming.

† For these species, the type locality is given in the current *AOU Check-list* (1998) as "Fort Vancouver, Washington" but it is more likely to be "Sauvie Island, near Portland, Oregon" (see main text, p. 177).

SPECIES COLLECTED, DESCRIBED AND NAMED BY TOWNSEND

Mountain Plover *Charadrius montanus* Townsend*

Charadrius montanus J.K. Townsend, 1837, *Journal of the Academy of Natural Sciences of Philadelphia*, 7, p. 192. (Central table-land of the Rocky Mountains = near Sweetwater River, Wyoming)

Vaux's Swift *Chaetura vauxi* (Townsend)

Cypcelus [sic] *Vauxii* J.K. Townsend, 1839, *Narrative of a Journey Across the Rocky Mountains etc.*, p. 348. (Columbia River = Fort Vancouver, Washington)

Chestnut-backed Chickadee *Parus rufescens* Townsend*

Parus rufescens J.K. Townsend, 1837, *Journal of the Academy of Natural Sciences of Philadelphia*, 7, p. 190. (Forests of the Columbia River = Fort Vancouver†, Washington)

Bushtit *Psaltriparus minimus* (Townsend*)

Parus minimus J.K. Townsend, 1837, *Journal of the Academy of Natural Sciences of Philadelphia*, 7, p. 190. (Forests of the Columbia River = probably near Fort Vancouver†, Washington)

Sage Thrasher *Oreoscoptes montanus* (Townsend)

Orpheus montanus J.K. Townsend, 1837, *Journal of the Academy of Natural Sciences of Philadelphia*, 7, p. 192. (Plains of the Rocky Mountains = Sandy Creek, lat 42° N., long 109° 30′ W., Wyoming)

Black-throated Gray Warbler *Dendroica nigrescens* (Townsend*)

Sylvia nigrescens J.K. Townsend, 1837, *Journal of the Academy of Natural Sciences of Philadelphia*, 7, p. 191. (No locality given = near Fort William, Portland, Oregon)

Townsend's Warbler *Dendroica townsendi* (Townsend*)

Sylvia Townsendi (Nuttall.) J.K. Townsend, 1837, *Journal of the Academy of Natural Sciences of Philadelphia*, 7, p. 191. (Forests of the Columbia river = Fort Vancouver†, Washington)

Hermit Warbler *Dendroica occidentalis* (Townsend*)

Sylvia occidentalis J.K. Townsend, 1837, *Journal of the Academy of Natural Sciences of Philadelphia*, 7, p. 190. (Forests of the Columbia river = Fort Vancouver, Washington)

MacGillivray's Warbler *Oporornis tolmiei* (Townsend)

Sylvia Tolmæi J.K. Townsend, 1839, *Narrative of a Journey Across the Rocky Mountains etc.*, p. 343. (The Columbia = Fort Vancouver, Washington)

Lark Bunting *Calamospiza melanocorys* Stejneger

Fringilla bicolor (not Linnaeus, 1766) J.K. Townsend, 1837, *Journal of the Academy of Natural Sciences of Philadelphia*, 7, p. 189. (Plains of the Platte River = western Nebraska)

Calamospiza melanocorys Stejneger, 1885, *Auk* 2, p. 49. New name for *Fringilla bicolor* Townsend, preoccupied.

Chestnut-collared Longspur *Calcarius ornatus* (Townsend*)

Plectrophanes ornata J.K. Townsend, 1837, *Journal of the Academy of Natural Sciences of Philadelphia*, 7, p. 189. (Prairies of Platte River = near forks of Platte River, western Nebraska)

BIRDS DESCRIBED AND NAMED BY OTHER NATURALISTS FROM TOWNSEND'S SPECIMENS

Black-footed Albatross *Diomedea nigripes* Audubon

Diomedea nigripes Audubon, 1839, *Ornithological Biography*, 5, p. 327. (Pacific Ocean, lat. 30° 44 N., long. 146° [W])

Swainson's Hawk *Buteo swainsoni* Bonaparte

Buteo vulgaris (not Swainson, 1832) Audubon, 1837, *Birds of America* (folio), 4, pl. 372. (Near the Columbia River = Fort Vancouver, Washington)

Buteo Swainsoni Bonaparte, 1838, *Geographical and Comparative List of the Birds of Europe and North America*, p. 3. New name for *Buteo vulgaris* Audubon, preoccupied.

American Black Oystercatcher[1] *Haematopus bachmani* Audubon

Hæmatopus Bachmani Audubon, 1838, *Birds of America* (folio), 4, pl. 427, fig. 1 (1839, *Ornithological Biography*, 5, p. 245). (Mouth of the Columbia River)

Western Gull *Larus occidentalis* Audubon

Larus occidentalis Audubon, 1839, *Ornithological Biography* 5, p. 320 (Cape Disappointment, [Washington])

Townsend's Solitaire[2] *Myadestes townsendi* (Audubon)

Ptilogony's [sic] *Townsendi* Audubon, 1838, *Birds of America* (folio) 4, pl. 419, fig. 2 (1839, *Ornithological Biography* 5, p. 206.) (Columbia River = Astoria, Oregon)

Swainson's Thrush *Catharus ustulatus* (Nuttall)

Turdus cestulatus [printer's error = ustulatus] Nuttall, 1840, *Manual of the Ornithology of the United States and Canada*, 2nd edn., 1, pp. vi, 400, 830. (Forests of the Oregon = Fort Vancouver, Washington)

Green-tailed Towhee *Pipilo chlorurus* (Audubon)

Fringilla chlorura Audubon, 1839, *Ornithological Biography* 5, p. 336. (No locality given = Ross's Creek, ca. 20 miles southwest of Blackfoot, Bingham County, Idaho)

Brewer's Sparrow *Spizella breweri* Cassin

Spizella Breweri Cassin, 1856, *Proceedings of the Academy of Natural Sciences of Philadelphia* 8, p. 40. (Western North America, California, and New Mexico = Laramie Hills, Wyoming)

Harris's Sparrow *Zonotrichia querula* (Nuttall[3])

Fringilla querula Nuttall, 1840, *Manual of the Ornithology of the United States and Canada*, 2nd edition, 1, p. 555. (A few miles west of Independence, Missouri)

BIRDS FROM PACIFIC ISLANDS

Hawaiian Hawk[4] *Buteo solitarius* Peale

Buteo solitarius Peale, 1848, *U.S. Exploring Expedition* 8, p. 62. (Island of Hawai'i)

Hawaiian Crow[5] *Corvus hawaiiensis* Peale

Corvus tropicus (not Gmelin, 1788) Kerr, 1792, *Animal Kingdom* 1 (2), p. 640 (Hawai'i)

Corvus hawaiiensis Peale, 1848, *U.S. Exploring Expedition* 8 p. 106. (A few miles inland from the village of Kaawaloa, Hawai'i)

Kaua'i 'O'o *Moho braccatus* Cassin

Mohoa [sic] *braccata* Cassin, 1855, *Proceedings of the Academy of Natural Sciences of Philadelphia* 7, p. 440. (Sandwich Islands = Kaua'i, Hawaiian Islands)

Gray-green Fruit Dove[6] *Ptilinopus purpuratus* (T.R. Peale)

Ptilinopus furcatus Peale, 1848. *Mammalia and Ornithology, US Exploring Expedition*, pp. 191-192. (Island of Tahiti)

1. Although Townsend collected a Black Oystercatcher specimen that Audubon named "Haematopus Townsendi", it was another specimen, collected by W.F. Tolmie, which was described one page earlier in the same publication that became the type specimen.

2. The type specimen was given to Townsend by Captain W. Brotchie of the Hudson's Bay Company.

3. Nuttall, if he used a specimen at all when writing his description, would have used the Townsend specimen at the Academy of Natural Sciences of Philadelphia (long since lost or destroyed).

4 & 5. The type specimens were sent from Hawai'i to Townsend by the Reverend Forbes.

6. Deignan (1961, p. 105) suggests that Peale would have used Townsend's specimens, in addition to his own, when writing his description of the Gray-green Fruit Dove. There is no evidence for his suggestion other than the fact that Peale and Townsend were both working at the National Institute in Washington, D.C. when the description was written.

MAMMALS

MAMMALS COLLECTED, NAMED AND DESCRIBED BY TOWNSEND

[**Giant Wolf** *Lupus gigas* **Townsend** is not a valid species.]

MAMMALS DESCRIBED AND NAMED BY OTHER NATURALISTS FROM TOWNSEND'S SPECIMENS

Townsend's Mole *Scapanus townsendi* (**Bachman**)

Scalops townsendii Bachman, 1839, *Journal of the Academy of Natural Sciences of Philadelphia* 8, p. 58. (Banks of the Columbia river = Fort Vancouver, Clark County, Washington)

Western Big-eared Bat *Plecotus townsendi* **Cooper**

Plecotus townsendi Cooper, 1837, *Annals of the Lyceum of Natural History of New York* 4, p. 573. (Columbia river = Fort Vancouver, Clark County, Washington)

Townsend's Ground Squirrel *Citellus townsendii* (**Bachman**)

Spermophilus townsendii Bachman, 1839, *Journal of the Academy of Natural Sciences of Philadelphia* 8, p. 61. (Prairies near the Walla-walla = near Wallula, Washington)

Least Chipmunk *Eutamias minimus* (**Bachman**)

Tamias minimus Bachman, 1839, *Journal of the Academy of Natural Sciences of Philadelphia* 8, p. 71. (Green River, near mouth of Big Sandy Creek [Sweetwater County, Wyoming])

Townsend's Chipmunk *Eutamias townsendii* (**Bachman**)

Tamias Townsendii Bachman, 1839, *Journal of the Academy of Natural Sciences of Philadelphia* 8, p. 68. (Lower Columbia River = near lower mouth of Willamette, Oregon)

Chickaree (Douglas Squirrel) *Tamiasciurus douglasii* (**Bachman**)

Sciurus douglasii Bachman, 1839, *Proceedings of the Zoological Society, London*, for 1838, p. 99. (Shores of the Columbia River, locality restricted to Mouth of Columbia River)

Townsend's Pocket Gopher *Thomomys townsendi* (**Bachman**)

Geomys townsendii Bachman, 1839, *Journal of the Academy of Natural Sciences of Philadelphia* 8, p. 105. (Columbia River = Nampa, Canyon County, Idaho)

Townsend's Vole *Microtus townsendi* (**Bachman**)

Arvicola townsendi Bachman, 1839, *Journal of the Academy of Natural Sciences of Philadelphia* 8, p. 60. (Columbia River = near mouth of Willamette, on or near Sauvie Island, Oregon)

Oregon Vole *Microtus oregoni* (**Bachman**)

Arvicola oregoni Bachman, 1839, *Journal of the Academy of Natural Sciences of Philadelphia* 8, p. 60. (Astoria [Clatsop County, Oregon])

White-tailed Jackrabbit *Lepus townsendi* **Bachman**

Lepus townsendi Bachman, 1839, *Journal of the Academy of Natural Sciences of Philadelphia* 8, p. 90. (Walla Walla = near Wallula, Walla Walla County, Washington)

Note: Harris's Antelope Squirrel *Ammospermophilus harrisii* (Audubon & Bachman, 1854). When naming this species Audubon and Bachman say that it was described from a Townsend specimen received from Edward Harris. It is stated to have probably come from west of the Rocky Mountains, on the route followed by Messrs Nuttall and Townsend. The specimen cannot have been personally collected by Townsend because it is a species of the south western US; a type locality of Santa Cruz Valley, Arizona, has been designated for it.

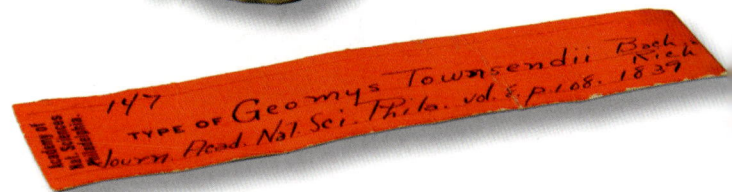

Townsend's Pocket Gopher (ANSP 147).

APPENDIX 3. TOLMIE LIST. Dr W.F. Tolmie's list of Townsend's birds sent to William Jackson Hooker, extracted from a letter dated 15th November 1836 from Tolmie, at Fort Vancouver, to W. J. Hooker, at Glasgow (Public Record Office, Kew MS LXII. Transcription of the unpublished MS and the addition of current identifications by B & R Mearns).

In Box no 2 for you are the birds enumerated
in the first leaf - I have rec^d them from a Phila _{Mr Townsend}
delphia ornithologist, who has been here for some time
& I send to my old friend William to whom I hope
they will be acceptable - On the envelope I will
copy some notes regarding the birds by Townsend …

List of Birds in Box N°. 2 –   *current identification*

1	*Tetrao Urophasianus – Young male from Walla walla*	Greater Sage-Grouse
2	_____ *Obscurus*	Sooty and Dusky Grouse
3	*Columba Fasciata*	Band-tailed Pigeon
4	_____ *Caroliniensis [sic]*	Mourning Dove
5	*Strix Cunicularia*	Burrowing Owl
6	*Garrulus Ultramarinus?*	Western Scrub-Jay
7	*Sturnus Ludovicianus*	Western Meadowlark
8	*Turdus Meruloides*	Varied Thrush
9	*Hirundo – new species*	Cliff Swallow
10	*Fringilla Vespertina*	Evening Grosbeak
11	*(Sialia) Saxicola new species*	Western Bluebird
12	*Fringilla Purpurea*	Purple or House Finch
13	*Muscicapa grisea?*	*Empidonax sp.*
14	*Bombycilla Caroliniensis [sic]*	Cedar Waxwing
15	*Fringilla Arctica*	Mountain Bluebird
16	*Sylvia – new species*	"Audubon's" Yellow-rumped Warbler
17	*Sterna Nigra*	Black Tern

Quadrupeds
Lepus   hare species
Arctomys (Spermophilus)   ground squirrel species

Lepus   *lives in the dry plains amongst the wormwood bushes*

Arctomys   *Killed at Fort Simpson - [illegible = ?Rum No 77]*

Hirundo   *new species – The nest most closely resembles that of Hirundo Fulva[1], but wants the gourdlike*
             mouth of the latter's[.] it is built of similar materials and in like situations

Sylvia   *new species Its note much resembles that of Regulus Cristatus[2] but is sharper. It arrives in April*
             and departs early in June.

Sialia   *(Saxicola) new species – The note is a plaintive warble a good deal resembling but not so*
             prolonged as that of the common species, to whom in its habits it is very similar

1. Cliff Swallow, rather than Cave Swallow.

2. Golden-crowned Kinglet.

A — Catalogue of Birds or submitted to my selection by Mr. Audubon on the part of Dr. Townsend from various quarters this 15 June 38

No	Name		Specimens	No
No 80	Crimson Icterus ?		30	No 8
79	Certhia coccinea		24	No 9
	Yellow head Sylvicola ?			
75	Muscicapa			
	Cinnamon Fringilla ?			No 23
232	Strepsilas interpres		3	
	Unequal billed Certhia ?	Sandwich	11	No 10
	Sylvia ?		18	
	Small yellow Icterus ?		2	No 12
87	Fulica		1	
88	Gallinula chloropus		1	
84	Strix Brachyotus		1	
82	Corvus		1	No 11
	Trochilus cyanocephalus		40	
240	Rufous necked Finch		1	
194	Siskin		3	
193	White throated Finch		1	No 4
206	Fringilla luteoventris		1	No 5
209	Small crested Titmouse		1	
	Phytotoma rara		1	
	Scolopax Wilsonii	Chili	1	
203	Grey Muscicapa		2	
211	Gallinula		2	No 6
223	Phalacrocorax		1	No 7
212	Podiceps		1	
217	Small Muscicapa		1	
186	Black Do		8	
187	Cinnamon Do		2	
182	Sterna ? (alba)	Tahiti	1	

195

Audubon sold some of Townsend's duplicate specimens in Britain. Lord Derby made a three-page list of all the items and marked up the ones he wanted in the right hand columns.

APPENDIX 4. DERBY LIST. Lord Derby's catalogue of birds. He wrote on the outside of the document: *"Copy of a List of Skins (chiefly Birds) | sent by Dr. Townshend [sic] to Mr. Audubon | for disposal, communicated to me 16th June [18]38".* **(Transcription of the unpublished MS and the addition of current identifications by B & R Mearns).**

Catalogue of Birds submitted to my Selection by Mr. Audubon
on the part of Dr. Townshend [sic] from various quarters this 15th June /[18]38

[Townsend field number]	[MS name]	[Locality]	[Number of specimens]	[Derby's selection]	[Current identification]
No. 80	Crimson Icterus ?		39	No. 8	'Apapane
79	Certhia coccinea		24 & 5	No. 9	'I'iwi
-	Yellowhead Pyrrhula ?		7		'O'u
75	Muscicapa		7		'Elepaio
-	Cinnamon Fringilla ?	Sandwich Is.	1	No. 13	'Akepa
232	Strepsiles interpres		3		Ruddy Turnstone
-	Unequal billed Certhia ?		11	No. 10	Lesser 'Akialoa / Nukupu'u
-	Sylvia ?		18		O'ahu Creeper?
-	Small yellow Icterus ?		2	No. 12	'amakihi sp.
87	Fulica		1		Hawaiian Coot
88	Gallinula chloropus		1		Common Moorhen
84	Strix Brachyotos		1		Short-eared Owl
82	Cassicus		1	No. 11	'o'o sp.
-	Trochilus cyanocephalus		40		hummingbird sp.
210	Rufous necked Finch		1		Rufous-collared Sparrow
194	Siskin		3		[Black-chinned?] Siskin
193	White throated Finch		1	No. 4	Common Diuca-Finch?
206	Fringilla luteoventris		1	No. 5	Grassland Yellowfinch?
209	Small crested Titmouse	Chili	1		Tufted Tit-Tyrant
-	Ichyotoma rara		1		Rufous-tailed Plantcutter
-	Scolopax Wilsoni		1		Magellan Snipe
203	Grey Muscicapa		2		flycatcher sp.
211	Gallinula		2	No. 6	Spot-flanked Gallinule
223	Phalacrocorax		1	No. 7	Red-legged Cormorant
212	Podiceps		1		grebe sp.
217	Small Muscicapa		1		flycatcher sp.
186	Black D^r [ditto]		8		Black Monarch
187	Cinnamon D^r [ditto]	Tahiti	2		Black Monarch
182	Sterna ? (alba)		1	*196*	Fairy Tern

	No.	Name	Whence	No of specimens	196	
No.	178	Columba		4		pigeon sp. [= Gray-green Fruit Dove or Pacific Pigeon]
	177	Puffinus		6		shearwater sp.
	183	Yellow Turdus ?	Tahiti	2	16	Tahiti Reed Warbler
	184	Dusky Dº [ditto]		1		Tahiti Reed Warbler
	191	Hirundo		1		Pacific Swallow
	185	Cypselus ?		1		Tahiti Swiftlet
	179	Aludo ?		4	15	Venerated Kingfisher
	181	Totanus		1	14	Wandering Tattler
	49	Falco fuscus		1		Sharp-shinned Hawk
	114	Siala occidentalis		1		Western Bluebird
	67	Turdus naevia		2		Varied Thrush
	109	_____ Wilsonii		1		Veery
	45	Sylvia Dº [ditto]		4		Wilson's Warbler
	57	Sitta Canadensis		2		Red-breasted Nuthatch
	99	Sylvia Celata		2		Orange-crowned Warbler
	120	Hirundo lunifrons	Columbia River	1		Cliff Swallow
	92	Vireo solitarius		2		Cassin's Vireo
	-	Sylvia Philadelphia ?		1		Mourning/MacGillivray's Warbler
	116	Certhia familiaris		1		Brown Creeper
	70	Scolopax Wilsonii		2		Wilson's Snipe
	95	Trochilus rufus		1		Rufous Hummingbird
	60	Picus redbreasted		1	No. 3	Red-breasted Sapsucker
	115	Rallus Carolinus		1		Sora
	130	Anas acuta ? young		1		Northern Pintail
	157	Loons [or Larus?]		1		loon sp. [or gull sp.]
	153	Tetrao Urophasianus young		1		Greater Sage-Grouse
	152	_____ phasianellus young		1		Sharp-tailed Grouse
	-	Phalacrocorax		1		cormorant sp.
	159	Larus		2		gull sp.
	156	Larus		1	259	gull sp.

No.			259	My purchases	
187	*Colymbus*	*Columbia River*	*1*		loon sp.
128	*Anser albifrons*		*1*		White-fronted Goose
162	*Charadrius pluvialis*		*1*		Golden Plover
48	*Garrulus Stelleri*		*3*		Steller's Jay
-	*Colaptes Mexicanus*		*1*		flicker sp.
97	*Picus pileatus*		*1*		Pileated Woodpecker
68	*Falco Columbarius*		*1*		Merlin or Prairie Falcon
101	*Muscicapa*		*1*		flycatcher sp.
105	*Dᵒ. [ditto]*		*1*		flycatcher sp.
140	*Fuligula Clangula*	*NW Coast*	*1*		Barrow's or Common Goldeneye
166	*Sterna nigra*		*1*		Black Tern
119	*Strix*	*California*			owl sp. [Burrowing Owl?]
125	*Procellaria Capensis*	*Atlantic Ocean*	*1*		Cape Pigeon
....	*Fulmar Petrel*		*1*		Southern Fulmar
....	*Puffinus*	*Pacific Dᵒ*	*1*		shearwater sp.
....	*Diomedea chlororhynca*	-------	*2*		Gray-headed Albatross (see appendix 10)
			278		
	Quadrupeds				
34	*Lepus small*	*Columbia River*	*1*	No. 1	Mountain Cottontail
19	*Neotoma Drumondii*	" "	*2*	No. 2	Bushytail Woodrat
-	*Meles Labradoria*	" "	*1*		American Badger
			4	2	

Townsend specimens currently at Liverpool Museum (following the numbering of Lord Derby's selection, above).

1. Bushy-tailed Woodrat *Neotoma cinerea* (LM D353a), 29th October 1836, Columbia River.
2. Mountain Cottontail *Sylvilagus nuttalli* (LM D573a), 20th August 1836, Walla Walla.
3. Red-breasted Sapsucker *Sphyraptes ruber* (no specimen traced).
4. ? Common Diuca-Finch *Diuca diuca* (no specimen traced).
5. ? Grassland Yellowfinch *Sicalis luteola* (no specimen traced).
6. Spot-flanked Gallinule *Gallinula melanops* (LM D3281d), June 1837, Valparaiso, Chile.
7. Red-legged Cormorant *Phalacrocorax gaimardi* (LM D2882b), no date, Valparaiso, Chile.
8. 'Apapane *Himatione sanguinea sanguinea* (LM D5359c), no date, Hawaiian Islands.
9. ' I'iwi *Vestiaria coccinea* (LM D511a), no date, Hawaiian Islands.
   I'iwi *Vestiaria coccinea* (LM D511b), no date, Hawaiian Islands.
10. 'Akialoa *Hemignathus obscurus* (LM D511c), no date, Hawaiian Islands.
11. 'O'o sp. Neither of the two *Moho* specimens in the Derby collection can be attributed to Townsend (C. Fisher *in litt.*, June 2002).
12. Common 'Amakihi *Hemignathus virens chloris* [O'ahu 'Amakihi *Hemignathus flavus*] (LM D5361), no date, O'ahu, Hawaiian Islands.
    Common 'Amakihi *Hemignathus virens virens* [Hawai'i 'Amakihi *Hemignathus virens virens*] (LM D5361a), no date, Hawai'i, Hawaiian Islands.
13. 'Akepa *Loxops coccinea rufa* ['Akepa *Loxops wolstenholmei*] (LM D4872), no date, O'ahu, Hawaiian islands.
14. Wandering Tattler *Heteroscelus incanus* (LM D3096), April 1837, Tahiti, Society Islands.
15. Tahitian Kingfisher *Halcyon venerata* (LM D2367), April 1837, Tahiti, Society Islands.
16. Tahiti Reed Warbler *Acrocephalus caffra* (no specimen traced).

Additional Townsend specimens at Liverpool Museum but not marked on Lord Derby's list:

17. Common 'Amakihi *Hemignathus virens stejnegeri* [Kaua'i 'Amakihi *Hemignathus kauaiensis*] (LM D5360), no date, Kaua'i (collector doubtful, see Medway 1981: 129).
18. Nukupu'u *Hemignathus lucidus hanapepe* (no number), no date, Kaua'i (collector doubtful, see Medway 1981: 129).
19. Varied Thrush *Zoothera naevia* (LM D1704 S), 25th January 1836, Columbia River.
20. Western Bluebird *Sialia mexicana occidentalis* (LM D1454 S), 21st May 1836, Columbia River.
21. Eastern Bluebird *Sialia sialis* (LM D1455 S), 21st May 1836, Columbia River. Eastern Bluebird does not occur on Columbia River. Although the specimen is credited to Townsend, it has a 1958 note by R. Wagstaffe stating that "Data obviously transferred from D1454 in error" and that "Only one <u>Sialia</u> received from Townsend and that is the cotype of <u>occidentalis</u>." (M. Largen *in litt.*, October, 2001).

Notes.
1. Our identification of the Hawaiian specimens is partly based on Medway (1981: 128-130) who matched some Liverpool specimens with the MS names. However, it would seem likely to us that "Unequal billed Certhia" referred to a range of long, curve-billed Hawaiian birds, such as Nukupu'u and Greater and Lesser 'Akialoa. Some other groups, such as 'Sylvia' may also have included more than one species.
2. Medway *op. cit.* reads No. 82 as 'Cossicus' but says it is a misspelling of Copsychus; he has perhaps been led astray by the old style double 's'. Townsend himself referred to 'Cassicus', seen on Hawai'i (see main text and appendix 11).

APPENDIX 5. McEUEN LIST. Compiled by Thomas McEuen and J.K. Townsend on 25th October 1839 (ANSP No 1. Coll. 404. Transcription of the unpublished MS and the addition of current identifications by B & R Mearns).

List of Townsend's birds lent to J.J. Audubon by Dr M'uhern

from Nos in Pickering list –		*names corresponding to nos in Townsend catalog*	*current identification*	
No	34	White chin Buzzard	x	Swainson's Hawk
√	33 2 spe	Clarke's crow	x	Clark's Nutcracker
	126	Red breasted snipe --- ?		dowitcher sp.
√	27	Ash coloured plover		Mountain Plover
√	12 √	Cream colored sparrow		Clay-colored Sparrow
√	64 √	Black crowned Bunting		Golden-crowned Sparrow
	31 ?	Say's flycatcher ---- ?		Say's Phoebe
	41	Merlin	x	Merlin or Prairie Falcon
√	13√	Black bellyed Bunting		Chestnut-collared Longspur
√	39 √	Rocky Mountain Antcatcher		Spotted Towhee
√	111√ 2 spe	Sylvia occidentalis		Hermit Warbler
√	100	Sylvia Townsendi		Townsend's Warbler
	32	Loggerhead	x	Loggerhead Shrike
√	28	Orpheus montanus		Sage Thrasher
√	117 √	Cypselus Vauxi		Vaux's Swift
	96 2spe	Parus minimus … [illegible]	x	Bush-tit
	15	same as 12		Brewer's Sparrow?
	37	Wren (Q Parkmani) ?		House Wren
√	113 2 spe	Fringilla amaena		Lazuli Bunting
√	11 √ 2 spe	Fringilla grammaca		Lark Sparrow

Birds lent to Mr Audubon by J K Townsend immediately
after his return from the Journey, with the approbation of Dr. Morton

Nos	68 *in J.K.T. list*	Falco columbarius	x	Merlin
	73	Diomedea fusca	x	Black-footed Albatross
	71	Pyrrhula (frontalis?)	x	Purple, House, or Cassin's Finch
	43	Buteo vulgaris	x	Swainson's Hawk

Hawk sent to the Acad. by Dr Jenkins from Mississippi   x   Harris's Hawk
lent by Dr Morton to Mr Audubon – Buteo Harrisi

of the above list, those marked thus x being 10 specimens we
believe are in the possession of Mr J.J. Audubon
those marked thus √ before the numbers are now in the Acad.
those 3 marked thus ? doubtful

This list was made up by Messrs Townsend & McEuen
October 25th 1839 from the Documents above referred to
That is list in the Acad. in Dr Pickering's hand writing
& from Townsend's original cataloge.

[on reverse side]
List of Birds sent
to Mr Audubon
from the Academy of Nat
Sc. of Philadᵃ

APPENDIX 6. TOWNSEND'S NUMBERED CATALOGUE OF BIRD SPECIMENS

Townsend's catalogue of the bird specimens that he obtained during his expedition to the west has disappeared, having long ago been destroyed or lost. We know that it once existed because the catalogue is referred to by Townsend in his 1836 bird report given to the Rev Parker (see appendix 9) and it is mentioned at the foot of the McEuen list of 1839 (see appendix 5). Townsend gave field numbers to his specimens and must have used a catalogue to keep track of his discoveries. Each species that was added to the collection was given a field number, so all specimens of that species had the same number, e.g. all the Clark's Nutcrackers were No. 33.

We have pieced together the following list from field numbers gleaned from a variety of sources. As one would expect, the species in the list reflect the chronology of the journey showing a westerly progression across the Rocky Mountains to the Pacific, time in the Hawaiian Islands, return to the Columbia River, and the journey home via the Hawaiian Islands, Tahiti and Chile.

There are still numerous gaps, now seemingly impossible to fill because of the loss of so many of Townsend's original specimen labels. There are also a few anomalies, some of them due to identification difficulties of that period (eg. there are two entries for Swainson's Hawk) and others due perhaps to errors of transcription from current labels and lists. A few are more difficult to account for, such as the early inclusion of Cape Petrel at No. 125 (see notes below). The Derby list includes some specimens without field numbers that may have simply been because they were duplicates or he may never have included these particular species in the catalogue.

The original catalogue would, of course, have been much more than a chronological list of birds and numbers. He would probably have noted down the dates, locations, colour of eyes and bare parts, and notes on plumage, calls and song, behaviour and habitat, and perhaps given them English and scientific names if he thought that they were new species. For some species he is known to have made a record of the names given to them by the local peoples.

Our initial hope was that the catalogue would help throw some light on the origins of Townsend's controversial seabirds and clarify the type locality of Swainson's Hawk, but there are still too many gaps to draw any definite conclusions on these two matters. The catalogue can perhaps still be added to if more information comes to light.

Field No.	Species name, as in original source	Current identification (by R&B Mearns)	Reference for field no.	Suggested collecting locations
1-7	*Carolina parakeet*	Carolina Parakeet	Townsend	St Louis to
	small Finch	unidentified	Townsend	Independence
	larger Finch	Harris's Sparrow	Townsend	24th March
	Small woodpecker	Yellow-bellied Sapsucker	Townsend	to 28th April 1834
8*	*another Finch*	unidentified	Townsend	
9				Independence
10				to Green River
11	*Fringilla grammaca*	Lark Sparrow	McEuen	29th April
12*	*Cream colored sparrow*	Clay-colored Sparrow?	McEuen	to 27th June 1834
13	*Black bellyed Bunting*	Chestnut-collared Longspur	McEuen	
14				
15*	*same as 12*	Brewer's Sparrow?	McEuen	
16				
17				
18				
19				
20				
21				
22	*[Northern Flicker]*	Northern Flicker	Brydges	
23				
24	*[Black-headed Grosbeak]*	Black-headed Grosbeak	USNM 2873	
25				
26				
27	*Ash coloured plover*	Mountain Plover	McEuen	
28	*Orpheus montanus*	Sage Thrasher	McEuen	
29				
30				
31	*Say's flycatcher*	Say's Phoebe	McEuen	
32	*Loggerhead*	Loggerhead Shrike	McEuen	
33*	*Clarke's crow*	Clark's Nutcracker	McEuen	
34*	*White chin Buzzard*	Swainson's Hawk	McEuen	
35*				
36				Green River to
37	*Wren (Q Parkmani)*	House Wren	McEuen	Columbia River
38				28th June to
39	*Rocky Mountain Antcatcher*	Spotted Towhee	McEuen	11th December 1834
40		Lewis's Woodpecker	USNM 2795	
41	*Merlin*	Merlin or Prairie Falcon	McEuen	
42				
43*	*Buteo vulgaris*	Swainson's Hawk	McEuen	
44				
45	*Sylvia Dⁱ [= Wilsonii]*	Wilson's Warbler	Derby	
46				
47		Band-tailed Pigeon	USNM 2825	
48	*Garrulus Stelleri*	Steller's Jay	Derby	
49	*Falco fuscus*	Sharp-shinned Hawk	Derby	
50				

51				
52				
53				
54	*an Eagle*	Golden Eagle or Bald Eagle	Rhoads	
55		Western Scrub-Jay	USNM 2841	
56				
57	*Sitta Canadensis* [or 97]	Red-breasted Nuthatch	Derby	
58				
59				
60	*Picus redbreasted*	Red-breasted Sapsucker	Derby	
61				
62				
63				
64	*Black crowned Bunting*	Golden-crowned Sparrow	McEuen	
65	*Picus Gairdneri*	Downy Woodpecker	Brydges	
66				
67	*Turdus naevia*	Varied Thrush	Derby	
68	*Falco Columbarius*	Merlin or Prairie Falcon	Derby, McEuen	
69				
70	*Scolopax Wilsonii*	Wilson's Snipe	Derby	
71	*Pyrrhula (frontalis?)*	House Finch?	McEuen	
72	*Parus rufescens*	Chestnut-backed Chickadee	Brydges	
73	*Diomedea fusca*	Black-footed Albatross	McEuen	Voyage to Hawaiian
74				Islands and visits to
75	*Muscicapa*	'Elepaio	Derby, Cambridge	Oahu and Kauai
76				12th December 1834
77				to 15th April 1835
78				
79	*Certhia coccinea*	'I'iwi	Derby	
80	*Crimson Icterus*	'Apapane	Derby	
81				
82*	*Cassicus*	'O'o sp.	Derby	
83				
84	*Strix Brachyotus*	Short-eared Owl	Derby	
85				
86				
87	*Fulica*	Hawaiian Coot	Derby	
88	*Gallinula chloropus*	Common Moorhen	Derby	
89				
89				
90				
91	*Picus pileatus* [or 97]	Pileated Woodpecker	Derby, USNM 2792	Return to the
92	*Vireo solitarius*	Cassin's Vireo	Derby	Columbia River
93				16th April 1835
94				to 30th November 1836
95	*Trochilus rufus*	Rufous Hummingbird	Derby	
96	*Parus minimus*	Bush-tit	McEuen	
97	*Picus pileatus* [or 91]	Pileated Woodpecker	Derby, USNM 2792	
98				
99	*Sylvia Celata*	Orange-crowned Warbler	Derby, USNM 1912, USNM 2929	
100	*Sylvia Townsendi*	Townsend's Warbler	Derby	
101	*Muscicapa*	flycatcher sp.	Derby	
102				
103				
104				
105	*Ditto [= Muscicapa]*	flycatcher sp.	Derby	
106				
107				
108				
109	------- *Wilsonii [=Turdus]*	Veery	Derby	
110				
111	*Sylvia occidentalis*	Hermit Warbler	McEuen	
112				
113	*Fringilla amaena*	Lazuli Bunting	McEuen	
114	*Siala occidentalis*	Western Bluebird	Derby	
115	*Rallus Carolinus*	Sora	Derby	
116	*Certhia familiaris*	Brown Creeper	Derby	
117	*Cypselus Vauxi*	Vaux's Swift	McEuen	
118				
119*	*Strix*	owl sp.	Derby	
120	*Hirundo lunifrons*	Cliff Swallow	Derby	
121				

122*	[California Gull]	California Gull	ANSP 4810	
123	[Hairy Woodpecker]	Hairy Woodpecker	USNM 1869	
124				
125*	Procellaria Capensis	Cape Pigeon	Derby	
126	Red breasted snipe	dowitcher sp.	McEuen	
127	Colymbus [or 187]	grebe sp.	Derby	
128	Anser albifrons	White-fronted Goose	Derby	
129				
130	Anas acuta ? young	Northern Pintail	Derby	
131				
132				
133				
134				
135				
136				
137				
138				
139				
140	Fuligula Clangula	Barrow's or C. Goldeneye	Derby	
141				
142	[Black Oystercatcher]	Black Oystercatcher	AMNH 45404	
143				
144				
145				
146				
147				
148				
149				
150				
151				
152	[Tetrao] phasianellus young	Sharp-tailed Grouse	Derby	
153	Tetrao Urophasianus young	Greater Sage-Grouse	Derby	
154				
155				
156	Larus	gull sp.	Derby	
157*	Loons [= Larus?]	loon sp. [or gull sp.]	Derby	
158				
159*	Larus	gull sp.	Derby	
	Phalacrocorax	Brandt's Cormorant	USNM 2742	
160				
161				
162*	Charadrius pluvialis	Golden Plover	Derby	
		Black-bellied Plover	USNM 2775	
163				
164				
165				
166	Sterna nigra	Black Tern	Derby	
167*				
168				Voyage and second
169				visit to the
170				Hawaiian Islands
171				31st November 1836
172				to 7th April 1837
173				
174				
175				
176*				
177	Puffinus	shearwater sp.	Derby	Tahiti
178	Columba	Gray-green Fruit Dove	Derby	8th April 1837
179	Aludo ?	Venerated Kingfisher	Derby	to 4th May 1837
180				
181	Totanus	Wandering Tattler	Derby	
182	Sterna ? (alba)	Fairy Tern	Derby	
183	Yellow Turdus ?	Tahiti Reed Warbler	Derby	
184	Dusky Dᵒ [= Turdus ?]	Tahiti Reed Warbler dark morph	Derby	
185	Cypselus	Tahiti Swiftlet	Derby	
186	Black Dᵒ [= Muscicapa]	Black Monarch male	Derby	
187	Cinnamon Dᵒ [= Muscicapa]	Black Monarch immature	Derby	
188				
189				
190				

An original Townsend specimen label with locality, sex and date, and field number on reverse (No.22) - Northern Flicker OUM3299

335

191	*Hirundo*	Pacific Swallow	Derby	
192				Valparaiso, Chile
193	*White throated Finch*	Common Diuca-Finch?	Derby	12th June 1837
194	*Siskin*	[Black-chinned?] Siskin	Derby	to 22nd August 1837
195				and voyage to
196				Philadelphia
197				
198				
199				
200				
201				
202				
203	*Grey Muscicapa*	flycatcher sp.	Derby	
204				
205				
206	*Fringilla luteoventris*	Grassland Yellowfinch	Derby	
207				
208				
209	*Small crested Titmouse*	Tufted Tit-tyrant	Derby	
210	*Rufous necked Finch*	Rufous-collared Sparrow	Derby	
211	*Gallinula*	Spot-flanked Gallinule	Derby	
212	*Podiceps*	grebe sp.	Derby	
213				
214				
215				
216				
217	*Small Muscicapa*	flycatcher sp.	Derby	
218				
219				
220				
221				
222				
223	*Phalacrocorax*	Red-legged Cormorant.	Derby, LMD2882b	
224				
225*	[instead of 125?]	Cape Pigeon	Derby	
226				
227				
228				
229				
230				
231				
232	*Strepsiles interpres*	Ruddy Turnstone	Derby	

* Notes to selected field numbers:

1-8. Townsend sent a batch of eight birds to Philadelphia while at Independence (see pp. 47-50, Townsend letter to McEuen, 23rd April 1834).

12 & 15. See note on Brewer's Sparrow (p. 82).

33. Although Townsend's *Narrative* suggests that Clark's Nutcracker was not added until a few days after leaving the Rendezvous near Bear River, 8th July 1834, the USNM card catalogue has details of a Clark's Nutcracker specimen that was collected on 16th June 1834 (USNM 1929).

34 & 43. First Swainson's Hawk found just before the Rendezvous. Second Swainson's Hawk may have been taken at the "Rough-legged Buzzard" nest found on the Banks of the Bear River (Nuttall 1840, p. 100). The Rough-legged Hawk does not breed as far south as Idaho (see appendix 8).

35. A Townsend letter to his family from Green River Rendezvous says that exactly 35 species were collected up to 27th June 1834.

82. Old style double 's' indicates that this is best read as "Cassicus", not "Copsicus", cf. Medway 1981 (see also appendix 4).

119. The Derby list (appendix 4) says that this owl was obtained in California (not visited by Townsend). It may have come from Thomas Nuttall.

122. Not recognised as California Gull at the time (first described in 1854 by G.N. Lawrence). Date on specimen is 18th August 1835.

125. Cape Pigeon is a bird of the southern oceans and is well out of the expected sequence at No. 125. This could be due to a transcription error by Lord Derby, i.e. it should be 225. Or, Townsend may simply have acquired his first specimen from the crew of a visiting ship or directly from Dr Tolmie or Dr Gairdner both of whom had arrived at Fort Vancouver via Cape Horn.

157. Handwriting of "Loons" could be read as "Larus" (see appendix 4).

159. Two species for one number. USNM database gives Brandt's Cormorant with No. 159 as a field number. Derby list gives "Larus" [gull sp.] under No. 159 but no number for "Phalacrocorax" [cormorant sp.]; in the Derby list "Larus" and "Phalacrocorax" are next to each other so there may have been a transcription error.

162. The Black-bellied plover specimen is in winter plumage, so would appear very similar to American Golden Plover.

167-176? No field numbers were traced for any birds obtained during Townsend's second visit to the Hawaiian Islands. If he did add some new Hawaiian birds to the catalogue it is unlikely to have been quite as many species as this.

APPENDIX 7A. BIRDS SEEN AND COLLECTED BY TOWNSEND AND NUTTALL ON THE OVERLAND JOURNEY FROM MISSOURI TO THE PACIFIC, THE VOYAGES BETWEEN THE COLUMBIA RIVER AND THE HAWAIIAN ISLANDS, AND NUTTALL'S VISIT TO CALIFORNIA.

KEY

Townsend a: "List of the Birds Inhabiting the Region of the Rocky Mountains, the Territory of the Oregon, and the North West Coast of America," *Journal of the Academy of Natural Sciences Philadelphia* 8: 151-158 (Townsend 1839f). This is a list of birds without any notes and includes species seen as far eastward as Independence, Missouri. **b**: Townsend's *Narrative of a Journey Across the Rocky Mountains* (1839c) contains a scientific appendix that includes a list or "Catalogue of Birds, Found in the Territory of the Oregon", as well as more detailed descriptions of 24 new, or supposedly new, species, pp. 331-352. The geographical scope of this list extends from the Pacific coast eastwards to Wyoming and beyond.

Audubon: Audubon's commentary on western species derived from Townsend and Nuttall that were included in Audubon's *Ornithological Biography* (in effect the text for *The Birds of America*). Plate numbers refer to original folio plate numbers but volume and page numbers refer to the octavo edition 1840-44 (Dover Reprint 1967).

Nuttall: Nuttall's second edition of his *Manual of the Ornithology of the United States and of Canada* (1840) contained his new observations on western birds, but only for the "Land Birds".

Notes: Miscellaneous notes on range, MS material, etc. For Tolmie, Derby and McEuen lists see appendices 3,4 & 5.

Specimens: All bird specimens are skins or mounts except for 32 skulls presented by Townsend to S.G. Morton (Morton 1840). The Morton collection is at ANSP but has not been searched for surviving Townsend specimens. Specimen numbers and notes are chiefly derived from museum databases. For **museum codes:** see p. ix.

GREATER WHITE-FRONTED GOOSE *Anser albifrons* (Scopoli, 1769)
Townsend a: ANSER albifrons; b: White-fronted Goose, *Anser albifrons*.
Notes: Derby list No. 128 *Anser albifrons*, Columbia River, 1 specimen.

Specimens: ANSP 6056, female, Columbia River, no date, *A. a gambeli*.

SNOW GOOSE *Chen caerulescens* (Linnaeus, 1758)
Townsend a: ANSER *hypoboreus [sic]*; b: White, or Snow Goose, *Anser hyperboreus*.

CACKLING GOOSE *Branta hutchinsii* Richardson, 1832
Townsend a: ANSER *Hutchinsii*; b: Hutchins's Brant Goose, *Anser Hutchinsii*.
Notes: see report given to Rev Parker (appendix 9).
Specimens: ANSP 6025, Oregon, *B. canadensis hutchinsii*.

CANADA GOOSE *Branta canadensis* (Linnaeus, 1758)
Townsend a: ANSER *canadensis*; b: Black-headed Goose, *Anser canadensis*.
Notes: see report given to Rev Parker (appendix 9).

TRUMPETER SWAN *Cygnus buccinator* Richardson, 1832
Townsend a: CYGNUS *buccinator*; b: Trumpeter Swan, *Cygnus buccinator*, (RICHARDSON.)
Notes: see report given to Rev Parker (appendix 9).

TUNDRA SWAN *Cygnus columbianus* (Ord, 1815)
Townsend a: CYGNUS *americanus* (Sharpless.); b: Bewick's Swan, *Cygnus Bewickii*.
Audubon: AMERICAN SWAN. Audubon examined a skin from Columbia River, see Trumpeter Swan, 6: 219-220.
Notes: Townsend skin ANSP 6106 of "Whistling swan" is cleaned (Huber 1930).
Specimens: ANSP 6106, female, Columbia River,10th Feb 1836. ANSP may have skull, No. 340 Cygnus Bewickii, Columbia river (Morton 1840, p. 43). USNM 2765, female. Columbia River, 23rd Oct 1835 (in hand register but no specimen traced).

WOOD DUCK *Aix sponsa* (Linnaeus, 1758)
Townsend a: ANAS *sponsa*; b: Summer, or Wood Duck, *Anas sponsa*.
Notes: see report given to Rev Parker (appendix 9).

GADWALL *Anas strepera* Linnaeus, 1758
Townsend a: [= FULIGULA *dispar?*]; b: not listed.
Audubon: GADWALL DUCK. "I have a fine male procured by Mr. TOWNSEND on the Columbia river", 6: 256.

AMERICAN WIGEON *Anas americana* Gmelin, 1789
Townsend a: ANAS *americana*; b: American Widgeon, *Anas Americana*.
Audubon: AMERICAN WIDGEON. Townsend states that it is common on the Columbia River, 6: 261.

AMERICAN BLACK DUCK *Anas rubripes* Brewster, 1902
Townsend a: ANAS *obscura*; b: Dusky Duck, *Anas obscura*. Seen Bear River, 9th July 1834.
Audubon: DUSKY DUCK. Reported from Columbia River, 6: 249.
Notes: Although Jobanek (1994) considered most Black Duck records for Oregon to be dubious, there seems to be no good reason not to accept Townsend's record for Idaho.

MALLARD *Anas platyrhynchos* Linnaeus, 1758
Townsend a: ANAS *boschas*; b: Mallard Duck, *Anas boschas*.

BLUE-WINGED TEAL *Anas discors* Linnaeus, 1766
Townsend a: not listed; b: not listed.
Audubon: BLUE-WINGED TEAL. Reported from the Columbia River, 6: 291.

NORTHERN SHOVELER *Anas clypeata* Linnaeus, 1758
Townsend a: ANAS *clypeata*; b: Shoveller Duck, *Anas clypeata*.

NORTHERN PINTAIL *Anas acuta* Linnaeus, 1758
Townsend a: ANAS *acuta*; b: Pintail Duck, *Anas acuta*. Townsend and local chief killed 26 when both fired at once, 18th October 1836.
Notes: Derby List No. 130 *Anas acuta?* young, Columbia River, 1 specimen.

GREEN-WINGED TEAL *Anas crecca* Linnaeus, 1758
Townsend a: ANAS *crecca*; b: Green-winged Teal, *Anas crecca*.

CANVASBACK *Aythya valisineria* (Wilson, 1814)
Townsend a: FULIGULA *valisneria*; b: Canvass-back Duck, *Fuligula valisneria*
Notes: Redhead *Aythya americana* was not described until 1838 by T.C. Eyton in his *Monograph of Anatidae*, p. 155. Townsend appeared to be familiar with the Redhead but says (11th April 1836) that he did not see it on the Columbia (Jobanek and Marshall 1992).

RING-NECKED DUCK *Aythya collaris* (Donovan, 1809)
Townsend a: FULIGULA *rufitorques*; b: Tufted, or Ring-necked Duck, *Fuligula rufitorques*.

GREATER SCAUP *Aythya marila* (Linnaeus, 1761) and **LESSER SCAUP** *Aythya affinis* (Eyton, 1838)
Townsend a: FULIGULA *marila*; b: Blue-bill, or Scaup Duck, *Fuligula marila*.
Notes: Lesser Scaup not described until 1838, by T.C. Eyton in his *Monograph of the Anatidae*, p. 157.

HARLEQUIN DUCK *Histrionicus histrionicus* (Linnaeus, 1758)
Townsend a: FULIGULA *histrionica*; b: Harlequin Duck, *Fuligula histrionica*.

SURF SCOTER *Melanitta perspicillata* (Linnaeus, 1758)
Townsend a: FULIGULA *perspicillata*; b: Black, or Surf Duck, *Fuligula perspicillata*.
Audubon: Townsend says that it is found on the Columbia River.

LONG-TAILED DUCK *Clangula hyemalis* (Linnaeus, 1758)
Townsend a: FULIGULA *glacialis*; b: Long-tailed Duck, *Fuligula glacialis*.

BUFFLEHEAD *Bucephela albeola* (Linnaeus, 1758)
Townsend a: not listed; b: not listed.
Audubon: BUFFEL-HEADED DUCK. Townsend "found this species on the streams of the Rocky Mountains", 6: 371.

COMMON GOLDENEYE *Bucephala clangula* (Linnaeus, 1758) and **BARROW'S GOLDENEYE** *Bucephala islandica* Gmelin, 1789
Townsend a: FULIGULA *clangula*; b: Golden-eye Duck, *Fuligula clangula*.
Audubon: Although he painted Barrow's Goldeneye (Plate CCCCIII) he thought that it was the summer plumage of Common Goldeneye and chose not to believe Swainson and Richardson who had included Barrow's Goldeneye in their *Fauna Boreali-Americana* (1832). Because there is no Audubon text for this species the origin of his specimen is not known, though it could have been from Townsend.
Notes: Derby List No. 140 Fuligula Clangula, N.W Coast, 1 specimen.

HOODED MERGANSER *Lophodytes cucullatus* (Linnaeus, 1758)
Townsend a: MERGUS *cucullatus*; b: Hooded, or Crested Merganser, *Mergus cucullatus*.

COMMON MERGANSER *Mergus merganser* Linnaeus, 1758
Townsend a: MERGUS *merganser*; b: Goosander, *Mergus merganser*.

RUFFED GROUSE *Bonasa umbellus* (Linnaeus, 1766)
Townsend a: TETRAO *umbellus*; b: Ruffed Grouse, or Pheasant, *Tetrao umbellus*.
Audubon: RUFFED GROUSE. Townsend "observed it on the Missouri and along the Columbia River", 5: 80.

GREATER SAGE-GROUSE *Centrocercus urophasianus* (Bonaparte, 1827)
Townsend a: TETRAO *urophasianus*; b: Cock of the plains, *Tetrao urophasianus*.
Audubon: PHEASANT-TAILED GROUSE. Notes received from Nuttall and Townsend, 5: 106-110. Audubon may have used Townsend's specimens for Plate CCCLXI.
Nuttall: COCK OF THE PLAINS. First seen north branch of the Platte; also abundant near Walla Walla, pp. 803-806.
Notes: Derby list No. 153 Tetrao Urophasianus young, Columbia River, 1 specimen. Tolmie list: "Tetrao Urophasianus - Young male from Walla walla".
On 12th and 16th August 1836 Townsend shot Sage Grouse near Fort Walla Walla (MS Journal).
Specimens: ANSP may have skull No. 338, Tetrao urophasianus, Columbia river (Morton 1840, p. 43).

AMERICAN SPRUCE GROUSE *Falcipennis canadensis* (Linnaeus, 1758)
Townsend a: TETRAO *canadensis*; b: Spotted Grouse, *Tetrao canadensis*.
Audubon: SPOTTED OR CANADA GROUSE. Information and skins supplied by
Townsend. Audubon considered Franklin's Spruce Grouse and Spruce Grouse the same
species, 5: 83-89.
Nuttall: SPOTTED GROUSE, or SPRUCE PARTRIDGE. "met with … towards the sources of
the Missouri and Oregon," pp. 811-813.
Notes: All Spruce Grouse seen by Townsend and Nuttall would have belonged to race
franklinii.

"pheasant sp." [= ?FRANKLIN'S SPRUCE GROUSE *Falcipennis canadensis
franklinii* (Douglas, 1829)]
Townsend a: PHASIANUS *americanus* (Aud.); b: Long-tailed Black Pheasant, *(not in the
collection.)*
Audubon: Listed as species not fully characterized, Royal octavo edition (1839) 5: 335.
Nuttall: AMERICAN PHEASANT, or GUAN *Penelope borealis*, Nobis. Nuttall repeats
Townsend's observations, states that it cannot be a true pheasant and places it in a
Mexican genus.
Notes: Graustein (1967: 443) identifies pheasants as Sharp-tailed Grouse - see discussion
p. 110.

WHITE-TAILED PTARMIGAN *Lagopus leucura* (Richardson, 1831)
Townsend a: TETRAO *leucurus*; b: White-tailed Grouse, *Tetrao leucurus*.
Audubon: WHITE-TAILED PTARMIGAN. Audubon says not observed by Nuttall
and Townsend, 5: 125.
Notes: Townsend seems never to have been in suitable habitat and included this species
in his lists perhaps on the basis of David Douglas's (1829) claim that it occurred as far
south as Mount Hood, or from reports of trappers he met.

DUSKY GROUSE *Dendragapus obscurus* (Say, 1823) and **SOOTY GROUSE** *D.
fuliginosus* (Ridgway, 1873)
Townsend a: TETRAO *obscurus*; b: Dusky Grouse, *Tetrao obscurus*.
Audubon: DUSKY GROUSE. Female received from Townsend, shot on the Columbia
River on 26th September 1834. Quotes by Townsend and Nuttall. 5: 89-93.
Nuttall: DUSKY GROUSE. "Breeds in the forests of the Oregon" and seen in the Blue
Mountains, pp. 809-811.
Notes: In 2006 Blue Grouse were split into Dusky and Sooty Grouse (AOU 47th
supplement to AOU *Check-list*). Sooty Grouse occurs on lower Columbia River. Large
flocks in Blue Mountains, some shot, would be Dusky Grouse (Townsend's MS Journal,
1836). Tolmie list: "Tetrao Obscurus."
Specimens: ANSP 24321, unsexed, Columbia River, no date, *D. o. fuliginosus*; collected
by JKT, ex Harris collection. ANSP 24320, male, Columbia River, March 1838. ANSP
may have skull No. 341, Tetrao obscurus (Morton 1840, p. 43).

SHARP-TAILED GROUSE *Tympanuchus phasianellus* (Linnaeus, 1758)
Townsend a: TETRAO *phasianellus*; b: Sharp-tailed Grouse, *Tetrao phasianellus*.
Audubon: SHARP-TAILED GROUSE. Seen by Townsend near Laramie's Fork, 5:
110-113.
Nuttall: SHARP-TAILED GROUSE. Found breeding near Laramie's Fork in June, 1834, pp.
806-808.
Notes: Derby list No. 152 [Tetrao] phasianellus, young, Columbia River,1 specimen.
On 9th and 27th July 1836 Townsend shot this species with Mr Pambrun near Fort
Walla Walla (MS Journal).
Specimens: ANSP 24304, male, Walla Walla, no date: *T. p. columbianus*; collected by
JKT, ex Harris collection.

GREATER PRAIRIE-CHICKEN *Tympanuchus cupido* (Linnaeus, 1758)
Townsend a: not listed; b: not listed but in main text for March and April 1834,
Missouri.
Notes: Seen several dates March and April, 1834 (MS Journal).

LESSER PRAIRIE-CHICKEN *Tympanuchus pallidicinctus* (Ridgway, 1873)
Townsend a: not listed; b: not listed.
Notes: Seen 1st April 1834; reddish-orange neck pouches as described by Townsend
refer to this species (MS Journal). Not separated from Greater Prairie Chicken at that
time.

WILD TURKEY *Meleagris gallopavo* Linnaeus, 1758
Townsend a: not listed; b: not listed.
Notes: Seen between St Louis and Independence (*Narrative*, 4th April 1834).

MOUNTAIN QUAIL *Oreortyx pictus* (Douglas, 1829)
Townsend a: PERDIX *plumifera*; b: Plumed Partridge, *Perdix plumifera*, (GOULD)
Audubon: PLUMED PARTRIDGE. Three specimens given to Townsend "by others",
one of which passed to Audubon. Female in Plate CCCCXXIII from Townsend; male
from Zoological Society of London. 5: 69-70.
Nuttall: PLUMED QUAIL *Lophortyx plumifera*, Nobis. "Mr. Townsend met with small coveys
of this fine species of 10 to 15 individuals each, in the woods of the Wahlamet, not far
from the Columbia," p. 791.
Specimens: ANSP 24332, female, Columbia River, ex Harris collection [= female in
Audubon Plate CCCCXXIII?]. USNM 2831, female, no locality

CALIFORNIA QUAIL *Callipepla californica* (Shaw, 1798)
Townsend a: not listed; b: not listed.
Audubon: CALIFORNIAN PARTRIDGE. Audubon says male (for Plate CCCCXIII)
procured near Santa Barbara on 6th March 1837, was sent to him by Townsend, 5: 67-
68. On that date Townsend was in Hawaiian Islands and Nuttall was back on the east
coast, though he had been at Santa Barbara, California, in March 1836. Since Nuttall
supplied notes about this species to Audubon he is the one most likely to have supplied
the specimen.
Specimen: USNM 2829, male, Santa Barbara, California, 6th March 1847 [=1837?] is
credited to Townsend [= Nuttall?].

NORTHERN BOBWHITE *Colinus virginianus* (Linnaeus, 1766)
Townsend a: not listed; b: not listed.
Notes: Seen between St Louis and Independence (MS Journal, 30th March 1834).

PACIFIC LOON *Gavia pacifica* (Lawrence, 1858)
Townsend a: not listed; b: not listed.
Audubon: BLACK-THROATED DIVER. "I find it enumerated in a list of the birds
observed by Mr. J.K. TOWNSEND on the Columbia river, where he also met with *Colymbus
glacialis* [Common Loon]", 7: 295.
Notes: Audubon must have been working from an unknown list, supplied directly to
him by Townsend. The loons seen on the Pacific coast are now designated Pacific Loon
having recently been split from Arctic Loon/Black-throated Diver *Gavia arctica*.

COMMON LOON *Gavia immer* (Brunnich, 1764)
Townsend a: COLUMBUS *glacialis*; b: Loon, or Great Northern Diver, *Colymbus glacialis*.
Notes: Derby list No. 157 Loons [could be read as Larus?], Columbia River; No. 187
Colymbus, Columbia River.

PIED-BILLED GREBE *Podilymbus podiceps* (Linnaeus, 1758)
Townsend a: not listed; b: not listed.
Notes: Jobanek and Marshall (1992) and Jobanek (1994) considered Townsend's
erroneous reference to the Little Grebe *Tachybaptus ruficollis* of Eurasia to be the Pied-
billed Grebe. Other authors have opted for Eared Grebe or Horned Grebe (Jobanek
1994).

RED-NECKED GREBE *Podiceps grisegena* (Boddaert, 1783)
Townsend a: PODICEPS *rubricollis*; b: Red-necked Grebe, *Podiceps rubricollis*.

EARED GREBE *Podiceps nigricollis* C.L.Brehm, 1831
Townsend a: PODICEPS *minor*; b: Little Grebe, *Podiceps minor*.
Notes: see notes for Pied-billed Grebe, above.

WESTERN GREBE *Aechmophorus occidentalis* (Lawrence, 1858) and **CLARK'S
GREBE** *Aechmophorus clarkii* (Lawrence, 1858)
Townsend a: not listed; b: not listed.
Notes: Townsend MS, 11th April 1836, refers to a grebe that will "probably prove
a new species, with an Unusually long neck" (appendix 9). This was probably the
Western Grebe and / or the similar Clark's Grebe, that winter at the mouth of the
Columbia River. These species were not described until 1858 and this may have given
rise to Townsend's erroneous inclusion of the Great Crested Grebe *Podiceps cristatus*
of Eurasia in the same manuscript. See Jobanek and Marshall (1992), Jobanek (1994)
for further discussion. Townsend says (11th April 1836) that specimens were almost
impossible to acquire (Jobanek and Marshall 1992).

BLACK-FOOTED ALBATROSS *Phoebastria nigripes* Audubon, 1839
Townsend a: DIOMEDEA *nigripes*, (Aud.); b: Brown Albatross, *Diomedea fusca*.
Audubon: BLACK-FOOTED ALBATROSS. "For a specimen of this Albatross, I am
indebted to Mr. TOWNSEND, who procured it on 25th December, 1834, on the Pacific
Ocean, in lat. 30°, 44', N. long. 146°", 7: 198. No plate.
Notes: Another "one of the brown Albatrosses of these seas" was taken in the vicinity
of "30° 45' N. Longitude 131° 37' W." (MS Journal,13th December 1836). 2. Three
albatross species are listed in Townsend (1839f) but only *D. fusca* is listed in Townsend
(1839c), the latter probably intended for Black-footed Albatross. McEuen list: No. 73
Diomedea fusca.
Specimens: Stone (1899) says type lost.

[**GRAY-HEADED ALBATROSS** *Thalassarche chrysostoma* J.R.Forster, 1785]
Townsend a: DIOMEDEA chlororhynchos, (Aud.); b: not listed.
Audubon: YELLOW-NOSED ALBATROSS. "A skin of this bird was sent to me by
Mr. TOWNSEND, who procured it in the Pacific Ocean, not far from the mouth of the
Columbia river", 7: 196. No plate.
Notes: 1. Locality error (see appendix 10). 2. The specimen described (but not figured)
by Audubon was in fact the similar Gray-headed Albatross.
Specimens: Derby list includes another two specimens of "Diomedea chlororhynca"
from "Pacific", not now traceable. They may possibly have been Yellow-nosed Albatross
D. chlorrhynchos as Townsend sailed through their range in the south Atlantic but they
are more likely to have been Grey-headed Albatrosses obtained in the Pacific (see
appendix 10).

[LIGHT-MANTLED SOOTY ALBATROSS *Phoebetria palpebrata* (J.R.Forster, 1785)]
Townsend a: DIOMEDEA fusca, (Aud.); b: not listed [see Black-footed Albatross].
Audubon: DUSKY ALBATROSS. "The skin from which I made my drawing of this
species was prepared by Mr. TOWNSEND, who procured the bird near the mouth of the
Columbia river", 7: 200-201. This specimen used in preparation of Plate CCCCVII.
Notes: 1. Locality error (see appendix 10). 2. Townsend persisted in using the name *D.*
fusca for Black-footed Albatross (Townsend 1847: 88-113).
Specimens: USNM 2718, adult; Audubon's type (Stone 1899).

[GIANT PETREL *Macronetes sp.*]
Townsend a: PROCELLARIA gigantea; b: not listed.
Audubon: GIGANTIC FULMAR. "A specimen of the Gigantic Fulmar, shot at some
distance from the mouth of the Columbia river, has been sent to me by Mr. TOWNSEND
…", 7: 202. No plate. Locality error (see appendix 10).
Specimens: USNM 2743, no date, catalogued as *M. giganteus.*

NORTHERN FULMAR *Fulmaris glacialis* (Linnaeus, 1761)
Townsend a: PROCELLARIA *pacifica*, (Aud.); b: not listed.
Audubon: PACIFIC FULMAR. "Three skins transmitted to me by Mr. TOWNSEND
… [from] North-west coast of America", 7: 208-209. No plate.
Specimens: ANSP 5190, *Fulmaris glacialis rodgersii*. Stone (1899) says type lost. USNM
2750, adult, *rodgersii* light phase; USNM 2751 adult *rodgersii* dark phase.

[SOUTHERN FULMAR *Fulmaris glacialoides* (Smith, 1840)]
Townsend a: PROCELLARIA tenuirostris, (Aud.); b: not listed.
Audubon: SLENDER-BILLED FULMAR. The following note from Mr TOWNSEND
was appended to this specimen: "Within a day's sail from the mouth of the Columbia
river. Its habits are very similar to those of Procellaria capensis [Cape Petrel], keeping
constantly around the vessel, and frequently alighting in her wake for the purpose of
feeding. They are easily taken with a hook baited with pork", 7: 210. No plate. Locality
error (see appendix 10).
Specimens: USNM 2032 adult. Stone (1899) says type lost.

WILSON'S STORM-PETREL *Oceanites oceanicus* (Kuhl, 1820)
Townsend a: THALASSIDROMA *Wilsonii*; b: Wilson's Stormy Petrel, *Thalassidroma*
Wilsonii.
Notes: Townsend caught petrels in the southern hemisphere, perhaps Wilson's Storm-
Petrel. Jobanek (1994) considered this to be a dubious record for Oregon. It is only
casual off the Oregon coast and has perhaps been confused with Leach's Petrel.

LEACH'S STORM-PETREL *Oceanodroma leucorhoa* (Vieillot, 1818)
Townsend a: THALASSIDROMA Leachii; b: not listed.
Notes: "upwards of twenty" petrels caught 13th December 1836 (MS Journal); probably
this species but could have included other small dark, white-rumped petrels (Wilson's
Storm-Petrel or Band-rumped Storm-Petrel *O. castro*).

RED-TAILED TROPICBIRD *Phaethon rubricauda* Boddaert, 1783
Townsend a: not listed; b: in main text but not listed in appendix.
Notes: Seen on way to Hawaiiian Islands (*Narrative*, 23rd Dec 1834), "I procured several
specimens" (Townsend 1847: 90).
Specimens: ANSP may have skull No. 593 from N.W. Coast, named as *Phaeton aethereus*
[= Red-billed or Red-tailed Tropicbird] (Morton 1840, p. 45).

AMERICAN WHITE PELICAN *Pelecanus erythrorhynchos* Gmelin, 1789
Townsend a: PELECANUS *onocrotalus*; b: White Pelican, *Pelecanus onocrotalus.* Seen Bear
River/Soda Springs, 9th July 1834.

BROWN PELICAN *Pelecanus occidentalis* Linnaeus, 1766
Townsend a: PELECANUS *fuscus*; b: Brown Pelican, *Pelecanus fuscus.* Seen off Oregon
coast.

BRANDT'S CORMORANT *Phalacrocorax penicillatus* (Brandt, 1838)
Townsend a: PHALACROCORAX *Townsendi (AUD.)*; b: Townsend's Cormorant,
Phalacrocorax Townsendi, (AUDUBON). "… inhabits the Columbia river, and is not
uncommon", p. 351-352.
Audubon: TOWNSEND'S CORMORANT. PLATE CCCCXII. Two males collected
by Townsend at Cape Disappointment, October 1836, 6: 438-439.
Notes: Derby list: Phalacrocorax 1 specimen, Columbia River; is perhaps this species.
Specimens: ANSP may have skull No. 478 (Morton 1840, p. 43; Morton 1849, p. 10.).
USNM 2742 Cape Disappointment, 8th Oct 1836.

DOUBLE-CRESTED CORMORANT *Pholacrocorax auritus* (Lesson, 1831)
Townsend a: not listed; b: Black Cormorant, *Phalacrocorax carbo.*
Notes: This cannot be the Great Cormorant *P. carbo* of Eurasia and is undoubtedly a
reference to the Double-crested Cormorant. See also Jobanek 1994.

PELAGIC CORMORANT *Phalacrocorax pelagicus* Pallas, 1811
Townsend a: *Phalacrocorax splendens (Townsend)*; b: Violet-green Cormorant, *Phalacrocorax*
splendens (TOWNSEND.)."inhabits in considerable numbers the Rocky Cape
at the entrance of the Columbia river", pp. 350-351. Townsend says he only collected
one skin.
Audubon: VIOLET-GREEN CORMORANT. Townsend quoted, 6: 440-441. Plate
CCCCXII prepared from Townsend's specimen.
Notes: Stone (1899) says type lost but USNM 2004 must be type if only one specimen
collected.
Specimens: USNM 2004, Mouth of Columbia River, Oct 1836.

Cormorants *Phalacrocorax sp.*
Townsend a: *Phalacrocorax leucurus*, (Aud.) and *Phalacrocorax leuconotus*, (Aud.); b: not listed
Audubon: Two species described from Townsend's notes were probably Pelagic
Cormorants – see p. 237.

GREAT BLUE HERON *Ardea herodias* Linnaeus, 1758
Townsend a: ARDEA *herodias*; b: Great Blue Heron, *Ardea herodias.*

SNOWY EGRET *Egretta thula* (Molina, 1782)
Townsend a: not listed; b: not listed.
Notes: Two skins rescued from Harris collection (Street 1948).
Specimens: ANSP 162372, Walla Walla, 3rd July 1836, *E.t.brewsteri*, presented by J.F. and
P.B. Street, 1948. Second rescued specimen not traced.

BLACK-CROWNED NIGHT HERON *Nycticorax nycticorax* (Linnaeus, 1758)
Townsend a: ARDEA *nycticorax*; b: Night Heron, *Ardea nycticorax.*

TURKEY VULTURE *Cathartes aura* (Linnaeus, 1758)
Townsend a: CATHARTES *aura*; b: Turkey Buzzard, or Vulture, *Cathartes aura.*
Audubon: TURKEY-BUZZARD. Townsend reported to have seen it on the Columbia
River, 1: 15.
Notes: One of only four species included in Townsend's abortive book (1839g).

CALIFORNIA CONDOR *Gymnogyps californianus* (Shaw, 1798)
Townsend a: CATHARTES *californianus*; b: Californian Vulture, *Cathartes californianus.*
Audubon: CALIFORNIAN VULTURE. Long quote from Townsend, 1: 12-13.
Nuttall: THE CONDOR, *Cathartes gryphus.* Nuttall does not distinguish between Andean
and California Condors; nor does he report seeing it or include any of Townsend's
observations, pp. 35-39.
Notes: See Townsend (1842) for account of hunting and collecting USNM 78005. Text
and plate included in Townsend (1839g).
Specimens: USNM 78005 (2717 and possibly 73053 are earlier numbers for same
specimen).

OSPREY *Pandion haliaetus* (Linnaeus, 1758)
Townsend a: FALCO *haliaetus*; b: Fish hawk, *Falco haliaetus.*

SWALLOW-TAILED KITE *Elanoides forficatus* (Linnaeus, 1758)
Townsend a: not listed; b: not listed.
Nuttall: SWALLOW-TAILED HAWK. Gives reports of birds reaching as far as 44° N, pp.
97-98.
Notes: Several seen shortly after leaving Independence (MS Journal,1st May 1834); a
rare spring vagrant in that area.

BALD EAGLE *Haliaeetus leucocephalus* (Linnaeus, 1758)
Townsend a: FALCO *leucocephalus*; b: White-headed, or Bald Eagle, *Falco leucocephalus.*
Nest seen 25th May 1834, Platte River.

NORTHERN HARRIER *Circus cyaneus* (Linnaeus, 1766)
Townsend a: FALCO *cyaneus*; b: Hen Harrier, or Marsh Hawk, *Falco cyaneus.*
Audubon: MARSH HAWK. Seen by Townsend "on the plains of the Columbia river
and prairies bordering the Missouri", 1: 105.
Notes: One shot a few miles west of St Charles, Missouri (MS Journal,1st April 1834)
has not been traced.

SHARP-SHINNED HAWK *Accipiter striatus* Vieillot, 1808
Townsend a: FALCO *fuscus*; b: Sharp-shinned Hawk, *Falco velox.*
Audubon: SHARP-SHINNED OR SLATE-COLOURED HAWK. Townsend
obtained specimens from the Columbia River, 1: 101.
Notes: Derby List No. 49 *Falco fuscus*, Colorado river, 1 specimen.

COOPER'S HAWK *Accipiter cooperii* (Bonaparte, 1828)
Townsend a: FALCO *Cooperii*; b: Cooper's Hawk, *Falco Cooperii.*
Specimens: ANSP 1317, female, Columbia River, no date.

RED-SHOULDERED HAWK *Buteo lineatus* (Gmelin, 1788)
Townsend a: FALCO *hyemalis*; Winter Hawk, or Falcon, *Falco hyemalis.*

SWAINSON'S HAWK *Buteo swainsoni* Bonaparte, 1838
Townsend a: FALCO *buteo*; b: Common Hawk, or Buzzard, *Falco (buteo) vulgaris*.
Audubon: COMMON BUZZARD. Audubon painted Swainson's Hawk from Townsend specimen (Plate CCCLXXII), and saw another Townsend specimen from the plains of the Snake River, 1: 30-32.
Nuttall: WHITE THROATED BUZZARD *Buteo montana*. Townsend found nest near Green River, Wyoming, p. 112. Also included under COMMON BUZZARD, p. 105.
Notes: McEuen list: No. 34 "White chin Buzzard" and No. 43 "Buteo vulgaris" – see appendix 8. No surviving specimens.

RED-TAILED HAWK *Buteo jamaicensis* (Gmelin, 1788)
Townsend a: FALCO *borealis*; b: Red-tailed Hawk, *Falco borealis*.
Nuttall: RED-TAILED HAWK OR BUZZARD. Occurs "far westward up the Missouri", pp. 102-104.
Specimens: ANSP 1524, unsexed, Rocky Mountains, no date, *B. j. calurus*.

?FERRUGINOUS HAWK *Buteo regalis* (G.R.Gray, 1844)
Townsend a: FALCO *sancti-johannis [sic]*; b: Black Hawk, *Falco sancti-johannes*.
Notes: Identity uncertain. Black Hawk could be dark morph of either Ferruginous Hawk or Red-tailed Hawk (or both). Or even dark adult Rough-legged Hawk (see Nuttall 1840. p 99). Cassin (1856: 103-105) considered *Falco sancti-johannis* to be "Rough Legged Buzzard" but included *Archibuteo ferrugineus* as the Western "Rough-legged Buzzard". Cassin (1856: 105) says that E.M. Kern was the first to bring back specimens of the Ferruginous Hawk.

ROUGH-LEGGED HAWK *Buteo lagopus* (Pontoppidan, 1763)
Townsend a: FALCO *lagopus*; b: Rough-legged Hawk, or Falcon, *Falco lagopus*.
Nuttall: ROUGH-LEGGED FALCON. "Mr. Townsend found its nest on the banks of Bear River," p. 100. [Out of range: = Swainson's Hawk? See appendix 8]

GOLDEN EAGLE *Aquila chrysaetos* (Linnaeus, 1758)
Townsend a: FALCO *albicilla*? [= immature Golden Eagle?]; b: Golden Eagle, *Falco chrysaetos*.
Notes: Townsend lost an eagle specimen (Rhoads 1903).

AMERICAN KESTREL *Falco sparverius* Linnaeus, 1758
Townsend a: FALCO *sparverius*; b: Sparrow Hawk, *Falco sparverius*.

MERLIN *Falco columbarius* Linnaeus, 1758 and **PRAIRIE FALCON** *Falco mexicanus* Schlegel, 1851
Townsend a: FALCO *aesalon* and FALCO *columbarius*; Merlin, b: *Falco aesalon* and Pigeon Hawk, *Falco columbarius*.
Audubon: PIGEON HAWK. Townsend reported to have found Merlin in Rocky Mountains and on shores of the Columbia River, 1: 88.
Nuttall: MERLIN *Falco aesalon*. "Two specimens … obtained by Mr. Townsend in the forests of Bear river in the Rocky Mountains where they were breeding in a cedar tree in the month of July," p. 62.
Notes: McEuen list: No. 41 Merlin; No. 68 Falco columbarius. There was obviously some confusion here: between the male and female Merlins because of their sexual dimorphism, or perhaps with much larger Prairie Falcons.
Specimens: ANSP 2157, female. ANSP may have skull No. 476, Falco columbarius? Columbia River (Morton 1840, p. 43).

PRAIRIE FALCON *Falco mexicanus* Schlegel, 1851
Townsend a: FALCO *aesalon*? (see above); b: *Falco aesalon*? (see above)
Nuttall: (see Peregrine Falcon below)
Notes: Described as a new species by Cassin in *Illustrations of the Birds of California* (1853-56) using a Townsend specimen from upper section of the Platte River. (It had already been described by H. Schlegel in 1851 from specimen from Mexico.)
Specimen; ANSP 2175, Source of Platte River, no date; type of Cassin's *Falco polyagrus* (Stone 1899).

PEREGRINE FALCON *Falco peregrinus* Tunstall, 1771
Townsend a: FALCO *peregrinus*; b: Great-footed, or Duck Hawk, *Falco peregrinus*.
Nuttall: COMMON OR WANDERING FALCON *Falco peregrinus*. "Mr. Townsend obtained a specimen on Big Sandy river of the Colorado of the West [Green River] in the month of July," p. 56. This was possibly Prairie Falcon ANSP 2175; the headwaters of the Platte and Big Sandy are adjacent to each other. See above.

SORA *Porzana carolina* (Linnaeus, 1758)
Townsend a: RALLUS *carolinus*; b: Soree, or Rail, *Rallus carolinus*.
Notes: Derby list No. 115 Rallus Carolinus, Columbia River, 1 specimen.

AMERICAN COOT *Fulica americana* Gmelin, 1789
Townsend a: FULICA *americana*; b: Common Coot, *Fulica Americana*.

SANDHILL CRANE *Grus canadensis* (Linnaeus, 1758)
Townsend a: not listed; b: Brown, or Sandhill Crane, *Grus Canadensis*.
Notes: Seen and heard in Missouri and on Platte River (*Narrative* and MS Journal).
Specimens: ANSP may have skull No. 339, Grus canadensis, Columbia river (Morton 1840, p. 43.).

WHOOPING CRANE *Grus americana* (Linnaeus, 1758)
Townsend a: GRUS *americana*; b: Hooping Crane, or Stork, *Grus Americana*.
Notes: Graustein (1967: 300) suggests that Whooping Cranes seen by Townsend near Soda Springs (9th July 1834) were Sandhill Cranes. But Whooping Cranes may well have bred in that area at that time (nearby Gray's Lake was chosen for reintroduction of the species in late 1900s). When Townsend saw Sandhill Cranes in Missouri he called them Sandhill Cranes. When he saw these birds near Soda Springs he called them "Hooping Cranes". In addition Nuttall had experience of Whooping Cranes in 1811 when he saw hundreds and would have kept Townsend right if need be. However, it is less likely that Townsend saw Whooping Cranes on the Columbia River (Jobanek 1994).

BLACK-BELLIED PLOVER *Pluvialis squatarola* (Linnaeus, 1758)
Townsend a: not listed; b: not listed.
Specimens: USNM 2775, male, Columbia River, 21st Oct 1836.

AMERICAN GOLDEN PLOVER *Pluvialis dominica* (Muller, 1776)
Townsend a: not listed; b: not listed.
Notes: Numerous in early part of journey (*Narrative* and MS Journal, 31st March 1834).
Derby list: No.162 Charadrius pluvialis, Columbia River, 1 specimen.

MOUNTAIN PLOVER *Charadrius montanus* Townsend, 1837
Townsend a: CHARADRIUS *montanus* (Towns.); b: Rocky Mountain Plover, *Charadrius montanus*, (TOWNSEND). "Inhabits the table land of the Rocky Mountains", p. 349.
Audubon: ROCKY-MOUNTAIN PLOVER. Quote by Nuttall, 5: 213-214. Plate CCCL drawn from Townsend's only specimen.
Notes: Original description in *Journal ANSP* 1837, p. 142. In McEuen's list as No. 27 Ash coloured Plover.
Specimens: ANSP 24353, female, Rocky Mountains; type.

KILLDEER *Charadrius vociferus* Linnaeus, 1758
Townsend a: CHARADRIUS *vociferus*; b: Killdeer Plover, *Charadrius vociferus*.

BLACK OYSTERCATCHER *Haematopus bachmani* Audubon, 1838
Townsend a: HAEMATOPUS *Townsendii*, (Aud.) and HAEMATOPUS *Bachmanii*, (Aud.); b: White-legged Oystercatcher, *Haematopus Bachmani*, (AUD.) "shot near Puget's sound, by my friend William Fraser Tolmie", p. 348-349.
Audubon: BACHMAN'S OYSTER-CATCHER and TOWNSEND'S OYSTER-CATCHER. Two species described from specimens received from Townsend, 5: 243-246. Plate CCCCXXVII prepared from these specimens.
Notes: Harrison and Walker (1997: 502) suggest that one specimen was Blackish Oystercatcher *H. ater* from South America; impossible to investigate now that one specimen is missing.
Specimens: AMNH 45404, unsexed, "Sitka – N.W.Coast America, June 1836, J.K. Townsend" is probable type specimen. Stone (1899) says both types lost but unlabelled skin in Harris collection (now missing, perhaps AMNH 45404) probably from Townsend (Street 1948.). ANSP may have skull No. 521, Haematopus ——, N.W. Coast (Morton 1840, p. 45).

AMERICAN AVOCET *Recurvirostra americana* Gmelin, 1789
Townsend a: RECURVIROSTRA *americana*; b: American Avocet, *Recurvirostra americana*. Seen near Independence Rock June 1834.
Audubon: AMERICAN AVOSET [sic]. Nuttall found it breeding "on islands of shallow ponds throughout the Rocky Mountains", 6: 28.
Specimens: Unlabelled skin in Harris collection possibly from Townsend (Street 1948) – untraced.

SPOTTED SANDPIPER *Actitis macularia* (Linnaeus, 1766)
Townsend a: TOTANUS *macularis*; b: Spotted Sandpiper, *Totanus macularis*.

SOLITARY SANDPIPER *Tringa solitaria* Wilson, 1813
Townsend a: not listed; b: not listed.
Audubon: SOLITARY SANDPIPER OR TATLER. Townsend not mentioned but recorded on the Columbia River, 5: 311.

GREATER YELLOWLEGS *Tringa melanoleuca* (Gmelin, 1789)
Townsend a: TOTANUS *flavipes*; b: Yellow-shanks Snipe, *Totanus flavipes*.
Audubon: TELL-TALE TATLER. Townsend found it on the Columbia River, 5: 319.
Notes: Seen between St Louis and Independance (MS Journal, 4th April 1834). No mention anywhere of Lesser Yellowlegs, even though first described in 1789.

WILLET *Catoptrophorus semipalmatus* (Gmelin, 1789)
Townsend a: TOTANUS *semipalmatus*; b: Semipalmated Snipe, or Willet, *Totanus semipalmatus*.

UPLAND SANDPIPER *Bartramia longicauda* (Bechstein, 1812)
Townsend a: not listed; b: not listed.
Seen near St Louis (MS Journal, 26th March 1834).

ESKIMO CURLEW *Numenius borealis* (J.R. Forster, 1772)
Townsend a: *Numenius borealis*; b: Esquimaux Curlew, *Numenius borealis.*
Notes: Not mentioned by Townsend in any MS Journal, letter or text (other than his two lists). It is most unlikely that Townsend saw Eskimo Curlew in Oregon or Washington (Jobanek 1994) but he could certainly have seen them in Kansas, Nebraska or Missouri as he was passing through that area at peak migration time (Gollup et al 1986, p.98). He would probably have been familiar with this species on the east coast and as it was then so common it is not surprising that he made no special mention of it. It is odd that he does not mention the Whimbrel *Numenius phaeopus* in his lists as they do occur on the west coast. Townsend would surely have known the difference between the two species.

LONG-BILLED CURLEW *Numenius americanus* Bechstein, 1812
Townsend a: NUMENIUS *longirostris*; b: Long-billed Curlew, *Numenius longirostris.*
Notes: Seen on the Platte River (MS Journal, 17th May 1834).

MARBLED GODWIT *Limosa fedoa* (Linnaeus, 1758)
Townsend a: LIMOSA *fedoa*; b: Great Marbled Godwit, *Limosa fedoa.*
Notes: "immense swarms of a bird which I have never before seen in this country. A Godwit, of a species with which I am not unacquainted [i.e. Townsend had seen it on the east coast]," (MS Journal, 19th October 1836).

SURFBIRD *Aphriza virgata* (Gmelin, 1789)
Townsend a: APHRIZA *Townsendii* (Aud.); b: Townsend's Sandpiper, *Frinca* [sic] *Townsendi* (AUDUBON.) "I shot one specimen of this curious bird on the base of the rocky cape at the entrance of the Columbia river, in November 1836", pp. 349-350.
Audubon: TOWNSEND'S SURF-BIRD. Townsend's female specimen from Cape Disappointment painted twice, in different poses (for Plate CCCCXXVIII), 5: 228-230.
Specimens: AMNH 156652, female; formerly mounted and almost certainly once part of the Jacob Giraud collection and was listed as type by Orton (1871). Stone (1899) says type lost. See also p. 233.

SEMIPALMATED SANDPIPER *Calidris pusilla* (Linnaeus, 1766)
Townsend a: TRINGA *semipalmata*; b: Semipalmated Sandpiper, *Tringa semipalmata.*
Notes: Western Sandpiper may also have been seen but was not described until 1857.

LEAST SANDPIPER *Calidris minutilla* (Vieillot, 1819)
Townsend a: TRINGA *Wilsonii*; b: Little Sandpiper, *Tringa Wilsonii.*
Audubon: LITTLE SANDPIPER. Townsend reported to have seen it on the Columbia River, 5: 285.

DUNLIN *Calidris alpina* Linnaeus, 1758
Townsend a: TRINGA *alpina*; Red-backed Sandpiper, *Tringa alpina.*

SHORT-BILLED DOWITCHER *Limnodromus griseus* (Gmelin, 1789) and **LONG-BILLED DOWITCHER** *L. scolopaceus* (Say, 1823)
Townsend a: SCOLOPAX grisea; b: Red-breasted Snipe, *Scolopax grisea.*
Audubon: RED-BREASTED SNIPE. Recorded on the Columbia River [by Townsend], 6: 10-15.
Notes: McEuen list No. 126 Red breasted Snipe. The differences between Long-billed and Short-billed Dowitchers were imperfectly known during Townsend's day. He probably saw both species.

WILSON'S SNIPE *Gallinago delicata* (Ord, 1825)
Townsend a: SCOLOPAX *Wilsonii* and SCOLOPAX *Drummondii*; b: Common American Snipe, *Scolopax Wilsonii.*
Notes: Derby list No. 70 Scolopax Wilsonii, Columbia River, 2 specimens.
Specimens: ANSP may have skull No. 475 Scolopax Drummondii (Wilsonii?), Columbia river (Morton 1840, p. 43.).

WILSON'S PHALAROPE *Phalaropus tricolor* (Vieillot, 1819)
Townsend a: PHALAROPUS *Wilsonii*; b: Wilson's Phalarope, *Phalaropus Wisonii.*

RED-NECKED PHALAROPE *Phalaropus lobatus* (Linnaeus, 1758)
Townsend a: PHALAROPUS *hyperboreus*; b: Hyperborean Phalarope, *Phalaropus hyperboreus.*

RED PHALAROPE *Phalaropus fulicarius* (Linnaeus, 1758)
Townsend a: PHALAROPUS *fulicarius*; b: Red Phalarope, *Phalaropus fulicarius.*

MEW GULL *Larus canus* Linnaeus, 1758
Townsend a: LARUS *canus*; b: Common Gull, *Larus canus.*

RING-BILLED GULL *Larus delawarensis* Ord, 1815
Townsend a: not listed; b: Ring-billed Gull, *Larus zonorrhynchus.*
Specimens: USNM 2771, male, 22nd Oct 1836. [Townsend was at Astoria, Oregon, on that date.]

CALIFORNIA GULL *Larus californicus* Lawrence, 1854
Townsend a: not listed; b: not listed.
Specimens: ANSP 4810, unsexed,18th Aug 1835; on that date JKT at Fort Vancouver (Washington) or Sauvie's Island (Oregon). USNM 2772, Cape Disappointment, 6th Oct 1836.

WESTERN GULL *Larus occidentalis* Audubon, 1839
Townsend a: LARUS *occidentalis*, (Aud.); b: not listed.
Audubon: WESTERN GULL. Two specimens received from Townsend. A male from Cape Disappointment, 7th October 1836 and young male Cape Disappointment, 6th October 1836, 7: 161-162. No plate.
Notes: One of the specimens from Cape Disappointment was probably a California Gull (see below).
Specimens: Stone (1899) says type at USNM but does not give specimen number. USNM 2767 Cape Disappointment, Washington, 7th Oct 1836 (in hand register but no specimen traced). USNM 2769, Cape Disappointment, Washington, 6th Oct 1836 (in hand register but no specimen traced).

GLAUCOUS GULL *Larus hyperboreus* Gunnerus, 1767 and **GLAUCOUS-WINGED GULL** *Larus glaucescens* Naumann, 1840
Townsend a: not listed; b: Glaucous Gull, *Larus glaucous.*
Notes: Townsend may have seen both species on the Pacific Coast but Glaucous-winged Gull is more numerous (not described until 1840, by J.F. Naumann).

gull *Larus* sp.
Notes: Derby list: No. 159 Larus, Columbia River, 2 specimens; No. 156 Larus, Columbia River, 1 specimen.

BLACK TERN *Chlidonias niger* (Linnaeus, 1758)
Townsend a: STERNA *nigra*; b: Black Tern, *Sterna nigra.*
Notes: Derby list: No. 166 Sterna nigra, N.W. coast, 1 specimen. Tolmie list: "Sterna Nigra."
Specimens: USNM 2020, no locality, [= Columbia River], July 1836.

PIGEON GUILLEMOT *Cepphus columba* Pallas, 1811
Townsend a: URIA *grylle*; b: Black Guillemot, *Uria grille.*
Notes: Black Guillemot does not occur on west coast; it forms a superspecies with Pigeon Guillemot.
Specimens: ANSP 4668, Straits of Juan de Fuca, Canada (donated to Townsend).

MARBLED MURRELET *Brachyramphus marmoratus* (Gmelin, 1789)
Townsend a: URIA *Townsendii*, (Aud.); b: Slender-billed Guillemot, *Uria Townsendi.* (AUDUBON.) "Inhabits the bays of N.W. Coast of America, in latitude 38° to 40°. The specimens were shot and presented to me by Captain W. Brotchie", p. 352.
Audubon: SLENDER-BILLED GUILLEMOT. Four specimens from Townsend shot by Captain Brotchie, 7: 278-280. Audubon painted two of them for Plate CCCCXXX. Stone (1899) says types lost.

ANCIENT MURRELET *Synthliboramphus antiquus* (Gmelin, 1789)
Townsend a: URIA *antiqua*; b: not listed.

CASSIN'S AUKLET *Ptychoramphus aleuticus* (Pallas, 1811)
Townsend a: PHALERIS *nodirostris* [= Least Auklet *Aethia pusilla*]; b: not listed
Notes: Least Auklet does not occur off the Oregon coast. Cassin's Auklet is regular off the Oregon coast so is more likely.

CRESTED AUKLET *Aethia cristatella* (Pallas, 1769)
Townsend a: PHALERIS *cristatella*; b: not listed.
Notes: Mainly an Alaskan species; considered accidental in southern waters.

RHINOCEROUS AUKLET *Cerorhinca monocerata* (Pallas, 1811)
Townsend a: CERATORHYNCHA *occidentalis*; b: not listed.
Notes: Mainly an Alaskan species; considered accidental in southern waters.

BAND-TAILED PIGEON *Patagioenas fasciata* Say, 1823
Townsend a: COLUMBA *fasciata*; b: Band-tail Pigeon, *Columba fasciata.* Very abundant, flocks of 50-60 birds, 21st May 1835, p. 320.
Audubon: BAND-TAILED DOVE, OR PIGEON. Specimens of three pairs and some young birds received from Townsend; quotes by Townsend and Nuttall. 4: 312-315. Two Townsend specimens used for Plate CCCLXVII.
Nuttall: BAND TAILED PIGEON. Very numerous on lower Columbia River and Willamette. Townsend reported to have seen nests on ground under small bushes, pp. 753-755.
Notes: Tolmie list: "Columba Fasciata."
Specimens: ANSP 13264, unsexed, Columbia River, no date. ANSP 175964, unsexed, Columbia River, no date. ANSP may have skull No. 479 Columba fasciata, Columbia river (Morton 1840, p. 44.). USNM 1933, male, Columbia River, 30th July 1835 (in hand leger but no specimen traced). USNM 2825, female, Columbia River, 16th May 1835 (in hand ledger but no specimen traced).

MOURNING DOVE *Zenaida macroura* (Linnaeus, 1758)
Townsend a: COLUMBA *carolinensis*; b: Turtle Dove, *Columba carolinensis.*
Notes: Tolmie list: " Columba Caroliniensis" [sic].

PASSENGER PIGEON *Ectopistes migratorius* (Linnaeus, 1766)
Townsend a: COLUMBA *migratoria*; b: Passenger Pigeon, *Columba migratoria.*
Notes: Townsend reported "wild pigeons" near Florisant, Missouri, (*Narrative*, 31st March 1834 and MS Journal, 6th April 1834). The *Narrative* appendix list is said to be for the "Territory of the Oregon" but also covers birds seen further *east* than Independence, Missouri, and this was probably why he included the Passenger Pigeon. Jobanek (1994) notes the nuisance caused by Townsend's inclusion of the Passenger Pigeon in this list.

CAROLINA PARAKEET *Conuropsis carolinensis* (Linnaeus, 1758)
Townsend a: not listed; b: not listed. Mentioned in text for 9th April (Boonville) and 20th April 1834 (Independence).
Notes: "I soon procured as many as I had occasion for" (MS Journal, 5th and 9th April 1834). Specimens are referred to in Hannah Townsend's diary entry dated 29th November 1835.
Specimens: USNM 12272, Illinois, 1834; more likely from Illinois, 1840 – see p. 288).

YELLOW-BILLED CUCKOO *Coccyzus americanus* (Linnaeus, 1758)
Townsend a: COCCYZUS *americanus*; b: Yellow-billed Cuckoo, *Coccyzus americanus*.
Audubon: YELLOW-BILLED CUCKOO. Seen on the Columbia River by Townsend, 4: 293-299.
Notes: Townsend reported this species as "abundant during the summer" along the lower Columbia River; now very rare there (Jobanek and Marshall 1992: 8,11).

EASTERN SCREECH-OWL *Megascops otus* (Linnaeus, 1758) and **WESTERN SCREECH-OWL** *Megascops kennicottii* (Elliot, 1867)
Townsend a: STRIX *naevia* [Gray morph] and STRIX *asio* [Red morph]; b: Mottled Owl, *Strix naevia* [Gray morph]. Red Owl, *Strix asio* [Red morph].
Notes: *Strix naevia* probably includes Western Screech Owl that was not described as separate species until 1867.

GREAT HORNED OWL *Bubo virginianus* (Gmelin, 1788)
Townsend a: STRIX *virginiana*; b: Great Horned Owl, *Strix virginiana*.

NORTHERN PYGMY OWL *Glaucidium gnoma* Wagler, 1832
Townsend a: STRIX *passerinoides*; b: Passerine Owl, *Strix passerinoides*.
Audubon: LITTLE COLUMBIAN OWL. Single specimen received from Townsend (from which two figures drawn for Plate CCCCXXXII) from Columbia River, November (no year), 1: 117-118.
Nuttall: SMALL SPARROW OWL. "Mr. Townsend and myself met with it near Fort Vancouver … According to Mr. T. the specimen he killed had been preying on a Ruby-crowned Wren", pp. 148-149. Specimen not traced.

BURROWING OWL *Athene cunicularia* (Molina, 1782)
Townsend a: STRIX *cunicularia*; b: Burrowing Owl, *Strix cunicularia*.
Audubon: BURROWING OWL. Some information supplied by Townsend, 1: 119-121. Plate CCCCXXXII includes Burrowing Owl *Strix cunicularia* and Large-headed Burrowing Owl *Strix californica*, [from Townsend = Nuttall?]
Notes: Tolmie list: Strix Cunicularia. Derby list: No. 119 Strix California.
Specimens: ANSP 2510, juv male, Columbia River, no date, *S. c. hypugaea*. Collected by Townsend, ex Harris collection.

GREAT GRAY OWL *Strix nebulosa* J.R. Forster, 1772
Townsend a: STRIX *cinerea*; b: Great Cinereous Owl, *Strix cinerea*.
Audubon: GREAT CINEREOUS OWL Audubon saw specimens from Rocky Mountains collected by Townsend, 1: 130-131.
Nuttall: GREAT GREY OR CINEREOUS OWL: "found in the Oregon Territory by my friend Mr. Townsend", p. 135.
Specimens: ANSP 2696, unsexed, Rocky Mountains, no date.

LONG-EARED OWL *Asio otus* (Linnaeus, 1758)
Townsend a: STRIX *otus*; b: Long-eared Owl, *Strix otus*.

SHORT-EARED OWL *Asio flammeus* (Pontoppidan, 1763)
Townsend a: STRIX *brachyotus*; b: Short-eared Owl, *Strix brachyotus*.
Audubon: SHORT-EARED OWL. Audubon received a Townsend specimen from 'the valley of the Columbia River", 1: 141.
Nuttall: SHORT-EARED OWL. "we have observed it … in the territory of Oregon", p. 141.
Specimens: The only two Townsend specimens at ANSP are both from the Hawaiian Islands.

BOREAL OWL *Aegolius funereus* (Linnaeus, 1758)
Townsend a: STRIX *Tengmalmi*; b: Tengmalm's Owl, *Strix Tengmalmi*.
Audubon: TENGMALM'S OWL. Found by Townsend and Nuttall near Malade River, Idaho, very tame, 1: 122.
Nuttall: TENGMALM'S OWL: "It inhabits the woods along the streams of the Rocky Mountains down to the Oregon", p. 147.
Notes: Jobanek (1994) considered reports for the lower Columbia as dubious.

NORTHERN SAW-WHET OWL *Aegolius acadicus* (Gmelin, 1788)
Townsend a: STRIX *acadica*; b: Little Owl, *Strix acadica*.

COMMON NIGHTHAWK *Chordeiles minor* (J.R. Forster, 1771)
Townsend a: CAPRIMULGUS *virginianus*; b: Night Hawk, *Caprimulgus virginianus*.

COMMON POORWILL *Phalaenoptilus nuttallii* (Audubon, 1844)
Townsend a: CAPRIMULGUS *Nuttallii*, (Aud.); b: not listed.
Audubon: Described and named from Nuttall's notes as NUTTALL'S WHIP-POOR-WILL *Caprimulgus Nuttallii*, Royal octavo edition (1839) 5: 335. [Described again from a specimen collected during Audubon's 1843 Missouri expedition]
Nuttall: NUTTALL'S GOAT-SUCKER: Seen Upper Platte 10th June, Sandy River 16th June, and Three Butes 7th August 1834, p. 747.

VAUX'S SWIFT *Chaetura vauxi* (Townsend, 1839)
Townsend a: CYPSELUS *Vauxii* (Towns.); b: Vaux's Chimney Swallow, *Cypselus Vauxi*, (TOWNSEND.) "… common on the Columbia river; breeds in hollow trees", p. 348.
Nuttall: VAUX'S CHIMNEY SWIFT *Chaetura Vauxii*, Nobis. Discovered by Townsend on Columbia River, p. 738.
Notes: Original description in *Narrative* (1839) in appendix, p. 348. McEuen list: No. 117 Cypselus Vauxi.
Specimens: ANSP 24169, female, Columbia River, no date; type specimen.

ANNA'S HUMMINGBIRD *Calypte anna* (Lesson, 1829)
Townsend a: TROCHILUS *anna*; b: not listed.
Audubon: ANNA HUMMING-BIRD. Nuttall supplied female and nest (used for Plate CCCCXXV), 4: 188-189.
Nuttall: YELLOW-CROWNED HUMMING-BIRD, pp. 712-713. Nuttall thought it was a new species but the yellow crown was only pollen, acquired when feeding from flowers. He collected the female and the nest at Santa Barbara, California. Not seen by Townsend as this species does not occur on the Columbia River.

RUFOUS HUMMINGBIRD *Selasphorus rufus* (Gmelin, 1788)
Townsend a: TROCHILUS *rufus*; b: Nootka Humming Bird, *Trochilus rufus*.
Audubon: RUFF-NECKED HUMMING-BIRD. Description from Townsend's specimens; quotes by Nuttall and Townsend, 4: 200-203.
Nuttall: NOOTKA HUMMING-BIRD. First seen in Blue Mountains; nest found by Nuttall; notes on display by Townsend, pp. 714-716.
Notes: Derby list: No. 95 Trochilus rufus, Columbia River, 1 specimen.
Specimens: OUM 4978, female, Columbia River, May 1835. USNM 12598, female, Columbia River, no date (in hand register but no specimen traced). USNM 1943, Columbia River, 29th May 1835. USNM 12598, female, Columbia River (in hand register but no specimen traced).

BELTED KINGFISHER *Ceryle alcyon* (Linnaeus, 1758)
Townsend a: ALCEDO *alcyon*; b: American Kingfisher, *Alcedo alcyon*.
Audubon: BELTED KINGFISHER. Found by Townsend "on the Missouri, the Rocky Mountains and the Columbia river", 4: 207.

LEWIS'S WOODPECKER *Melanerpes lewis* (G.R. Gray, 1849)
Townsend a: PICUS *torquatus*; b: Lewis Woodpecker, *Picus torquatus*.
Audubon: LEWIS' WOODPECKER. Quotes by Nuttall and Townsend, 4: 280-282. Specimens for Plate CCCCXVI probably from Townsend.
Nuttall: LEWIS'S WOODPECKER. First seen near Bear River; also on the Willamette in flocks, p. 679-680.
Specimens: ANSP 19106, juv female, Columbia River, no date. ANSP 24204, Columbia River, 16th June 1836. ANSP may have skull No. 480 Picus torquatus, Columbia river (Morton 1840, p. 44.). OUM 3523, male, Columbia River, 22nd Sept 1834. USNM1909, female, Columbia River, 22nd Sept 1834 (on hand register but no specimen traced). USNM 2795, female, Rocky Mts, no date, (in card catlogue but no specimen traced).

? ACORN WOODPECKER *Melanerpes formicivorus* (Swainson, 1832)
Townsend a: PICUS *pyrrhonotus*, (Aud.); b: Black, Red-Backed Woodpecker, *(not in the collection.)*
Audubon: RED-BACKED WOODPECKER. Quote by Townsend but no specimen collected, Royal octavo edition (1839) 5: 335-336.
Notes: From Townsend's description appears to be Acorn Woodpecker or White-headed Woodpecker *Picoides albolarvatus*; both are right size and colours, but red is on the back of the head not the back. Seen at Fort Vancouver, which suggests that Acorn Woodpecker is the more likely species.

RED-BELLIED WOODPECKER *Melanerpes carolinus* (Linnaeus, 1758)
Townsend a: not listed; b: not listed.
Seen between St Louis and Independence (MS Journal, 30th March 1834).

RED-BREASTED SAPSUCKER *Sphyrapicus ruber* (Gmelin, 1788) and **RED-NAPED SAPSUCKER** *S. nuchalis* Baird, 1858
Townsend a: PICUS *ruber*; b: Red-breasted Woodpecker, *Picus ruber*.
Audubon: RED-BREASTED WOODPECKER. Quote by Nuttall; several specimens procured by Townsend on Columbia River, 4: 261-268. Plate CCCCXVI probably from these specimens.
Nuttall: RED-BREASTED WOODPECKER. First seen "in the forests of the Oregon and the Blue Mountains [latter probably includes the similar Red-naped Sapsucker]", pp. 681-682.
Notes: Derby list: No. 60 Picus redbreasted, Columbia River, 1 specimen.
Specimens: ANSP 19184, Columbia River [= unlabelled skin in Harris collection probably from Townsend (Street 1948, p. 182)]. ANSP 24216, unsexed, Columbia River, no date. USNM 1938, female, Columbia River, 8th July 1835. USNM 2796, male, Columbia River, 28th Oct 1835 (in hand register but no specimen traced).

DOWNY WOODPECKER *Picoides pubescens* (Linnaeus, 1766)
Townsend a: PICUS *Gairdnerii*, (Aud.); b: Downy Woodpecker, *Picus pubescens*.
Audubon: GAIRDNER'S WOODPECKER. Not figured. Specimens from the Columbia River, source not given, 4: 252-254.
Notes: Gambel donated specimen of Gairdner's Woodpecker to ANSP on 2nd March 1841 (Graustein, p. 360 – ANSP 19254). One specimen noted to be in USNM (Hall 1937, p. 12) but is not on current database.
Specimens: ANSP 19254, male, Columbia River, no date, *P.p.gairdnerii*. OUM 3658, male, Columbia River, October 1834, *P.p.gairdnerii*. OUM 3659, female, Columbia River, 7th November [?1834] *P.p.gairdnerii*.

HAIRY WOODPECKER *Picoides villosus* (Linnaeus, 1766)
Townsend a: PICUS *Harrisii*, (Aud.); b: Harris' Woodpecker, *Picus Harrisi*, (AUDUBON.)
"… abundant in the forests on the Columbia river", p. 347-348.
Audubon: HARRIS' WOODPECKER. Discovered by Townsend who sent Audubon
a pair: male 18th January 1836, female 7th September 1834, 4: 242-244. Townsend
specimens used for Plate CCCCXVII.
Nuttall: HARRIS'S WOODPECKER. "This species was observed and collected in the forests
of the Oregon by Mr. Townsend", p. 687.
Specimens: ANSP 24246, female, Columbia River, no date, *P.v. harrisi*. ANSP 19230,
male, Columbia River, no date, *P. v. harrisi*. Both probably used by Audubon (Stone
1899). USNM 1869, female, Columbia River, 27th Sept 1834.

NORTHERN FLICKER *Colaptes auratus* (Linnaeus, 1758)
Townsend a: PICUS *mexicanus*; b: Red-shafted Woodpecker, *Picus mexicanus*.
Audubon: RED-SHAFTED WOODPECKER. Female from Townsend from
Columbia River, 1st April 1836, 4: 290-292.
Nuttall: RED-SHAFTED FLICKER. First seen near Laramie's Fork, and westwards to the
Pacific, p. 667.
Notes: Derby list: [No number] Colaptes Mexicanus, Columbia River,1 specimen.
Specimens: ANSP 24190, Columbia River, no date, *C.a.cafer*. ANSP 18821, female, no
locality or date, *C. a.collaris*. ANSP may have skull No. 473 Coloptes [sic] mexicanus,
Columbia river (Morton 1840, p. 43). OUM 3299, no locality, 19th June 1835,
C.a.collaris. USNM 1886, unsexed, Columbia River, 18th Oct 1834 (in hand register but
no specimen traced).

[LINEATED WOODPECKER *Dryocopus lineatus* (Linnaeus, 1766)]
Townsend a: PICUS *lineatus*; b: not listed.
Audubon: LINEATED WOODPECKER. Not figured. Audubon saw specimen from
Dr Gairdner at Edinburgh, 4: 233.
Nuttall: Nuttall noted that "a single specimen of [the Lineated Woodpecker] was sent
from Oregon by the late Dr. Gairdner, to Edinburgh".
Notes: Error for Pileated Woodpecker. Lineated Woodpecker occurs from Mexico to
central South America.

PILEATED WOODPECKER *Dryocopus pileatus* (Linnaeus, 1758)
Townsend a: PICUS *pileatus*; b: Pileated Woodpecker, *Picus pileatus*.
Audubon: PILEATED WOODPECKER. Specimen received from Townsend, 4: 230.
Notes: Derby list: No. 97 Picus pileatus, Columbia River, 1 specimen. Seen between St
Louis and Independence (MS Journal, 30th March 1834).
Specimens: USNM 2792, Colorado [= Columbia] River, 9th May 1835, *D.p.picinus*.

[IMPERIAL WOODPECKER *Campephilus imperialis* (Gould, 1832)]
Townsend a: PICUS *imperialis*; b: Black, White-Banded Woodpecker, *(not in the collection.)*
Audubon: IMPERIAL WOODPECKER. Not figured. Supposedly seen by Townsend
on 14th August 1834, 4: 212-213.
Nuttall: IMPERIAL WOODPECKER. "My friend Townsend … shot [but did not retrieve] a
bird of this species in the Rocky Mountains," pp. 667-668.
Notes: Imperial Woodpecker is a Mexican species – error for Pileated Woodpecker (see
p. 110).

OLIVE-SIDED FLYCATCHER *Cantopus cooperi* (Nuttall, 1832)
Townsend a: MUSCICAPA *Cooperii*; b: Cooper's Flycatcher, *Muscicapa Cooperi*.

WESTERN WOOD-PEWEE *Cantopus sordidulus* Sclater 1859
Townsend a: MUSCICAPA *virens* and MUSCICAPA *Richardsonii*, (Sw.); b: Wood Pewee
Flycatcher, *Muscicapa virens*.
Nuttall: WOOD PEWEE *Tyrannus virens*. "Mr. Townsend and myself frequently saw them in
the dark forests of Oregon", p. 317.
Notes: Eastern Wood Pewee *C. virens* could have been seen in early part of trip, in
Missouri and eastern Nebraska.
Specimens: USNM 2962, egg of *C. s. saturatus*; listed in S.F. Baird's original catalogue as
"Tyrranula virens, Columbia River" from the Trudeau collection (S. Olson *pers comm*),
possibly originating from Townsend or Nuttall.

ACADIAN FLYCATCHER *Empidonax virescens* (Vieillot, 1818)
Townsend a: MUSCICAPA *acadica*; S b: mall Green-crested Flycatcher, *Muscicapa
acadica*.
Nuttall: SMALL PEWEE. Collected in the Oregon Territory by Townsend, p. 321.
Notes: Acadian Flycatcher is an eastern species, perhaps seen in Missouri. The bird
collected in Oregon would have been Pacific-slope Flycatcher *E. difficilis* or Cordillerean
Flycatcher *E. occidentalis*.

WILLOW FLYCATCHER *Empidonax traillii* (Audubon, 1828)
Townsend a: MUSCICAPA *Traillii*; b: Traill's Flycatcher, *Muscicapa Trailli*.
Audubon: TRAILL'S FLYCATCHER. "Many specimens" procured by Townsend on
Columbia River; Audubon gave one to Bonaparte and one to Derby, 1: 234-235.
Nuttall: TRAILL'S PEWEE. Seen by Nuttall and Townsend on edge of forests of
Willamette and Columbia Rivers, p. 324.
Notes: Specimen from Audubon to William Swainson (Herrick 1938, 2: 176) was
possibly a Townsend specimen.
Specimens: LM D4473a, a Derby specimen without data is possibly from Townsend.

LEAST FLYCATCHER *Empidonax minimus* (Baird and Baird, 1843)
Townsend a: MUSCICAPA *pusilla*, (Sw.); b: Little Flycatcher, *Muscicapa pusilla*,
(SWAINSON.)
Audubon: LEAST PEWEE FLYCATCHER. Specimen and notes from Nuttall, 1: 236-
238.
Nuttall: LITTLE PEWEE. Seen by Nuttall on Sauvie's Island, Oregon, pp. 324-325.
Notes: Least Flycatcher does not breed on the lower Columbia River; more likely to
have been a more western *Empidonax* species, perhaps *E. hammondii*.

PACIFIC-SLOPE FLYCATCHER *Empidonax difficilis* Baird, 1858
Townsend a: not listed; b: not listed.
Specimens: ANSP 35440, unsexed, Columbia River, no date; marked "B. Hoopes" with
note "J.K. Townsend and J.G. Bell" on back of label. The specimen is from the Hoopes
and Turnbull collection, donated by E.D. Cope [Specimen collected by Townsend *or*
Bell but not both]. OUM 14179, male, Columbia River, June 1835; identified at Oxford
as Yellow-bellied Flycatcher *E. flaviventris* but must be Pacific-slope Flycatcher *E. difficilis*
because Townsend on lower Columbia River on that date.

Flycatchers *Empidonax sp*
Notes: Derby list: No. 101 Muscicapa, Columbia River,1 specimen; No. 105 Muscicapa,
Columbia River,1 specimen. Tolmie list: "Muscicapa grisea?" [= Gray Flycatcher
Empidonax griseus or White-eyed Vireo *Vireo griseus*].
Specimens: MCZ 17286, male, Columbia River, June 1836 (ex Brown University);
identified only as *Empidonax sp*.

BLACK PHOEBE *Sayornis nigricans* (Swainson, 1827)
Townsend a: MUSCICAPA *nigricans*, (Sw.); b: not listed.
Audubon: ROCKY-MOUNTAIN FLYCATCHER. Specimens from Nuttall from
California (used in Plate CCCCXXXIV), 1: 218.

EASTERN PHOEBE *Sayornis phoebe* (Latham, 1790)
Townsend a: MUSCICAPA *fusca*; b: Pewit Flycatcher, *Muscicapa fusca*.
Notes: Only a rare vagrant to the west coast states. Included in Townsend's list probably
because of sightings in Missouri, Kansas and Nebraska. Confusion with Black Phoebe
has been suggested but this too is a vagrant on the Columbia (see Jobanek 1994).

SAY'S PHOEBE *Sayornis saya* (Bonaparte, 1825)
Townsend a: MUSCICAPA *Saya*; b: Say's Flycatcher, *Muscicapa Saya*.
Audubon: SAY'S FLYCATCHER. First seen Laramie Mountains about 14th June
1834; only Nuttall's observations are quoted, 1: 217-218. Plate CCCLIX.
Nuttall: SAY'S FLYCATCHER. First seen about 14th June 1834 in Laramie Mountains,
Wyoming, p. 311.
Notes: McEuen list: No. 31 Say's flycatcher. Audubon's original watercolour (NYHS
1863.17.359) is inscribed with date of 16th June.

WESTERN KINGBIRD *Tyrannus verticalis* Say, 1823
Townsend a: MUSCICAPA *verticalis*; b: Arkansas Flycatcher, *Muscicapa verticalis*.
Audubon: ARKANSAW FLYCATCHER. Some "remarkably fine skins" received from
Townsend; quotes by Nuttall and Townsend, 1: 199-200.
Nuttall: ARKANSA FLYCATCHER. First seen on the Platte River, and westwards as far as the
Columbia and Willamette Rivers, p. 306-307.

EASTERN KINGBIRD *Tyrannus tyrannus* (Linnaeus, 1758)
Townsend a: MUSCICAPA *tyrannus*; b: Tyrant Flycatcher, or King Bird, *Muscicapa
tyrannus*.

LOGGERHEAD SHRIKE *Lanius ludovicianus* Linnaeus, 1766
Townsend a: LANIUS *ludovicianus*; b: Loggerhead Shrike, *Lanius ludovicianus*.
Audubon: LOGGERHEAD SHRIKE. Specimens of both sexes received from
Townsend from Rocky Mountains and Columbia River, 4: 136.
Nuttall: LOGGER-HEAD SHRIKE. "Mr. Townsend found it in the Rocky Mountain range
and in the territory of Oregon," p. 289.
Notes: McEuen's list: No. 32 Loggerhead [Shrike].
Specimens: ANSP 23902, no date, Columbia River, *L .l .gambeli*.

NORTHERN SHRIKE *Lanius excubitor* Linnaeus, 1758
Townsend a: LANIUS *borealis*; b: Great American Shrike, *Lanius septentrionalis*.

WHITE-EYED VIREO *Vireo griseus* (Boddaert, 1783)
Townsend a: VIREO *noveboracensis*; b: White-eyed Flycatcher, or Vireo, *Vireo
noveboracensis*.
Audubon: WHITE-EYED VIREO, OR GREENLET. Townsend saw the species on
the Columbia River, 4: 146-149.
Nuttall: WHITE-EYED VIREO, OR FLYCATCHER. "Found, in Oregon, by my friend Mr.
Townsend", p. 347.
Notes: This is an eastern species. Although seen near Independence, Missouri, (MS
Journal, April 1834) it is only a vagrant on the Columbia. White iris of adult is unique,
but possible confusion of non-adults with Cassin's Vireo, see below.

YELLOW-THROATED VIREO *Vireo flavifrons* (Vieillot, 1808)
Townsend a: not listed; b: not listed.
Seen near Independence (MS Journal, April 1834).

BLUE-HEADED VIREO *Vireo solitarius* (Wilson, 1810) *and* **CASSIN'S VIREO** *Vireo cassinii* Xantus, 1858
Townsend a: VIREO *solitarius*; b: Solitary Flycatcher, or Vireo, *Vireo solitarius*.
Audubon: SOLITARY VIREO, OR GREENLET. Several specimens received from Townsend. Nuttall says seen in May on the Columbia River, 4: 145.
Nuttall: SOLITARY VIREO, OR FLYCATCHER. Found breeding in the oak forests on the Columbia River, p. 346.
Notes: Solitary Vireo now split into above two species, Cassin's Vireo occurring Washington, Oregon and Idaho. Derby list: No. 92 Vireo solitarius, Columbia River, 2 specimens.
Specimens: OUM 12721, male, Columbia River, May 1835, *V.s.cassinii* [= Cassin's Vireo].

WARBLING VIREO *Vireo gilvus* (Vieillot, 1808)
Townsend a: *Vireo gilvus*; b: Warbling Flycatcher, or Vireo, *Vireo gilvus*.

RED-EYED VIREO *Vireo olivaceus* (Linnaeus, 1766)
Townsend a: not listed; b: not listed.
Seen near Independence (MS Journal, April 1834).

GRAY JAY *Perisoreus canadensis* (Linnaeus, 1766)
Townsend a: GARRULUS *canadensis*; b: Canada Jay, *Garrulus canadensis*.
Audubon: CANADA JAY. Specimens and notes from the Columbia River from Townsend, 4: 121-126. Plate CCCCXIX.
Nuttall: CANADA JAY. Nuttall saw family party on Willamette River, Oregon, p. 248.
Notes: "Oregon Jay", male, Columbia River, 11th Oct 1836, in Harris collection (Street 1948, p. 182.)
Specimens: ANSP 162471, Columbia River [= Fort George, Astoria, Oregon],11th Oct 1836, *P.c.obscurus*; donated by J.F. and P.B. Street 1948.

STELLER'S JAY *Cyanocitta stelleri* (Gmelin, 1788)
Townsend a: GARRULUS *Stelleri*; b: Steller's Jay, *Garrulus Stelleri*.
Audubon: STELLER'S JAY. Nuttall and Townsend quoted; 2 eggs received from Nuttall, 4: 107-109.
Nuttall: STELLER'S JAY. First seen "in the Blue Mountains of the Oregon, east of Walla-Walla", pp. 243-245.
Notes: Derby list: No. 48 Garrulus Stelleri, 3 specimens.
Specimens: ANSP may have skull No. 482 Garrulus stelleri, Columbia river (Morton 1840, p. 44.). There must have been, or still may be, specimens at ANSP; Stone (1891, p. 444) mentions skins at ANSP with no locality or collector information.

WESTERN SCRUB-JAY *Aphelocoma californica* (Vigors, 1839)
Townsend a: GARRULUS *ultramarinus*; b: Ultramarine Jay, *Garrulus ultramarinus*.
Audubon: ULTRAMARINE JAY. Seen in California and at Fort Vancouver, quote from Nuttall, specimens probably from Townsend, 4: 115-117.
Nuttall: ULTRAMARINE JAY. Seen near Fort Vancouver, p. 245.
Notes: A misidentification – not Gray-breasted Jay *Aphelocoma ultramarina* that occurs Arizona to central Mexico. Tolmie list: "Garrulus Ultramarinus ?" 1 specimen.
Specimens: Stone (1891, p. 444) mentions one *Aphelocoma californica* "collected by J.K. Townsend, Columbia river" at ANSP but not found there in 2003. ANSP may have skull No. 477 Garrulus ultramarinus, Columbia river (Morton 1840, p. 43.) USNM 2841, male, Columbia River, 11th Oct 1834 (in hand register but no specimen traced).

CLARK'S NUTCRACKER *Nucifraga columbiana* (Wilson, 1811)
Townsend a: CORVUS *(Nucifraga) columbiana*; b: Clark's Crow, *Corvus columbianus*.
Audubon: CLARKE'S NUTCRACKER. First seen by Townsend and Nuttall near Bear River, July 1834; quotes from Nuttall and Townsend, 4: 127-129.
Nuttall: AMERICAN NUT-CRACKER. First seen in pine grove near Bear River and afterwards in Blue Mountains, pp. 251-253.
Notes: McEuen's list: No 33 Clarke's crow, 2 specimens.
Specimens: ANSP may have skull No. 525 Corvus Columbianus Clarke's Crow (Morton 1840, p. 45.) USNM 1929, male, Rocky Mts, 16th June 1834 (in hand register but no specimen traced).

AMERICAN MAGPIE *Pica hudsonia* (Sabine, 1823)
Townsend a: CORVUS *pica*; b: Magpie, *Corvus pica*.
Audubon: COMMON MAGPIE. First seen by Townsend and Nuttall near Snake River; quote by Nuttall, 4: 99-106.
Nuttall: MAGPIE. First seen by Nuttall 15th July 1834 near Fort Hall, pp. 231-233.
Notes: Stone (1892, p. 442) says "several specimens from the Columbia river by J.K. Townsend" at ANSP; one specimen is dated 29th May 1836 (Jobanek 1994: 13). Catalogue entry at ANSP lists a Townsend skin via Harris (Street 1948, p. 184).
Specimens: ANSP 2932, male, Columbia River, no date; donated by J.F. and P.B. Street in 1948. ANSP 3304, unsexed, Columbia River, [29th May 1836?].

YELLOW-BILLED MAGPIE *Pica nuttalli* (Audubon, 1837)
Townsend a: not listed; b: not listed.
Audubon: YELLOW-BILLED MAGPIE. Nuttall provided a specimen from Santa Barbara, California (used for Plate CCCLXII), 4: 104.
Nuttall: YELLOW-BILLED MAGPIE. Nuttall saw them at Santa Barbara, California, p. 236.
Notes: In 1841 Nuttall donated a specimen, through Townsend, to ANSP (Graustein, p. 360). Stone (1892, p. 442) says specimens at ANSP were from California and Columbia River, the latter from the Townsend collection. All these birds were presumably collected in California by Nuttall (unless brought back from there in 1845 by William Gambel or another collector); they were certainly not collected in the field by Townsend because the Yellow-billed Magpie is endemic to California, not visited by Townsend.
Specimens: ANSP 3303, unsexed, Columbia River [= California], no date. ANSP 3306, unsexed, Columbia River [= California], no date. USNM 2845, unsexed, Santa Barbara, California, April 1836, collected by Nuttall; type.

AMERICAN CROW *Corvus brachyrhynchos* C.L. Brehm, 1822
Townsend a: CORVUS *americanus*; b: Common Crow, *Corvus carone*.
Notes: see also Northwestern Crow, below.

? NORTHWESTERN CROW *Corvus caurinus* Baird, 1858
Townsend a: CORVUS *ossifragus*; b: Fish Crow, *Corvus ossifragus*.
Audubon: FISH CROW. Townsend reported Fish Crow "plentiful on the Columbia River", and brought back specimens, 4: 98.
Nuttall: FISH CROW. "common on the banks of the Oregon, where they were nesting in the month of April [1835], p. 300.
Notes: The Fish Crow *Corvus ossifragus* does not occur on the west coast. The birds seen were either Northwestern Crow (not described until 1858, by S.F. Baird, from specimens from Fort Steilacoom, Washington) or more likely, as stated by Jobanek (1994), the small Oregon race of American Crow *C. b. hesperis*. It is possible that both species were seen at the mouth of the Columbia and in Willapa Bay; the two west coast crows can spend a lot of time scavenging on the shore though Northwestern Crow does not normally come as far south as the Columbia River.
Specimens: None traced. Stone (1891, p. 446) does not refer specifically to any Townsend specimens.

COMMON RAVEN *Corvus corax* Linnaeus, 1758
Townsend a: CORVUS *corax*; b: Raven, *Corvus corax*. Reported from around Fort Hall and other locations.
Audubon: RAVEN. Townsend reported it common on salmon fisheries on the Columbia River, 4: 85.
Notes: Reported from Platte River westward (MS Journal). Stone (1891, p. 445) does not refer to any ANSP specimens from Townsend.

HORNED LARK *Eremophila alpestris* (Linnaeus, 1758)
Townsend a: ALAUDA *cornuta* (Wils.); b: Shore Lark, *Alanda* [sic] *cornuta*, (WILSON.)
Audubon: SHORE LARK. Nuttall found nest on 10th June 1834, 3: 48.
Nuttall: SHORE LARK. Nest found near Sweetwater River, 10th June 1834; and seen later near Bear River, pp. 523-525.

PURPLE MARTIN *Progne subis* (Linnaeus, 1758)
Townsend a: HIRUNDO *purpurea*; b: Marten [sic], *Hirundo purpurea*.
Nuttall: PURPLE MARTIN. "Mr. Townsend met with it on the Rocky Mountains", p. 723.

TREE SWALLOW *Tachycineta bicolor* (Vieillot, 1808)
Townsend a: HIRUNDO *bicolor*; b: White-bellied Swallow, *Hirundo bicolour*.

VIOLET-GREEN SWALLOW *Tachycineta thalassina* (Swainson, 1831)
Townsend a: HIRUNDO *thalassina*; b: Violet-green Swallow, *Hirundo thalassina*.
Audubon: VIOLET-GREEN SWALLOW. First seen Ham's Fork and later on Columbia River; quotes by Nuttall and Townsend, 1: 186. Plate CCCLXXXV.
Nuttall: VIOLET-GREEN SWALLOW. First observed Ham's Fork and later on Columbia River; quote by Townsend on nest sites, p. 725.
Specimens: AMNH 40224, female, Columbia River, 14th July 1838 [=1835?]; has label in G.N.Lawrence's handwriting "Presented by J.W. Audubon from the collection of J.J. Audubon". USNM 1895, male, Columbia River 12th July 1835 (in hand register but no specimen traced). USNM 1945, female, Columbia River, 12th July 1835 (in hand register but no specimen traced).

BANK SWALLOW *Riparia riparia* (Linnaeus, 1758)
Townsend a: HIRUNDO *riparia*; b: Bank Swallow, *Hirundo riparia*
Notes: Probably includes Northern Rough-winged Swallow *Stelgidopteryx serripennis*.

CLIFF SWALLOW *Petrochelidon pyrrhonota* Vieillot, 1817
Townsend a: HIRUNDO *fulva*; b: Cliff, or Republican Swallow, *Hirundo fulva*.
Nuttall: CLIFF SWALLOW. Nest sites on Pillar Rock near Chinook, mouth of the Columbia, pp. 729-731.
Notes: Derby list: No. 120 Hirundo lunifrons, Columbia River, 1 specimen. Tolmie list: "Hirundo – new species" 1 specimen. Seen on Platte River (MS Journal, 17th May 1834). Nests seen on Columbia River (MS Journal, 26th June 1836).

BARN SWALLOW *Hirundo rustica* Linnaeus, 1758
Townsend a: HIRUNDO *rufa*; b: Barn Swallow, *Hirundo rufa*.

BLACK-CAPPED CHICKADEE *Poecile atricapilla* Linnaeus, 1758
Townsend a: PARUS *atricapillus*; b: Black-capt Titmouse, *Parus atricapillus*.
Nuttall: Nuttall claimed, in error, that he and Townsend had seen mixed flocks of Carolina Chickadees *Poecile carolinensis* and Chestnut-backed Chickadees. Carolina Chickadee is an eastern species.

CHESTNUT-BACKED CHICKADEE *Poecile rufescens* Townsend, 1837
Townsend a: PARUS *rufescens*, (Towns.) Original description, p. 190; b: Chestnut-backed titmouse, *Parus rufescens*, (TOWNSEND.) "Inhabits the Columbia river; common, gregarious", p. 337.
Audubon: CHESTNUT-BACKED TITMOUSE. Nuttall and Townsend quoted, 2: 158-159. Source of specimens for Plate CCCLIII not given but known to be from Townsend.
Nuttall: CHESTNUT-BACKED TITMOUSE. Seen throughout the year in forests of the Columbia, pp. 267-269.
Specimens: ANSP 23665, female, Columbia River, no date, *Pt. rufescens*; type specimen according to Stone (1899). USNM 1924, Forests of Columbia River, no date. USNM 2931, Forests of Columbia River, no date. OUM 11571, female, Columbia River, 12th May 1835.

BUSH-TIT *Psaltriparus minimus* (Townsend, 1837)
Townsend a: PARUS *minimus*, (Towns.); b: Brown-headed Titmouse, *Parus minimus*, (TOWNSEND.) "I first observed this little species on the Columbia river in May, 1835, and procured a pair", p. 337.
Audubon: CHESTNUT-CROWNED TITMOUSE. Nuttall supplied nest and a quote, 2: 160-161. Source of specimens for Plate CCCLIII not given but known to be from Townsend.
Nuttall: CHESTNUT CROWNED TITMOUSE. First seen on or near Sauvie's Island and at Fort Vancouver, pp. 269-270.
Notes: Original description in *Journal ANSP* 1837 p. 190. Townsend text in his appendix p. 337. McEuen's list: No. 96 Parus minimus, 2 specimens.
Specimens: Type now lost? Stone (1899) says no specimens at ANSP or USNM; still missing 2003.

RED-BREASTED NUTHATCH *Sitta canadensis* Linnaeus, 1766
Townsend a: SITTA *canadensis*; b: Red-bellied Nuthatch, *Sitta canadensis*.
Audubon: RED-BELLIED NUTHATCH. Found by Townsend on the Columbia, 4: 179-181.
Nuttall: RED-BELLIED NUTHATCH. Mid-October 1834 found "in the woods of the Oregon" near the Willamette River, p. 698.
Notes: Derby list: No. 57 [or No. 97] Sitta Canadensis, Columbia River, 2 specimens.
Specimens: ANSP 9154, male, no further data.

WHITE-BREASTED NUTHATCH *Sitta carolinensis* Latham, 1790
Townsend a: SITTA *carolinensis*; b: White-breasted Nuthatch, *Sitta carolinensis*.
Audubon: WHITE-BREASTED NUTHATCH. Townsend brought back specimens from the Columbia River, 4: 175-178.

BROWN CREEPER *Certhia americana* Bonaparte, 1838
Townsend a: CERTHIA *familiaris*; b: Common Creeper, *Certhia familiaris*.
Notes: Derby list: No. 116 Sylvia familiaris, Columbia River, 1 specimen [= Brown Creeper?].

ROCK WREN *Salpinctes obsoletus* (Say, 1823)
Townsend a: MYIOTHERA *obsoleta*; b: Rocky-mountain Wren, *Troglodytes obsoleta*.
Audubon: ROCK WREN. Quote by Nuttall; adult female collected on 21st June 1834 by Nuttall at Ham's Fork, 2: 113-115. See Plate CCCLX.
Nuttall: ROCK WREN. Family party seen Ham's Fork; and seen westwards almost to Fort Vancouver, pp. 490-492.

CAROLINA WREN *Thyrothorus ludovicianus* (Latham, 1790)
Townsend a: not listed; b: not listed.
Audubon: GREAT CAROLINA WREN. Seen by Townsend on the Missouri, 2: 118.

BEWICK'S WREN *Thyromanes bewickii* (Audubon, 1827)
Townsend a: TROGLODYTES *Bewickii*; b: Bewick's Wren, *Troglodytes Bewicki*.
Audubon: BEWICK'S WREN. Found by Nuttall and Townsend on the Columbia River; examined their specimens, 2: 120-122.
Nuttall: BEWICK'S WREN. Frequently seen marshy meadows of the Willamette River, pp. 489-490. SPOTTED GREY WREN. *Troglodytes maculosa*, NOBIS. [= grayish form of Bewick's Wren]. Seen 4th July 1835, Point Chinook, p. 492.
Specimens: ANSP may have skull No. 595 Troglodytes Bewickii, Columbia river (Morton 1840, p.45).

HOUSE WREN *Troglodytes aedon* (Vieillot, 1808)
Townsend a: TROGLODYTES *Parkmanii*, (Aud.); b: Common Wren, *Troglodytes aedon*.
Audubon: PARKMAN'S WREN. Collected by Townsend on the Columbia River, 2: 127, 133-135 and plate 122. Not figured in original folio.
Nuttall: PARKMAN'S WREN. Specimens obtained "in the forests of the Oregon" by Townsend, p. 483.
Notes: McEuen's list: No. 37 Wren (Q Parkmani).
Specimens: ANSP 13618, unsexed, no date, Columbia River, *T.a.parkmanii*. USNM 66644; Stone (1899) says this is a Townsend specimen, the type for *T. parkmani*. MCZ 140279, *T.a.parkmanii*, is the Parkman's Wren from Townsend that Audubon donated to Parkman; it was later in the J.E. Thayer collection (Thayer 1916) that went to Harvard University in the 1930s.

WINTER WREN *Troglodytes troglodytes* (Linnaeus, 1758)
Townsend a: TROGLODYTES *fulvus* [= Winter Wren?]; b: Winter Wren, *Troglodytes hyemalis*.
Audubon: WINTER WREN. Seen by Nuttall on Columbia River, 2: 129.
Nuttall: WINTER WREN. "Mr. Townsend obtained specimens … forests of the Columbia", p. 482.

MARSH WREN *Cistothorus palustris* (Wilson, 1810)
Townsend a: not listed; b: not listed.
Nuttall: MARSH-WREN. Nuttall saw this species mid-July 1834 in a marsh near Fort Hall, pp. 496-497.

AMERICAN DIPPER *Cinclus mexicanus* Swainson, 1827
Townsend a: CINCLUS *americanus*, (Sw.); b: Morton's Water Ouzel, *Cinclus Mortoni*, (TOWNSEND.) "It was shot by Captain W. Brotchie, near Fort McLoughlin", p. 339. Townsend shot one near Fort Vancouver, *Narrative* Appendix p. 339-340 and Columbian Water Ouzel, *Cinclus Townsendi*, (AUDUBON.) "… inhabits the swiftly running streams of fresh water in the vicinity of Fort Vancouver", p. 340.
Audubon: AMERICAN DIPPER. Quote by Townsend, 2: 182-186. Royal octavo edition (1839) 5: 303-304. Plate CCCCXXXV.
Nuttall: AMERICAN WATER-OUSEL, OR DIPPER. Inhabits clear mountain streams near Fort Vancouver. "Mr. Townsend obtained a specimen at [= from] Fort McLaughlin", pp. 407-409.
Notes: Unlabelled skin in Harris collection probably from Townsend (Street 1948, p. 182) – specimen untraced.
Specimens: USNM 2861, N.W. coast, Oregon, no date; type of *Cinclus Townsendi*. USNM 2862, Oregon, no date; type of *C.Mortoni*.

GOLDEN-CROWNED KINGLET *Regulus satrapa* Lichtenstein, 1823
Townsend a: REGULUS *cristatus*; b: Golden-crested Wren, *Regulus cristatus*.

RUBY-CROWNED KINGLET *Regulus calendula* (Linnaeus, 1766)
Townsend a: REGULUS *calendula*; b: Ruby-crowned Wren, *Regulus calendula*.
Specimens: ANSP 16848, male, Oct (no year), Columbia River.

EASTERN BLUEBIRD *Siala sialis* (Linnaeus, 1758)
Townsend a: not listed; b: not listed.
Audubon: COMMON BLUE-BIRD. Does not extend to the west of the Rocky Mountains, though "it was observed by Dr TOWNSEND on the headwaters of the Missouri," 2: 173. Elsewhere this species was listed for the Columbia River (Jobanek 1994).
Notes: This is an eastern species that does not normally occur in the Pacific Northwest – origin doubtful (see appendix 4).
Specimens: LM D1455(S), 21st May 1836 [museum labelling error = 1834, in Missouri?].

WESTERN BLUEBIRD *Siala mexicana* Swainson 1832
Townsend a: SIALIA *occidentalis*, (Towns.) Original description, p. 188; b: Western Blue Bird, *Sialia occidentalis*, (TOWNSEND.) "Common on the Columbia river in the spring", p. 344.
Audubon: WESTERN BLUE-BIRD. Noted as a discovery of Townsend (though first specimens were from Mexico, described by Swainson); quote by Nuttall, 2: 176-178. Plate CCCXCIII.
Nuttall: WESTERN BLUE-BIRD. First seen "in the small rocky prairies of the Columbia"; also at Fort Vancouver and Santa Barbara, pp. 513-514.
Notes: Tolmie list: "(Sialia) Saxicola new species".
Specimens: USNM 1930, male?, Columbia River, no date. USNM 1931, male, Columbia River, 3rd July 1835 (in hand register but no specimen traced). USNM 2949, female, Columbia River, no date. LM D1454(S), 21st May 1836. [MCZ 16394, Columbia River, June 1835 – catalogued only as *Sialia sp?*]

MOUNTAIN BLUEBIRD *Siala currucoides* (Bechstein, 1798)
Townsend a: SIALIA *arctica*, (Sw.); b: Arctic Blue Bird, *Sialia arctica*.
Audubon: ARCTIC BLUE-BIRD. Townsend's observations quoted in error (1839a: 343-344) under Western Bluebird, 2: 178-179.
Nuttall: ARCTIC BLUE-BIRD. First seen near Fort Laramie, nest found Sandy River, and seen in winter at Fort Vancouver, pp. 514-515.
Notes: Tolmie list: "Fringilla Arctica".
Specimens: ANSP 16481, no date, Rocky Mountains. ANSP may have skull No. 481 Fringilla arctica, Columbia river (Morton 1840, p. 44). AMNH 39327, female, Columbia River, 26th June 1835 – has label in unknown hand "Presented by V.G. Audubon from J.J. Audubon's collection".

TOWNSEND'S SOLITAIRE *Myadestes townsendi* (Audubon, 1838)
Townsend a: PTILIOGONYS *Townsendii*, (Aud.) and *Turdus Townsendii*, (Aud.); b:
Townsend's Thrush, *Ptiliogonys Townsendi*, (AUDUBON.) "shot by my friend Captain W.
Brotchie, of the Honorable Hudson's Bay Company, in a pine forest near Fort George,
(Astoria.)" p. 338-339. "White tailed Thrush, *(not in the collection.)*" is also listed.
Audubon: TOWNSEND'S PTILIOGONYS. One specimen received from Townsend
(included in Plate CCCCXIX), 1: 243.
Nuttall: TOWNSEND'S PTILOGONYS. "The only individual known was shot by Captain
Brotcher [sic] in the Hudson's Bay Service in the neighborhood of Ft. George (Astoria)
and presented to Mr. Townsend", pp. 361-362.
Notes: Type specimen shot by Capt Brotchie *Narrative* appendix, p. 338-339. Audubon
included White-tailed Thrush in original edition of *Ornithological Biography* (1839) 5: 336.
Specimens: Stone (1899) says USNM 2922 is type but not listed under Townsend
specimens on current USNM database. Type not traced.

VEERY *Cathurus fuscescens* (Stephens, 1817)
Townsend a: TURDUS *Wilsonii*; b: Wilson's Tawny Thrush, *Turdus Wilsonii*.
Nuttall: WILSON'S THRUSH, or VEERY. Heard at Ham's Fork and in thickets of Snake
River, pp. 396-397.
Notes: Derby list: No. 109 Turdus Wilsonii, 1 specimen.

SWAINSON'S THRUSH *Catharus ustulatus* (Nuttall, 1840)
Townsend a: not listed; b: not listed.
Nuttall: WESTERN THRUSH *Turdus cestulatus [= ustulatus]*, NOBIS. Original description from
Townsend specimen shot on banks of Columbia 10th June [1835 or 1836]; nest found
prairies of Willamette, pp. 400-401.
Specimens: ANSP 23644, Washington State, no date; identified as type by Stone (1899).
USNM 2040, Columbia River, no date; also claimed as type.

HERMIT THRUSH *Catharus guttatus* (Pallas, 1811)
Townsend a: TURDUS *minor* and TURDUS *nanus*, (Aud.); b: Hermit Thrush, *Turdus
minor*.
Nuttall: DWARF THRUSH. "obtained in the forests of the Oregon by Mr. Townsend", p.
396.
Specimens: ANSP 16090, female, May [no year], Columbia River, *C. g nanus*. MCZ
16298, Columbia River, no date, "Turdus nanus" probably *C.g nanus*.

WOOD THRUSH *Hylocichla mustelina* (Gmelin, 1789)
Townsend a: not listed; b: not listed.
Notes: An eastern species. Seen near Independence, Missouri (MS Journal, 10th April
1834).

AMERICAN ROBIN *Turdus migratorius* Linnaeus, 1766
Townsend a: TURDUS *migratorius*; b: American Robin, *Turdus migratorius*.
Nuttall: AMERICAN ROBIN. "a denizen of the territory of Oregon", p. 383.

VARIED THRUSH *Ixoreus naevius* (Gmelin, 1789)
Townsend a: TURDUS *naevia*; b: Varying Thrush, *Turdus naevia*.
Audubon: VARIED THRUSH. Quotes by Nuttall and Townsend, 3: 22. Townsend
probably supplied specimens for Plate CCCLXIX.
Nuttall: VARIED THRUSH. Seen in Oregon only in winter, arriving in October, p. 389.
Notes: Derby list: No. 67 Turdus naevia, Columbia River, 2 specimens. Tolmie list:
"Turdus Meruloides". Unlabelled skin in Harris collection probably from Townsend
(Street 1948, p.182.)
Specimens: ANSP 23631, female, Columbia River, donated by J.F. and P.B. Street in
1948. ANSP may have skull No. 524 Turdus meruloides, Columbia river (Morton 1840,
p. 45; Morton 1849, p. 10). LM D1704(S), 25th Jan 1835 [on this date Townsend in
Hawaiian Islands, therefore not 1835 = 1836, Fort Vancouver]. OUM 11091, male,
Columbia River, 1st Jan 1836. USNM 1844, male, Columbia River,18th Jan 1836 (in
hand ledger but no specimen traced).

SAGE THRASHER *Oreoscoptes montanus* (Townsend, 1837)
Townsend a: TURDUS *(orpheus) montanus*, (Towns.); b: Mountain Mocking Bird, *Orpheus
montanus*, (TOWNSEND.) "Inhabits the banks of the Platte river, west [= east] of the
Rocky Mountains", p. 338.
Audubon: MOUNTAIN MOCKING-BIRD. Quote by Nuttall who found a nest, 2:
194-195. Townsend's specimen used for Plate CCCLXIX.
Nuttall: MOUNTAIN MOCKING-BIRD. Seen on arid plains between upper Platte and Green
Rivers, June 1834, p. 371-372.
Notes: Original description *Journal ANSP* (1837) 7:192. McEuen list: No. 28 Orpheus
montanus.
Specimens: ANSP 23728 [42° N 109° 30′ W.], no date; type specimen.

AMERICAN PIPIT *Anthus rubescens* (Tunstall, 1771)
Townsend a: ANTHUS *spinoletta*; b: Brown Lark, *Anthus spinoletta*.

CEDAR WAXWING *Bombycilla cedrorum* Vieillot, 1808
Townsend a: BOMBYCILLA *carolinensis*; b: Cedar Bird, or Cherry Bird, *Bombycilla
carolinensis*.
Audubon: CEDAR WAXWING, OR CEDAR-BIRD. Townsend obtained specimens
on the Columbia River, 4: 169-173.
Notes: Tolmie list: "Bombycilla Caroliniensis" [sic].

ORANGE-CROWNED WARBLER *Vermivora celata* (Say, 1823)
Townsend a: SYLVIA *celata*; b: Orange-crowned Warbler, *Sylvia celata*.
Nuttall: ORANGE-CROWNED VERMIVORA. "in the month of May we saw this species
abundant in the forests of the Oregon", pp. 473-474.
Notes: Derby list: No. 99 Sylvia Celata, Columbia River, 2 specimens.
Specimens: USNM 2929, male, Colorado [= Columbia] River, 15th May 1835. USNM
1912, Columbia River, 16th May 1835.

?NASHVILLE WARBLER *Vermivora ruficapilla* (Wilson, 1811)
Townsend a: not listed; b: Ash-headed Warbler, *Sylvia (not described.)*
Nuttall: NASHVILLE VERMIVORA, OR WARBLER: not mentioned for the far West, pp. 472-
473.
Notes: Nashville Warbler may equate to the "Ash-headed Warbler". Nashville Warbler
is a regular breeder on the lower Columbia River, but was evidently not seen well by
Townsend and not at all by Nuttall. Virginia's Warbler is another possibility; Townsend
passed through south-east Idaho where it breeds.

NORTHERN PARULA *Parula americana* (Linnaeus, 1758)
Notes: Seen near Independence, Missouri (MS Journal, 5th and 25th April 1834).

YELLOW WARBLER *Dendroica petechia* (Linnaeus, 1766)
Townsend a: SYLVIA *aestiva*; Blue-eyed Yellow Warbler, *Sylvia aestiva*.

CHESTNUT-SIDED WARBLER *Dendroica pensylvanica* (Linnaeus, 1766)
Townsend a: not listed; not listed.
Notes: Seen near Independence, Missouri (MS Journal, April 1834).

YELLOW-RUMPED WARBLER *Dendroica coronata* (Linnaeus, 1766)
Townsend a: SYLVIA *Auduboni*, (Towns.); b: Audubon's Warbler, *Sylvia Auduboni*,
(TOWNSEND.) "Very common on the Columbia river in the spring", p. 342-343.
Audubon: AUDUBON'S WOOD-WARBLER. Quotes by Nuttall and Townsend, 2:
26-27. In Plate CCCXCV from Townsend's specimens.
Nuttall: AUDUBON'S WARBLER. Common breeder on the Columbia River, pp. 414-415.
Notes: Original description in *Journal ANSP* (1837) 7: 191. Tolmie list: "Sylvia - new
species" is possibly this warbler.
Specimens: ANSP 23826, male, Columbia River, no date, *D.c. auduboni*; noted as type by
Stone (1899). USNM 2909, male, no locality [= near Fort Vancouver], 31st May 1835.
USNM 2911, female, no locality [= Fort Vancouver], April 1836. Stone (1899) says 3
specimens at USNM.

BLACK-THROATED GRAY WARBLER *Dendroica nigrescens* (Townsend, 1837)
Townsend a: SYLVIA *nigrescens*, (Towns.); b: Black-throated Gray Warbler, Sylvia
nigrescens, (TOWNSEND.) "This species is not uncommon in the forests of oak on the
Columbia river", p. 341.
Audubon: BLACK-THROATED GREY WOOD-WARBLER. Discovered by
Townsend; quote by Nuttall, 2: 62-63. Townsend's specimens used in Plate CCCXCV.
Nuttall: BLACK-THROATED GREY VERMIVORA. First heard 23rd May [1835]; abundant in
the forests of the Columbia, pp. 471-472.
Specimens: USNM 1908, male, Fort William, 16th June 1835; cotype. USNM 2915,
male, Fort William, 14th May 1835; cotype.

TOWNSEND'S WARBLER *Dendroica townsendi* (Townsend, 1837)
Townsend a: SYLVIA *Townsendii*, (Nutt.); b: Townsend's Warbler, *Sylvia Townsendi*,
(NUTTALL.) Only two specimens collected, p. 341-342.
Audubon: TOWNSEND'S WOOD-WARBLER. Nuttall quote, 2: 59. Townsend's
specimens from Columbia River used in Plate CCCXCIII.
Nuttall: TOWNSEND'S WARBLER. Inhabits tall conifers on the Columbia River, Townsend
collected specimen 28th October 1835, pp. 446-447.
Notes: McEuen list: No. 100 Sylvia Townsendi.
Specimens: USNM 2918, male, no locality [= Fort Vancouver], 28th October 1835;
type specimen. Second specimen missing?

HERMIT WARBLER *Dendroica occidentalis* (Townsend, 1837)
Townsend a: SYLVIA *occidentalis*, (Towns.); b: Hermit Warbler, *Sylvia occidentalis*,
(TOWNSEND.) "I shot a single pair of these birds in a pine forest on the Columbia
river, on the 28th of May, 1835", pp. 340-341.
Audubon: HERMIT WOOD-WARBLER. Pair shot Fort Vancouver 28th May 1835
and Nuttall quote, 2: 60. Townsend's specimens from Columbia River used in Plate
CCCXCV.
Nuttall: HERMIT WARBLER. Difficult to see high in trees, Townsend collected two
specimens at Fort Vancouver, pp. 445-446.
Notes: Original description *Journal ANSP* 1837, p. 190-191. McEuen list: No. 111 Sylvia
occidentalis, 2 specimens.
Specimens: ANSP 162369, male, Columbia River, 28th May 1837 [= 1835]; type
specimen. Second specimen missing?

BLACKPOLL WARBLER *Dendroica striata* (J.R. Forster, 1772)
Townsend a: not listed; b: not listed.
Seen near Independence, Missouri (MS Journal, April 1834).

CERULIAN WARBLER *Dendroica cerulea* (Wilson, 1810) and **MAGNOLIA WARBLER** *D. magnolia* (Wilson, 1811)
Townsend a: SYLVIA *azurea*; b: Caerulean Warbler, *Sylvia azurea*
Audubon: COERULIAN WOOD-WARBLER. Audubon says seen by Townsend at Fort Vancouver, 2: 46.
Notes: This is an eastern species. Possible confusion with another *Dendroica* warbler with double wingbars: e.g. Magnolia Warbler *D. magnolia* female or fall Blackpoll Warbler *D. striata*.
Specimens: Magnolia Warbler ANSP 8680, unsexed, Columbia River. This specimen may be the source of this strange record of a far western Cerulian Warbler. However, the locality for this Magnolia Warbler specimen may also be erroneous as this species is only accidental in Columbia River area.

BLACK-AND-WHITE WARBLER *Mniotilta varia* (Linnaeus, 1766)
Townsend a: not listed; b: not listed.
Notes: Seen between St Louis and Independence (MS Journal, 5th and 25th April 1834).

OVENBIRD *Seiurus aurocapillus* (Linnaeus, 1766)
Townsend a: TURDUS *aurocapillus*; b: Golden-crowned Thrush, *Turdus aurocapillus*.
Audubon: GOLDEN-CROWNED WAGTAIL (THRUSH). Audubon says Townsend collected specimens from Columbia River, 3: 35-36.
Notes: Although included in Townsend's list for the Territory of Oregon this would have been because of sightings westwards from Independence to Wyoming; the Ovenbird does not normally occur in Oregon or Washington.

LOUISIANA WATERTHRUSH *Seiurus motacilla* (Vieillot, 1808) and **NORTHERN WATERTHRUSH** *S. noveboracensis* (Gmelin, 1789)
Townsend a: not listed; b: not listed.
Audubon: AQUATIC WOOD-WARBLER (LOUISIANA WATER THRUSH). Townsend says breeds on Missouri River [= Louisiana Waterthrush], and seen on Columbia [= Northern Waterthrush], 3: 37-39.
Nuttall: NEW YORK, OR AQUATIC THRUSH: Seen in Oregon as well as in Missouri, p. 402.
Notes: Not generally recognised as two species in Townsend's day. A waterthrush [Louisiana Waterthrush] noted as breeding beside the Missouri River [MS Journal, 11th April 1834].

MACGILLIVRAY'S WARBLER *Oporornis tolmiei* (Townsend, 1839)
Townsend a: SYLVIA *Tolmoei*, (Towns.); b: Tolmie's Warbler, *Sylvia Tolmiei*, (TOWNSEND.) "common in spring on the Columbia ... I dedicate the species to my friend W.F. Tolmie", p. 343.
Audubon: MACGILLIVRAY'S GROUND-WARBLER. Quotes by Townsend and Nuttall, 2: 74-75. Townsend's specimens from Columbia River used in Plate CCCXCIX under name of Mourning Warbler.
Nuttall: TOLMIE'S GROUND WARBLER. Common in woods and plains of the Columbia, pp. 460-461.
Specimens: ANSP 23765, Columbia River, no date; Stone (1899) considered this specimen as the type. MCZ 35008, Columbia River, 31st May 1835. USNM 1861, male, no locality [= Fort Vancouver], 4th May 1836. USNM 1910, male, no locality [= Fort Vancouver], May 1835. USNM 2907, male, no locality [= Fort Vancouver], May 1835.
Notes: Derby list: [no number] Sylvia Philadelphia, Columbia River, 1 specimen [probably MacGillivray's Warber rather than Mourning Warbler].

COMMON YELLOWTHROAT *Geothlypis trichas* (Linnaeus, 1766)
Townsend a: SYLVIA *trichas*; b: Maryland Yellow-throat, Sylvia *trichas*.
Audubon: MARYLAND YELLOW-THROAT. Townsend supplied specimens from Columbia River but not used in plate, 2: 79.
Nuttall: MARYLAND YELLOW-THROAT. "also exists in the territory of Oregon, where Mr. Townsend obtained specimens", p. 455.

[MASKED YELLOWTHROAT *Geothlypis acquinoctialis* (Gmelin, 1789)]
Townsend a: SYLVIA *Delafieldii*, (Aud.); not listed.
Audubon: DELAFIELD'S GROUND-WARBLER. Single specimen from Townsend from "California" [= error of collector and locality], 2: 81-82.
Nuttall: MASKED YELLOW-THROAT. Occurs Oregon Territory, Audubon described Townsend's single specimen, p. 458.
Notes: The Masked Yellowthroat is a Central and South American species. It does not occur at Valparaiso, Chile, (Townsend's only South American landfall) so could not have been collected in the wild by Townsend. See also Jobanek (1994).

WILSON'S WARBLER *Wilsonia pusilla* (Wilson, 1811)
Townsend a: SYLVIA *Wilsonii*; b: Wilson's Green-black Capt Warbler, *Sylvia Wilsoni*.
Nuttall: WILSON'S SYLVAN FLYCATCHER. Arrival in Oregon noted about first week of May [1835], pp. 335-336.
Notes: Derby list: No. 45 Sylvia [Wilsonii], Columbia River, 4 specimens.
Specimens: OUM 12671, male, Columbia River, May 1835.

YELLOW-BREASTED CHAT *Icteria virens* (Linnaeus, 1758)
Townsend a: ICTERIA *viridis*; b: Yellow-breasted Chat, *Icteria viridis*.
Nuttall: YELLOW-BREASTED ICTERIA: "Mr. Townsend saw it at Walla-Walla on the Columbia, breeding in the month of June", p. 339.
Notes: Seen near Independence, Missouri (MS Journal, 10th May 1834).

WESTERN TANAGER *Piranga ludoviciana* (Wilson, 1811)
Townsend a: TANAGRA *ludoviciana*; b: Louisiana Tanager, *Tanagra ludoviciana*.
Audubon: LOUISIANA TANAGER. First seen Laramie's Fork 4th June 1834; Nuttall and Townsend quotes, 3: 231. Townsend supplied male specimens for Plate CCCLIV.
Nuttall: LOUISIANA TANAGER: First seen 4th June 1834, Laramie's Fork, and downstream from Fort Vancouver, pp. 543-545.
Specimens: ANSP may have skull No. 474 Tanagra ludoviciana, Columbia river (Morton 1840, p. 43).

GREEN-TAILED TOWHEE *Pipilo chlorurus* (Audubon, 1839)
Townsend a: FRINGILLA *chlorura*, (Aud.); b: Green-tailed Finch, *Fringilla (not described.)*
Audubon: GREEN-TAILED SPARROW. Townsend quote; original description from Townsend's only specimen, Royal octavo edition (1839), 5: 336. No plate.
Specimens: USNM 1896, male, [Ross's Creek, near Blackfoot, Idaho], 12th July 1834; type specimen.

SPOTTED TOWHEE *Pipilo maculatus* Swainson, 1827
Townsend a: FRINGILLA *arctica* [Eastern Towhee not listed]; b: Arctic Ground Finch, *Fringilla arctica* [Eastern Towhee not listed].
Audubon: ARCTIC GROUND-FINCH. Nuttall quoted, 3: 166. Male and female specimens from Townsend used in Plate CCCXCIV.
Nuttall: ARCTIC GROUND FINCH. *Fringilla arctica*, NOBIS. "confined to the western side of the Rocky Mountains", pp. 610-611.
Notes: McEuen list: No. 39 Rocky Mountain Antcatcher.
Specimens: ANSP 24001, female, Columbia River, no date, catalogued as *P.erythrophthalmus ferox*. USNM 2867, male, Fort Vancouver, 27th May 1835.

AMERICAN TREE SPARROW *Spizella arborea* (Wilson, 1810)
Townsend a: FRINGILLA *canadensis*; b: Tree Sparrow, *Fringilla canadensis*.

CHIPPING SPARROW *Spizella passerina* (Bechstein, 1798)
Townsend a: FRINGILLA *socialis*; b: Chipping Sparrow, *Fringilla socialis*.
Nuttall: CHIPPING SPARROW. "Mr. Townsend found it to be a common species in the territory of the Oregon", p. 574.

CLAY-COLORED SPARROW *Spizella pallida* (Swainson, 1832)
Townsend a: EMBERIZA *pallida* (Sw.); b: not listed.
Audubon: CLAY-COLOURED BUNTING. Townsend specimen (for Plate CCCXCVIII) collected 15th June 1834, 3: 71-72. [error = Brewer's Sparrow, see below]
Nuttall: CLAY-COLORED SPARROW. *Fringilla pallida*, NOBIS. Seen on the plains of the Platte and Green Rivers, pp. 579-580 [= Brewer's Sparrow].
Notes: McEuen list: No. 12 Cream-colored sparrow; No. 15 "same as 12" [One of these, at least, probably was a Clay-colored Sparrow as they appear early in Townsend's field list, i.e. too far east to be Brewer's Sparrow – see p. 82].

BREWER'S SPARROW *Spizella breweri* Cassin, 1856
Townsend a: not listed; b: not listed.
Audubon: included under CLAY-COLOURED BUNTING. Specimen for plate collected 15th June 1834, 3: 72.
Notes: Misidentified as *S. pallida*; in 1852 John Cassin described it as new species from Townsend specimens (see Mearns and Mearns 1995).
Specimens: ANSP 24050, Black Hills [= Laramie Mts, Wyoming], no date. USNM 1905, Sweetwater, Wyoming, 15th June 1834; cotype. USNM 2890, Sweetwater, Wyoming, 15th June 1834; cotype.

FIELD SPARROW *Spizella pusilla* (Wilson, 1810)
Townsend a: FRINGILLA *pusilla*; b: Field Sparrow *Fringilla pusilla*.
Notes: An eastern species, not occurring west of the Rocky Mountains.

VESPER SPARROW *Pooecetes gramineus* (Gmelin, 1789)
Townsend a: FRINGILLA *graminea*; b: Bay-winged Bunting, or Grass Finch, *Fringilla graminea*.
Audubon: BAY-WINGED BUNTING. Procured by Townsend on the Columbia River, 3: 65.

LARK SPARROW *Chondestes grammacus* (Say, 1823)
Townsend a: FRINGILLA *grammaca*; b: Lark Finch, *Fringilla grammaca*.
Audubon: LARK BUNTING. Quote from Nuttall, 3: 63-64. Townsend quote included in error under Song Sparrow (Stone 1906).
Nuttall: LARK FINCH: seen on the Plains of the Platte, Snake and Columbia Rivers, pp. 557-558.
Notes: McEuen list: No. 11 Fringilla grammaca, 2 specimens.
Specimens: ANSP 24077, Rocky Mountains, no date, *C.g.strigatus*.

LARK BUNTING *Calamospiza melanocorys* (Stejneger, 1885)
Townsend a: FRINGILLA *bicolor*, (Towns.) Original description from specimen from "plains of the Platte river", p. 189; b: Omitted from list. In text as Prairie Finch, *Fringilla bicolor*, (TOWNSEND.) "... inhabits a portion of the Platte country, east of the first range of the Rocky Mountains", pp. 346-347.
Audubon: PRAIRIE LARK-FINCH. Quotes from Nuttall and Townsend, 3: 195-196. Townsend specimens used in Plate CCCXC.
Nuttall: WESTERN OR PRAIRIE REED BIRD: first seen 24th May 1834 on north branch of

the Platte, p. 203.
Notes: Stone (1899) also refers to ANSP 23953 female; not now listed on database.
Specimens: ANSP 23951, male, Rocky Mountains, no date; Stone (1899) gives this specimen as the type. ANSP 23952, female, Rocky Mountains, no date. USNM 2869, male, Platte River, Nebraska, 24th May-1st June 1834, credited to Nuttall [=Townsend]; cotype.

SAVANNAH SPARROW *Passerculus sandwichensis* (Gmelin, 1789)
Townsend a: FRINGILLA *savanna*; b: Savannah Finch, *Fringilla savanna*.
Audubon: SAVANNAH BUNTING. Found by Townsend in Rocky Mountains and on Columbia River, 3: 68-70.
Notes: Seen shortly after leaving Independence, Missouri (MS Journal, 10th May 1834).
Specimens: ANSP 24091, Rocky Mountains, no date, *P.s.spodocephala*.

GRASSHOPPER SPARROW *Ammodramus savannarum* (Gmelin, 1789)
Townsend a: not listed; b: not listed.
Audubon: YELLOW-CROWNED BUNTING. Audubon says found by Townsend in Oregon, 3: 73-74.
Nuttall: Summer breeder "in the distant territory of the Oregon".
Notes: Early records of this species on the lower Columbia are questionable (Jobanek 1994), though it does occur there now.

FOX SPARROW *Passerella iliaca* (Merriam, 1786)
Townsend a: FRINGILLA *Townsendii*, (Aud.); b: Brown Longspur, *Plectrophanes Townsendi*, (AUDUBON.) In text as Townsend's Ground Finch, or Longspur *Plectrophanes Townsendi*. "… common in the neighborhood of Fort Vancouver", p. 345.
Audubon: TOWNSEND'S FINCH. Townsend quoted, but with text intended for Clay-colored Sparrow [= Brewer's Sparrow (Stone 1906)], 3: 143-144. Female collected 15th February 1836 used in Plate CCCCXXIV.
Nuttall: TOWNSEND'S FINCH. "collected in the Oregon Territory by Mr. Townsend on 15th of February [1836]," p. 583.
Specimens: USNM 2874, female, no locality, 15th Feb 1836; type for *Pi.townsendi*.

SONG SPARROW *Melospiza melodia* (Wilson, 1810)
Townsend a: FRINGILLA *cinerea* (Gm.); b: Ash-colored Finch, *Fringilla cinerea*, (GM.)
Audubon: BROWN SONG-FINCH. Nuttall and Townsend quoted [latter in error for text intended for Lark Sparrow (Stone 1906)]; specimen from Columbia River, 3: 145.
Nuttall: BROWN SONG-SPARROW. *Fringilla guttata*, NOBIS. Woody districts of the Columbia, pp. 581-583.
Specimens: ANSP 24028 [Fort Vancouver], no date, *M.m. morphna* type for *M. fasciata guttata* (Nuttall). ANSP 24030, Columbia River, no date, *M.m.morphna*. USNM 1860, female, Fort Vancouver, 18th Jan 1836. USNM1942, unsexed, Fort Vancouver, no date. USNM 83644, male, Columbia River, 18th Oct 1834, *morphna* (in hand register but no specimen traced).

[RUFOUS-COLLARED SPARROW *Zonotrichia capensis* (Muller, 1776)]
Townsend a: FRINGILLA *Mortoni*, (Aud.); b: not listed.
Audubon: MORTON'S FINCH. "Procured by Townsend in upper California", 3: 151. Townsend specimen used in Plate 190 of octavo edition.
Nuttall: "*Fringilla Mortonii* of Audubon … is a native of Chili, where Mr. Townsend procured the specimen described by Audubon, and inadvertently attributed to the Rocky Mountains", p. 557.
Notes: Included in error – Rufous-collared Sparrow is a South American species (see appendix 15 and Jobanek 1994.)
Specimens: ANSP 10614, [Valparaiso, Chile], no date.

HARRIS'S SPARROW *Zonotrichia querula* (Nuttall, 1840)
Townsend a: not listed; b: not listed.
Audubon: HARRIS' FINCH. Found by Audubon on his Missouri River expedition of 1843, 7: 331-332.
Nuttall: Nuttall gives first published description, p. 555. Nuttall's type now lost.
Notes: MS description by Townsend pre-dates Nuttall's publication (see p. 48-50.).

WHITE-CROWNED SPARROW *Zonotrichia leucophrys* (J.R. Forster, 1772)
Townsend a: FRINGILLA *leucophrys*; b: White-crowned Bunting, or Finch, *Fringilla leucophrys*.
Nuttall: WHITE-CROWNED FINCH. Seen Santa Barbara, California, pp. 553-554.
GAMBEL'S FINCH. *Fringilla Gambelii*, NOBIS. Single male specimen obtained by Mr. Townsend near Fort Walla Walla, and perhaps seen Big Vermillion, Missouri, 1st May 1834, pp. 556-557.
Specimens: Stone (1899) says type of *gambelii* now lost; still missing 2003. ANSP may have skull No. 527 Fringilla Gambeli [sic], Oregon (Morton 1840, p. 45; Morton 1849, p. 10.).

GOLDEN-CROWNED SPARROW *Zonotrichia atricapilla* (Gmelin, 1789)
Townsend a: FRINGILLA *atricapilla*, (Gm.); b: Black-headed Finch, *Fringilla artricapilla*, (GM.)
Audubon: BLACK-AND-YELLOW-CROWNED FINCH. Quote by Nuttall; seen Rocky Mountains and Columbia River. Townsend supplied specimens of young male shot 12th July 1834 [= Green-tailed Towhee collected on same date?], 3: 162.
Nuttall: YELLOW-CROWNED FINCH. *Fringilla aurocapilla* NOBIS: seen in Rocky Mountains and at Fort Vancouver, pp. 555-556.
Notes: McEuen list: No. 64 Black-crowned Bunting.
Specimens: ANSP 24067, no date [Fort Vancouver]. Stone (1899) says that ANSP 24067, a male collected by Townsend on 22nd Sept1836, Columbia River, is Nuttall's type.

DARK-EYED JUNCO *Junco hyemalis* (Linnaeus, 1758)
Townsend a: FRINGILLA *oregona*, (Towns.) Original description pp. 188-189; b: Oregon Snow Finch, *Fringilla Oregona*, (TOWNSEND.) "Common on the Columbia river in winter", p. 346.
Audubon: OREGON SNOW-BIRD. Specimen procured 5th October 1834, 3: 91-92. Townsend specimens from Columbia River used in Plate CCCXCVIII.
Nuttall: OREGON SNOW-BIRD. Seen in the forests of Oregon, p. 587.
Notes: Stone (1899) gives ANSP 24048 as type. Not listed on current ANSP database.
Specimens: ANSP 23952, unsexed, Columbia River, no date, *J.h.oreganus*. OUM 13516, male, Columbia River, October 1835. OUM 13517, female, Columbia River, 5th Oct 1834. USNM 1947, male, Fort Vancouver, 5th Oct 1834. USNM 1948, male?, no locality, 16th Oct 1834.

CHESTNUT-COLLARED LONGSPUR *Calcarius ornatus* (Townsend, 1837)
Townsend a: FRINGILLA (plectrophanes) *ornata*, (Towns.); b: Omitted from list. In text as CHESTNUT-COLORED [SIC] FINCH, *Plectrophanes ornata*, (Townsend,) "Inhabits the plains of the Platte river, near the first range of Rocky Mountains", p. 344-345.
Audubon: CHESTNUT-COLLARED LARK-BUNTING. Short quote by Townsend, 3: 53. Townsend specimen used in Plate CCCXCIV.
Nuttall: BROWN COLLARED LARK BUNTING. Seen in early May 1834 "on the wide grassy plains of the Platte", p. 537.
Notes: McEuen list: No. 13 Black-bellied Bunting.
Specimens: ANSP 24099, male, no locality, no date; type specimen.

SNOW BUNTING *Plectrophenax nivalis* (Linnaeus, 1758)
Townsend a: EMBERIZA *nivalis*; b: Snow Bunting, *Emberiza nivalis*.

ROSE-BREASTED GROSBEAK *Pheucticus ludovicianus* (Linnaeus, 1766)
Townsend a: not listed; b: not listed.
Notes: Seen near Independence, Missouri (MS Journal, 11th May 1834).

BLACK-HEADED GROSBEAK *Pheucticus melanocephalus* (Swainson, 1827)
Townsend a: FRINGILLA *maculata*; b: Mottled, or Spotted Grosbeak, *Fringilla maculata*.
Audubon: BLACK-HEADED GROSBEAK. Long quote by Nuttall, 3: 214. Townsend probably supplied specimens for Plate CCCLXXIII.
Nuttall: BLACK-HEADED GROSBEAK. *Coceothraustes melanocephalus*, NOBIS. Seen often near Green River and westwards to Columbia River, pp. 625-627.
Notes: Male specimen from Black Hills [= Laramie Mts], 3rd June 1834, at Charleston Museum (Sass 1908); may not still exist. An Audubon original watercolour (NYHS 1863.17.373) of a male Black-headed Grosbeak is inscribed "Nov. 27th 1836"; this may be from a JKT specimen but on that date he was at sea (error for 1835?).
Specimens: USNM 2873, Columbia River, 28th July 1835.

BLUE GROSBEAK *Passerina caerulea* (Linnaeus, 1758)
Townsend a: not listed; b: not listed.
Nuttall: BLUE GROSBEAK. "Mr. Townsend and myself met with them in May in the borders of the Platte near Scott's Bluffs, when they were already mated and breeding", p. 624.

LAZULI BUNTING *Passerina amoena* (Say, 1823)
Townsend a: FRINGILLA *amoena*; b: Luzuli Finch, *Fringilla amoena*.
Audubon: LAZULI FINCH. Nest presented by Nuttall. Male and female shot 3rd June 1836 by Townsend (used in Plate CCCXCVIII and CCCCXXIV), 3: 100.
Nuttall: LAZULI FINCH. "Common in the Oregon Territory", nests found, pp. 546-547.
Notes: McEuen list: No. 113 Fringilla amoena, 2 specimens.
Specimens: OUM 13664, male, Columbia River, June 1836.

DICKCISSEL *Spiza americana* (Gmelin, 1789)
Townsend a: not listed; b: not listed.
Notes: Seen near Independence, Missouri (MS Journal, 10th May 1834).

BOBOLINK *Dolichonyx oryzivorus* (Linnaeus, 1758)
Townsend a: not listed; b: not listed.
Audubon: RICE-BIRD. Observed in Rocky Mountains by Townsend, 4: 13.
Notes: Considerable numbers seen on the Platte River (MS Journal, 17th May 1834).

RED-WINGED BLACKBIRD *Agelaius phoeniceus* (Linnaeus, 1766)
Townsend a: ICTERUS *phoeniceus* and ICTERUS *gubernator*; b: Red-winged Starling, or Oriole, *Icterus phoeniceus*.
Audubon: RED-WINGED STARLING. Procured on the plains, Rocky Mountains and Columbia River, 4: 29-30. CRIMSON-WINGED TROOPIAL. Two specimens procured Oregon, 4: 29-30.
Nuttall: TWO COLORED TROOPIAL. "obtained in Oregon by my friend Mr. T. [sic] K. Townsend", p. 187.

TRICOLORED BLACKBIRD *Agelaius tricolor* (Audubon, 1837)
Townsend a: ICTERUS *tricolo* [sic], (Aud.); b: not listed.
Audubon: RED-AND-WHITE-WINGED TROOPIAL. Nuttall supplied specimen from California (for Plate CCCLXXXVIII) and quote, 4: 27-29.
Nuttall: THREE COLORED TROOPIAL. *Icterus tricolor*, NOBIS. Seen around Santa Barbara and Monterey, California, p. 186.
Notes: 2nd March 1841 William Gambel donated specimen received from Nuttall to ANSP (Graustein 1967, p. 360).
Specimens: USNM 2836, male, Santa Barbara, California, April 1836 (collected by Nuttall); type.

WESTERN MEADOWLARK *Sturnella neglecta* Audubon, 1844 and **EASTERN MEADOWLARK** *S. magna* (Linnaeus, 1758)
Townsend a: STURNUS *ludovicianus*; b: Meadow Lark, or Starling, *Sturnus ludovicianus*.
Notes: On Tolmie list as "Sturnus Ludovicianus".

YELLOW-HEADED BLACKBIRD *Xanthocephalus xanthocephalus* (Bonaparte, 1826)
Townsend a: ICTERUS *zanthrocephalus*; b: omitted from list.
Audubon: YELLOW-HEADED TROOPIAL. Quotes by Nuttall and Townsend; seen Missouri and banks of the Platte, 4: 24-26.
Nuttall: YELLOW-HEADED TROOPIAL. Seen near the Platte, Townsend found nest, pp. 187-189.
Notes: Seen near Independence, Missouri, and "Kansa Agency" (MS Journal, April-May 1834).

RUSTY BLACKBIRD *Euphagus carolinus* (Muller, 1776) and **BREWER'S BLACKBIRD** *E. cyanocephalus* (Wagler, 1829)
Townsend a: QUISCALUS *ferrugineus*; b: Rusty Blackbird, or Grakle, *Quiscalus ferrugineus*.
Audubon: RUSTY GRAKLE. Procured by Townsend on the Columbia River, 4: 65-68.
Notes: Both species would have been seen; but sightings west of Wyoming would have been Brewer's Blackbirds. The latter was first described from a Mexican specimen by Wagler in 1829 but it was not fully appreciated as a separate species in North America until after 1843 when Audubon named it Brewer's Blackbird. See also Jobanek 1994.

COMMON GRACKLE *Quiscalus quiscula* (Linnaeus, 1758)
Townsend a: not listed; b: not listed.
Notes: Seen frequently near Independence, Missouri (MS Journal, 28th April 1834).

BROWN-HEADED COWBIRD *Molothrus ater* (Boddaert, 1783)
Townsend a: not listed; b: not listed.
Notes: Seen shortly after leaving Independence, Missouri (MS Journal, 28th April and 2nd May 1834).

BULLOCK'S ORIOLE *Icterus bullocki* (Swainson, 1827)
Townsend a: ICTERUS *Bullockii*; b: Bullock's Oriole, *Icterus Bullocki*.
Audubon: BULLOCK'S ORIOLE. Descriptions taken from two Townsend specimens dated 21st June 1836, 4: 43-45. Plate CCCLXXXVIII.
Nuttall: WESTERN TROOPIAL. *Icterus Bullockii*: "westward along the woody borders of the Platte, beyond Laramie's Fork, to the shores of the Oregon and the Pacific", pp. 174-175.
Specimens: ANSP 3501, male, Columbia River, 3rd June 1836; ex Harris collection (see also Street 1948, p. 184).

PURPLE FINCH *Carpodacus purpureus* (Gmelin, 1789), **HOUSE FINCH** *C. mexicanus* (Muller, 1776) and **CASSIN'S FINCH** *C. cassinii* Baird, 1854
Townsend a: FRINGILLA *purpurea* [Purple Finch]; b: Crimson-fronted Bullfinch, *Fringilla frontalis* [House Finch].
Notes: McEuen's list: No. 71 Pyrrhula (frontalis?). Tolmie list: "Fringilla Purpurea". Townsend may have seen all three species.
Specimens: ANSP 10382, unsexed, Columbia River, no date, *C.p.californicus*.

RED CROSSBILL *Loxia curvirostra* Linnaeus, 1758
Townsend a: LOXIA *curvirostra*; b: American Crossbill, *Loxia curvirostra*.
Specimens: ANSP 10344, unsexed, Columbia River, no date, *L.c.sitkensis*; [ex G.A. McCall collection = Townsend specimen?].

PINE SISKIN *Carduelis pinus* (Wilson, 1810)
Townsend a: FRINGILLA *pinus*; b: Pine Finch, *Fringilla pinus*.
Specimens: ANSP 10289, female, Columbia River, no date.

[BLACK-CHINNED SISKIN *Carduelis barbata* (Molina, 1782)]
Audubon: STANLEY GOLDFINCH, "found in upper California", 3: 137-138. Plate 185 in octavo edition.
Specimen: USNM 2035, male, [Valparaiso, Chile, June-Aug 1834]. USNM 2036, female, [Valparaiso, Chile, June-Aug 1834].
Notes: Not from California; perhaps collected by Townsend in Chile.

AMERICAN GOLDFINCH *Carduelis tristis* (Linnaeus, 1758)
Townsend a: FRINGILLA *tristis*; b: American Goldfinch, *Fringilla tristis*.

EVENING GROSBEAK *Coccothraustes vespertinus* (Cooper, 1825)
Townsend a: FRINGILLA *vespertina*; b: Evening Grosbeak, *Fringilla vespertina*.
Audubon: EVENING GROSBEAK. Long quote by Townsend, 3: 217-221. Specimen from Laramie Mountains dated 3rd June 1824 [= 1834] used in Plate CCCCXXIV.
Nuttall: EVENING GROSBEAK. "In the pine woods of the Oregon … they are seen abundant about the middle of May", pp. 620-621.
Notes: Tolmie list: "Fringilla Vespertina".
Specimens: ANSP 9696, female, Columbia River, no date, *C.v.brooksi*.

Chestnut-collared Longspur.

APPENDIX 7B. MAMMALS SEEN AND COLLECTED BY TOWNSEND ON THE OVERLAND JOURNEY FROM MISSOURI TO THE PACIFIC.

KEY

Townsend a: Catalogue of Quadrupeds found in the Territory of Oregon, in the appendix to Townsend's *Narrative of a Journey* (1839).

Townsend b: List of Quadrupeds procured by Mr. Townsend and sent to the Academy of Natural Sciences, *Journal of the Academy of Natural Sciences of Philadelphia* 1839 **8**: 73-74.

Townsend c: Letter to J.E. Gray; a list of specimens that Townsend wanted to sell to BMNH, dated 13th January 1849.

Townsend d: Fifty-six specimens in skins, of North American Mammalia, collected by Dr. J.K. Townsend in Oregon and the Rocky Mountains, and presented by him to the Academy. *Proceedings of the Academy of Natural Sciences of Philadelphia* (Anon 1850, p. 13).

Bachman: Mammal papers by John Bachman, mainly *Journal of the Academy of Natural Sciences of Philadelphia and Proceedings of the Zoological Society of London*.

VQ: *Viviparous Quadrupeds*– Audubon and Bachman's work on the mammals of North America, with illustrations by both John James Audubon and his son, J.W. Audubon.

Notes: Any additional references or notes.

Specimens: Current location of Townsend's mammals. Specimens are mostly skins and mounts but an additional twenty-seven skulls (Morton 1840, 1849) were presented by Townsend to the S.G. Morton collection, now mostly at ANSP.

SHREW sp. [Vagrant Shrew *Sorex vagrans* Baird, 1857 and Dusky Shrew *Sorex monticolus* Merriam, 1890 amongst those possible]
Townsend a: Columbia Shrew, *Sorex*, *(undescribed.)*
Notes: 1 Sorex (new) Columbia River (*Bulletin of the Proceedings of the National Institute* 1841 p. 104). Sorex (undescribed) Shrew Mouse, Columbia River (*Bull. Proc. National Institute* 1841 p. 134). This may well have been a new species.
Specimens: Above specimen(s) from Columbia River donated to National Institute (USNM), 13th September 1841. Not traced.

TOWNSEND'S MOLE *Scapanus townsendii* (Bachman, 1839)
Townsend a: Townsend's Shrew Mole, *Scalops Townsendii*, (BACHMAN.) Listed and in text, pp. 314-315. b: *Scalops Townsendii*. c: Scalops Townsendii, Bach. 2 [specimens] [Oregon]. d: Scalops Townsendii (2 spec.).
Bachman: Original description by Bachman, one specimen received from Nuttall and one from Townsend – latter from "Banks of the Columbia river, May 9th, 1835" (*JANSP* 1839 **8**: 58-60).
VQ: Plate CXLV, "Townsend's Shrew Mole" by J.W. Audubon from Townsend and Nuttall specimens.
Notes: Specimen from Fort Vancouver, Clark Co., Washington (see True 1898 *Proc. US Nat. Mus. 19*: 63). True gives date of Townsend specimen as 9th May 1835 – but at that time Townsend was at Warrior's Point not Fort Vancouver. He had not yet been to Fort Vancouver yet that spring but on 2nd May 1835 (Townsend, letter to his father) he said that he "expected to do so in about 2 weeks or more." Restricted type locality should therefore be near St. Helens, Columbia Co., Oregon.
Specimens: ANSP 449 mounted skin, skull inside. Collected by Townsend on "the banks of the Columbia river" 9th May 1835, paratype. The true type of Bachman's *T. townsendii* (from Nuttall) does not appear to exist (Koopman 1976).

STAR-NOSED MOLE *Condylura cristata* (Linnaeus, 1758)
Townsend a: 1. Thick-tail Star-nose Mole, *Condylura macroura*. 2. Long-tail Star-nose [Mole], *Condylura longicaudata*. b-c: not listed.
Notes: Thick-tail Star-nosed Mole described by Richard Harlan on basis of thick tail but this is Star-nosed Mole during rutting season when tail becomes greatly enlarged. Townsend did not pass through the current range of the Star-nosed Mole after leaving Independence, Missouri.

WESTERN SMALL-FOOTED MYOTIS *Myotis ciliolabrum* (Merriam, 1886)
Townsend a: Say's Bat, *Vespertillio subulatus*, (SAY.). b: not listed. c: Vespertilio subulatus, Say 3 [specimens] [Oregon]. d: Vespertilio subulatus, Say (3 spec.).
VQ: Bats not included, though Audubon did prepare some bat paintings (Ford, 1951, p. 62).
Specimens: ANSP 1126 skin, skull inside, 11th August 1835 Oregon, Columbia River [= Sauvie's Island or Fort Vancouver].

YUMA MYOTIS *Myotis yumanensis* (H. Allen, 1864)
Townsend a: ?Little Bat, *Vespertilio*, *(undescribed.)*. b: not listed. c: *Vespertilio (nondescript?) 1* [specimen] *R.Mts.* d: V. ———.
VQ: Bats not included, though Audubon did prepare some bat paintings (Ford, 1951, p. 62).
Notes: "Little Bat" and *Vespertillio (nondescript?")* probably refer to specimens below. Yuma Myotis was not described until 1864 so Townsend's specimen would have been a new discovery had it been written up at the time.
Specimens: ANSP 1125 skin, skull inside. ANSP 2604 skull only is skull No 466 (Morton 1840: 35).

TOWNSEND'S BIG-EARED BAT *Corynorhinus townsendii* (Cooper, 1837)
Townsend a: Great-eared Bat, *Plecotus Townsendii*, (COOPER.). Listed as Townsend's Great-eared Bat in text, pp. 324-325. b: not listed. c: *Plecotus Townsendii, Cooper. 1 [specimen] [Oregon] "There is a mutilated specimen in the box, in addition to the other."* d: Plerotus [sic] Townsendii.
VQ: Bats not included, though Audubon did prepare some bat paintings (Ford, 1951, p. 62). An original Audubon water colour at New-York Historical Society (NYHS Z3333[s-30]) includes Western Big-eared Bat from Columbia River (Olson and Mazzinelli 2003).
Notes: First described by Cooper from three Townsend specimens from the Columbia River (*Annals of the Lyceum of Natural History of New York* 1837 4: 73. Specimens: ANSP 2605 is skull No. 467 (Morton 1840, p. 35.). The holotype is "not known to be in existence" (Handley 1959).

AMERICAN PIKA *Ochotona princeps* (Richardson, 1828)
Townsend a: Little Chief Hare, *Lagomys princeps*, (RICHARDSON.). b: not listed. c: *Lagomys princeps, Rich. 1* [specimen] *R.Mts.* d: Lagomys princeps.
Bachman: Describes male taken by Townsend on 13th August 1834 (*JANSP* 1837 **7**: 354-358).
VQ: Quote by Nuttall "we were not fortunate enough to obtain any good specimens." But text says later "The specimen of Mr. Townsend was procured in latitude 42° [N]" and seems to have been used for Plate LXXXIII.
Specimens: ANSP 372 body mount, skull inside, N America, 1834.

?BRUSH RABBIT *Sylvilagus bachmani* (Waterhouse, 1839)
Townsend a: Marsh Hare, *Lepus palustris*, (BACHMAN.)
Notes: Not Marsh Rabbit because Townsend did not enter its range (south east US). He did pass through range of Pygmy Rabbit *Brachylagus idahoensis* (a species not described until 1891) and may have entered northernmost part of range of Brush Rabbit.

MOUNTAIN (or NUTTALL'S) COTTONTAIL *Sylvilagus nuttallii* (Bachman, 1837)
Townsend a: 1. Nuttall's little Hare, *Lepus Nuttallii*, (BACHMAN.) Listed and in text, pp. 313-314. 2. Wormwood Hare, *Lepus artemesia*, (BACHMAN.). b: *Lepus artemesia*, - seven specimens. [*Lepus nuttalli* not listed]. c: 1. Lepus Nuttalii [sic], Bach. 1 [specimen] RMts. 2. *Lepus artemesia, Bach. 3* [specimens] [R.Mts.]. d: L. Nuttallii, L. Artemesia (3 spec.).
Bachman: 1. *L. nuttalli* original description by Bachman, with quote by Nuttall. Nuttall's type from "banks of several small streams which flow into Shoshone and Columbia rivers" [Designated type locality = near mouth of Malheur River] (*JANSP* 1837 **7**: 345-348). Townsend quote (*JANSP* 1839 8: 79-80). 2. *Lepus artemesia* described by Bachman from Townsend specimen with quote by Townsend, but proved to be same species as *L. nuttalli* (*JANSP* 1839 8: 94-96).
VQ: 1. Plate XCIV of "Lepus Nutallii. Nuttall's Hare" by J.W. Audubon from specimen collected by Nuttall. 2. Plate LXXXVIII "Lepus artemisia [sic]. Worm-wood Hare" from Townsend specimen.
Notes: "1 Lepus Nuttalii [sic] (unique specimen) Rocky Mountains …on north branch of the Platte River" amongst Townsend donation to National Institute on 14th February 1842 (*Proceedings of National Institution* Feb 1842, p.147).
Specimens: ANSP 382 mounted skin, skull separate, juvenile, holotype. Early museum label says "Lepus nuttalli – TYPE – J.K. Townsend – Rocky Mts." Type locality determined by Nelson (1909) as "eastern Oregon, near mouth of Malheur River," with date of August 1834 (see Koopman 1976). LM D573a skin and skull 20th August 1836, Fort Walla Walla [2005 LM database gives date of 29th January 1836].

SNOWSHOE HARE *Lepus americanus* Erxleben, 1777
Townsend a-c: not listed, though Townsend says he shot several on the Columbia River (see Bachman "Remarks on Genus Lepus" *JANSP* 1839 8: 82).
Specimens: no specimen prepared (see *JANSP* 1839 8: 82).

WHITE-TAILED JACKRABBIT *Lepus townsendii* Bachman, 1839
Townsend a: Townsend's Hare, *Lepus Townsendii*, (BACHMAN.) Listed and in following text, pp. 325-327. b: *Lepus Townsendii*, – two specimens. c: *Lepus Townsendii, Bach 2 [specimens] [Oregon]*. d: Lepus Townsendii (2 spec.).
Bachman: Original description by Bachman from specimens from "Walla-walla" with quote by Townsend (*JANSP* 1839 8: 90-94).
VQ: Plate III of "Lepus Townsendii. Townsend's Rocky Mountain Hare" presumably painted from Townsend's specimens because this plate was produced before 1843 when Audubon obtained his additional specimens.
Specimens: ANSP 378 mounted skin, skull inside, holotype, marked "type" on old label. ANSP 2086 Oregon, is skull No. 538 (Morton 1840, p. 36.).
[Skull No. 537 not found in 2003.]

YELLOW-PINE CHIPMUNK *Neotamias amoenus* J.A. Allen, 1890
Townsend a: Four-lined Squirrel, *Tamias quadrivittatus* [sic], (SAY.). b: *Tamias quadrivittatus*. (Say.). c: *Tamias quadrivittatus 3 [specimens] R.Mts.* d: *T. grandivittatus* [sic].
VQ: Yellow-Pine Chipmunk not included in text. Plate XXIV of "Four Striped Ground-Squirrel" is of Yellow-Pine Chipmunk. Original watercolour inscribed "August 12th 1834. Rocky Mountains. J.K.T." (see Boehme 2000, p. 64.) On that date Townsend near Hyndman Peak, Idaho, outside the range of Four-lined Chipmunk *Tamias quadrivittatus*. Townsend not mentioned in text for *T. quadrivittatus*.
Notes: Plate XXIV has also been erroneously identified as Colorado Chipmunk *Tamias quadrivittatus* in Cahalane 1967.
Specimens: ANSP 245 body mount, skull inside, Rocky Mountains, 1834; catalogued as *T. quadrivittatus*. ANSP 247 body mount, skull inside, Rocky Mountains, 1834 (see below); catalogued as *T. quadrivittatus*. Photograph of specimen in Peck 2000, p. 93. ANSP 2116 skull No. 464 (Morton 1840, p. 35). Catalogued as *T. quadrivittatus*. USNM 93 skin only, Columbia River, no date but field number possibly 114; catalogued as *E. amoenus*.

LEAST CHIPMUNK *Neotamias minimus* (Bachman, 1839)
Townsend a: Little Ground Squirrel, *Tamias minimus*, (BACHMAN.) Least Ground Squirrel in text, pp. 323-324. b: *Tamias minimus*. c: *Tamias minimus, Bach. 2 [specimens] [R. Mts]*. d: T. minimus (2 spec.).
Bachman: Original description by Bachman from Townsend specimen from "the Rio Colorado" [Green River, Wyoming] with Townsend quote (JANSP 1839 8: 71-73).
Specimens: ANSP 248 mounted skin, skull inside, ?holotype; oldest label reads "Rocky Mountains – Dr. Townsend." Type locality restricted to Green River, near mouth of Big Sandy Creek, Sweetwater Co., Wyoming (J.A. Allen 1890 *Bulletin American Museum of Natural History* 3: 110-112) – see Koopman 1976.

TOWNSEND'S CHIPMUNK *Neotamias townsendii* (Bachman, 1839)
Townsend a: Townsend's Ground Squirrel, *Tamias Townsendii*, (BACHMAN.) Listed and in text, pp. 321-323. b: *Tamias Townsendii*. c: Tamias Townsendii, Bach. 2 [specimens] Oregon. d: Tamias Townsendii (3 spec.).
Bachman: Original description by Bachman with quote from Townsend – no locality given = near lower mouth of Willamette, Columbia River (*JANSP* 1839 8: 68-71).
VQ: Plate XX of "Tamias Townsendii. Townsend's Ground Squirrel" from Townsend specimens, with Townsend notes.
Specimens: ANSP 241 mounted skin, skull inside, lectotype; catalogue entry reads "Tamias townsendii – Oregon – 1834 – Dr. Townsend." Type locality restricted to "Lower Columbia River, near lower mouth of Willamette River, Oregon," by Howell (1929) – see Koopman 1976. USNM 92 mounted skin with skin inside (later dismounted, skin now lost but skull re-numbered as A38797), Columbia River, 4th October 1834.

HARRIS'S ANTELOPE SQUIRREL *Ammospermophilus harrisii* (Audubon and Bachman, 1854)
VQ: "Spermophilus Harrisii. Harris's Spermophile". J.W. Audubon figured it as Harris's Antelope Ground Squirrel. J.J. Audubon and Bachman say: "Mr. J.K. TOWNSEND who gave the specimen to our esteemed friend, EDWARD HARRIS Esq, from whom we received it, and with whose name we have honoured this pretty little animal from the West." (Vol 3: p. 268). The locality given was "Probably west of the Rocky Mountains, on the route followed by Messrs NUTTALL and TOWNSEND".
Notes: Not personally collected by Townsend. This species is resident in Arizona and Northern Mexico, so could not have been collected in the field by Townsend. It may be a specimen from the Fremont expedition, collected by Edward Kern. Type locality now restricted to Santa Cruz Valley, Arizona (A.H. Howell, 1938. *North American Fauna* 56: 167). No specimens.

CALIFORNIA GROUND SQUIRREL *Spermophilus beecheyi* (Richardson, 1829)
Townsend a: Douglass' Marmot, *Arctomys Douglassi* [sic]. Collected at the Dalles 30th June 1836 – see *Narrative*. b: *Spermophilus Douglassii* – four specimens of various sizes. c: Spermophilus Douglasii 4 [specimens] Oregon. d: Spermophilus Douglasii (4 spec.).
VQ: Plate XLIX of "Douglasses [sic] Spermophile" from four Townsend specimens. One marked "Falls of the Columbia River", one marked "Walla-walla".
Notes: Original Audubon watercolour indicates specimens collected 22nd June 1835 (Boehme 2000, p. 127) when Townsend at or near Fort Vancouver.
Specimens: ANSP 329, 330, 331, 332, 333, 334 body mounts, some with skulls removed.

BELDING'S GROUND SQUIRREL *Spermophilus beldingi* Merriam, 1888
Townsend a: Richardson's Marmot, *Arctomys Richardsonii*. b: *Spermophilus Richardsonii*. c: *Spermophilus Richardsonii 2 [specimens] [Oregon]*. d: S. Richardsonii (2 spec.).
VQ: Plate L of "Spermophilus Richardsonii. Richardson's Spermophile" from two Townsend specimens shows black tip to tail of one, while the other has lost the hair off its tail. Text says specimens from "Rocky Mountains, (about latitude 45° [N],)". Townsend did not pass through range of Richardson's Ground Squirrel. The plate almost certainly shows Belding's Ground Squirrel because of black tip to tail.
Notes: Belding's Ground Squirrel first described by Merriam (1888) from Donner, Placer Co., California.

FRANKLIN'S GROUND SQUIRREL *Spermophilus franklinii* (Sabine, 1822)
Townsend a: Franklin's Marmot, *Arctomys Franklinii*. b: *Spermophilus Franklinii*. c: *Spermophilus Franklinii, 1 [specimen] Missouri*. d: S. Franklinii.
VQ: "Mr. Townsend … observed it in Oregon". Plate LXXXIV perhaps from Townsend specimens.
Specimens: ANSP 348 body mount, skull inside, Rocky Mountains, 1834.

TOWNSEND'S GROUND SQUIRREL *Spermophilus townsendii* (Bachman, 1839)
Townsend a: Townsend's Marmot, *Spermophilus Townsendii*, (BACHMAN.) Listed and in text, pp. 316-317. b: *Spermophilus Townsendii*. c: *Spermophilus Townsendii*, *Bach*. 1 [specimen] Oregon.
Bachman: Original description by Bachman from Townsend specimen from "prairies near Walla-walla" (*JANSP* 1839 8: 61-62). See also Scheffer 1946.
VQ: Plate CXLVII, fig. 1 of "Spermophilus Townsendii. American Souslik" by J.W. Audubon from Townsend specimen, plus notes from Townsend. Plate depicts Washington Ground Squirrel (see below).
Specimens: ANSP 344 mounted skin, skull separate, holotype; earliest museum label reads "Spermophilus townsendii Bach. – TYPE – J.K. Townsend – Columbia River." Catalogue entry gives 1834 as date of collection (see Koopman 1976). ANSP 345 body mount, skull inside, Rocky Mountains, 1834. ANSP 346 skin and skull, Rocky Mountains, 1834; catalogued as *S. citellus*.

THIRTEEN-LINED GROUND SQUIRREL *Spermophilus tridecemlineatus* (Mitchell, 1821)
Townsend a: Hood's Marmot, *Spermophilus tridecemlineatus*. b: *Spermophilus tridecem-lineatus*. c: *Spermophilus tridecemlienatus [sic] 1 [specimen] Missouri*. d: S. tridecimlineatus [sic].
VQ: Townsend not mentioned (nor on original watercolour).
Specimens: ANSP 852 body mount, upper Missouri [between Independence and Fort Hall, 1834].

WASHINGTON GROUND SQUIRREL *Spermophilus washingtoni* (A.H. Howell, 1938)
Townsend a - c: not listed.
VQ: This species is depicted in Plate CXLVII, fig 1, as "American Souslik." Notes: Townsend seems to have acquired both dappled *S.w.washingtoni* and undappled *S.w.townsendi* (see also Scheffer 1946. *J. Mammalogy* 27: 395).
Specimens: none traced (see Townsend's Ground Squirrel, above).

BLACK-TAILED PRAIRIE DOG *Cynomys ludovicianus* (Ord, 1815)
Townsend a: Prairie Dog, or Marmot, *Arctomys ludovicianus*.
"Small-pouched Marmot"
Townsend a: Small-pouched Marmot, (the opening of the pouches within the mouth,) not in my collection.
Not listed in other references. Identity not known; ground squirrels and chipmunks both have internal cheek pouches.

EASTERN FOX SQUIRREL *Sciurus niger* Linnaeus, 1758
Townsend a-c: Not listed, but mentioned in *Narrative* for early part of trip, east of Independence, Missouri.

DOUGLAS'S SQUIRREL (or CHICKAREE) *Tamiasciurus douglasii* (Bachman, 1839)
Townsend a: Douglass' Tree Squirrel [sic], *Sciurus Douglassii*, (BENNETT.) Douglass' Squirrel in text, pp. 317-318. b: *Sciurus Douglasii*, (Bennett,). c: *Sciurus Douglasii 3 [specimens] [Oregon]*. d: Sciurus Douglasii (3 spec.).
Bachman: Bachman quotes Townsend's observations and describes his specimen from "the Columbia river" (*JANSP* 1839 8: 63-64).
VQ: Plate XLVIII of "Sciurus Douglassii. Douglass' Squirrel" from Townsend specimens, with Townsend notes. Audubon had four Townsend specimens, one from "Falls of the Columbia", another from "Walla Walla." The original watercolour gives date of specimen as "Sepr 26th (no year)" (see Boehme 2000, p. 124).
Notes: Bachman originally named this species after Townsend but Gray had earlier named it after Douglas (without actually publishing a description), so Bachman changed to Douglas.
Specimens: ANSP 286 mounted skin, skull inside, holotype; collected by Townsend on "Columbia River" in 1834. Allen (1898, p. 284) says "probably from the lower part of the Columbia, – doubtless from [Fort] Vancouver or below." He restricts type locality for *douglasii* "to the coast phase of the group, as represented at the mouth of the Columbia River". He also records that the specimen label reads "286. Type of *Sciurus douglasii* Bach. Columbia River, J.K. Townsend. The specimen is in full summer pelage and Allen thought that it must have been taken in August or September (see note concerning original watercolour). This could have been in 1834, or 1835, or 1836, though earliest year is most likely as it is a common species and he was known to be out collecting in those months. See also Koopman 1976. ANSP 287 body mount, Oregon, 1834. Skull No. 470 "SCIURUS TOWNSENDI" (Morton 1840, p. 35.) but identified as "S. DOUGLASII" in Morton 1849, p. 5, was not found at ANSP in 2003.

RED SQUIRREL *Tamiasciurus hudsonicus* (Erxleben, 1777)
Townsend a: 1. Downy Squirrel, *Sciurus lanuginosus*, (BACHMAN.). Listed and in text, pp. 320-321. 2. Richardson's Squirrel, *Sciurius Richardsonii*, (BACHMAN.) Columbia Pine Squirrel in text, pp. 318-319. b: *Sciurus lanuginosus*. *Sciurus Richardsonii*. c: 1. *Sciurus lanuginosus*, *Bach*. 1 [specimen] Oregon. 2. *Sciurus Fremontii*, *Townsend* 1 [specimen] Upper California. d: S. Richardsonii, S. lanuginosus, S. Fremontii.
Bachman: *S. Richardsoni* and *S. lanuginosus* described by Bachman, both with quotes from Townsend. *S. Richardsonii* from Townsend specimen from the "Rocky Mountains, August 12th, 1834" [when Townsend at Big Lost River, south central Idaho]. *S. lanuginosus* from specimen from "coast near Sitka" given to Townsend by W.F. Tolmie (*PZS* 1839, pp. 100-103 and *JANSP* 1839 8: 64-68). Townsend's *Narrative* (Appendix p. 321) says from Tolmie from Fort McLoughlin.
VQ: 1. Plate XXV of "Sciurus Lanuginosus. Downy Squirrel" from Townsend specimen obtained from W.F. Tolmie. 2. Plate V of "Sciurus Richardsonii. Richardson's Columbian Squirrel" from Townsend specimen, with Townsend notes. 3. Plate CXLIX of "Sciurus Fremonti. Fremont's Squirrel" by J.W. Audubon from specimen from "the Rocky Mountains" collected on a Fremont expedition. Townsend not mentioned except as describer. See ANSP 285 below.
Specimens: 1. *S. lanuginosus* ANSP 295 mounted skin, skull separate, holotype; collected by W.F. Tolmie. Type locality restricted to Fort McLoughlin, Campbell Island, British Columbia. Marked "type" on original museum label (see Koopman 1976). 2. *S. richardsoni* ANSP 293 mounted skin, skull inside, holotype; collected by Townsend in "Rocky Mountains". Type locality restricted to Head of Big Lost River, Custer Co., Idaho by Merriam (1891) (see Koopman 1976). 3. *S. fremontii* ANSP 285 mounted skin, skull inside, said to have been collected by Townsend at "South Pass, Rocky Mountains". Marked "type" on museum label and in original catalogue entry. Back of label says, "The data on the reverse of this label was transcribed by me from the base of old stand – W. Stone" (see Koopman 1976). Possibly collected during a Fremont expedition; Hall (1981) suggests that it was probably collected in the park region of central Colorado.

NORTHERN FLYING SQUIRREL *Glaucomys sabrinus* (Shaw, 1801)
Townsend a: 1. Hudson's Bay Flying Squirrel, *Pteromys sabrinus*. 2. Oregon Flying Squirrel, *Pteromys oregonensis*, (BACHMAN.) Listed and described in text, pp. 329-330. b: 1. *Pteromys Sabineus* [sic] – one specimen. 2. *Pteromys oregonensis*, – one specimen. c: *Pteromys Oregonensis*, *Bach*. 1 [specimen] [Oregon]. *Pteromys sabrinus* 1 [specimen] [Oregon]. *Pteromys alpinus*, *Bach*. 1 [specimen] [Oregon]. d: Pteromys oregonensis, P. salerinus [sic].
Bachman: *P. oregonensis* described by Bachman from Townsend specimen from "pine woods of the Columbia, near the sea" with information from Townsend (*JANSP* 1839 8: 101-102). See Rhoads (1897, p. 324) who says probably collected St. Helens, Columbia Co., Oregon [with no real evidence – Eds.].
VQ: 1. Plate CXLIII fig 1 [not fig 2] is of "Pteromys Alpinus. Rocky Mountain Flying Squirrel," from Townsend (or Drummond) specimen. 2. Plate XV of "Pteromys Oregonensis. Oregon Flying Squirrel" is presumably from Townsend specimen; Bachman refers to Townsend's notes.
Specimens: ANSP 252 body mount, Columbia River, 1834. ANSP 253 skin and skull, adult, ?holotype; museum label reads "Pteromys oregonensis – TYPE – Oregon (Columbia River), J.K. Townsend." Catalogue entry gives 1834 as collection date. Type locality restricted to "probably near St. Helen, Columbia Co., Oregon" (see Koopman 1976).

NORTHERN POCKET GOPHER *Thomomys talpoides* (Richardson, 1828)
Townsend a: Gopher, or Kamas Rat, *Geomys borealis*. b: *Geomys borealis*. (Richardson.). c: 1. Pseudostoma borealis 1 [specimen] [Oregon]. 2. *Pseudostoma Richardsonii* 1 [specimen] [Oregon]. d: Pseudostoma Richardsonii.
Bachman: Two Townsend specimens described by Bachman (*JANSP* 1839 8: 103-104).
VQ: Four Townsend specimens described and probably used in Plate CXLII (though one specimen is a Townsend's Pocket Gopher).
Specimens: ANSP 143 body mount [missing in 2003]. ANSP 144 skin and skull, Columbia River (?), 1834.

TOWNSEND'S POCKET GOPHER *Thomomys townsendii* (Bachman, 1839)
Townsend a: Townsend's Gopher, *Geomys Townsendii, (Richardson's M.S.S.)*. Listed and described in text, pp. 330-331. b: *Geomys Townsendii* – two specimens. (Richardson's manuscripts.) c: *Pseudostoma Townsendii, Rich. M.S.S. 1 [specimen] R. Mts.* d: P. borealis, P. Townsendii (Rich.).
Bachman: Original description by Bachman from Townsend specimen from "the Columbia river" (*JANSP* 1839 8: 105).
VQ: Species included under "Pseudostoma Borealis" and in Plate CXLII (see above).
Notes: Type specimen now has no date. Bailey (1915 *North American Fauna*, 38: 42-46) says "very probably secured … near the present town of Nampa." Unless Bailey was able to assign the specimen to the western population on morphological differences, the type could equally well have come from eastern part of range near Fort Hall where Townsend spent far more time.
Specimens: ANSP 146 skin and skull, SW Wyoming, 1834. ANSP 147 skin and skull, holotype; oldest label reads "Pseudostoma richardsonii. Rocky Mountains. J.K. Townsend". Catalogue entry adds "1834 – Type of G. townsendii Bach et Rich." Type locality restricted to Nampa, Canyon Co., Idaho by Bailey (1915), (see Koopman 1976).

PLAINS POCKET GOPHER *Geomys bursarius* (Shaw, 1800)
Townsend a: Perhaps included as: Gopher, or Kamas Rat, *Geomys borealis*. b: *not listed.* c: *Pseudostoma bursarius 1 [specimen] Oregon.* d: P. borealis.
Notes: The range of this species does not extend to Columbia River, indeed current range only extends westwards into eastern Wyoming.
Specimens: ANSP 136 mounted skin, skull inside, ?holotype; original label reads "Pseudostoma bursarius, Columb. River. J.K. Townsend." The catalogue entry adds "Geomys oregonensis – 1834 – Type". Locality erroneous. USNM 91 skin only, Columbia River (dismounted May 1911). Locality erroneous (see above).

HISPID POCKET MOUSE *Chaetodipus hispidus* Baird, 1858
Townsend a-c: not listed.
Notes: Inhabits shortgrass prairies so would have been collected in the early stages of the expedition not long after leaving Independence, Missouri.
Specimens: ANSP 694 poorly made-over mounted skin, skull separate. Described by Rhoads in 1894, from specimen from "Rocky Mountains in the summer of 1834". Holotype of *Perognathus hispidus paradoxus* Merriam 1900 – see Koopman, 1976.

AMERICAN BEAVER *Castor Canadensis* Kuhl, 1820
Townsend a: Beaver, *Castor fiber.*
Skull No. 463 (Morton 1840, p. 34.) not found at ANSP in 2003.

WHITE-FOOTED MOUSE *Peromyscus leucopus* (Rafinesque, 1818)
Townsend a: White-footed Mouse, *Mus leucopus.* b: *Mus leucopus?* c: *Mus leucopus 1 [specimen] [Oregon].* d: *Mus cercopus* [sic].
Bachman: Townsend had "Several very striking varieties, if not new species" (*JANSP* 1839 8: 73).
VQ: "Mr. TOWNSEND brought us several skins [from the Columbia River]." No reference to source of the skins used in Plate XL.
Specimens: Skull No. "469. MUS LEUCOPUS?" Columbia River (Morton 1840, p. 35.) is perhaps at ANSP.

BUSHY-TAILED WOODRAT *Neotoma cinerea* (Ord, 1815)
Townsend a: Rocky Mountain Rat, *Neotoma Drummondii.* b: *Neotoma Drummondii.* c: *Neotoma Drummondii 3 [specimens] [Oregon].* d: Neotoma Drummondii (4 spec.).
VQ: Townsend's notes quoted. No information on source of specimens for Plate XXIX but possibly from Townsend.
Notes: Skull No. 516 (Morton 1840, p. 35) perhaps at ANSP (see below).
Specimens: ANSP 153 body mount, skull inside, Oregon, 1834. ANSP 154 skin and skull, Oregon, 1834. ANSP 2008 skull only, Columbia River. USNM 87 skin and skull, Columbia River (skull missing 30.11.1987). LM D353a skin and skull, Columbia River, 29th October 1836 (2005 LM database gives 29th January 1836).

HOUSE MOUSE *Mus musculus* Linnaeus, 1758
Townsend a: Common Mouse, *Mus musculus*

CREEPING (or OREGON) VOLE *Microtus oregoni* (Bachman, 1839)
Townsend a: Small Meadow Mouse, *Arvicola oregonii*, (BACHMAN.) Oregon Meadow mouse in text, pp. 315-316. b: *Arvicola oreganii* [sic]. c: *Arvicola Oregoni, Bach. 1 [specimen] Oregon.* d: Arvicola oregoni.
Bachman: Original description by Bachman from Townsend specimen from "the Columbia river" dated 2nd November 1836 (*JANSP* 1839 8: 60-61).
VQ: Plate CXLVII, fig 2, of "Arvicola Oregoni. Oregon Meadow Mouse" by J.W. Audubon from Townsend specimen dated 2nd November 1836.
Notes: Astoria, Clatsop Co., Oregon is correctly designated type locality based on date of specimen.
Specimens: Type not listed at ANSP by Koopman 1976.

TOWNSEND'S VOLE *Microtus townsendii* (Bachman, 1839)
Townsend a: Townsend's Meadow Mouse, *Arvicola Townsendii*, (BACHMAN.) Listed and in text, p. 315. b: *Arvicola Townsendii.* c: *Arvicola Townsendii, Bach. 1 [specimen] [Oregon].*
Bachman: Original description by Bachman from Townsend specimen from "the Columbia river" dated 21st July 1835 (*JANSP* 1839 8: 60).
VQ: Plate CXLIV, fig 1, of "Arvicola Townsendii. Townsend's Arvicola" by J.W. Audubon from Townsend specimen.
Notes: Bailey (1900. *North American Fauna* 18: 46) gives locality as "near mouth of Willamette on or near Wappatoo (or Sauvie) Island."
Specimens: No specimens. Type not listed at ANSP by Koopman 1976. Townsend said he did not receive the type back from Bachman (JKT to J.E. Gray 13th January 1849, BMNH DF 200/146/202).

COMMON MUSKRAT *Ondatra zibethecus* (Linnaeus, 1766)
Townsend a: Musk-rat, or Musquash, *Fiber zibethicus.*

JUMPING MOUSE *Zapus* sp.
Townsend a: Jumping Mouse, Meriones labradorius, (RICHARDSON.) b: Meriones labradorius [plus a note on tail length] c: Meriones Americanus 1 [specimen] Oregon. d: Meriones Americanus.
VQ: Plate LXXXV, "Meriones Hudsonicus. Jumping Mouse" "Mr. TOWNSEND brought specimens from Oregon." But there is no mention of source of the specimen used in plate.
Notes: There are three Jumping Mouse species that Townsend could have collected: Western *Z. princeps*, Meadow *Z. hudsonius* and Pacific *Z. trinotatus*.

NORTH AMERICAN PORCUPINE *Erethizon dorsatum* (Linnaeus, 1758)
Townsend a: American Porcupine, *Hystrix dorsata.*

COYOTE *Canis latrans* Say, 1823
Townsend a: Prairie Wolf, *Canas latrans.* b: Canis [species not specified]. c: not listed.
Notes: Prairie Wolf mentioned (Townsend, 1850).
Specimens: ANSP 602 skin only, Rocky Mountains, 1834. ANSP 2267 skull only, Oregon, 1834 is skull No. 532 or 535 (Morton 1840, p. 36). ANSP 2272 skull only, Oregon 1834 is skull No. 532 or 535 (Morton 1840, p. 36).

GRAY WOLF *Canis lupus* Linnaeus, 1758
Townsend a: 1. Common Wolf, *Canus lupus.* 2. Dusky Wolf, *Canus nubilis?*Townsend footnote: "*This is probably a new species. It is much larger than *nubilus*, as described, and differs much in its habits." 3. Cinereous Wolf, *Canus.* b: Canis [species not specified]. c: *Canis. Nondescript? [4 specimens, Oregon] Skins in Spirits.* d: Canis ——? (or great Oregon Wolf) (2 spec.), Canis ——? (lesser Oregon Wolf, Canis ——? (least Oregon Wolf).
Notes: Townsend finally got around to describing his wolf specimens as new species: *Canis gigas* Townsend *JANSP* 1850 **2**: 75-79 [= *Canis lupus fuscus* Richardson].
Specimens: 1. *Lupus gigas* Townsend ANSP 589 mounted skin, cotype; marked in catalogue "type, Rocky Mountains – 1834, J.K. Townsend, coll." 2. *Lupus gigas* Townsend ANSP 590 Skin. Cotype. Marked in catalogue "type, Rocky Mountains – 1834, J.K. Townsend, coll." 3. *Lupus gigas* Townsend ANSP 2266 skull, cotype, "Canis gigas Townsend" written on skull. "Rocky Mountains Dr. J.K. Townsend" in catalogue. Skull No. 531. (Morton 1840, p. 36). 4. ANSP 662 skin only, Rocky Mountains (on database, but not mentioned by Koopman 1976).

RED FOX *Vulpes vulpes* (Linnaeus, 1758)
Townsend a: 1. Red Fox, *Canus vulpes*. 2. Cross Fox, *Canus cinereo-argentatus*. 3. Black, or Silver Fox, *Canus cinereo-argentatus*. b: Canis [species not specified]. c: not listed.
Notes: Cross Fox, with dark 'cross' over shoulders and down middle of back is colour variation of Red Fox.

COMMON GRAY FOX *Urocyon cinereoargenteus* (Schreber, 1775)
Townsend a: Grey Fox, *Canus cinereo-argentatus*. b: Canis [species not specified].

AMERICAN BLACK BEAR *Ursus americanus* Pallas, 1780
Townsend a: Black Bear, *Ursus americanus*.
Skull No. 471 (Morton 1840, p. 35) not found at ANSP in 2003.

GRIZZLY BEAR *Ursus arctos* Linnaeus, 1758
Townsend a: Grizzly Bear, *Ursus ferox*. / White Bear / Brown Bear. Grizzly Bears mentioned several times in main text of *Narrative*.

HARBOR SEAL *Phoca vitulina* Linnaeus, 1758
Townsend a: Hair Seal, *Phoca vitulina*. Also mentioned in text for 6th September 1836.

NORTHERN RACCOON *Procyon lotor* (Linnaeus, 1758)
Townsend a: Racoon, *Procyon lotor*.

AMERICAN MARTEN *Martes americana* (Turton, 1806)
Townsend a: Pine Marten, *Mustela martes*.

FISHER *Martes pennanti* (Erxleben, 1777)
Townsend a: Fisher, *Mustela Pennanti*.

SHORT-TAILED WEASEL (or ERMINE) *Mustela erminea* Linnaeus, 1758
Townsend a: Ermine Weasel, Mustela erminea. b: Mustela (putorius) ermina? c: Mustela erminea? 1 [specimen] Oregon. d: Mustela erminea.
VQ: "Putorius Erminea. White Weasel" – Bachman says he "saw a specimen procured by TOWNSEND in Oregon."

LONG-TAILED WEASEL *Mustela frenata* Lichtenstein, 1831
Townsend a-c: not listed.
VQ: Specimen for Plate LX of "Putorius Frenata. Bridled Weasel" was captured by Townsend but no additional data is given.
Specimens: ANSP 2459 is skull No. 465 from Columbia River (Morton 1840, p. 35.).

AMERICAN MINK *Mustela vison* Schreber, 1777
Townsend a: Mink, *Mustela vison*.

WOLVERINE *Gulo gulo* (Linnaeus, 1758)
Townsend a: Wolverene [sic], or Glutton. *Gulo luscus*.

AMERICAN BADGER *Taxidea taxus* (Schreber, 1777)
Townsend a: American Badger, Meles labradoria. b: Meles labradoria. c: Meles Labradoria. 1 specimen Oregon. d: Meles Labradorica [sic].
VQ: Text refers to Townsend's specimens but does not say if they were used for Plate XLVII.
Notes: Townsend says he prepared two badger skins received from Indians at Fort Walla Walla, 16th July and 13th August 1836 (MS Journal).
Specimens: ANSP 524 skin & skull, juvenile, Oregon, 1834. ANSP 2447 skull only, female is skull No. 281 from Columbia River (Morton 1840, p. 33). ANSP 2448 skull only, male is skull No. 530 from Columbia River (Morton 1840, p. 36).

NORTHERN RIVER OTTER *Lontra canadensis* (Schreber, 1777)
Townsend a: Land, or River Otter, *Lutra canadensis*.

SEA OTTER *Enhydra lutris* (Linnaeus, 1758)
Townsend a: Sea Otter, *Lutra marina*.

MOUNTAIN LION *Puma concolor* (Linnaeus, 1771)
Townsend a: Cougar, or Panther, *Felis concolor** Townsend footnote: *There is a second species of Panther, of which, unfortunately, I possess only the skull and one foot. I believe it to be undescribed*. b-c: not listed.
Notes: Townsend's "second species" could be Mountain Lion or Bobcat or Lynx (the only cat species present in that region).
Specimens: ANSP 2240 skull only, adult male is skull No. 203 or 529 from Columbia River (Morton 1840, pp. 31, 36.).

CANADA LYNX *Lynx canadensis* Kerr, 1792
Townsend a: Hudson's Bay Lynx, *Felis hudsonicus*. b: not listed. c: *"a handsome spotted Lynx"* Lynx 2 [specimens] [Oregon] Skins in Spirits. d: Lynx canadensis?
Specimens: ANSP 2248 skull only, juvenile, Oregon (ex Morton collection, from Townsend).

ELK *Cervus canadensis* Linnaeus, 1758
Townsend a: Wapiti, or Red Deer, (Elk of the hunters,) *Cervus canadensis*.

BLACK-TAILED (OR MULE) DEER *Odocoileus hemionus* (Rafinesque, 1817)
Townsend a: Black-tailed Deer, *Cervus macrourus*.

WHITE-TAILED DEER *Odocoileus virginianus* (Zimmermann, 1780)
Townsend a: White-tailed Deer, *Cervus leucurus*. b: *Cervus leucurus*. A fawn, resembling very strongly that of CERVUS *virginianus*. c: *"the Oregon Deer"* Cervus leucurus 1 [specimen] Oregon Skins in Spirits. d: Cervus leucurus.
VQ: "There is now in the Academy of Sciences at Philadelphia a young male which was procured some years since by the late Mr. J.K. TOWNSEND on the Columbia River. Plate CXVIII not from Townsend specimen.
Specimens: ANSP 2356 is skull No. 533 (Morton 1840, p. 36.). [Skull No. 534 not found in 2003.]

MOOSE *Alces alces* (Linnaeus, 1758)
Townsend a: Moose, *Cervus alces*.
Skull No. 515 (Morton 1840, p. 35, and 1849, p. 5.) not found at ANSP in 2003.

PRONGHORN *Antilocapra americana* (Ord, 1815)
Townsend a: Prong-horned Antelope, *Antelope furcifer*.

AMERICAN BISON *Bos bison* (Linnaeus, 1758)
Townsend a: American Bison, or Buffalo, *Bos americanus*.

MOUNTAIN GOAT *Oreamnos americanus* (de Blainville, 1816)
Townsend a: Mountain Goat, *Capra americana*.

BIGHORN SHEEP *Ovis canadensis* Shaw, 1804
Townsend a: Mountain Sheep, *Ovis montana*.

Additional notes:

ANSP 170 Brown Rat *Rattus norvegicus* and ANSP 424 Big Brown Bat *Eptesicus fuscus* may or may not have been collected on Townsend's journey to the Pacific Northwest: both species occur in both eastern and western North America. The ANSP database also lists ANSP 143 *"Geomys"* with no further data; missing in 2003; ANSP 225 has no data of any kind; missing in 2003. A Townsend specimen of a shrew is claimed to have been used in the description of the Southern Short-tailed Shrew *Blarina carolinensis* (see Handley 1994). USNM database also lists a Meadow Vole *Microtus pennsylvanicus* (USNM 5153 skin only) as collected by Townsend in Pennsylvania.

For some of Townsend's mammals, the ANSP database gives vague collecting localities such as "Rocky Mountains", "Oregon", "Columbia River" and almost always gives 1834 as year of collection. Such localities, especially "Columbia River" and the year of collection may sometimes be incorrect assumptions.

Nomenclature follows Baker *et al* 2003.

APPENDIX 8. WHERE AND WHEN DID TOWNSEND COLLECT HIS SWAINSON'S HAWKS?

Swainson's Hawk *Buteo swainsoni* is a prairie hawk that hunts small mammals and insects in grassland, usually nesting in isolated trees or clumps of trees but occasionally on cliffs. Its current breeding range (with a relevance to Townsend's 1834 route) extends from eastern Nebraska, through Wyoming and Idaho, and into Oregon and Washington as far west as the Cascade Mountains; it is decidedly scarce on the Pacific side of the Cascades (Slipp 1947). The type specimen of Swainson's Hawk was collected by Townsend, but the location has long been a matter of contention. At one time it was thought to have come from west of the Cascades near Fort Vancouver, Washington, but the locality currently accepted is east of the Cascades near Fort Walla Walla, Oregon. Are either of these locations correct?

First we must consider the times in which Audubon and Townsend were working. In 1834 *Buteo* identification and the geographical distribution of each species was very imperfectly known. In addition to Red-tailed Hawks, Townsend and Nuttall both thought that they were seeing Common Buzzards *Buteo vulgaris* (now *Buteo buteo*, a European species that was then widely but erroneously believed to also occur in North America). In addition, they thought that there was another species present to which they gave the names "White throated Buzzard" and "White chin Buzzard". In 1838 Charles Bonaparte correctly believed this bird to be a different species and called it Swainson's Hawk *Buteo swainsoni*[1] – the English and scientific names that we still use today. Bonaparte's publication had no description of the bird and instead referred the reader to an Audubon painting in *The Birds of America* (Plate CCCLXII) – which was of a bird collected by Townsend. In the absence of the surviving specimen, the Audubon plate depicting Townsend's hawk has become the type (sometimes referred to as an iconotype). Audubon's accompanying text, in the *Ornithological Biography*, is not particularly helpful:

> "COMMON BUZZARD. BUTEO VULGARIS … The specimen from which the figure before you was taken, was shot by Mr. TOWNSEND on a rock near the Columbia River[2], on which it had a nest. Unfortunately he has not supplied me with any account of this species … (Audubon 1840-44, 1: 30-32).

Audubon goes on to mention a second specimen acquired by Townsend on the Snake River plain (see later) but this is not the one he painted. On the basis of Audubon's text, the type locality for the Swainson's Hawk was long accepted as Fort Vancouver, Washington (see early *AOU Checklists*). This seems to have been a rather arbitrary decision based on the fact that the fort lay beside the Columbia River and that Townsend collected there on and off for several months during 1834, 1835 and 1836. This area is well outside the current normal breeding range of Swainson's Hawk, and there are no rocky outcrops in the immediate vicinity (though there are prominent rocks upstream, e.g. Rooster Rock at 23 miles and Beacon Rock at 33 miles). It was later argued by Slipp (1947) that the type locality should be Fort Walla Walla that also lay beside the Columbia River but it is further west within the species' normal breeding range. Moreover, this is a relatively treeless area and just a few miles downstream there are terrific cliffs where the birds could have nested.

Slipp even postulated that the bird was collected on 4th September 1834, and if Slipp's theory is correct, it has to have been taken then, or within a few days of this date, because Townsend passed through Fort Walla Walla and the surrounding area in just a few days[3]. There is nothing in Townsend's *Narrative* or in any other sources to support or refute this exact date but the most recent edition of the *AOU Checklist* (1998) now accepts Fort Walla Walla as the type locality.

In fact we have misgivings about both Fort Vancouver and Fort Walla Walla as the type locality. Fort Vancouver because we share Slipp's doubts, mainly that the fort lies well outside the normal range of the species. Fort Walla Walla is certainly a more plausible locality, but if we accept Slipp's assertion that this fort is the correct locality we are also being asked to accept, without any evidence, that the bird was beside its nest on a rock in early September, a rather late nesting date; indeed, many hawks begin their southward migration in this month. So, if these localities are not correct, can we find the true locality by examining some additional sources, including unpublished material?

Unfortunately, although we have investigated the literature and searched museum collections for information on all the hawks that Townsend collected we have not found anything conclusive. We do not know how many Swainson's Hawks he prepared, though in practical terms he is unlikely to have skinned many because of a self-imposed restriction on keeping large and bulky birds that would have added to his transportation difficulties. There are no surviving skins and therefore no specimen labels with dates and localities. Moreover, there are no surviving notes by Townsend concerning Swainson's Hawks. He never mentions the species except in three lists: one in the appendix to his *Narrative* (1839), another in the *Journal of the Academy of Natural Sciences of Philadelphia* (1839), and, a little more helpfully, in McEuen's list where Swainson's Hawks are given the field numbers 34 and 43 (see appendix 5). However, by referring to Nuttall's observations and other sources, we can examine Townsend's few documented encounters with Swainson's Hawk, in supposed chronological order, east to west:

1. As Townsend and the rest of the Wyeth party travelled westwards the first Swainson's Hawks that we hear about were seen on Big Sandy River, or another stream in the vicinity, just before, during or after their arrival at the Trappers' Rendezvous near Ham's Fork, Wyoming. In his *Manual of Ornithology* Nuttall has this to say:

> "WHITE THROATED BUZZARD (*Buteo montana*) … Mr. TOWNSEND and myself observed this bird on the woody margins of the Rocky Mountain streams which pass into the Colorado of the West [Green River], about the month of July, when they were breeding, in a tree where Mr. T. found the nest containing 2 white eggs. Its habits closely resembled those of the Red Tailed Buzzard. From the *Buteo vulgaris* it appears to be sufficiently distinct, however, allied (Nuttall 1840: 112).

The party were near the Green River on 16th June and left its watershed on 5th July, crossing over to the Bear River by 8th July. Nuttall says the nest was in a tree, not on a rock.

2. The first entry of the "White chin Buzzard" (No 34) in Townsend's catalogue (a chronological list of field numbers, see appendix 6) appears immediately after Clark's Nutcracker (No 33). The nutcracker is first mentioned in his *Narrative* near the Bear River on 8th July 1834, but the USNM card catalogue includes details of a nutcracker specimen collected on 16th June 1834 (USNM 1929). So the hawk could have been taken between 16th June and 8th July, or a few days later. On 16th June Townsend was on the lower part of the Sandy River that leads into the Green River; this ties in well with Nuttall's observation on the "White Throated Buzzard". The reconstructed catalogue is not particularly useful in the present context. The catalogue is only useful in indicating when the first specimen of a particular species was obtained. There could have been any number of specimens with the numbers 34 and 43, possibly including specimens taken in other years.

3. Nuttall (1840: 99-100) mentions a Rough-legged Buzzard, described as "black … tail rounded with 5 bands of white, and terminated with dull white". He adds that "My friend, Mr Townsend found its nest on the banks of the Bear River, west of the Rocky Mountains. The nest formed of large sticks, was in a thick willow bush, about 10 feet from the ground and contained two young almost fledged." Since the Rough-legged Buzzard does not breed this far south it must have been some other species. A dark morph Swainson's Hawk is very likely, but other possibilities are a dark morph Ferruginous or Red-tailed Hawk. Even if it was a Swainson's Hawk the nest was in a tree.

4. Later in Townsend's catalogue there is an entry for "Buteo Vulgaris" (No 43) – a Swainson's Hawk. The first specimen was prepared earlier than the Steller's Jay (No 48) that Townsend says he first encountered in the Blue Mountains (about 31st August 1834) so the hawk would have been collected at some point further east than the Blue Mountains. This may have been the second hawk mentioned by Audubon: "Another specimen [of "Common Buzzard *Buteo vulgaris*"] in my possession, procured by Mr. TOWNSEND on the plains of the Snake River, has the upper parts brown, streaked and spotted with reddish-white; the upper tail-coverts white, barred with dusky, the lower parts as above described" (Audubon 1840-44, 1: 30-32). Townsend may have taken this bird on the Snake River plain at Kamas Prairie (about 16th August), still a favourite resort of Swainson's Hawks.

5. Finally, there is the adult female specimen taken "on a rock near the Columbia River", the one that Audubon painted. Many such specimens that Audubon painted were passed to the Edward Harris collection, or to S.F. Baird and subsequently to the Smithsonian Institution (USNM). As we have said, there is now no trace of any of Townsend's specimens of Swainson's Hawks.

None of this is very conclusive. All it tells us is that Townsend first saw Swainson's Hawks near the Green River, again near Bear River, and again on the Snake River plain, all between 16th June and 16th August. When nests were found they were in trees. No cliff sites are mentioned (though there are cliffs in these areas where Swainson's Hawks could have been nesting). Nothing disproves the possibility of a specimen being collected further west at Fort Walla, or Fort Vancouver.

There may, however, be a solution to our question about the true type locality. Audubon is known to be unreliable in his statements regarding the provenance of certain specimens. He (or his co-author William MacGillivray) could simply have mixed up Columbia River with the Colorado River, i.e. the Colorado of the West, a now obsolete name for the Green River, in Wyoming. Indeed, Townsend's label may have simply said "Col. R." or at some point this abbreviation may have been used by Audubon or MacGillivray in their various handwritten drafts for the *Ornithological Biography* and it was later construed as Columbia River instead of Colorado River. This is the explanation that we favour because this would set the collecting locality in the heart of the breeding range of Swainson's Hawk where we know that Townsend saw and collected this species. If we simply change Audubon's text to read "The specimen from which the figure before you was taken, was shot by Mr. TOWNSEND on a rock near the *Colorado River* [= Green River], on which it had a nest." Everything then fits neatly into place, including the time of year, right in the midst of the hawk nesting period.

Although there is no evidence for this idea we do know cases where this mistake has occurred in the reverse. A Pileated Woodpecker specimen (USNM 2792) is listed in the Smithsonian Institution database as being collected by Townsend on the Colorado River on 9th May 1835 – at a time when Townsend was certainly on the Columbia River. An Orange-crowned Warbler (USNM 2929) collected on 15th May 1835 is likewise mistakenly catalogued as originating from the Colorado River.

1. There is an illustration of Swainson's Hawk included in the *Fauna Boreali-Americana* (1832) that pre-dates Audubon's plate of this hawk. Richardson and Swainson, the co-authors of the volume on birds, considered it to be the same as the "*Buteo vulgaris*". Bonaparte had seen the illustration but does not refer to it in his 1838 list when he named the species Swainson's Hawk. Swainson's plate also has a claim to be the type, in which case the locality would be Carlton House (Houston 1998). This is a question for taxonomists but Swainson's illustration would at least have the merit of offering a more certain type locality.

2. Audubon describes it as a female in the *Ornithological Biography* (1840-44,1:30-32). The bird depicted is in a rather dark adult plumage and would be described by current field workers as "two-toned". Most "two-toned" individuals are males but not exclusively so (Bill Mattox *in litt.*, 2003).

3. In July and August 1836 Townsend spent several weeks at Fort Walla Walla, much earlier in the breeding season. But by then the Swainson Hawk specimen that Audubon was to paint had already been shipped to Philadelphia, so it would have to have been taken on the very brief 1834 visit to Walla Walla, *if at all*.

Typical nesting habitat for Swainson's Hawk (near Green River).

APPENDIX 9. THE COLUMBIA RIVER BIRD REPORT GIVEN BY TOWNSEND TO THE REV SAMUEL PARKER

Rediscovery of the report

In the summer of 1935 Clifford M. Drury, a Presbyterian minister and historian, was visiting the eastern United States researching the life of Marcus Whitman when he came across an important Townsend manuscript. He was staying in New Jersey at the home of a great grandson of the Reverend Samuel Parker when he discovered Townsend's *"Description of the Birds of the Columbia River Region"* drawn up at Fort Vancouver on 11th April 1836 and handed to the Reverend Parker just before Parker returned to the east. Parker used just a part of the bird report in his *Journal of an exploring tour beyond the Rocky Mountains* (1838).

Drury took photostat copies of the manuscript and from it made a typescript copy, because it would then be easier to read. It was not published at the time but carbon copies of the typescript were circulated; although Hall (1937) did not quote from it he used it while writing about Nuttall and Townsend on the Columbia. A few years later, while teenager David B. Marshall was at an Oregon Audubon Society meeting or field trip, a copy of the typescript was handed to him and he was advised to keep it in a safe place. Marshall put it away and did not find it again until January 1989, but this time he realised its significance and the whole document was soon afterwards published in full in *Northwestern Naturalist* with a commentary on the ornithological content (Jobanek and Marshall 1992).

There are a number of minor errors and changes in the copy that seem to be the work of the typist: for examples, the mistakes in the scientific names, the incorrect Christian name for Thomas Say, and the reference to the *"Cygnus bewickii of Mr. Garrell"*, an obvious mistake for (William) Yarrell, a well-known English naturalist of Townsend's day. Some of the spelling errors also seem atypical of Townsend. Although the original manuscript cannot be traced there is no doubt about the original authorship of the bird report. It was evidently prepared hurriedly and was probably written without any preliminary draft. There is no real attempt to include all the birds he had seen as there is no mention of the herons, shorebirds, or California Condor, and the warblers are dismissed in just a short paragraph. This is Townsend's report:

DESCRIPTION OF THE BIRDS OF THE COLUMBIA RIVER REGION

The birds of Oregon territory are not numerous, compared with the immense swarms which inhabit at all seasons the more civilized parts of our vast Continent: they are however sufficiently so to afford the ornithologist and collector an ample field for a year's study & a not unprofitable opportunity of procuring specimens. The birds of this region are particularly interesting from the fact, that until now it has never been visited by an ornithologist. The only naturalist who had previously explored the Country to any extent was the late talented and lamented Mr. David Douglas; but he was not particularly au fait in this science & collected such specimens as appeared to him curious & peculiar, neglecting by far the greater number to attend to his more immediate pursuits. Many, & perhaps the majority, (but not by any means the whole,) of the birds found here may probably be seen at times on the Continent of Mexico & in Alta California – Some of the spring visitors undoubtedly retire then to attend to the duties of reproduction, & many others which are not migratory are probably resident there. During the autumn [sic] & winter few birds are to be seen. As the season advances many of these seek more congenial latitudes & from Dec. to March the Ornithologist may rest his labors – Occasionally however, even at the more vigorous [rigorous] season, when the ground is covered with snow, when the water courses are choked by ice & the branches of the forest trees are encrusted with the same material, a solitary robin [**American Robin**] may be heard whistling from the depths of a tangled bush in which he had ensconsed [sic] himself, but his notes are feeble & his madrigal short, the chill weather represses his disposition to melody & the soaring & hungry hawk prevents his egress from the sheltered bramble. As I have said, there are but few birds which reside here during the whole year. Among the few we may name, first & chiefest of all the feathered tribes the noble & majestic White-headed Eagle, (Falco leucocephalus) [**Bald Eagle**]; there are also a few species of Hawks, the Red-tailed kind common throughout the states – (Falco borealis) [**Red-tailed Hawk**], the Hen Harrier (F. cyaneous) [**Northern Harrier**] & one or two others. There are also two species of Jay, one the **Steller's Jay** (Garrulus stelleri) exceedingly abundant at all seasons, & the other [**Western Scrub-Jay**], a new species probably, with a white belly & blue pectrol [sic] collar, somewhat resembling one

common to the Southern states of the Union the G. floridaneous [Florida Scrub Jay *A. coerulescens*]. There is also the Magpie (Garrulus pica) [**American Magpie**[1]], thousands of Ravens [**Common Raven**], Crows [**American Crow**[2]] & Several species of small Sparrows are abundant during the winter. One of these – a new one [**Dark-eyed Oregon Junco**] – very similar to our Common Snow Bird, only seen at this season, & another very plentiful at all times. Two species of Grous (Tetrao) are also abundant, the Common Partridge of the States (T. umbellus) [**Ruffed Grouse**], and another, the Dusky Grous of the Rocky Mountains (T. obscurus) [**Dusky Grouse**, and **Sooty Grouse**[3] near coast]. In December the beautiful & familiar little Crested Wrens (Regulus calendula & cristatus) [**Ruby-crowned** and **Golden-crowned Kinglets**] appear & glean about the trees for a few days in quest of their insect food. They remain now but a short time, before the end of the month they leave the country to make timely preparations for their early nidification. The few kinds that I have remarked are the chief that inhabit this part of the country, (Fort Vancouver on the Columbia River) during the winter. I cannot say positively as to what species may reside at this season in the interior, but I have reason to believe that my catalogue would not be much augmented, as the cold is there equally if not more severe, & the facilities of obtaining food &c, not greater. While treating this part of my subject however I must not omit to mention a curious & very interesting species of Dipper or Water Ousel (the Cinclus pallasii) [**American Dipper**[4]] hitherto very little known. I have observed it only in the autumn & winter. Its habits are very curious & peculiar, particularly that of its descending to the bottoms of swiftly running streams & foraging there for minute shell fish &c. I have seen it remain under water for at least two minutes during which time it coursed along the pebbly bottom with as much apparent ease & satisfaction as though it travelled upon the dry land.

About the first of Nov. the Swans, Geese & ducks are seen winging their way over us in the most peculiar manner. Their flight is then high, & the screams, – particularly of the Swans – are at times almost deafening: they first visit the Sea & contiguous bays & for a few weeks feed on the shell fish & spawn found abundantly on the shores: they then begin gradually to approach the interior of the country & before December the lakes & lagoons of the low lands in the vicinity of the river are literally swarming with them. The Swan, (I have not been able to indicate with certainty more than a single species) is not the one common to the States. It is the Cygnus bewickii of Mr. Garrell [Yarrell], about the size of our common species, (the C. americanus of Sharpless [**Tundra Swan**[5]]) & characterized by the same unsullied plumage. Its habits are very similar to those of our bird: – its attitudes & motions while sailing over the surface of the stream equally elegant & graceful, & its voice even louder & more sonorous. The great length & extreme tortuosity of the trachea, – forming several convolutions within the cavity of the sternum – satisfactorily accounts for their pre-eminence in this respect [**Trumpeter Swan**]. Unlike the Swans that frequent the Chesapeake Bay & its vicinity [i.e. Tundra Swans] their flesh is not exquisite [sic] tasted. This is in consequence of the inferior quality of the food upon which they here subsist, the lakes affording them only a supply of small thin shelled Mollusca with a few aquatic insects &c. I have however occasionally seen Swans having their crops filled with the Wappatoo or Indian potatoe [Broad-leaved Arrowhead *Sagittaria latifolia*]. Such are always finer flavoured than the ordinary ones & were this vegetable to constitute their sole diet, I have no doubt that this noble bird might vie in delicacy of flesh with our own, which is so widely & justly celebrated on that account. What I have said of the Swan applies to the whole family of water fowl which pass the winter in this region. They would not be considered worthy to ornament the tables of our pampered epicures, but they are nevertheless eatable & when properly dressed, even palatable.

Of Geese there are here four species: the Canada or Common Wild Goose (Anser canadensis) [**Canada Goose**], the White-fronted goose (A. albifrons) [**Greater White-fronted Goose**], the White or Snow G. (A. hyperborecus [sic]) [**Snow Goose**] & Hutchins' Barnacle (A. hutchinsii) [**Cackling Goose**[6]]. The three first are more or less common in the U. States, but the last is never found there. It is rather larger than the common Brant (A. bernicla) & in its markings resembles very much its relative the A. canadensis. There are about twelve species of true ducks, most of them common to various parts of the U. States: I have however discovered one new one closely allied to the [Northern] Pintail (Anas acuta)[7]. The celebrated Canvass-back (Faligula valisneria [sic]) [**Canvasback**] is common, but his, (with us,) inseparable companion, the Red-head (F. serina) [**Redhead**[8]] is never seen. The **Green-winged Teal** (Anas Crecca) is very abundant, & is I think the finest flavored of all the water fowl which visit the country. They appear early in Dec. in immense flocks – & are constantly seen searching with singular assiduity along the muddy flats of the river & inland lakes for the minute shell fish upon which they feed. The Indians destroy them in great numbers, & I have myself killed 8 by a single discharge of small shot. Of Mergansers there are only two species, the Goosander (Mergus merganser) [**Common Merganser**] & the Hooded or Crested one (M. cucullatus) [**Hooded Merganser**]. The former is only seen in the winter & the latter in the summer. They are both eaten & are scarcely inferior to most of the ducks. There are two species of Pelecan [sic] resident here, the white & brown one (Pelecanus onocrotalus & fuscus) [**American White Pelican**[9] and **Brown Pelican**]. During the autumn & winter they are found chiefly in the neighborhood of the sea, seldom advancing high on the river, but on the approach of spring they retire inland to breed, choosing the most solitary & retired places in the neighborhood of water courses. In the month of June 1834 while on my way to this country I observed great numbers of these birds on a small stream called the Blackfoot River in the Rocky Mountains, at that time engaged in rearing their young. The Black Cormorant (Phalacrocorax carbo) [not Great Cormorant, = **Double-crested Cormorant**[10]] is common on the river, & there are at least three more species of the same genus not yet indicated, which reside near the Cape [Cape Disappointment] & probably nest upon the surf washed sides. They never ascend the rivers & are in consequence very difficult to be procured. The Loon or Great Northern Diver (Columbus glacialis) [**Common Loon**] is very plentiful in the river, & I have reason to believe that there are other species of this genus scarcely less common, but from their extreme shyness & quickness in diving I have not as yet been able to procure them. The closely allied Genus Podiceps or Grebes have several representatives here: the Chief of them is the small Dobchick (P. minor) [not Little Grebe = **Pied-billed Grebe**[11]]. There is also the Red-necked G. (P. rubicollis) [**Red-necked Grebe**], the Crested G. (P. cristatus) [**Great Crested Grebe**[12]], & one whichmay [sic] probably prove a new species, with an Unusually long neck [**Western Grebe** or **Clark's Grebe**[13]].

The well known & remarkable shyness of this tribe of birds, & the amazing quickness with which they evade the shot by diving almost with the swiftness of light beneath the surface, forms a nearly insurmountable barrier to the aquisition [sic] of specimens for examinations. Even the Indians who are so proverbially expert in procuring game of all kinds & who resort to so many stratagems to deceive their victims, acknowledge their inability to kill the Grebes, & although I have repeatedly offered exorbitant prices for the more ordinary specimens I have received from them only two or three during a residence of eighteen months in the country.

Before taking leave of the Water Birds, I must mention as well worthy a place, the incomparable Summer or **Wood Ducks** (Aix sponsa). This is perhaps the most beautiful of all the tribe in the known world with the single exception of the Mandarin of China. They are very plentiful here in the summer, but are never seen in the winter. As the autumn approaches they become rather fat & are tolerable eating: they are then killed by the Indians in considerable numbers & brought to the fort to exchange for powder & shot. Although I am rather fond of their flesh, yet I cannot say that I ever saw an Indian carrying a string of these splendid birds & soiling & ruffling their beautiful plumage without an inadvertent feeling of regret that anything so supremely lovely should be sacrificed for a purpose so sensual & common place.

In addition to the Aquatic birds already mentioned, & which governed by the powerful instinct wisely given to all inferior animals, voluntarily take up their abode here during a limited period in each year, there are others which occasionally visit the rivers & bays impelled by the same instinct to seek shelter from the violent storms which sometimes agitate the the [sic] ocean during the winter season. Gulls, Terns, Auks, Petrels &c flock in innumerable hosts & even the gigantic Brown Albatross, (Diomedes [sic] fusca) [Light-mantled Sooty Albatross, error for **Black-footed Albatross**[14]] is sometimes seen – they remain only during the prevalence of the gale and seem while they are compelled to stay, uneasy & discontented. On the subsidence of the storm they wing their way out to beyond the dangerous breakers to pursue as before their roving & solitary life upon the trackless deep.

As in other Countries the spring brings with its fruits & flowers, hosts of the lovely & gladsome feathered creatures, – many of whom however tarry but a very short period. They remain, according to the forwardness of the season, from 2 to 3 weeks, & then pass away, (generally south) to breed. But there are a considerable number that continue with us until September, & in a fine April morning or May the woods & groves are vocal with their delightful carrolings [sic]. Hundreds of Warblers, Wrens, Titmice & Nuthatches are straining their throats for the mastery – some are engaged in wooing & endeavoring to render themselves agreeable to their affianced mates others, who are farther advanced in the important and interesting concerns of their being are constructing those beautiful & admirably adapted fabrics which have astonished philosophers of all ages, & compelled even the most infidel of them to exclaim, – Here, as in every part of his bountiful Creation hasGod [sic] made himself manifest, supplying in the stead of the principle of reason, a mysterious & unerring instinct always sufficient for the end intended to be accomplished.

About the end of March the splendid little Nootka Humming Bird (Trochilus rufus) [**Rufous Hummingbird**] makes his appearance. The gorgeous little messenger comes so suddenly that you are led to wonder where he could have sprung from, as the fact of his performing with his feeble & delicate little wings a long migration of weeks over a cold & flowerless country, & even sometimes across the sea, seems almost incredible. But here he is however, & who can look upon his exquisite plumage without delight? He flutters around you sipping the honey from every flower, & as tho the sun beams sticks upon his gorgeous throat you almost feel an inclination to veil your eyes from its dazzling splender [sic]. The neck of this most magnificent little creature differs from that of the common one at the east [Ruby-throated Hummingbird] by presenting even more variation of color according as the rays of the light are received upon it. If you look at it in front, in the direction of the grain of the scaly feathers, it is only red with the most brilliant metalic [sic] luster; turn the bird in the hand & examine it obliquely, and equally luminous sapphire is seen, while in some other points of view the tints vary from purple to violet & crimson, blended with minute & ever varying shades of orange & scarlet. This gorgeous little fellow inhabits this section of the country in great numbers during summer & I have been informed is equally abundant in the neighborhood of Fort Walla Walla, several days journey into the interior. The nest is built in the same kind of situation & composed of very nearly the same materials as that of our own. The eggs are two, pure white & about the size of small peas.

About the middle of April the Warblers arrive. Of these I have already noticed eleven species, six of which are new & the remaining five, more or less common in the States. Several of them are but transcient [sic] visitors, remaining with us but a few weeks, but the majority nest here & most of them produce two broods in a season.

In the train of this numerous body follows the Wrens, Tanagers, Thrushes, Flycatchers, Finches, &c. – a large & musical band, among which we may remark several of the finest songsters in the known world. The Wilson's Thrush (Turdus wilsonii) [**Veery**[15]] is pre-eminent in this respect. This fine species breeds here & is consequently, for a part of the year in full song. The only Tanager found here is a new species somewhat similar to the one [Summer Tanager] inhabiting the extreme western portion of the Union (the [**Western Tanager**] Tanagra ludovicianus), but sufficiently distinct. Its voice in breeding season is beautiful, resembling some of the thrushes. But in no instance have I seen combined more richness & delicacy of plumage with the most perfect harmony of voice than in the case of a large Bullfinch (also a new species) [**Black-headed Grosbeak**] which visits this section in the spring. I have made several unsuccessful attempts to domesticate this fine bird, but my failures I attribute to the young having been too mature when taken, & I have no doubt that were they procured early from the nest they might easily be raised, & they would certainly form a most valuable addition to any aviary. Of the genus to which this is attached, (Fringilla) there are within the territory of Oregon 17 species, 6 of which are new & several of the remainder common at home. The **Yellow-billed Cuckoo**[16] (Coccyzus americanus) is abundant during the summer. His singular voice may be heard repeatedly along the banks of the river. It is a very curious coincidence that the Indians here have imbibed the same notions with regard to the prognosticating qualities of this species, which have obtained so universally among their civilized brethern [sic]. They assert that the common cry of this bird, – resembling the words kow-kow-kow-kow – is a sure indication of rain, & that even in the dryer seasons this forbody [sic] call is sure to be followed by a plentiful dash of the element: but although apparently so I have often listened to the Cuckoo's monotonous notes when the earth has been suffering for want of moisture, and none has fallen for weeks afterward.

Of the Woodpeckers (Picus) there are several that are very interesting – Among these we may instance the very distinct & fine species discovered by Lewis & Clark in their memorable exploring expedition – the (P. torquatus)– [**Lewis's Woodpecker**]. This species is very little known at home, no subsequent traveller having met with it. It is common in this vicinity & breeds here. (P. pileatus) [**Pileated Woodpecker**[17]] is also plentiful & there are several more hitherto unknown to naturalists. –

Of the swallow tribe (Hirundo) we have the familiar Marten [sic] [**Purple Martin**], the Common Green-blue species (H. bicolor) [**Tree Swallow**] the Sand Marten (H. riparia) [**Bank Swallow**[18]], & a most lovely new one with a green back & purple rump, the most beautiful of all the genus hitherto discovered [**Violet-green Swallow**]. The common Chimney Bird or Swallow [not Chimney Swift, = **Vaux's Swift**[19]] is also abundant during the summer, but in this savage country he has to accomodate [sic] himself with the trunk of some hollow tree in which to rear his brood instead of the more congenial dwelling which he always selects in civilized countries.

I have now to speak of a fine and useful genus of Birds (Tetrao or Grous) abundant in this country, & with a few remarks on the varieties shall conclude this paper. In the neighborhood of Fort Walla Walla & east of the great dividing ridge of the Rocky Mountain Chain we find a most noble & very interesting species, the Cock of the Plains of Lewis & Clark (T. uropasianus [sic]) [**Greater Sage-Grouse**]. In size & gracefulness of manner he may vie with any of his class in Europe, but I am sorry to say that as an article of food the comparison will not hold. This inferiority is owing to a rank & most nauseous species of wormwood (Artemisia) [Sagebrush *Artemisia tridentata*] which forms their chief subsistence. This quality however contributes to their preservation, & being a most prolific species many parts of the country are literally swarming with them. In passing with our caravan across the mountains in 1834 we found these birds so numerous in some placcs that at times we even made use of our riding whips to prevent them being trodden to death by our horses.

On the next largest & finest species I have already spoken in the early [part] of this paper, the Dusky Grous, (T. obscurus of Wm Say [= Thomas Say]) [**Dusky Grouse**]. It inhabits, plentifully, the interior of the country & is also frequently found here [**Sooty Grouse**]. Its flesh is excellent, the color dark – but richer & more juicy than that of the Partridge [Ruffed Grouse]. The Sharp-tailed Grous (T. phasianellus) [**Sharp-tailed Grouse**[20]] is occasionally seen in this vicinity but it is considered by the Indians a rare bird. Many of them are indeed not acquainted with it. It inhabits abundantly the eastern declivity of the Mountains but is probably only a straggler here. The Spotted Grous [**Spruce Grouse**], the Ptarmigan, the Rock Grous & several others, with some new species which I have myself indicated reside in the upper country. They are mostly wild & difficult ti [sic] be procured, many of them affecting the elevated land & even living for a great part of the year above the region of perpetual snow, so that the naturalist must be a zealous & hardy one who succeeds in procuring specimens of the whole.

I have thus given [in] a very succinct manner an account of the birds of Oregon territory – It must be considered however only in the light of a sketch. Time is not allowed me, nor has it been consistent with my design to furnish an elaborate history of all the species, their habits & manners &c – My object has been simply to give some idea of the ornithological treasures of a new & unexplored country, & if any public use should ever be made of the present hasty production, I should wish that it might have the effect of inciting some of the young & enterprizing [sic] naturalists of our country to visit this most interesting region, with a view of adding to our stock of knowledge in one of the loveliest of the Natural Sciences & increasing & enriching the Cabinets of our already extensive Museum.

Fort Vancouver, Columbia River

April 11th, 1836

Furnished by Mr John K. Townsend, Ornithologist, from Philadelphia.

1. The American Magpie is now only an occasional visitor to the lower Columbia River but there are old breeding records and it may then have been more abundant (Jobanek and Marshall 1992: 10).

2. Jobanek (1994: 13-14) considered all crow records to refer to the small race of the American Crow *Corvus brachyrhynchos hesperis* rather than the Northwestern Crow. Townsend lists the Fish Crow among his sightings but this is an east coast species.

3. The Blue Grouse *Dendragapus obscurus* has recently been split into two species (*AOU Check-list, 47th Supplement*, 2006): Townsend saw Dusky Grouse *Dendragapus obscurus* in the Blue Mountains and Sooty Grouse *D. fuliginosus* on the lower Columbia River.

4. The dark brown *Cinclus pallasii* occurs in South and Central Asia, Himalayas, China and northern Indonesia. In 1836 Townsend seems to have been unaware that the American Dipper had been described in 1827 by William Swainson from a Mexican specimen.

5. Bewick's Swan (now considered a race of the Tundra Swan) does not occur in North America and though he says he saw only one species of swan he probably saw both Tundra and Trumpeter Swans. His description of the trachea must refer to the Trumpeter Swan.

6. The northern populations of "small Canada Geese" are now gathered into a single species: Cackling Goose *Branta hutchinsii* (Richardson). Hutchins's Goose, the nominate race, has the most easterly distribution and the races found on the lower Columbia River are more likely to have been *B. h. taverneri* , *B. h. minima* or *B. h. leucopareia* (*AOU Check-list, 45th Supplement*, 2004). See also Jobanek and Marshall 1992: 10.

7. Probably just a Northern Pintail in eclipse or immature plumage, perhaps even an aberrant individual.

8. The Redhead does occur on the lower Columbia in winter but it is rare.

9. The White Pelican is now considered rare on the lower Columbia.

10. It is not clear why Townsend should have mistaken any of the west coast cormorants for the Great Cormorant. Jobanek and Marshall (1992: 10) suggest that this was a mistake for the Double-crested Cormorant but maybe he meant the larger Brandt's Cormorant that he had perhaps not yet managed to see at close quarters. He certainly collected Brandt's Cormorants at some stage, yet he still included the Great Cormorant in his list of birds from the territory of Oregon in his scientific appendix to the *Narrative* (Townsend 1839c) but not in his list in the *Journal of the Academy of Natural Sciences of Philadelphia* (Townsend 1839f).

11. This is a curious error but must refer to the Pied-billed Grebe as Jobanek and Marshall suggest (1992: 9).

12. The Great Crested Grebe *Podiceps cristatus* is a Eurasian species and would have been familiar only to Nuttall. Jobanek and Marshall (1992: 9) considered this to be a reference to the Horned or Eared Grebe but the larger Red-necked Grebe in winter plumage is more similar to the Great Crested Grebe.

13. Grebes with "an Unusually long neck" must be a reference to Western Grebes and Clark's Grebes. If he had managed to collect specimens they would have been new species as they were not described until 1858.

14. The scientific name refers to the Light-mantled Sooty Albatross but this was the Black-footed Albatross, a common species off the Oregon coast that Townsend collected between the Columbia River and the Hawaiian Islands. Townsend's seabirds are discussed in appendix 10.

15. More likely to have been the Swainson's Thrush, described from a Townsend specimen in 1840 by Nuttall with an acknowledgment that on the Columbia River they had mistaken it for the Veery.

16. Although the Yellow-billed Cuckoo was once fairly common in Oregon it is now rare (see Jobanek and Marshall 1992: 11).

17. It is interesting that Townsend should refer to the Pileated Woodpecker since this is the species he probably saw in the Sawtooth Mountains, Idaho. Audubon suggested that the woodpecker seen in the mountains was the Imperial Woodpecker, a Mexican species, an error that Townsend appears not to have tried to correct (see p. 110).

18. Townsend would have seen both Bank Swallow and Northern Rough-winged Swallow, the latter being much more common but not described as a new species until 1838.

19. The Chimney Swift is an eastern species. Townsend later realised that this was a new discovery, naming it Vaux's Swift in 1839.

20. The Sharp-tailed Grouse seems to have been extirpated from the lower Columbia not long after 1836, as there are no other records from the area (Jobanek and Marshall 1992: 10).

APPENDIX 10. TOWNSEND'S CONTROVERSIAL SEABIRDS

During Townsend's voyages to and fro between the Columbia River and the Hawaiian Islands, and on his voyage home via Tahiti and Chile, he often passed the time by collecting seabirds. He shot Red-tailed Tropicbirds as they flew above the ship in the hope that they would fall on deck (Townsend 1848, p. 90), he trailed a baited hook in the water to catch albatrosses, and entangled the wings of petrels by letting out un-baited lines from the stern (see pp. 262, 264 - 265). He must have been among the first collectors to use *chum* for attracting seabirds (see pp. 264 - 265), though it was undoubtedly an age-old practice, the effects of which were common knowledge among seafarers.

Several seabirds that Townsend collected were described by Audubon in the *Ornithological Biography*, including one new species: the **Black-footed Albatross** that Townsend preserved on the way to the Hawaiian Islands on 25th December 1834. This is a common, dark-coloured albatross that breeds in the western Hawaiian Islands and is a regular autumn and winter visitor off the Oregon coast. Unfortunately, four other species of seabird collected by Townsend are rather problematic (Stone 1930a, 1931, 1934; Burns 1931, 1934a, 1934b; Bourne 1967, Fisher 1965, Jobanek 1986). The difficulty has arisen, not because of the details of the descriptions, nor because of Audubon's plate, or lack of a plate, in *The Birds of America*, but because the localities at which the seabirds were said to have been collected are extremely unlikely. The four species (current nomenclature) and their supposed collecting localities (as given by Audubon 1840-44, 7: 196-210) are as follows:

Gray-headed Albatross *Thalassarche chrysostoma*:
"in the Pacific Ocean, not far from the Columbia River."

Light-mantled Sooty Albatross *Phoebetria palpebrata*:
"near the mouth of the Columbia river."

Southern Fulmar *Fulmarus glacialoides*:
"within a day's sail from the mouth of the Columbia river."

Giant petrel sp. *Macronectes giganteus / M. halli*:
"shot at some distance from the mouth of the Columbia river."

All four species are birds of the southern oceans and rarely if ever stray into the northern hemisphere. The records were at first accepted but by the 1930s, as seabird distribution became better known, the unusual nature of Townsend's records became more apparent. The case for these birds all suddenly straying to the Oregon coast is extremely weak and none of them are now properly accepted into the *AOU Check-list*. So what went wrong? Why the mistakes?

Few of Townsend's seabird specimens still survive and of those that do, none have original labels. But the information supplied with Audubon's descriptions seems to consist of too many words to have been taken from a normal-sized specimen label, so either Townsend supplied extra details (Audubon says "a note from Mr. TOWNSEND was appended" to the specimen of Southern Fulmar), or Audubon made assumptions based on some labels that he saw. Perhaps the information about locality was based on verbal discussions with Thomas Nuttall who had made the same sea crossings between the Columbia River and the Hawaiian Islands as Townsend. There is no evidence for this latter possibility, nor indeed the previous supposition.

We know that communication between Audubon and Townsend was rather poor – they were sometimes on different continents and Townsend was not always very co-operative. We know that Audubon and his co-worker on the text, William MacGillivray, made many mistakes elsewhere in the *Ornithological Biography*. They said that some land birds were collected in California by Townsend, even though Townsend never went there, clearly demonstrating their degree of confusion about Townsend's actual route. Sometimes the dates they gave for specimens were obviously incorrect, as we happen to know that the collector was elsewhere at the time (see e.g. California Quail, appendix 14). And then there was the strange inclusion of Rufous-collared Sparrow (as 'Morton's Finch') that is not a North American species but had been collected by Townsend as far away as Chile (see appendix 13).

Townsend had a very incomplete knowledge of the seabirds of the Pacific Northwest, gleaned from his own limited experience of offshore species, from travellers he met while on the Columbia River, and by what he could read in bird books and papers at Philadelphia. He erroneously claimed to have discovered several new species of cormorant, Audubon irresponsibly describing two of them merely on the vaguest of descriptions, without any specimens (1840-44, 4: 336). In the *Journal of the Academy of Natural Sciences of Philadelphia* (1839) Townsend also lists Least Auklet and Crested Auklet as occurring on the "North West Coast of America", all species that are considered accidental south of Alaskan waters. These two auklets, and the Rhinoceros Auklet, Ancient Murrelet, Marbled Murrelet, and Pigeon Guillemot, were omitted from the *Narrative* Appendix, as though he had realised his earlier mistake. Audubon also says (*Ornithological Biography* 4: 304) that Townsend had informed him that the Little Auk "is found near the mouth of the Columbia River." Both Least Auklet and Little Auk may be references to Cassin's Auklet, a similar-sized species that does occur regularly all along the Oregon, Washington and British Columbia coastlines. At no time does there appear to have been any Townsend specimens for any of these auklets.

For the first time in this long-running debate, we have examined some overlooked MS material of Townsend's but have failed to find any supporting evidence for the occurrence off the Oregon coast of any of the four seabird species in question. On the contrary, Townsend's reconstructed catalogue of specimens, compiled in the field, sequentially as they were obtained, does not seem to have room for all the seabirds he is supposed to have collected near the Columbia River (see appendix 6). It therefore seems much more likely that they were collected at a later date in more southerly latitudes.

Possible explanation for the errors

From the current distribution and reports of seabirds off the Oregon coast it would be reasonably safe to assume that Townsend saw the following species on his voyages between the Columbia River and the Hawaiian Islands: Northern Fulmar (dark and light phase), Black-footed Albatrosses (adults and juveniles) and lesser numbers of Laysan Albatrosses. When he collected look-alikes in the southern Pacific and Atlantic he may have thought that they were the same species that he had seen off the Oregon coast and, when writing about them for Audubon, said that they occurred near the Columbia River. Thus Gray-headed Albatross from southern Pacific he equated with Laysan Albatross; Light-mantled Sooty Albatross, and giant petrels he equated with adult and juvenile Black-footed Albatross; and Southern Fulmar with Northern Fulmar. The case for the two 'white' albatrosses, Gray-headed and Laysan Albatross is fairly convincing, so too is that for the two fulmars that as far as we know Townsend thought were the same species. Less convincing is the idea that Townsend could have made three species (Light-mantled Sooty Albatross, a giant petrel and Black-footed Albatross) out of the highly variable Black-footed Albatross that is so abundant off the Oregon coast.

Conclusion

There now seems little doubt that the four contentious species said to have been collected from near the Columbia River or Oregon coast were from further south, collected between the Hawaiian Islands and Chile, or on the voyage round Cape Horn. Townsend seems to have been either *unable* to correct the errors because of confusion in his own mind, or he was *unwilling* to rectify the errors, perhaps because he did not want to publicly correct Audubon, or because he did not want to diminish his own contribution to the list of the birds of North America. Townsend was clearly out of his depth regarding the seabirds of the Pacific Northwest. The simplest explanation is that Townsend confused species seen and collected in the southern oceans with those he saw off the Oregon coast, believing them to be the same. We see no reason to change the decisions made by the AOU based on the arguments of Stone (1930a, 1931, 1934).

APPENDIX 11. TOWNSEND'S BIRD SPECIMENS FROM THE HAWAIIAN ISLANDS – A PROVISIONAL LIST

"If Townsend had but published a list of his captures, he would indeed have rendered a very good service; but of course the value of island-forms, to say nothing of the fact that many of them were threatened with extinction by colonization and civilization, had not then been appreciated, if ever entertained by naturalists."

Wilson and Evans (1890-99) *Aves Hawaiienses*, p. xiv.

"Most of those [Hawaiian specimens] from that excellent collector [Townsend] are in a capital state of preservation, though now over 50 years old."

Wilson and Evans (1890-99) *Aves Hawaiienses*, part II, p.3.

The bird specimens collected by Townsend in the Hawaiian Islands are amongst the most intriguing that he ever preserved. Although all of them were probably labelled when first obtained, few original labels now survive. Most labels would have simply said "Sandwich Islands", with the date, and perhaps the sex of the bird determined during preparation. He did not specify from which island the specimens had originated as he did not realise the importance of such information. He was in good company: when Charles Darwin collected in the Galapagos Islands in 1835 he stored finch specimens from the first two islands in the same bag and never managed to sort out where they were from (Weiner 1994: 21-30).

Townsend may have been less interested in the Hawaiian birds than those of North America because the islands had an exotic avifauna, well outside his normal experience, and because they were far removed from the bounds of his self-appointed study area. This reaction was certainly true of Thomas Nuttall who neglected most of the plants and concentrated instead on the shells and ferns. Townsend should, perhaps, have written up his discoveries of the new species and races, but he probably knew that he would be hampered by the lack of easily accessible reference material. Most Hawaiian specimens at that time were in European museums and it would have been too difficult for Townsend to work through his specimens properly without seeing other collections. As a result he failed to earn much recognition for collecting on the islands. Ferdinand Deppe's birds received more attention and were worked through by a variety of ornithologists, especially the German naturalists Lichtenstein and Cabanis.

Examples of most of the species collected in the Hawaiian Islands by Townsend were deposited in the collection of the Academy of Natural Sciences of Philadelphia. Other Hawaiian specimens that remained in the United States were in the Edward Harris collection that was passed down through the family until eventually given away, with other specimens from elsewhere, in order to settle a debt. Although some were used as target practice, a proportion of the collection and a handful of Hawaiian specimens were rescued and deposited in US museums (see appendix 16).

Townsend decided to sell his Hawaiian duplicates in Europe and gave them to Audubon to offer to Lord Derby (who made a list of the 121 specimens on offer, selecting about 17 skins for himself (see appendix 4). Audubon disposed of the remaining majority to an Edinburgh dealer named Macduff Carfrae, who sold them to Sir William Jardine and other collectors; with some no doubt ending up as decorations in Victorian living rooms. At the sale of the Jardine collection in 1886 the Hawaiian birds, "some of them bearing Townsend's labels", were bought by Scott B. Wilson for Cambridge University Museum. These were later examined by F.W. Frohawk when preparing the coloured plates for *Aves Hawaiienses* (Wilson and Evans, 1890-99); though the text does not clarify which of the skins were painted.

Townsend's success in the islands was limited by a variety of factors. On Maui and Hawai'i it was lack of time, on Kaua'i he had time but ran out of ammunition and preservative, on O'ahu there were other distractions and he failed to appreciate the importance of collecting at high altitude in order to find a range of additional species. Everywhere there was the physical difficulty of collecting on the islands, a problem that in time became notorious. In 1900 Alvin Seale wrote: "The difficulty of collecting in these islands, with their dense tropical jungles and knife-like mountain ridges, has been mentioned by all former collectors, and I can only add, that while I have collected in difficult places before, including the boggy tundra of Siberia, the high mountains of Alaska, the Tamerack swamps of Michigan, and the Everglades of Florida, I have found nothing that would discourage an Ornithologist so much as one of these islands." (in Berger 1972, p. 185). The exact difficulties were further itemised by George Munro:

"Moving on into the forest, conditions change and become more difficult. Trees increase, the perspective narrows, shadows darken, footing is less secure and moisture, fog and mist increase. The birds seem to reflect their environment. They are perpetually shaking the rain drops from their feathers as they slip about under the protecting canopy of leaves and vines.

"Finally, the innermost fastnesses of the mountain forest are reached. Here is the continuous drip, drip, drip, of the rain forest; aerial roots, creepers, tangled vines, fallen trees raise the floor of the forest and afford treacherous footing. Moss clothes the trees, ferns cling to rotting trunks, lianes web the branches in tough embrace. Below are mud, mire and bog, and often lava cracks and holes. Nor is this level country. Steep palis [cliffs] drop into abrupt ravines to imperil life and limb. Vagrant drifts of fog forever wind and twist among the spectral trees. Hours may pass without sight of birds and disappointment add further depression to the melancholy scene" (Munro 1960, p. 67-69).

No wonder that some species were hard to find. The Po'o-uli *Melamprosops phaeosoma*, endemic to Maui, was not discovered until 1973.

The history of collecting on each of the islands is well summarised by Olson (1996). Townsend was among the first to seriously collect birds on O'ahu and Kaua'i, and it is a matter of some regret that he had so little time on Maui and Hawai'i where he was also amongst the first in the field. The following provisional list gives the present location of Townsend's surviving Hawaiian specimens. There may be others that we do not know about, as well as those that with the passage of time cannot be linked back to the enthusiastic young collector from Philadelphia.

The taxonomy of the endemic birds of the Hawaiian Islands is continually under review. Hawaiian specialists do not agree in their classifications and many museum collections have not kept up with the latest ideas. On the whole, we have followed the *AOU Check-list* (1998) and changes to the *Check-list* up to the 47th Supplement (2006).

O'AHU

The earliest bird collecting took place on O'ahu May-June 1786 HMS Queen Charlotte; May 1825 HMS Blonde; September 1834 M. Botta (= Paola Botta? (Palmer 1928)). Townsend collected there during the following periods: **5th January – 10th February 1835 (with Thomas Nuttall); 16th – 26th March 1835 (with Thomas Nuttall); 23rd December 1836 – 9th February 1837 (with Ferdinand Deppe for part of this time); 27th February – 18th March 1837.** *The most productive collecting periods were during his first visit and later when Townsend and Deppe shared a cottage in the Nu'uano valley when Townsend reported that he had collected over 80 specimens (see p. 241). Specimens from the 1835 visit to O'ahu are not now separable from those collected in 1837.*

'Elepaio *Chasiempis sandwichensis* (Gmelin, 1789)
Endemic to Hawaiian Islands (O'ahu, Kaua'i, Hawai'i). The subspecies *gayi* is confined to O'ahu. Endangered.
Specimens: **Cambridge, UK: 27/Mas/7/b/1** (labelled 10 Jan 1834 [= 1835, or perhaps 1837]). **Cambridge UK: 27/Mas/7/b/2.** Both catalogued as *C.s.gayi*.
Note 1: Cambridge specimens are among those in the Derby list as: *No. 75, Muscicapa, Sandwich Is., 7 specimens.* Originally sold by Audubon to the dealer Carfrae, bought by Sir William Jardine, and then bought for Cambridge at the Jardine sale in 1886.

O'ahu 'O'o *Moho apicalis* Gould, 1860
Endemic to O'ahu. Not seen since 1837 (by Townsend and Deppe). Extinct.
Specimens: **MCZ 17598** (ex Brown University Museum in 1872) is referred to by Greenway (1967, p. 424)]; see also comments under Hawai'i 'O'o.
FMNH 306728 (ex Princeton University Museum).

O'ahu 'Amakihi *Hemignathus flavus* (Bloxam, 1827)
Endemic to O'ahu; previously known as *H. virens chloris*.
Specimens: **ANSP 3378** (received from or examined by Cambridge 1889).
ANSP 3381. Liverpool 5361. Cambridge UK: 27/12/c/14, 27/Dre/12/c/15, 27/Dre/12/c/16; a note on the card catalogue queries their identity and accepts only 27/Dre/12/c/16 as *H.v.chloris*.
Note 1: Liverpool and Cambridge specimens are among those in Derby list as: *[no field number] small yellow Icterus, Sandwich Is., 2 specimens*. There now seem to be more Townsend specimens of this species in the UK than listed by Derby.

Nukupu'u *Hemignathus lucidus* Lichtenstein, 1839
Endemic to the Hawaiian Islands; the O'ahu subspecies *H.l.lucidus* was last recorded in 1839.
Specimens: **Cambridge UK: 27/Dre/3/a/1, 27/Dre/3/a/2.**
Note 1: This species perhaps included in Derby list as: *[no number] Unequal billed Certhia, Sandwich Is., 11 specimens*. Species composition of the 11 specimens sent to Britain are unknown though Medway (1981: 129) identifies at least one specimen (LM D511c) as Lesser 'Akialoa *H.obscurus obscurus*.

O'ahu 'Alauahio or **O'ahu Creeper** *Paroreomyza maculata* (Cabanis, 1850)
Endemic to O'ahu; probably extinct (last reported 1986).
Specimens: **ANSP 375** (examined by Newton & Wilson 1889). **ANSP 18582. Cambridge UK: 27/Dre/8/a/59.** In 1975 **Cambridge UK 27/Dre/8/a/58** was sent to the National Museum, Bulawayo.
Note 1: Cambridge and Bulawayo specimens among those in Derby list as: *[no field number] Sylvia, Sandwich Is., 18 specimens.*

'Akepa *Loxops coccineus* (Gmelin, 1789)
Endemic to the Hawaiian Islands; the O'ahu group *L. wolstenholmei* Rothschild, 1893 (formerly *L. rufa*) is now extinct.
Specimens: **ANSP 19841. USNM 14689** (Although now missing this was perhaps the specimen donated to the National Institution by Townsend by 13th September 1841 as "Fringilla cinnamonea" (*Proceedings National Institution* 1841: 103). **Liverpool D4872**, labelled "Received from Mr. Townsend per Mr. Audubon June 1838" (see Medway 1981, p. 129). **Cambridge UK: 27/Dre/6/a/27** (ex ANSP in 1899).
Note: 1: Liverpool and Cambridge specimens among those in Derby list as: *[no field number] Cinnamon Fringilla, Sandwich Is., 1 specimen*. Liverpool specimen = Derby list No. 13 (see Medway 1981, p. 129).
Note 2: Two additional specimens (ANSP 156995, ANSP 156996) were donated to the Academy by West Chester Teachers School and may also have been collected by Townsend.

KAUA'I
Cook's ships stopped at Kaua'i in 1778 and specimens of 'I'iwi were received when bartering with the natives. No other species from Kaua'i are known to have made their way to Europe from that visit. Townsend was therefore the first naturalist to collect there, between **10th February – 16th March 1834 (with Thomas Nuttall).** *Unfortunately, few birds were collected and the following quote by Townsend explains the disappointing results for the five week period that he was there: "After we had been here about four days, however, a heavy S.W. wind sprung up, blowing steadily towards Oahu, in consequence of which the Avon could not leave her anchorage, and we were compelled to remain where we were. Under some other circumstances this detention would not have been irksome; but we had made provision for only a few days' residence, and in a very short time all our ammunition, poison for preserving specimens, &c. were exhausted, and it was impossible to obtain even substitutes for these indispensable articles at this place. So we had nothing for it, but to yield to our fate with what grace we could, and spend the remainder of our forced sojourn in collecting plants, shells, and such other matters as the "moth and rust would not corrupt" (Townsend 1847: 122). Worse still, those specimens that he did collect did not receive the attention they deserved and became widely scattered.*

Short-eared Owl *Asio flammeus* (Pontoppidan, 1763)
A.f.sandwichensis is resident on all the main Hawaiian Islands.
Specimens: Although Nuttall (1840, p. 141.) says that the owl was seen on Kaua'i the two Townsend specimens at ANSP have no positive link with this island.

Kaua'i 'O'o *Moho braccatus* Cassin, 1855
Endemic to Kaua'i. Considered extinct (not seen since 1988).
Specimens: **ANSP 18581** not on current data list but should be according to Wilson and Evans (1890-99) and Stone (1899, p. 43).

Note 1: Townsend's specimen at Philadelphia was used as the type when John Cassin described it as a new species. "The only specimen that I have ever seen of this species was brought by Dr. Townsend from the Sandwich Islands and is marked as a male. It has heretofore in the collection of the Academy been mistaken for *Certhia pacifica* Gmelin [Hawai'i Mamo *Drepanis pacifica*], but it is clearly not that bird and but little resembles it …" (*Proceedings ANSP* 1855, p. 440.)
Note 2: The 1853 illustration by Reichenbach (in *Handb. Spec. Orn.* II Abth., p. 33, pl. dcxiv, fig. 4099) may well be from a Townsend specimen of this species (Olson 1994, p. 96, col. 2).
Note 3: Greenway (1967, p. 424) says there is a Kaua'i 'O'o at MCZ but does not name collector.

'Anianiau or **Lesser 'Amakihi** *Hemignathus parvus* (Stejneger, 1887)
Endemic to Kaua'i, where still common in suitable habitat.
Specimens: **ANSP 18648. USNM 14686** is from Townsend via National Institution, yet this species was not described until 1887, from material supplied by another collector (Olson 1994). This is probably the "Nectarinia viridis" donated to the National Institution by 13th September 1841 (*Proceedings of National Institution* 1841: 103).
Note 1: ANSP 156994 was donated by West Chester Teachers School and may also have been collected by Townsend.

Nukupu'u *Hemignathus lucidus* Lichtenstein, 1839
Endemic to the Hawaiian Islands; the subspecies *H.l.hanappe* is restricted to Kaua'i, now considered extinct.
Specimens: **Liverpool [no number]** - see Medway 1981, p. 129. **FMNH 308799** is presumably the specimen rescued from the Harris collection and donated to Princeton University Zoology Museum (Street 1948, p. 148) and later transferred to the Field Museum, Chicago.
Note 1: This species perhaps included in Derby list as: *[no number] Unequal billed Certhia, Sandwich Is., 11 specimens*. Species composition of the 11 specimens sent to Britain are unknown though Medway (1981: 129) identifies at least one specimen (LM D511c) as Lesser 'Akialoa *H.obscurus obscurus*.

Kaua'i 'Amakihi *Hemignathus kauaiensis* Pratt, 1989
Endemic to Kaua'i, where still common in some areas.
Specimens: Liverpool D5360 is perhaps from Townsend but has no positive link to him (see Medway 1981, p. 129).
Note 1: Perhaps in Derby list: *[no number], small yellow Icterus, Sandwich Is., 2 specimens.*

MAUI
Although Maui is a relatively large island it was not thoroughly worked by collectors until after 1879. This was perhaps due to the steepness of the mountains immediately behind the village of Lahaina. Townsend was at Lahaina for less than two days, from the **morning of 14th February until the evening of 15th February 1837.** *Most of that time was apparently spent visiting the missionaries and the school. He does not make any mention of birds in his MS Journal during his visit.*

Nukupu'u *Hemignathus lucidus* Lichtenstein, 1839
Endemic to the Hawaiian Islands; the subspecies *H.l.affinis* is confined to Maui. Perhaps extinct.
Specimens: **ANSP 18760** [alternative number ANSP 3363]. This specimen has been attributed to the Maui race. It may have been collected by Townsend or sent to him by one of the missionaries he met there, or the subspecies designation may be incorrect. Olson 1994 says that no birds were collected on Maui until 1879.

HAWAI'I
Of all the islands that Townsend visited, this was the one that had already received the most attention by naturalists. Captain Cook's third great expedition called here in 1779, HMS Blonde arrived in 1825 (a few birds were collected by the Rev Bloxam), David Douglas botanised in 1832, and the French ship La Bonité was here from September to October 1836. Townsend's brief visit extended from **17th – 25th February 1837.** *He landed at the Bay of Kealakekua, and the following day went into the woods and noticed some birds he had not seen on the other islands (but for some reason had not taken his gun). On the 19th the ship tried to leave but they were detained by the winds. Hoping to take full advantage of the delay he went ashore early on the 20th and hunted in the neighbouring woods for most of the day - "but met with very little success." That evening and until the evening of the next day he was sailing up the west coast to Kawaihae, a more*

barren area. Though he went ashore there he does not mention any birds or further shooting expeditions.

The missionary at Kealakekua was the Reverend Forbes who later sent specimens of the Hawaiian Hawk and Hawaiian Crow to Townsend while he was working at Washington, D.C. These appear to be the only species sent, no doubt at the specific request of Townsend.

Nene or **Hawaiian Goose** *Branta sandvicensis* (Vigors, 1833)
Endemic to Hawaiian Islands; Hawai'i and Maui, recently re-introduced to Kaua'i and Moloka'i.
Specimens: The only two specimens at Philadelphia have no known link with Townsend: ANSP 6001 has no data; ANSP 6004 is from the Rivoli collection.
Note 1: Townsend's MS Journal mentions three or four live geese that he bought on Hawai'i and tried to take home with him (see p. 251).

'Io or **Hawaiian Hawk** *Buteo solitarius* Peale, 1848
Endemic to the island of Hawai'i. Occurs as a light and dark phase; still fairly common.
Specimens: **ANSP 2304.** Stone (1899, p. 45) confirms ANSP 2304 as type.
Note 1: The Rev Forbes sent one specimen to Townsend when he was at Washington, D.C. (W. Baird to S.F. Baird, 23rd November 1842). The hawk was first described by T.R. Peale from the specimen obtained "near Karakakoa [Kealakekua] Bay, by the Rev. Mr. Forbes, Presbyterian missionary on that station; he transmitted it to Mr. J.K. Townsend, who kindly loaned it to be drawn" (Peale 1948, pp. 62-63; plate XVI).

Short-eared Owl *Asio flammeus* (Pontoppidan, 1763)
A.f.sandwichensis is resident on all the main Hawaiian Islands.
Specimens: **ANSP 2638** is labelled "Hawaii, Sandwich Islands, J.K. Townsend". Possibly an original label [handwriting not seen by B&R Mearns]. No date.

'Alala or **Hawaiian Crow** *Corvus hawaiiensis* Peale, 1848
Endemic to the island of Hawai'i. Critically endangered (captive population only).
Specimens: **ANSP 2830. ANSP 2831.** Stone (1899, p. 46) lists ANSP 2830 as type.
Note 1: "These specimens were obtained a few miles inland from the (now long-abandoned) village Ka'awaloa, celebrated as being the spot where the renowned Captain Cook was killed ...Our specimens, collected by the Expedition, of the Hawaiian Crow, or Alala, with many important notes attached, were lost in the wreck of the U.S. Ship *Peacock*; but we are happy to acknowledge our obligations to Mr. J.K. Townsend, who has kindly loaned us others, collected at the same place, and sent to him [after Townsend's return to the eastern US] by the Rev. Mr. Forbes, missionary at Karakakoa [Kealekekua] Bay" (Peale 1848, p. 107). These birds were used by Titian Peale in making his description and plate XXVIII (Peale 1848, pp. 106-107). Peale "must be considered fortunate to have so readily obtained the loan of two others from Dr. J.K. Townsend, which were sent from Kaawaloa by Mr. Forbes, a missionary at Karakakoa Bay, and were afterwards deposited in the collection of the Philadelphia Academy" (Wilson and Evans (1890-99), part IV, page 1). William Baird mentions that Townsend had "just received" a box from the Hawaiian Islands with a male and female Hawaiian Crow (and Hawaiian Hawk) while Townsend was at Washington, D.C. (W. Baird to S.F. Baird, 23rd November 1842).

'Oma'o or **Hawaiian Thrush** *Myadestes obscurus* (Gmelin, 1789)
Endemic to the island of Hawai'i; fairly common at some locations.
Specimens: **ANSP 3355** is in the database as a Townsend specimen.

Hawai'i 'O'o *Moho nobilis* (Merrem, 1786)
Endemic to the island of Hawai'i. Extinct.
Specimens: John Cassin (*Proceedings ANSP* (1855) 7: 440) indicates that there was at least one Townsend specimen in the ANSP collection. There is a mounted specimen without data at ANSP that is possibly from Townsend but the other 13 specimens on the ANSP database have no known link with Townsend. **MCZ 17599** is on MCZ online database (2006) as received in exchange in 1872 [from Brown University]. Either MCZ 17598 or MCZ 17599 is probably ex Harris collection and Princeton University Museum - see Street 1848, pp. 182-184.
Note 1: Townsend spent a day near Kealakekua Bay hunting for birds: "One,

a Cassicus with yellow axillary feathers, (which are used by the natives in making the ornament called "Re", & worn on the head,) was pretty plentiful but shy" (MS Journal, 20th February 1837). This is surely a reference to the distinctive yellow axillaries of the Hawai'i 'O'o, though it might also include the Hawai'i Mamo *Drepanis pacifica* since its black and yellow feathers were also used by the islanders for making feather cloaks and helmets.
Townsend used the generic name *Cassicus*, referring to their similarity of the South American caciques *Cacicus*. Medway (1981: 129) read *Cassicus* in the Derby list as *Cossicus* and interpreted it as a misspelling of *Copsychus*, an Asian genus.
Note 2. Perhaps included in Derby list as *No. 82, Cassicus, Sandwich Is., 1 specimen*. This specimen was marked on the list for acquisition for Lord Derby's collection but there is no evidence of any Townsend link with the *Moho* or *Drepanis* specimens amongst his birds, now at Liverpool Museum (C. Fisher and M. Largen, *in litt.*, 2002). The *Cassicus* specimen in the Derby list may refer to *Moho nobilis* or to a *Moho* species from another island.

Hawaii 'Amakihi *Hemignathus virens* (Gmelin, 1788)
Endemic to Hawai'i, Moloka'i, Lana'i and Maui; the nominate race *H.v.virens* is confined to the island of Hawai'i.
Specimens: **Liverpool 5361a** bears an original Lord Derby label: *"12 Mr Townsend by Audubon June/38"*.
Note 1: Liverpool specimen is among those in Derby list as: *[no field number] small yellow Icterus, Sandwich Is., 2 specimens* [= No. 12 on Derby list – see Medway 1981, p. 129.]

Lesser 'Akialoa *Hemignathus obscurus* (Gmelin, 1788)
Formerly resident on Hawai'i; now extinct.
Specimens: **Liverpool D511c** identified as *H. obscurus obscurus*. This specimen is possibly No.10 on Derby list – see Medway 1981: 129.

Hawai'i Mamo *Drepanis pacifica* (Gmelin, 1788)
Endemic to the island of Hawai'i. Extinct.
Note 1. Perhaps seen by Townsend on Hawaii and perhaps included in Derby list as *No. 82, Cassicus, Sandwich Is., 1 specimen* (see notes under Hawai'i 'O'o, above).

SPECIMENS NOT ATTRIBUTABLE TO PARTICULAR ISLANDS

Common Moorhen *Gallinula chloropus* (Linnaeus, 1758)
Uncommon O'ahu, Kaua'i, Hawaii, now absent from Moloka'i and Maui.
Specimens: none traced.
Note 1: Among those in Derby list as: *No. 88, Gallinula chloropus, Sandwich Is., 1 specimen.*

Hawaiian Coot *Fulica alai* Peale, 1848
Endemic to main Hawaiian Islands. Fairly common only on Kaua'i, O'ahu and Maui.
Specimens: none traced.
Note 1: Among those in Derby list as: *No. 87, Fulica, Sandwich Is., 1 specimen.*

Ruddy Turnstone *Arenaria interpres* (Linnaeus, 1758)
Common winter visitor from Arctic breeding grounds.
Specimens: none traced.
Note 1: Among those in Derby list as: *No. 232, Strepsiles interpres, Sandwich Is., 3 specimens.* Considering the high field number, these specimens may have come from Tahiti and/or Chile (see appendix 6)

Short-eared Owl *Asio flammeus* (Pontoppidan, 1763)
A.f.sandwichensis is resident on all the main Hawaiian Islands.
Specimens: **ANSP 2635** has no data (but a second specimen (ANSP 2638) is attributed to the island of Hawai'i).
Note 1: Among those in Derby list as: *No. 84, Strix Brachyotus, Sandwich Is., 1 specimen.*
Note 2: Seen on Kaua'i by Nuttall and Townsend (Nuttall 1840, p. 141).

'O'o sp. *Moho sp.*
Various species were endemic to, Moloka'i. All are now extinct.
Note 1: Perhaps a species from O'ahu or Kaua'i is on Derby list as: *No. 82, Cassicus, Sandwich Is., 1 specimen* rather than Hawai'i 'O'o (see notes under Hawai'i 'O'o, above).

'O'u *Psittirostra psittacea* (Gmelin, 1789)

O'ahu, Kaua'i, Hawai'i, Lana'i, Maui, Moloka'i. Probably now extinct on all islands.

Specimens: **ANSP 3356. ANSP 3357. ANSP 3358.**

Note 1: Among those in Derby list as: *[no field number] Yellowhead Pyrrhula, Sandwich Is., 7 specimens.*

Note 2: A specimen, possibly of this species, was donated to the National Institution by Townsend by 13th September 1841 as "Pyrrhula psittacina" (*Proceedings National Institution* 1841: 103) but cannot now be traced.

Note 3: Liverpool T12303, from H.B. Tristram collection, ex Jardine sale 1886, is possibly a Townsend specimen but there is no proven link. It is in the Jardine sale catalogue as No. 6510 from Sandwich Islands, collector "Uncertain" ([Jardine] 1886, p. 151).

Note 4: In Morton skull catalogue as: "522. PYRRHULA? _____. Sandwich islands". The skull was presented by Townsend to S.G. Morton collection (Morton 1840, p. 45.).

O'ahu 'Amakihi *Hemignathus flavus* (Bloxam, 1827), **Kaua'i 'Amakihi** *H. kauaiensis* Pratt, 1989 and **Hawai'i 'Amakihi** *H. virens* (Gmelin, 1788),

Note 1: One of these species perhaps in Morton skull catalogue as: "486. ICTERUS _____. Yellow ICTERUS. Sandwich islands." The skull was presented by Townsend to S.G. Morton collection (Morton 1840, p. 44.).

Lesser 'Akialoa *Hemignathus obscurus* (Gmelin, 1788) or **Greater 'Akialoa** *H. ellisianus* (G.R. Gray, 1860)

Hawai'i (*H. obscurus*). O'ahu, Kaua'i, Lana'i (*H. ellisianus*). Probably now extinct on all islands.

Specimens: **ANSP 3360** (examined by Newton and Wilson, August 1889); no information on identity of ANSP 3360 as database does not follow recent changes in Hawaiian bird taxonomy.

'I'iwi *Vestiaria coccinea* (J.R. Forster, 1780)

Endemic to Hawaiian Islands: O'ahu, Kaua'i, Maui, Moloka'i, Lana'i, Hawai'i.

Specimens: **ANSP 3364** (Possibly one of the two specimens rescued from the Harris collection (Street 1948, pp. 182-184). **Liverpool D511a and D511b** (see Medway 1981, p. 129). In 1975 **Cambridge UK: 7/Dre/11/a/1** was sent to the National Museum, Bulawayo. **MCZ 72036**, no date or locality, has collector name as "J.K. Townsend?"

Note 1: Liverpool, Cambridge and Bulawayo specimens among those in Derby list as*: No. 79, Certhia coccinea, Sandwich Is., 24 + 5 specimens.*

Note 2: In Morton skull catalogue as: "483. CERTHIA VESTIARIA. Sandwich islands". The skull was presented by Townsend to S.G. Morton collection (Morton 1840, p. 44.).

Note 3: ANSP 156992 was donated by West Chester Teachers School may also have been collected by Townsend.

'Apapane *Himatione sanguinea* (Gmelin, 1788)

Endemic to Hawaiian Islands; nominate race *H.s.sanguinea* occurs on Kaua'i eastwards, including O'ahu, Maui, Hawai'i, Moloka'i.

Specimens: **USNM 14692** (currently missing). This was possibly the specimen donated to National Institution by Townsend by 13th September 1841 as "Nectarinia rubra" (*Proceedings National Institution* 1841: 103).

Liverpool D5359c (see Medway 1981, p. 129). **Cambridge UK: 27/Dre/4/a/1, 27/Dre/4/a/3, 27/Dre/4/a/18.** In 1975 **Cambridge UK: 27/Dre/4/a/2** was sent to the National Museum, Bulawayo.

Note 1: The Liverpool, Cambridge and Bulawayo specimens are among those in Derby list as: *No. 80, Crimson Icterus, Sandwich Is., 39 specimens.*

Note 2: One specimen in the Harris collection seems to have been sent to ANSP (Street 1948, p. 184) but it is not on the ANSP database (as a Townsend specimen).

Note 3: 27/Dre/4/a/18 is ex Carfrae and Strickland collection.

Note 4: In Morton skull catalogue as: "485. ICTERUS _____. CRIMSON ICTERUS. Sandwich islands." The skull was presented by Townsend to S.G. Morton collection (Morton 1840, p. 44.).

Note 5: ANSP 156993 was donated by West Chester Teachers School and may also have been collected by Townsend.

Hawaiian Honey-creeper *Drepanidinae*

Specimens: **USNM 14729** (currently missing) has card index notation: "Drepanis ?"

Note 1: At the moment the identity of this Townsend specimen cannot be determined. It is unlikely to be either of the Mamo *Drepanis* species because there is no mention of Mamo specimens connected with Townsend in the ornithological literature. It is more likely to be another species of the *Drepanidae*.

Honeyeater *Meliphagidae* **or Honeycreeper** *Drepanidinae*

In Morton skull catalogue as: "484. Dicoeum _____. Sandwich islands." The skull was presented by Townsend to S.G. Morton collection (Morton 1840, p. 44.). Dicoeum = *Dicaeum* = Flowerpeckers, a family not present on the Hawaiian Islands. This specimen (if it still exists) is probably one of the Hawaiian Honeyeaters or Honeycreepers.

I'iwis by F. W. Frohawk (from *Aves Hawaiiensis* by S.B. Wilson and A.H.Evans, 1890 - 99).

APPENDIX 12. TOWNSEND'S BIRD SPECIMENS FROM TAHITI

In his MS Journal Townsend says that on Tahiti he collected about 110 specimens, of 16 species (see pp. 254 - 256) of which at least 18 specimens (16%) of nine species are still known to survive. Townsend counted adult and immature Tahiti Monarch as two species, and the two colour phases of the Tahiti Reed Warbler as two species. This reduces his tally to 14 species. The list below contains a minimum of 13 species. It is unlikely that Townsend included the Red Junglefowl in his sixteen species but his six "puffinus" specimens could well have included more than one species.

The Tahiti Swiftlet was undescribed until 1848, when Titian Peale named it from specimens collected during the US Exploring Expedition. One of Townsend's Gray-green Fruit Dove specimens (USNM 14816) is listed as a co-type, described by Peale.

Thirty-one bird skins on the Derby list (see appendix 4) are from Tahiti but these would have been mostly duplicates that Townsend wished to sell – the total number of specimens from Tahiti was, of course, larger. Two specimens, of two species, listed in the *Proceedings of the National Institution* (September 1841: 102-104), were donated by Townsend to the National Institution and later transferred to the USNM. Townsend's bird skulls from Tahiti in the Morton collection (Morton 1840) may still exist at the ANSP but we have not searched for them; likewise, the bird skeletons from Tahiti on the USNM card catalogue.

Bird name as in text or MS (if any)	Reference & number of specimens	Identification (by B&R Mearns)	Surviving specimens
"puffinus"	Derby list No. 177	shearwater sp. *Procellariidae*	6 specimens
	no ref found	Pacific Reef-Heron *Egretta sacra*	ANSP 6667
"dunghill fowl"	*Narrative* "about a dozen killed"	Red Junglefowl *Gallus gallus*	
"Totanus"	Derby list No. 181 1 specimen	Wandering Tattler *Heteroscelus incanus*	ANSP 11245 LM D3096
"Sterna ? (alba)"	Derby list No. 182 1 specimen	Common Fairy Tern *Gygis alba*	ANSP 5011 USNM A02022
"594. STERNA? ____."	Morton 1840, p. 45. see also Morton 1849, p. 10.	Common Fairy Tern *Gygis alba*	
"Columba" "Dove" in MS	Derby list No. 178 4 specimens	Gray-green Fruit Dove *Ptilinopus purpuratus*	ANSP13139 USNM 6797 USNM A14816[1]
"Columba taitensis"	*Proc Nat Inst* 1841, p.103 1 specimen	?Gray-green Fruit Dove *Ptilinopus purpuratus*	
	USNM card catalogue	Pacific Pigeon *Ducula pacifica* (skeleton)	USNM 6797
"very small parroquet"	MS Journal	Blue Lorikeet *Vini peruviana*	ANSP 22118
"Cuckoo (Cuculus)"	MS Journal 3 specimens	Long-tailed Cuckoo *Eudynamis taitensis*	
"Cypselus ?" [2]	Derby list No. 185 1 specimen	Tahiti Swiftlet *Aerodramus leucophaeus*	ANSP 21799
"Aludo ?" [= Alcedo]	Derby list No. 179 4 specimens	Venerated Kingfisher *Halcyon venerata*	ANSP 21455 ANSP 21456 LM D2367
"526. ALCEDO? _____."	Morton 1840, p. 45 see also Morton 1849, p. 10. [3]	? Venerated Kingfisher *Halcyon venerata*	
"Hirundo"	Derby list No. 191 1 specimen	Pacific Swallow *Hirundo tahitica*	
"Black [Muscicapa]" [4]	Derby list No. 186 8 specimens	Tahiti Monarch *Pomarea nigra*, adult	ANSP 1100 ANSP 1101 ANSP 1102 ANSP 1103
	USNM card catalogue		USNM 14385
"Muscicapa nigra"	*Proc Nat Inst* 1841, p.103. 1 specimen	Tahiti Monarch *Pomarea nigra*, adult?	
"Cinnamon [Muscicapa]"	Derby list No, 187 2 specimens	Tahiti Monarch *Pomarea nigra*, immature	

	USNM card catalogue	Tahiti Monarch *Pomarea nigra* (skeleton)	USNM 6796
"Yellow Turdus ?"	Derby list No. 183 2 specimens	Tahiti Reed Warbler *Acrocephalus caffra*	ANSP 17347
"Dusky [Turdus ?]"	Derby list No. 184 1 specimen	Tahiti Reed Warbler *A. caffra*, dark morph	ANSP 18189
"Drepanis"[5]	USNM card catalogue	honeyeater sp. *Meliphagidae* sp. (skeleton)	USNM 6798

1. "USNM 14816 Adult or subadult male. Tahiti Island, Society Islands, Oceania. Received by the National Institute not later than September 13, 1841. Collected by John K. Townsend … while there can be no proofs adduced, it may easily have formed one of Peale's original series, and for that reason is here listed" (Deignan 1961, p. 105).

2. Handwriting seems to indicate "Cypselus" rather than "Cypcelus" that is now used. ANSP 21799 listed at ANSP as "*Callocalia leucophaea*".

3. Identified in Morton 1849, p. 10. as "*Todiramphus divinus*".

4. Entries in Derby list are in columns with frequent use of "do" = ditto. Names in square brackets are drawn down from above (see Derby list, appendix 4).

5. "Drepanis" is probably some form of honeyeater *Meliphagidae* (Storrs Olson *pers. comm.*)

Two Townsend specimens acquired on the voyage home: Wandering
Tattler (top) from Tahiti (LM D3096) and Red-legged Cormorant (below)
from Chile (LM D2882b).

APPENDIX 13. TOWNSEND'S BIRD AND MAMMAL SPECIMENS FROM CHILE

Townsend said that his illness "utterly incapacitated me from making collections in Chili" yet he evidently amassed a small collection of birds, some of which still survive (at least 23 specimens of 18 species). Fifty-six specimens recorded in the Derby list (see appendix 4) were from Chile but these would have been mostly duplicates that Townsend wished to sell – the total number of specimens (and species) must have been larger. Most of the birds listed are common and widespread species that under normal circumstances would have been easy for Townsend to collect himself, though he could, of course, have bought or been given some of them whilst resident in Valparaiso. Eight Chilean specimens, of three species, listed in the *Proceedings of the National Institution* (September 1841: 102-104; 1842: 146-147.) were donated by Townsend and later transferred to the USNM. Audubon's plate CCCCXXVII may include a South American Blackish Oystercatcher *Haematopus ater* (Harrison and Walker 1997: 502) but Townsend's specimen appears to be lost. Townsend's bird skulls from Chile in the Morton collection (Morton 1840) may still exist at the ANSP but we have not searched for them; likewise, the bird skeletons from Chile on the USNM card catalogue. There may be some duplication in the list below where there is no obvious link between existing specimens and those mentioned in the Derby list or in other references.

Bird name as listed in text or MS (if any)	Reference & number of specimens	Identification (by B&R Mearns)	Surviving specimens
"Podiceps"	Derby list No. 212 1 specimen	grebe sp. *Podicipedidae*	
"519. PROCELLARIA ___."	Morton 1840, p. 45.	petrel or shearwater sp. *Procellariidae*	
"520. PUFFINUS _____."	Morton 1840, p. 45.	shearwater sp. *Procellaridae*	
_____	no ref found	Red-legged Cormorant *Phalacrocorax gaimardi*	ANSP 5415
"Phalacrocorax"	Derby list No. 223 1 specimen	Red-legged Cormorant *Phalacrocorax gaimardi*	LM D2882b
_____	no ref found	Chilean Flamingo *Phoenicopterus chilensis*	ANSP 6118
"Condor of the Andes"	Townsend 1848 male and female	Andean Condor *Vultur gryphus*	ANSP 37 male
"Gallinula"	Derby list No. 211 2 specimens	Spot-flanked Gallinule *Gallinula melanops*	ANSP 6383 LM D3281d
"Scolopax Wilsoni"	Derby list, no number 1 specimen	Magellan Snipe *Gallinago paraguaiae*	
"523. STERNA INCA."	Morton 1840, p.45.	Inca Tern *Lorosterna inca*	
"329. PSITTACUS _____."	Morton 1840, p. 43.	Parrot sp. *Psittacidae*	
"Trochilus cyanocephalus"	Derby list, no number ? 40 specimens	Sparkling Violetear *Colibri coruscans*	
"596. TROCHEILUS [sic] CYNOCEPHALUS [sic]"	Morton 1840, p. 45. ?	Sparkling Violetear *Colibri coruscans*	
"Trochilus sephanoides"	*Proc Nat Inst* 1841, p. 102. 3 specimens	Green-backed Firecrown *Sephanoides sephanoides*	USNM 14217 USNM 14225 USNM 14232
"Trochilus sephanoides"	*Proc Nat Inst* 1842, p. 147. 3 specimens	Green-backed Firecrown *Sephanoides sephanoides* [same as the three 1841 specimens mentioned above?]	
_____	no ref found	Striped Woodpecker *Picoides lignarius*	ANSP 19275
_____	no ref found	Grey-flanked Cinclodes *Cinclodes oustaleti*	ANSP 6804
_____	no ref found	Plain-mantled Tit-Spinetail *Leptasthenura aegithaloides*	ANSP 6824
_____	no ref found	Dusky-tailed Canastero *Asthenes humicola*	ANSP 13858
_____	no ref found	Moustached Turca *Pteroptochos megapodius*	ANSP 18755
"Grey Muscicapa"	Derby list No. 203 2 specimens	flycatcher sp. *Tyrannidae*	

"Small Muscicapa"	Derby list No. 217 1 specimen	flycatcher sp. *Tyrannidae*	
	USNM card catalogue	flycatcher sp. *Tyrannidae* (skeleton)	USNM 6799
"Small crested Titmouse"	Derby list No. 209 1 specimen	Tufted Tit-Tyrant *Anairetes parulus*	ANSP 4212 BMNH 1886.6.24.397[1]
_____	no ref found	Many-coloured Rush-Tyrant *Tachuris rubigastra*	ANSP 4220
_____	no ref found	Fire-eyed Diucon *Xolmis pyrope*	ANSP 4049
_____	no ref found	Austral Negrito *Lessonia rufa*	ANSP 4143
"Ichyotoma rara"	Derby list, no number 1 specimen	Rufous-tailed Plantcutter *Phytotoma rara*	
_____	no ref found	Austral Thrush *Turdus falcklandii magellanicus*	ANSP 16197
_____	ex Boston Soc NH 1872	Lesser Seed-Finch *Oryzoborus angolensis* [= Brazil][2]	MCZ 75966
"Fringilla flavescens"	*Proc Nat Inst* 1841, p. 103.	?	
"Fringilla luteoventris"	Derby list No. 206 1 specimen	? Grassland Yellowfinch *Sicalis luteola*	
"White throated Finch"	Derby list No. 193 1 specimen	? Common Diuca-Finch *Diuca diuca*	
_____	USNM card catalogue	Grey-hooded Sierra-Finch *Phrygilus gayi*	USNM 15122
_____	no ref found	Grey-hooded Sierra-Finch *Phrygilus gayi*	ANSP 10702
"Fringilla atricapilla"	*Proc Nat Inst* 1841, p. 103.	? Black-throated Sierra-Finch *Phrygilus atriceps*	
_____	no ref found	Band-tailed Sierra-Finch *Phrygilus alaudinus*	ANSP 10682
"Rufous necked Finch"	Derby list No. 210 1 specimen	Rufous-collared Sparrow *Zonotrichia capensis*[3]	ANSP 10613 ANSP 10614
"Siskin"	Derby list No. 194 3 specimens	Black-chinned Siskin *Carduelis barbata*[4]	ANSP 10275 USNM A02035 USNM A02036

1. Sir William Jardine's catalogue of his collection includes "Euscarthmus parulus 2747 Valparaiso - Townsend" amongst the *Tyrannidae* (Jardine 1874, p. 67.), presumably purchased in Edinburgh from the dealer Carfrae from whom he bought other specimens. In the National Museums of Scotland library in Edinburgh there is a sale catalogue ([Jardine]1886) with handwritten notes in the margins, made during the sale on 17th June 1886, indicating the purchaser and prices for each of the lots. The Townsend specimen was among lots 141 and 142, bought by Edward Gerrard for £3.15s., that went to the British Museum (Natural History) in London and ended up in its Sub-department of Ornithology at Tring, Hertfordshire. The original label for this specimen is pasted to the back of the Jardine label and says "Valparaiso, June 1837 J.K. Townsend". Sharpe (1906, p. 360) says that 516 Jardine specimens purchased by Gerrard went to the BMNH in June 1886 so it is probable that there were a few more Townsend specimens in this batch. (The BMNH collection is not computerised so additional Townsend specimens would be hard to find).

2. The Lesser Seed-Finch MCZ 75966 is labelled as being from Chile but this species does not occur there naturally; it may have originated from Brazil (or it may be misidentified).

3. "Fringilla Mortonii" Audubon, see *Ornithological Biography* (1839) 5: 312 [= *Zonotrichia capensis*]. Joseph Grinnell (1932, p. 323) reported that ANSP 10614 was "still in fairly good condition, tip of upper mandible broken; plumage somewhat worn and faded; "apparently an adult male"; one label, not an original, says: "Columbia R. J.K.T."; another says, in part, "original specimen from Dr. J.K. Townsend Collection. Chili. Pres.[ented] by Dr.Woodhouse".

4. "Carduelis Stanleyi" Audubon, see *A Synopsis of the Birds of America* (1839) p. 118 [= *Carduelis barbatus*]. Joseph Grinnell (1932, p. 323) reported that USNM 2035 was an "adult male in full fresh plumage; skin in excellent condition, though slight soiling and disarrangement of barbs at ends of wings and tail would seem to show that the bird had been kept alive in confinement … The old label, in S.F. Baird's hand-writing says "2035. [male] California." An "n" indicates that Baird received the bird from Audubon." This is a southern South American species and the locality "California" is erroneous, unless it was obtained there as a cage bird by someone other than Townsend (who did not visit California). There is no positive link of this specimen with Townsend but we do know that he could have collected (or purchased caged birds) in Chile. Some South American specimens passed by Audubon to Baird were originally from William Swainson and this may be one of them.

TOWNSEND'S MAMMAL SPECIMENS FROM CHILE

The only known Townsend mammal specimens from Chile appear to be an otter "Lutra chilensis" donated by Townsend to the National Institution some time prior to 13th September 1841 (*Proceedings of the National Institution* 1841: 104) and a skull listed in the S.G. Morton catalogue as "536. Canis _____. Fox of Chili." (Morton 1840, p. 36.). The Morton skull collection is in the ANSP.

APPENDIX 14. NUTTALL'S BIRD SPECIMENS FROM THE WEST

Not only did Thomas Nuttall provide Audubon with notes on the birds seen on the journey across the Rocky Mountains and on the Columbia River but he also made a significant contribution to the early study of the birds of California, despite his short stay there and his usual preoccupation with plants. In volume four of Audubon's *Ornithological Biography* (1838) five species are ascribed to California; in the next volume (1839) a further 19 are included; and in the octavo edition of *The Birds of America* (1840-44) the total rises to 33 species. Most of them were included on the authority of Nuttall, and some are erroneously attributed to Townsend (Grinnell 1909: 8-10). Although Nuttall's observations were a significant contribution to Audubon's works, the specimens from the west that he gave to the artist are relatively few and worth examining in more detail.

Nuttall left the Columbia River in September 1835 to spend the winter in the Hawaiian Islands. After four months there he took a ship bound for California, landing at Monterey. After a while he took a small boat further down the coast to Santa Barbara where he found accommodation in a house with a tree-bordered garden (Graustein 1967). Although he later went down to San Diego and boarded the *Alert* for a passage round Cape Horn to Boston, most of his bird observations from this period seem to have been at Santa Barbara. We know that Nuttall did not usually carry a gun, so he either borrowed a gun or got someone to go out shooting with him (the wording in his notes on the acquisition of the Yellow-billed Magpie do not make it clear which method he used). His specimen of Anna's Hummingbird was caught in his hat.

When Nuttall arrived in Boston in September 1836 Audubon happened to be in the city: "Nuttall has arrived – he breakfasted with me the other day – gave me 6 new species of Birds and tells me that he will urge both Townsend and the Society at Philadelphia to allow me to portray all the species which they have procured within the limits of our Territories" (Audubon to E. Harris, 25th September 1836). With the aid of the *Ornithological Biography* it is easy to identify five of the six species mentioned above but the remaining one is more difficult because of Audubon's carelessness. Hall (1938: 5) suggests that it was a Gairdner's Woodpecker, a supposed new species that Audubon named without referring to the precise origin of the single specimen he possessed. Another possibility is that Nuttall acquired a South American cage bird in one of the California ports.

Nuttall seems later to have given Audubon an additional bird or two, and several nests. Even so, Nuttall's entire collection was evidently a very small one. It probably contained only the few species that Nuttall considered to be novel, as he would not have wanted to burden himself with specimens that he knew Townsend had already collected. In addition to the bird skins there were nests of **Anna's Hummingbird**, **Rufous Hummingbird**, **Lazuli Bunting**, **Bush-tit** (and perhaps **Chestnut-backed Chickadee**), two eggs of **Steller's Jay** (Audubon 1840-44), and perhaps a few more items. It is not clear whether Nuttall brought all of these back himself or simply handed some of them over to Townsend while they were both on the Columbia River. Either way, Audubon acquired both skins and nests from Nuttall and painted them for *The Birds of America* (1827-38) as well as describing them in the *Ornithological Biography*. The 'only' new species were the Yellow-billed Magpie and the Tricolored Blackbird – pretty astute of Nuttall considering the small amount of time he spent in California and the fact that he was on or near the coast, travelling in areas that had been worked by earlier naturalists. The following quotations are from the octavo edition of *The Birds of America* and refer to the minimum seven species that we believe Nuttall provided for Audubon. Very few of Nuttall's western bird specimens have survived: all those that we have traced are listed.

California Quail *Callipepla californica* (Shaw, 1798)
"CALIFORNIAN PARTRIDGE … Mr. TOWNSEND has lately sent me a beautiful specimen of the male, which he procured on the 6th of March, 1837, near Santa Barbara in California. I have to regret, however, that he has not furnished me with any account of its habits." Audubon then gives some notes on the gentle and confiding behaviour of the quail supplied by Nuttall (Audubon 1840-44, 5: 67). See Plate CCCCXIII.

Audubon has allowed several errors to appear here: Townsend never went to California and this location cannot be a mistake for the Columbia River because this species does not appear on any of Townsend's bird lists; moreover, on the specified date Townsend was on Oʻahu in the Hawaiian Islands. This specimen was no doubt collected by Nuttall at Santa Barbara in

1836 (on 6th March 1837 Nuttall was back on the east coast, but it was not unusual for Nuttall to be a year out in his dating!)

Specimens: USNM 2829, male, Santa Barbara, California, 6th March 1847 [= 1836?]; credited to J.K. Townsend [= T. Nuttall?].

Anna's Hummingbird *Calypte anna* (Lesson, 1829)
"ANNA HUMMING-BIRD … My good friend THOMAS NUTTALL, while travelling from the Rocky Mountains toward California, happened to observe on a low oak bush a Humming-bird's nest on which the female was sitting. Having cautiously approached, he secured the bird with his hat. The male in the meantime fluttered angrily around, but as my friend had not a gun, he was unable to procure it.

The nest, which he has presented to me, is attached to a small branch, and several leaves from a twig issuing from it, which have apparently been bent down for the purpose. It is very small, even for the size of the bird, being an inch and a half in depth, and an inch and a quarter in breadth …

The figures of the nest and female are taken from the specimens presented to me by Mr. NUTTALL. Those of the male I made from specimens, for the use of which I am indebted to Mr. LODDIGE, of London, whose collection of Humming-birds is unrivalled. This species is the fourth now found within the limits of the United States" (Audubon 1840-44, 4: 188). See Plate CCCCXXV.

Downy Woodpecker *Picus pubescens* (Linnaeus, 1766)
Nuttall may possibly have given Audubon a specimen of Gairdner's Woodpecker, a specimen Nuttall had received from Dr Gairdner while on the Columbia River in 1835 (Stone 1899: 16). It could also have come from, or via, Townsend. Gairdner's Woodpecker was not illustrated in *The Birds of America*.

Black Phoebe *Sayornis nigricans* (Swainson, 1827)
"ROCKY-MOUNTAIN FLYCATCHER … The only specimen of this Flycatcher in my possession was given to me by my esteemed friend THOMAS NUTTALL, Esq., who procured it in North California, but was unable to give me any account of its habits. It has been briefly characterised by Mr. SWAINSON in his Synopsis of the Birds of Mexico" (Audubon 1840-44, 1: 218). See Plate CCCCXXXIV.

Yellow-billed Magpie *Pica nuttalli* (Audubon, 1837)
"YELLOW-BILLED MAGPIE … I have conferred on this beautiful bird the name of a most zealous, learned, and enterprising naturalist, my friend THOMAS NUTTALL, Esq., to whom the scientific world is deeply indebted for the many additions to our zoological and botanical knowledge which has resulted from his labours. It is to him alone that we owe all that is known respecting the present species, which has not hitherto been portrayed. In a note inserted by him in my journal, he says:

"As we proceed to the south in Upper California, around the village of Sta. Barbara, we find the Common Magpie [American Magpie] substituted by this remarkable species, which is much more shy and cautious, as well as more strictly insectivorous. It utters, however, nearly if not quite the same chatter. In the month of April they were everywhere mated, and had nearly completed their nests in the evergreen oaks of the vicinity (*Quercus agrifolia*). The only one I saw was situated on a rather high tree, towards the summit, and much concealed among the thick and dark branches. Their call was *pait*, *pait*; and on approaching each other, a low congratulatory chatter was heard. After being fired at once, it seemed nearly impossible again to approach them within gun-shot. When alighted in the thick oaks, they remained for a considerable time silent, and occasionally even wholly hid themselves; but after a while the call of recognition was again renewed, and if the pair then met, they would often fly off a mile or more without stopping, in quest of insects. We often saw them on the ground, but never near the offal of the oxen, so attractive to the Crows and Ravens around" (Audubon 1840-44, 4: 104). See Plate CCCLXII.

Specimens: ANSP 3303, unsexed, Columbia River, no date [credited to J.K. Townsend = T. Nuttall, California, April 1836?]. ANSP 3306, unsexed, Columbia River, no date [credited to J.K. Townsend = T. Nuttall, California, April, 1836?]. USNM 2845, unsexed adult, Santa Barbara, California, April 1836.

Rock Wren *Salpinctes obsoletus* (Say, 1823)

Audubon drew a specimen of an adult female Rock Wren "given to me by Mr. NUTTALL", presumably the one collected on 21st June 1834 while at the Trappers' Rendezvous (Audubon 1840-44, 2: 113-115). It would be surprising if Nuttall carried this specimen around with him for a couple of years. He may simply have added it to Townsend's collection and it would have arrived in Philadelphia on 12th July 1836. See Plate CCCLX.

Tri-colored Blackbird *Agelaius tricolor* (Audubon, 1837)

"RED-AND-WHITE-WINGED TROOPIAL, OR MARSH BLACKBIRD … This beautiful species was discovered in Upper California by my friend THOMAS NUTTALL, Esq., from whom I received the specimen represented in the plate, together with the following account. "Flocks of this vagrant bird, which, in all probability, extends its migrations into Oregon, are very common around Santa Barbara in Upper California, in the month of April. Their habits are similar to those of the Red-winged Starling … They are seldom seen but in the near suburbs of the town, feeding at this time almost exclusively on the maggots or larvae of the blow-flies, which are generated in the offal of the cattle constantly killed around the town for the sake of the hides … They are also common around Monterey …" (Audubon 1840-44, 4: 27-28). See Plate CCCLXXXVIII.

Specimens: USNM 2836, male, Santa Barbara, California, April 1836.

Plate CCCLXII. Yellow-billed Magpie at centre.

APPENDIX 15. AUDUBON'S CENTRAL AND SOUTH AMERICAN LAND BIRDS IN *THE BIRDS OF AMERICA*

The following seven species are among those that Audubon included in error in *The Birds of America*, considering each of them to be birds of western North America, though none of them occurs north of Mexico. Audubon was confused not just about their distribution, but also about their scientific status. He thought that several were new discoveries but in fact only one had not been described.

These mistakes may have been due to Audubon's lack of thorough research in museums and libraries, or forgetfulness or confusion as to who had given him the specimens, or they may have been the fault of his co-author William MacGillivray. In those days it was extremely difficult to know when a species was undescribed because new discoveries were being published at a tremendous rate in a huge variety of books and journals, many of them in languages other than English, so some of their errors are excusable. Nevertheless, it seems incredible that neither of the two authors seems to have had a clear idea of the itineraries of either Townsend or Nuttall. Even though Audubon had met both Nuttall and Townsend after they had returned from the West he still had not grasped that Townsend never went to California (even in its wider geographical sense of those times). Nor did Audubon know enough about their movements to be able to correct some obvious dating errors (mistakes that also appeared in the texts for several genuine North American species). However, the responsibility for all these errors must rest with Audubon since he did not give co-author status to MacGillivray.

The Lineated Woodpecker, Rufous-collared Sparrow and Masked Yellowthroat all appeared in Townsend's list of birds from "the Rocky Mountains, the territory of the Oregon and the north west coast of America" (Townsend 1839a), but only after Audubon's descriptions of these birds had appeared in the *Ornithological Biography* (see also Jobanek 1994).

Lineated Woodpecker *Dryocopus lineatus* (Linnaeus, 1766)
This bird occurs from Mexico to central South America. Audubon wrote about it in his *Ornithological Biography* (1839) (as the Lineated Woodpecker) having seen a supposed specimen at Edinburgh said to have been collected by Dr Gairdner at Fort Vancouver, but he did not paint it. Townsend (1839a) included this woodpecker in his list of birds found in the Rocky Mountains and westwards to the Pacific. Gairdner's bird was no doubt the Pileated Woodpecker (see also Jobanek 1994).

Imperial Woodpecker *Campephilus imperialis* (Gould, 1832)
Audubon included this large Mexican woodpecker in the *Ornithological Biography* (as the Imperial Woodpecker) on the strength of a vague description of a bird seen by Townsend in the Sawtooth Mountains, Idaho. Audubon could not paint it because no specimen was taken. The bird seen by Townsend was probably a Pileated Woodpecker (see 14th August 1834 in *Narrative*; Jobanek 1994).

Magpie-Jay *Calocitta formosa* (Swainson, 1827)
"COLUMBIA MAGPIE, OR JAY ... The specimen from which the drawings were taken was presented to me by a friend who had received it from the Columbia river" (Audubon 1840-44, 4: 105).
This species occurs only in Mexico and other parts of Central America and could not have come from the Columbia River (nor from Colombia). It cannot, therefore, have been given to him by either Nuttall or Townsend and this is supported by Audubon's statement that these two naturalists had missed it on the Columbia River (see p. 279). The specimen may have come from William Swainson via William Bullock who collected in Mexico in 1825-26 (see e.g. Mearns and Mearns 1992: 419-427), though why Audubon described it as new after Swainson had already done so is unclear, as he must surely have seen Swainson's well-known 'Synopsis of Birds discovered in Mexico', published in the *Philosophical Magazine* in 1827.

Masked Yellowthroat *Geothlypis aequinoctialis* (Gmelin, 1789)
"DELAFIELD'S GROUND-WARBLER ...The only specimen in my possession was obtained from Mr. TOWNSEND, who procured it in California" (Audubon 1840-44, 2: 81).
This species occurs from Costa Rica southwards into central South America but not in Chile. Townsend could not have collected it himself and there

must also be doubts that he obtained it from other sources before giving it to Audubon. It seems more likely to have come from William Swainson, one of Audubon's English correspondents. Swainson visited Brazil in 1816-18 and later described specimens received from William Bullock junior who collected in Mexico in 1825-26 (see e.g. Mearns and Mearns 1992: 419-427). Swainson also had birds from many other sources.

Rufous-collared Sparrow *Zonotrichia capensis* (Müller, 1776)
Audubon described this species as "MORTON'S FINCH ... A single specimen of this pretty little bird, apparently an adult male, has been sent to me by Dr. TOWNSEND, who procured it in Upper California. Supposing it to be undescribed, I have named it after my excellent and much esteemed friend Dr. MORTON of Philadelphia, Corresponding Secretary of the Academy of Natural Sciences of that city" (Audubon 1840-44, 3: 151). There are three mistakes here: Townsend never went to Upper California, the species does not occur there, and it had already been described. This sparrow is widespread and common in much of Central and South America, occurring in open country, brush, gardens, etc., and there is little doubt that Townsend brought the bird back from Valparaiso, Chile. The type, ANSP 10614, has a label that says, in part, "original specimen from Dr. J.K. Townsend Collection. Chili. Pres.[presented] by Dr. Woodhouse" (Grinnell 1932: 323). Townsend had at least one duplicate that he asked Audubon to sell in Europe (see Derby list, appendix 4). Whitmer Stone (1899) blamed Townsend for mislabelling the specimen but was later of the opinion that Audubon "was the one at fault" (Stone 1934).

Yellow-faced Siskin *Carduelis yarrellii* (Audubon, 1839)
This was a new species described and named by Audubon as "YARRELL'S GOLDFINCH. CARDUELIS YARRELLII ... found in Upper California, it may be considered as forming part of our Fauna" (Audubon 1840-44, 3: 136-137). He says that the specimen was presented to him by William Swainson, but the locality of "Upper California" is not correct as this species breeds only in the arid zone of eastern Brazil. There can be no direct connection with Townsend or Nuttall.

Black-chinned Siskin *Carduelis barbata* (Molina, 1782)
Audubon described this species as new, calling it the "STANLEY GOLDFINCH. CARDUELIS STANLEYI ... named in honour of the illustrious Earl of Derby ... found in Upper California" (Audubon 1840-44, 3: 137-138). Audubon does not say any more about the origin of this bird but there are two specimens that Audubon is believed to have used (USNM 2035, male; USNM 2036, female. Both are now catalogued as being from Valparaiso, Chile).

This siskin is a South American species, inhabiting Chile and western Argentina. Examination of one of the specimens used by Audubon (USNM 2035) showed that the tail and wing tips were worn, indicating that the bird may have been kept in captivity; and it had an old label in S.F. Baird's handwriting with a mark to show that before Baird donated it to the Smithsonian Institution he had received it from Audubon (Grinnell 1932: 323).

There is no certain evidence that Audubon received these specimens from Townsend, though it is a species that Townsend may have acquired at Valparaiso, since three "Siskin" specimens from Chile were included in the Derby list (see appendix 4) and it is possible that one, or all, of these may have been Black-chinned Siskins. If it was indeed a captive bird that Audubon described then Townsend could have bought it from a dealer in the streets of Valparaiso, but it might equally well have arrived in Upper California by ship as a cage-bird and been acquired by Audubon from another source.

There are a few additional species from Central and South America included by mistake in *The Birds of America* but none of them have any connection with Townsend, neither real connections nor erroneous ones. These include: the Crested Bobwhite *Colinus cristatus* that Audubon figured as the Welcome Partridge from a specimen collected during the Beechey voyage from the "northwest coast of America" – though Guatemala is the northern extent of its range; and Trudeau's Tern *Sterna trudeaui* that Audubon figured from a specimen received from Dr James Trudeau, supposedly collected in New Jersey but no doubt from South America.

APPENDIX 16. THE EDWARD HARRIS COLLECTION OF BIRDS

In October 1937 a man by the name of Henry Makin walked into the office of J. Fletcher Street, and because he knew Street had an interest in birds, mentioned that his parents had in a barn, near Moorestown, New Jersey, a collection of birds stored away in old boxes and trunks. Makin's father had received the birds, as well as items of furniture, guns and a mineral collection, in payment for a plumbing bill. The donor was Edward Harris III, a son of the Edward Harris who had been such a staunch supporter of Audubon, who had travelled with him up the Missouri in 1843, and who had financed the purchase of many of Townsend's bird and mammal specimens for Audubon.

Realising the possible significance of this news, Street went to Moorestown to talk to Makin's parents and learnt that the birds had originally arrived in a large and beautiful chest of drawers. It was not long before Mrs Makin thought the chest would be just the place to store the family linen and insisted on the removal of the birds. Mr Makin found an old trunk and some packing boxes, shoved in the birds and finished off the job by pressing down on them with his feet in order to close the lids. Worse was to follow. When their son began to grow up he became interested in Harris's guns, went to the barn loft and took some of the birds out of the trunks to use as targets!

When Street looked at the remains of the collection he was horrified: there were headless birds, loose heads, wings and feet all mixed up with complete skins in better condition, some with labels, some without. Street took away "twelve bushel baskets" of skins, some 400 identifiable birds of 228 species. From the labels it was easy to see that some skins were from the Missouri River expedition, and that the remainder had originally been collected by a French naturalist by the name of August Lefevre, by Spencer Baird (from the vicinity of Carlisle, Pennsylvania), and by Townsend (from his western journey, including the Hawaiian Islands). None of them had been collected locally by Edward Harris.

Only three of the Townsend skins still had labels, each with a locality and a precise date. Eleven other birds have been attributed to Townsend (Street 1948), but there were presumably others that were not worth saving, or already lost.

Townsend skins found in the Harris Collection (based on Street 1948)

Species	Collecting locality, date	Current location
1. Snowy Egret *Egretta thula*	Walla Walla, Columbia River, 3rd July 1836.	ANSP 162372
2. Snowy Egret *Egretta thula*	Walla Walla, Columbia River, 13th August 1836.	
3. Black Oystercatcher *Haematopus bachmani*		
4. American Avocet *Recurvirostra americana*		
5. Red-breasted Sapsucker *Sphyrapicus ruber*		ANSP 19184 ?
6. Gray Jay *Perisoreus canadensis*	[Astoria], Columbia River, 11th October 1836.	ANSP 162471
7. American Dipper *Cinclus mexicanus*		
8. Varied Thrush *Ixoreus naevius*		ANSP 23631
9. Hermit Warbler *Dendroica occidentalis*	[Columbia River]	ANSP 162369
10. Oʻahu ʻŌʻo *Moho apicalis*	[Oʻahu, Hawaiian Islands]	MCZ 17599
11. Kauaʻi Nukupuʻu *Hemignathus lucidus hanapepe*	[Kauaʻi, Hawaiian Islands]	FMNH 308799
12. ʻApapane *Himatione sanguinea*	[Hawaiian Islands]	
13. ʻIʻiwi *Vestiaria coccinea*	[Hawaiian Islands]	ANSP 3364 ?
14. ʻIʻiwi *Vestiaria coccinea*	[Hawaiian Islands]	

Street donated the majority of the Townsend skins to the Philadelphia Academy but some of them are not now easily traced, or may have been discarded. The ANSP database does not include any Black Oystercatcher or American Dipper collected by Townsend. Those that did not go to Philadelphia include the *Hemignathus lucidus hanapepe* that went to Princeton University (and later to the Field Museum, Chicago); some others may have been included in the 41 skins that went to Newark Museum, Delaware, 23 of which were soon disposed of because of their condition. The remaining specimens at Newark could not be traced in 2002.

APPENDIX 17. TOWNSEND'S BAT SPECIMENS FROM THE WEST

Townsend's Big-eared Bat *Corynorhinus townsendii* (Cooper, 1837)
Townsend collected this bat on the lower Columbia River, within the confines of Fort Vancouver, presumably with the permission of the chief factor because normally the bats at the fort were protected. Some specimens of this bat seem to have been included in the first consignment of natural history material sent back by ship that arrived in Philadelphia on 12th July 1836. In Townsend's absence (or by agreement at a later date), Dr Morton at the Philadelphia Academy of Natural Sciences sent the bats to William Cooper, a New York naturalist who moved to a farm at Guttenberg on the New Jersey side of the Hudson River in 1837. Cooper published the first description of the Western Big-eared Bat in the *Annals of the Lyceum of Natural History of New York:*

> "PLECOTUS TOWNSENDII … Three specimens of this very distinct new species were brought from the Columbia river by Mr. John K. Townsend, where he procured them on his late journey in company with Mr. Nuttall … I regret being obliged to describe these … Bats from dried specimens, in which state the most characteristic marks especially about the head, are often difficult to detect" (Cooper 1837: 73-75).

After his return home, Townsend included the bat in his scientific appendix to the *Narrative*, explaining why the bats were protected on the Columbia River:

> "TOWNSEND'S GREAT-EARED BAT … Inhabits the Columbia river district, rather common. Frequents the store houses attached to the forts, seldom emerging from them even at night. This, and a species of *Verpertilio [sic], (V. subulatus,)* [**Western Small-footed Myotis**] which is even more numerous, are protected by the gentlemen of the Hudson's Bay Company, for their services in destroying the dermestes [beetles] which abound in their fur establishments" (Townsend 1839a: 324-325).

Much to Townsend's frustration, it took him years to retrieve anything from Cooper – and when he eventually did so, he was far from pleased. Only one of the three original specimens had been returned and it was in a very poor state. Morton had sent the bats to Cooper, so Townsend thought it only reasonable that Morton should get them back and asked him to do so by letter:

> "Did I inform you that this package, (which of course purported to contain the animals sent by you to Cooper,) had in it, but a very few of the Specimens which I sent to Philadª. from Oregon? The least valuable of them have been returned, but of the species which I prized most, (the new Plecotus described & named after me by Cooper [**Townsend's Big-eared Bat** *Plecotus townsendi*]) but <u>one</u> of the <u>three perfect</u> specimens sent by you to Cooper, & that one so mutilated as to be worth absolutely nothing, was found in the package. You are probably aware of the extreme rarity of this species – mine, I believe to be the only specimens extant. They are therefore extremely valuable – & the species is so rare, even at [Fort] Vancouver, where alone I have ever seen it, that it is not likely to be collected in future. Now there must be <u>two</u> perfect specimens somewhere, belonging to me, & as they are my property, & as I value them very highly, I surely have a <u>right</u> to ask what has become of them. Will you therefore do me the favor to write to Cooper on the subject, & make enquiries about the missing specimens. I do not know Cooper & have never seen him, or corresponded with him, & as you sent these specimens to him for description, I trust you will not object to write to him on the subject. If I can recover the <u>two</u> specimens of <u>Plecotus</u>, I shall be satisfied, though I might very properly inquire why the third – the one sent me – was destroyed (JKT to Morton, 8th May 1844, Library Company MSS at HSP).

It is unclear if Townsend ever got all the specimens back. Audubon was living in New York at that time and may have got hold of them from Cooper because he painted one of the Western Big-eared Bats, inscribing his watercolour: "Plecotus townsendii, Cooper – Natural Size / Townsend's Plecotus - / J.J.A. / 16th Feby / [18]46" (see Olson and Mazzitelli 2003). Unfortunately, none of Audubon's bat paintings were included in the *Viviparous Quadrupeds* despite Audubon and Bachman's original intentions. The original watercolour of the bat (NYHS Z3333 [S-30], fig 17) is now housed at the New-York Historical Society along with the majority of Audubon's other bird and mammal watercolours.

The mutilated specimen that Townsend got back from Cooper (and one other Western Big-eared Bat) were among nearly 60 mammals that Townsend offered to sell to J.E. Gray at the British Museum (Natural History) (JKT to J.E. Gray, 13th January 1849, see appendix 18). Although the majority of these specimens later went instead to the Academy of Natural Sciences of Philadelphia, the two bat specimens seem not to have survived. There is a Townsend's Big-eared Bat skull at Philadelphia that was donated by Townsend to the Morton collection (ANSP 2605; see Morton 1840: 35) but this is not necessarily from one of the two missing specimens.

Western Small-footed Myotis *Myotis ciliolabrum* (Merriam, 1886)
A Townsend specimen of this species survives, collected at Sauvie Island, Columbia River, on 11th August 1835 (ANSP 1126, skin with skull inside). It is probably one of the three "*Vespertillio [sic] subulatus*" specimens offered to J.E. Gray at the British Museum (Natural History) that later went instead to the Academy of Natural Sciences of Philadelphia. This was not a new discovery: in 1823 it had been described by Thomas Say and was long known as *Myotis subulatus* before being separated into two species, *M. leibii* being the eastern form.

Yuma Myotis *Myotis yumanensis* (H. Allen, 1864)
There is a Townsend specimen of this species, with no supporting data (ANSP 1125, skin with skull inside). It is perhaps the species referred to in Townsend's appendix as "? Little Bat *Vespertilio (undescribed.)*" (Townsend 1839a: 313). It may also be the specimen listed in a letter to J.E. Gray at the British Museum (Natural History) as "Vespertillio (nondescript?) 1 [specimen] R.Mᵗˢ" that later went instead to the Academy of Natural Sciences of Philadelphia. There is also a skull in the Morton collection that may be this species (ANSP 2604; see Morton 1840: 35, where listed as "Vespertilio _____ ").

This was a new discovery by Townsend but was not described at the time. The first scientific description was published in1864, from a specimen from Old Fort Yuma, California (opposite the present town of Yuma, Arizona), hence the bat's scientific name. Its range extends northwards and Townsend could have collected it anywhere in his travels between South Pass, Wyoming, and the mouth of the Columbia River.

APPENDIX 18. TOWNSEND'S ATTEMPT TO SELL MAMMAL SPECIMENS IN EUROPE. Townsend letter to John Edward Gray 13th January 1849 (BMNH DF 200/146/202) – unpublished MS, transcribed by B & R Mearns.

Philadelpia, Jan. 13th 1849.

To/
J.E.Gray, F.R.S.

Dear Sir.

I have in my possession, a small collection of Quadruped skins, brought by me from the Rocky Mountains & the Oregon Territory, some years since, which I am anxious to dispose of; & my friend Prof. Baird, of Dickinson College, advised me to write to you on the subject, supposing that you might desire these skins for your Museum or for yourself.

My great desire has been to place them among the collections of our Academy of Natural Sciences, of which I have been a member for fifteen years; but I cannot, at this time, afford to <u>present</u> them, as I have done my birds, & various other interesting & rare objects of Nat. History; & the exchequer of the Society is now so low, that it cannot compensate me adequately for them. Under these circumstances, I have concluded to offer them to you. I have not fixed a price upon them, nor can I do so at present, having no data to guide me. Objects of this character possess no determined, or market value, being regulated & governed wholly by their rarity, & the difficulty of procuring them. Besides, this is the first time I have been compelled to sell specimens of Nat. History, having always hitherto, for nearly twenty years, been enabled to enjoy the luxury of giving to the Museums & private collections in which I have felt an interest.

This collection was made exclusively by my own hands, & the skins are, generally, in excellent order. Some of the largest of them are preserved in poisoned whiskey, but the smaller specimens, & by far the greatest number, are stuffed dry. The skins composing this latter part were entrusted for description to my friend, the Rev. Dr. Bachman, who, some few years since, carried them with him to Europe, & from thence sent on his descriptions, which were published in the Journal of the Academy of Natural Sciences, of this City. He compared the skins with those in the various collections in the Museums of the old world; & I believe he stated that most of them, of which single specimens only existed in the collection, were unique.

Messrs Audubon & Bachman subsequently had these specimens in their possession, & from them the figures were made, & the descriptions drawn up which appear in their fine work, entitled, the "Quadrupeds of America."

In a letter received by me from Audubon, accompanying the return of the skins, he says. "I trust the Academy will give you a good price for these specimens, for there are several that are unique; not found in any collection in Europe, & of course very valuable as establishing species published by us only."

In the spirits, if my recollection serves me, are Seven or eight specimens, in perfect order, & at least three, - possibly four, - new species; - absolute <u>nondescripts</u>. This is the opinion of several scientific gentlemen who have recently examined them; among whom are, Dr. Bachman, Dr. Trudeau, & Mr. Titian R. Peale. One of these is a Wolf, of, I think, unprecedentedly large size. Of this there are two specimens. There are also two other Wolves of different species, of less size. These Wolves we think new, & they have never been described. There are two specimens of a handsome spotted Lynx, from the Oregon, which, I think, without having compared them carefully, - differ from any described by Dr. Richardson, & may possibly prove to be new. There is also the skin of the Oregon Deer, <u>Cervus leucurus</u>. I append a list of the specimens. My object at present, is to ascertain whether you would like these skins, & what sum you would be willing to give me for them. I presume that they will command a higher price in England than in this Country, because you are more capable of appreciating the value of such rarities than our more plodding & money-loving people.

I confess I feel great regret in offering to dispose of this collection out of my own beloved Country, but it is sufficiently evident that I cannot realise half its value here; & necessity compels me, most reluctantly, to offer it for sale where I have a prospect of realizing an adequate sum for it.

<u>List of Skins in Spirits</u>

Canis	Nondescript?	2	specimens	Hab. Oregon.	
Dᵒ· [= ditto]	Dᵒ·	1	Dᵒ·	"	dᵒ·
Dᵒ·	Dᵒ·	1	Dᵒ·		Dᵒ·
Cervus leucurus		1	Dᵒ·		Dᵒ·
Lynx		2	Dᵒ·		Dᵒ·
		7	Dᵒ·		

List of dry stuffed Skins

		Hab.			Hab
Meles Labradoria	1 specimen	Oregon	Brought over	22 specimens	
Lepus *Townsendii, Bach.	2 Dᵒ·	Dᵒ·	Meriones Americanus	1 Dᵒ·	Oregon
" *Nuttalii, Bach	1 Dᵒ·	R.Mᵗˢ	Mus leucopus	1 Dᵒ·	Dᵒ·
" *artemesia, Bach.	3 Dᵒ·	Dᵒ·	Scalops *Townsendii, Bach.	2 Dᵒ·	Dᵒ·
Lagomys princeps, Rich.	1 Dᵒ·	R.Mᵗˢ.	Neotoma Drummondii,	3 Dᵒ·	Dᵒ·
Spermopilus [sic] Douglasii,	4 Dᵒ·	Oregon	Sciurus Douglasii	3 Dᵒ·	Dᵒ·
" Richardsonii	2 Dᵒ·	Dᵒ·	" *Richardsonii, Bach	1 Dᵒ·	R.Mᵗˢ
" Franklinii,	1 Dᵒ·	Missouri	" *lanuginosus, Bach	1 Dᵒ·	Oregon
Tamias *Townsendii, Bach	2 Dᵒ·	Oregon	Pteromys *Oregonensis, Bach	1 Dᵒ·	Dᵒ·
" quadrivittatus	3 Dᵒ·	R.Mᵗˢ.	" Sabrinus	1 Dᵒ·	Dᵒ·
" *minimus, Bach	2 Dᵒ·	Dᵒ·	" *alpinus, Bach	1 Dᵒ·	Dᵒ·
	22 Dᵒ· Carried up			37 Dᵒ· carried over	

		Hab.			Hab.
Brought over	37 specimens		Brought up	46 specimens	
Mustela erminea?	1 Dᵒ·	Oregon	Vespertilio (nondescript?)	1 Dᵒ·	R.Mᵗˢ
Pseudostoma *Richardsonii	1 Dᵒ·	Dᵒ·	Sciurus *Fremontii, Townsend	1 Dᵒ·	Upper California
" borealis	1 Dᵒ·	Dᵒ·	Spermophilus *Townsendii,Bach	1 Dᵒ·	Oregon
" *Townsendii, Rich. M.S.S.	1 Dᵒ·	R.Mᵗˢ.	" tridecemlienatus [sic]	1 Dᵒ·	Missouri
" bursarius	1 Dᵒ·	Oregon	Arvicola *Oregoni, Bach.	1 Dᵒ·	Oregon
Plecotus *Townsendii, Cooper	1 Dᵒ·	Dᵒ·	" *Townsendii, Bach.	1 Dᵒ·	Dᵒ·
Vespertilio subulatus, Say	3 Dᵒ·	Dᵒ·	52 specimens, dry skins		
	46 Do, carried up			7 Dᵒ· in spirits	

59 Dᵒ· total.

Of these there are two species in the possession of Dʳ. Bachman, & which have not yet been returned to me. They are the curious Plecotus Townsendii*, of Cooper (of which I believe there is no other specimen extant,) & the Arvicola Townsendii. These specimens he neglected to return to me with the collection, & I have hitherto refrained from writing to him respecting them on account of terrible distress in his family, resulting from recent heavy bereavements. I will however write to him soon, & I have no doubt I shall receive them before it may become necessary to send them away. I am however, extremely anxious to have this matter attended to with as little delay as possible. Will you oblige me by writing as soon as you receive this, telling me what price you will give for the skins, & they shall be shipped as you may direct, provided, of course, I may think the sum offered to be adequate.

I am, dear Sir, with great respect,

and esteem,

Your most obᵗ. servant

John. K. Townsend, M.D

*There is a mutilated specimen
[of Plecotus Townsendii] in the box, in addition to the other.

P.S. I enclose a slip [see p. 310] from Professor Baird, of Dickinson College, as a sort of voucher in support of what I have said in reference to the specimens. He says that all the skins are in "first rate order." Most of them were so when they were entrusted to Dʳ. Bachman for description; but during the time, a few were slightly injured by injudicious handling (not by insects). I believe, however, that those injured are confined to the duplicates.

John K. Townsend.

BIBLIOGRAPHY

UNPUBLISHED SOURCES *Listed in sections, chronologically.*

Family papers: this all-embracing title covers original family letters, papers and documents in the possession of Donald Townsend Little and Dana Dunbar King. This includes genealogical charts and 'person sheets' that they, and other members of the family, have compiled, as well as material extracted from Philadelphia city directories 1785-1875, Priscilla Kirk Townsend's diaries 1848-1860, biographical sketches of Elisha Townsend, 'Memoirs of Edward Townsend' (6 pp., circa 1890, with a few additional incidents and anecdotes by Elizabeth S. Hoadley, 6 pp., July 1949), and from Jordan (1911). Many details about the family may be attributed to additional research carried out by D.T. Little in Philadelphia at the American Philosophical Society, College of Physicians, Library Company, Free Library and the Philadelphia Archives of the Historical Society of Philadelphia, and from elsewhere including Cape May Court House and Princeton University Library.

MS SOURCES BY TOWNSEND (JKT)

Journal

JKT, written on journey to Independence, Missouri, and slightly beyond, westwards, 13th March – 20th May 1834. ANSP Coll. 404 III (2).

JKT, written between Blackfoot River and Fort Vancouver, 10th July 1834 to 16th September 1834. Original journal lost, extracts appeared in *Waldie's Select Circulating Library* 1835, pt. 2, pp. 427-432 (reproduced in Hulbert 1934: 184-226).

JKT, written on the Columbia River, Hawaiian Islands etc., 18th December 1835 – 15th October 1837. ANSP Coll. 404 III (3).

Letters to family at Phildelphia

JKT, at Philadelphia, to Hannah Townsend, 7th April 1826. Family papers.
A letter of encouragement and advice to his twelve-year-old sister who is at Westtown Boarding School.

JKT, at Philadelphia, to Hannah Townsend, 4th August 1830. Family papers.
Family news for his sisters Hannah and Mary who are staying in Port Elizabeth, New Jersey.

JKT, at Pittsburg, to "My Dear Sister", 17th March 1834. ANSP Coll. 404 I: 1.
Description of trip by stage coach to Pittsburg

JKT, at Louisville, Kentucky, to "Dear Sister", [23rd March 1834]. ANSP Coll. 404 I: 2.
Description of trip from Pittsburg to Louisville.

JKT, at St Louis, to Charles Townsend, 27th March 1834. ANSP Coll. 404 I: 3
Mississippi River, Jefferson Barracks, Wyeth in St Louis.

JKT, at Hickory Grove (50 miles above St Louis), to Charles Townsend, 1st April 1834. ANSP Coll. 404 I: 4.
Townsend and Nuttall travel on foot to Independence. Some birds.

JKT, at Boonville, Missouri, to Charles Townsend, 9th April 1834. ANSP Coll. 404 I: 5.
Carolina Parakeets. Local people.

JKT, at Independence landing, to Charles Townsend, 21st April 1834. ANSP Coll. 404 I: 6.
Has received letters from home. Western manners. Capt Sublette joins the party.

JKT, at Kansas Agency, Missouri, to Charles Townsend, 3rd May 1834. ANSP Coll. 404 I: 7.
Extracts from his journal 28th April to 3rd May 1834. Kansas Indians. Some birds.

JKT, at Trappers' Rendezvous, Rocky Mountains, to Charles Townsend, 27th June 1834. ANSP Coll. 404 I: 8.
Account of his trip since last letter, of 3rd May 1834.

JKT, at O'ahu, [Hawaiian Islands], to Charles Townsend, 6th January 1835. ANSP Coll. 364.
The voyage from the Columbia River, and hospitality on O'ahu.

JKT, on board brig *May Dacre*, Columbia River, to Charles Townsend, 2nd May 1835. ANSP Coll. 404 I: 9.
Report of first trip to Hawaiian Islands. Poverty and sickness among tribes on the Columbia.

JKT, at Fort Vancouver, Columbia River, to Charles Townsend, 10th September 1835 (and addition dated 26th September 1835). ANSP Col 404 I: 11.
Discusses ways of coming home. Tribes of the Columbia and losses to diseases. Capt Thing's adventures with Blackfeet tribe.

JKT, at Fort Vancouver, Columbia River, to Charles Townsend, 11th April 1836. ANSP Coll. 404 I: 12.
Discusses return trip. Nuttall has left for Hawaii.

JKT, "State of Arkansas, 20 miles from the hot Springs of Washita", to Charles Townsend, 4th March 1840. Family papers.
Amusing account of a trip to the hot springs with a friend, John Jackson.

JKT, at Little Rock, Arkansas, to Charles Townsend, 14th March 1840. Family papers.
Continuation of his account of a trip to Washita hot springs with John Jackson. They lodge with a widow who has heard of JKT through her subscription to 'Waldie's Circulating Library'. They plan to travel to Independence, Missouri, via St Louis.
JKT, at Washington, to Priscilla Townsend, 29th December 1841. Family papers.
Family news. Account of an unsuccessful and rather miserable Christmas trip that he and his wife undertook to see cousins at Sandy Spring, Maryland. He is working for no pay at the National Institute and is seeking funds owed to his wife.

JKT, at Washington, to Hannah Townsend, 5th June 1843. Family papers.
His wife's mother and sister are staying. Discusses the heat in Washington, huge numbers of small beetles, his baby son and the joys of parenthood, and his thoughts on original sin and christenings.

JKT, at Washington, to Hannah Townsend, 6th June [1843]. Family papers.
Mentions the illness that has troubled him for years, the same illness that afflicts his mother. His displeasure about the sale of chapters from his Narrative *in newspaper form. Cousins from Maryland have recently visited. Report on his first meeting with John Bachman. The National Institute still lacks funds.*

JKT, at Washington, to Hannah Townsend, 12th April 1844. Family papers.
Concern for Hannah's health. Refers to insect book his sister Mary is preparing and his growing network of entomological correspondents. His financial crisis has eased a little.

JKT, at Washington, to Hannah Townsend, 16th April 1844. Family papers.
Working as a clerk copying letters. Recent eight-day scientific meeting under the auspices of the National Institute he considers a great success and boasts of being mentioned in the opening address and by other speakers. Offers advice to his sister Mary about her insect book. He has decided to concentrate on beetles and mentions some collecting techniques.

Other Townsend letters and miscellaneous items

JKT, at Independence, to Thomas McEuen, 23rd April 1834. ANSP Coll. 3. Coll. 567.
Concerning JKT's first consignment of specimens (and includes a description of a new bird – see below).

JKT, [at Independence, no date], ANSP Coll. 247.
Bird description [Harris's Sparrow]; believed to be an insert for above letter.

JKT, at Fort Vancouver, for Rev Samuel Parker, 11th April 1836. Original MS untraced; a typescript copy is reproduced by Jobanek and Marshall 1992.
A rather hastily prepared and incomplete "Description of the birds of the Columbia River region" that he gave to Parker.

JKT, at New Garden, Philadelphia, to Dr Samuel Morton, 6th September 1838. APS.
JKT ensconced at Dr Michener's house. Skulls.

JKT, at Philadelphia, to John Vaughan, 20th January 1838. APS. [Not seen]
Indian vocabularies. Specimens and shells selected by Titian Peale.
JKT, at Cape May, to Dr Samuel Morton, 4th August [1839]. APS.
JKT's 'Ornithology of United States'. Certificate requested from Morton. JKT's illness.

JKT, at Philadelphia, to John Vaughan, 30th August 1839. APS. [Not seen].
JKT presents unnamed volume to APS.

JKT (with Thomas McEuen), at Philadelphia, 25th October 1839. ANSP Coll. 404.
"List of birds lent to J.J. Audubon by Dr M'uhern [sic]."

JKT, at Philadelphia, to Spencer F. Baird, 10th October 1840. SI Box 34.
JKT has been rail shooting with George Leib.

JKT, [at Washington, D.C.], 13th December 1841. Huntingdon Library.
List of specimens presented to National Institute.

JKT, at Washington, D.C., to S.S. Haldeman, 17th February 1844. ANSP Coll. 73.
JKT's Washington address. Invitation to a meeting.

JKT, at Washington, D.C., to S.S. Haldeman, 7th March 1844, ANSP Coll. 73.
Haldeman cannot come, asks for papers to be read. Mentions treatment of JKT by the US Exploring Expedition.

JKT, at Washington, D.C., to S.S. Haldeman, 25th March 1844. ANSP MS211c.
Concerning insects and best equipment for catching and preserving them.

JKT, at Washington, D.C., to Dr Samuel Morton, 8th May 1844. HSP: Library Company Manuscripts, Morton Collection No. Yi2 7388 ff 22.
Townsend asks Morton to retrieve Oregon bat specimens from William Cooper. Complains about losses from his insect collection.

JKT, at Washington, D.C., to J. Varden, 3rd July 1844. Huntingdon Library.
Asks Varden to have the hand barrow ready to transport Castlenau's insects.

JKT, at Washington, D.C., to Spencer. F. Baird, 7th June 1844. SI Box 34.
JKT busy at National Institute, hummingbird exchanges, naming of species.

JKT, at Washington, D.C., to Dr Samuel Morton, 3rd & 28th April 1845. HSP: Library Company Manuscripts, Morton Collection No. Yi2 7388 ff 53. *Two letters concerning Mexican Indian skull presented by Baron Gerolt.*

JKT, at Washington, D.C., to Edward Harris, 4th July 1845. ADAH.
Asks for live Summer Ducks [Wood Ducks], wants to know distribution of Cliff Swallow in eastern US. Hunting dogs.

JKT, at Philadelphia, to Dr Samuel Morton, 9th June 1846. ANSP. [Not seen]
Mummy presented to Mr Hodge.

JKT, at Philadelphia, to Spencer F. Baird, 13th January 1849. SI Box 34.
Asks Baird for certificate so JKT can sell mammals to Gray or Schlegel.

JKT, at Philadelphia, to John Edward Gray, 13th January 1849. BMNH DF 200/146/202.
JKT lists 59 mammal specimens that he wants to sell.

JKT, at Philadelphia, to John Edward Gray, 21st March 1849. BMNH DF 200/146/203a.
Asks for $250 for the mammal specimens.

JKT, at Philadelphia, to John Edward Gray, 15th June 1849. BMNH DF 200/146/204.
No reply to first letter. Repeats rarity value and asks for $250 for them.

JKT, Indian Vocabularies, 5 volumes. APS. [Not seen]

JKT, undated MS comments in the ANSP copy of volume 5 of Audubon's *Ornithological Biography*. ANSP.
Townsend's comments are discussed by Stone (1906).

MS SOURCES TO TOWNSEND (JKT)

From family

Hannah Townsend, at Philadelphia, to JKT, commences 29th November 1835. ANSP Coll. 404 II: 1. [20pp. Entries for various dates 1835 and 1836.]
Family events and local news.

Priscilla and Charles Townsend, at Philadelphia, to JKT, 3rd March 1836. ANSP Coll. 404 II: 2.
News of family and friends. Both parents hope that JKT's long journey will cure his wanderlust for ever.

From others

John Bachman, at Charleston, to JKT, 7th February 1838. PUL (see Rice 1960: 72).
Bachman advises JKT to include his mammal and bird discoveries in an appendix to his forthcoming book. He considers it to have been wise of JKT to send his birds to Audubon.

John Bachman, at Charleston, to JKT, 24th May 1838. PUL (see Rice 1960: 72).
Thanks JKT for mammal specimens and asks for a prompt reply to his questions about them because he plans to sail to Europe soon.

John James Audubon, at London, to JKT, 31st May 1838. Fugitive leaves, see Alice Ford 1964, p. 456.
Asks for measurements and more information about JKT's birds.

Edward Stanley, at Liverpool, 16th June 1838. LM.
"Catalogue of Birds submitted to my Selection by Mr. Audubon on the part of Dr. Townshend [sic] from various quarters."
Lord Derby makes a list of his acquisitions for his own records.

Victor Audubon, [no address], to JKT, 2nd June 1840. Houghton Library, Harvard University.
Negative response from Harpers regarding a new edition of JKT's "interesting work."

Levi Woodbury, at Washington, D.C., to JKT, 31st January 1841. Huntingdon Library.
Acknowledgement of Woodbury's election as president of National Institute.

A.D. Bache, at Washington, D.C., to JKT, 31st January 1841. Huntingdon Library.
Acknowledgement of Bache's election as director of National Institute.

John Torrey, at New York, "to "My Dear Sir", 2nd February 1841. APS [filed under JKT].
Refers to possible trip across the Rockies supported by [Harford Jones] Brydges; Brydges would prefer addressee to accompany him to Hudson Bay.

Thomas Nuttall, at Philadelphia, to JKT, 7th April 1841, at 6th Street near Pennsylvania Avenue, Washington D.C. HSP: Gratz Collection: American Scientists, case 7, box 24.
Nuttall says he can only arrange the plants in the National Institute herbarium if the collection is sent to Philadelphia where he has access to reference books and comparative material.

George Leib, at Philadelphia to JKT, 20th January 1842. HSP: Society Collection, filed under Leib.
A short list of birds and birds' eggs for presentation to the National Institute. Encourages JKT to describe his large wolf skins in order to establish priority.

John Bachman, at Charleston, to JKT, 2nd November 1842. PUL (see Rice 1960: 72).
Apologizes for not collecting certain birds for JKT. Hopes to be in Washington, D.C. next May.

C.A. Hasler, at Washington, D.C. to JKT, 16th April 1843. Huntingdon Library.
Specimens from Amazon and Rio de Janiero.

Miscellaneous MS sources

J.J. Audubon at London, to S.G. Morton, 26th December 1837. APS.
JKT should have received a better welcome from the citizens of Philadelphia and would have done better if collections disposed of in London. Audubon wants more information and specimens from JKT, and complains that neither Nuttall nor Morton have written to him. A postscript asks for eggs and nests from JKT.

J.J. Audubon at London, to S.G. Morton, 19th January 1838. APS.
Requires more bird notes from JKT. Sends Morton latest bird plates many of which depict JKT specimens.

J.J. Audubon at New York, to S.G. Morton, 9th September 1839. APS.
Tells of his plans to publish The Quadrupeds *and the octavo edition of the birds. Asks if JKT still has his mammal specimens from the west.*

Spencer F. Baird to William Baird correspondence (25 letters examined, 16th June 1841 to 2nd July 1853). S.F. Baird Papers 1833-1889. SI Record Unit 7002, box 38, folder 25.

William Baird to Spencer F. Baird correspondence (101 letters examined, 2nd July 1841 to 9th February 1852). S.F. Baird Papers 1833-1889. SI Record Unit 7002, box 38, folders 15-20.

Harris, T.W. 1838. Description of the Insects collected by Mr. John Kirk Townsend in Oregon, near the mouth of the Columbia river. 30pp. MCZ 6Mn1308.41.4.
An incomplete assessment of JKT's insect collection, mainly beetles.

Samuel Morton, at Philadelphia, to Alex Bache, 21st July 1836. APS.
First JKT specimens have arrived at the Academy.

Samuel Morton, at Philadelphia, to George Ord, 19th December 1837. APS.
JKT's collections.

William Fraser Tolmie, at Fort Vancouver, to W.J. Hooker, 15th November 1836. Public Record Office, Kew LX11 152. *Some plants and all the birds (17 species) that Tolmie sent to Hooker were collected by JKT.*

Committee of the American Philosophical Society, at Philadelphia, to Samuel Morton, 21st December 1837. APS.
APS Committee wants meeting with ANSP in order to obtain JKT's specimens.

Committee of APS (signed Ord, Patterson, and Peale), at Philadelphia, 30th April 1838. APS.
No specimens yet received from JKT.

Minutes of APS Historical and Literary Committee, 1840-43.
JKT's Indian Vocabularies received.

Charles Pickering, at Washington, D.C., 22nd November 1842. SI Record Unit 7058 Box 12.
Report on the state of the National Institute collections. JKT employed as taxidermist.

John Cassin to Spencer F. Baird correspondence (66 letters examined, 13th February 1843 to 15th May 1851). S.F. Baird Papers 1833-1889. SI Record Unit 7002, box 17, folders 3-5.

S.S. Haldeman, at New York, to John Le Conte, 15th August 1844. APS.
JKT's insects from the Pacific North West.

S.J. Tilden, [no address], to C.W. Lawrence, 30th July 1845. APS.
Recommendations for a John Townsend for employment [probably not JKT].

T.R. Peale, c.1855. "The South Seas Surveying and Exploring Expedition, Its Organisation, Equipment, Purposes, Results and Termination." A typed copy, pp. 10-16. AMNH.
Criticism of the aftermath of the expedition.

Mahlon Kirk, Woodburn Farm, Oakland, Maryland, to Witmer Stone, 27th February 1904. ANSP.
Resting place of JKT's Journal etc. His devotion to family. A postscript refers to William Baird.

Mahlon Kirk, Woodburn Farm, Oakland, Maryland, to Witmer Stone, 26th May 1904. ANSP.
Wants a copy of Cassinia article on JKT by Stone (1903).

Mahlon Kirk, Woodburn Farm, Oakland, Maryland, to Witmer Stone, 4th June 1904. ANSP.
Thanks Stone for a copy of Cassinia and copies of JKT portrait. Points out error in Cassinia *article (that JKT's wife was Charlotte Holmes not Harriet Holmes).*

Mahlon Kirk, Woodburn Farm, Oakland, Maryland, to Witmer Stone, 19th July 1904. ANSP.
Invites Stone to his home. Would Academy want to buy artefacts from the Columbia and Hawaiian Islands?

Mahlon Kirk, Woodburn Farm, Oakland, Maryland, to Witmer Stone, 28th July 1904. ANSP.
Thanks Stone for the portrait.

Mahlon Kirk, Woodburn Farm, Oakland, Maryland, to Witmer Stone, 15th September 1905. ANSP.
Regarding JKT's son, Frank Holmes Townsend.

Mahlon Kirk's son (Mahlon Kirk III), Rockville, Maryland, to Witmer Stone, 12th October 1911. ANSP.
Wants to sell a fish skin coat collected by JKT to the Academy of Natural Sciences of Philadelphia.

Sandwich Island collection of Bird Skins [at Cambridge, England]. Typescript, [no date, but post 1902]. University Museum of Zoology, Cambridge, UK.
Includes some JKT specimens from the Hawaiian Islands.

McDaniel, W.B. (Ed.) 1963. [Letters of John James Audubon to Richard Harlan and John Kirk Townsend, of Philadelphia; letters of John Bachman to Harlan]. The College of Physicians of Philadelphia: Fugitive Leaves from the Library. 12 mimeographed pages, FL 196-210.

PUBLISHED SOURCES

Alden, R.H. and Ifft, J.D. 1943. *Early Naturalists in the Far West. Occasional Papers of the California Academy of Sciences,* No. XX, 60 pp. San Francisco.

American Ornithologists' Union. 1998. *Check-list of North American Birds*, 7th ed. AOU, Washington, D. C. [and up to *47th Supplement to the AOU Check-list* (2006)].

Anon. 1847. *Annual announcement of the Medical Department of Pennsylvania College, session 1847-48.* J.H. Gihon, Printer, Philadelphia.

Anon. 1849. *Annual catalogue of the Officers and Students in Pennsylvania College, Gettysburg.* Printed by H.C. Neinstedt, Gettysburg.

Anon. 1850. Donations to Museum. Fifty-six specimens, in skins, of North American Mammalia, collected by Dr. J.K. Townsend in Oregon and the Rocky Mountains, and presented by him to the Academy. *Proceedings of the Academy of Natural Sciences of Philadelphia*, p. 13.

Audubon, J.J. 1827-38. *The Birds of America,* double elephant folio, 4 vols. London.

Audubon, J.J. 1831-39. *Ornithological Biography*, 5 vols. Edinburgh.

Audubon, J.J. 1840-44. *The Birds of America,* octavo edition, 7 vols. New York and Philadelphia. [Quoted extracts are from the 1967 Dover Publications reprint.]

Audubon, J.J. 1966. *The original water-color paintings by John James Audubon for The Birds of America*, 2 vols. [Introduction by M.B. Davidson, pp. xi-xxxi.] American Heritage Publishing Co., New York.

Audubon, J.J. and Bachman, J. 1845-48. *Viviparous Quadrupeds of North America*, 3 vols. New York.

Bachman, J. 1837. Observations on the different species of Hares (genus LEPUS) inhabiting the United States and Canada. *Journal of the Academy of Natural Sciences of Philadelphia* 7: 282-361, 403.

Bachman, J. 1839a. [Skins of the genus *Sciurus* exhibited]. *Proceedings of the Zoological Society for 1838*, pp. 85-105.

Bachman, J. 1839b. Description of several New Species of American Quadrupeds. *Journal of the Academy of Natural Sciences of Philadelphia* 8: 57-74.

Bachman, J. 1839c. Additional Remarks on the Genus *Lepus*, with corrections of a former paper, and descriptions of other species of Quadrupeds found in North America. *Journal of the Academy of Natural Sciences of Philadelphia* 8: 75-105.

Bailey, V. 1900. [re Townsend Vole]. *North American Fauna* 17: 46-47.

Bailey, V. 1915. [re Townsend Pocket Gopher]. *North American Fauna* 39: 42-43.

Baker, R.J., Bradley, L.C., Bradley, R.D., Dragoo, J.W., Engstrom, M.D., Hoffman, R.S., Jones, C.A., Reid, F., Rice, D.W., and Jones, C. 2003. *Revised Checklist of North American Mammals North of Mexico.* Occasional Papers of the Museum of Texas Tech University, No. 229, pp. 1-24.

Beidleman, R.G. 1957. Nathaniel Wyeth's Fort Hall. *Oregon Historical Quarterly* 58: 197-250.

Berger, A.J. 1972. *Hawaiian Birdlife*. University Press of Hawaii, Honolulu.

Blaugrund, A. and Stebbins, T.E. (eds.) 1993. *John James Audubon. The Watercolours for the Birds of America.* The Herbert Press for The New-York Historical Society.

Boehme, S.E. (Ed,) *John James Audubon in the West. The Last Expedition. Mammals of North America.* Harry N. Abrams, Inc., Publishers, in association with the Buffalo Bill Historical Center.

Buchanan, R. (Ed.) 1868. *The Life and Adventures of John James Audubon, the Naturalist.* London.

Bulletin of the Proceedings of the National Institute 1841-1846, vols 1-4. [*National Institution* in title until part way through 1842.]

Burchsted, F.F. 1999. *Edward Charles Pickering.* In Garraty, J.A and Carnes, M.C. (Eds.) *American National Biography*, vol. 17, pp. 475-476. OUP, New York and Oxford.

Burns, F.L. 1919. *The Ornithology of Chester County, Pennsylvania*, pp. 13-15. [Not seen]

Burns, F.L. 1931. "Townsend's Oregon Tubinares." *Auk* 48: 106-109. [with comment by W.S. Stone]

Burns, F.L. 1934a. Townsend's Sooty Albatross. Auk 51: 225-226 [with comment by W.S. Stone].

Burns, F.L. 1934b. Type Localities of Townsend's "Columbia River" Birds. *Auk* 51: 403-404.

Cahill, E. 1999. *The Life and Times of John Young. Confidant and Advisor to Kamehameha the Great.* Island Heritage Publishing, Aiea, Hawai'i.

Calahane, V.H. (Ed.) 1967. *The Imperial Collection of Audubon Animals. The Quadrupeds of North America.* Hammond Incorporated, Maplewood, New Jersey.

Cassin, J. 1855. Notices of some new and little known birds in the collection of the U.S. Exploring Expedition in the *Vincennes* and *Peacock*, and in the collection of the Academy of Natural Sciences of Philadelphia. *Proceedings of the Academy of Natural Sciences of Philadelphia* 7: 438-441.

Cassin, J. 1856. *Illustrations of the Birds of California, Texas, Oregon, British and Russian America.* J.B. Lippincott, Philadelphia (1991 facsimile edition by The Texas State Historical Association, Austin, Texas; with introduction by R.M. Peck).

Cassin, J. 1858. *United States Exploring Expedition during the years 1838, 1839, 1840, 1841, 1842 … Mammalogy and ornithology.* J.B. Lippincott & Co., Philadelphia.

Chalmers, J. 2003. *Audubon in Edinburgh and his Scottish Associates.* NMS Publishing, a division of NMS Enterprises Ltd, Edinburgh.

Conrad, T.A. 1837. Descriptions of new Marine Shells from Upper California. Collected by Thomas Nuttall, Esq. *Journal of the Academy of Natural Sciences of Philadelphia*, 7: 227-268.

Cooper, W. 1837. Two Species of Plecotus. *Annals of the Lyceum of Natural History of New York* 4: 73-75 and plate.

Corning, H (Ed.). 1930. *Letters of John James Audubon, 1826-1840.* 2 vols. Club of Old Volumes, Boston, Massachesetts.

Coues, E. 1900. Date of Discovery and Type Locality of the Mountain Mockingbird. *Auk* 17: 68-69.

Cutright, P.R. 1969. *Lewis and Clark: Pioneering Naturalists.* University of Nebraska Press, Lincoln & London.

Daily National Intelligencer [Washington, D.C.], 7th February 1851. [J.K. Townsend death notice.]

Dall, W.H. 1915. *Spencer Fullerton Baird. A Biography.* J.B. Lippincott, Philadelphia and London.

Deane, R. 1909. Some original manuscript relating to the history of Townsend's Bunting. *Auk* **26**: 269-272.

Deignan, H.G. 1961. Type specimens of birds in the USNM. *United States National Museum Bulletin* **221**: 105, 638, 656-657.
Dictionary of American Biography 1928-36, vol XVIII, pp. 617-618 [J.K. Townsensd]; vol XII, pp. 596-597 [Ezra Michener]. New York.

Douglas, D. 1829. Observations on the *Vultur Californianus* of Shaw. *Zoological Journal* **4**: 328-330.

Douglas, D. 1914. *Journal kept by David Douglas during his travels in North America, 1823-1827*. William Wesley and Son, London.

Edwards, R.L. 1834. [Letter of 23rd June 1834 published in the *Missouri Enquirer*, reproduced in *Niles Register* 11th October 1834.]

Evans, H.E. 1997. *The Natural History of the Long Expedition to the Rocky Mountains, 1819-1820*. Oxford University Press, New York and Oxford.

Ewan, J. and Ewan, N.D. 1981. *Biographical Dictionary of Rocky Mountain Naturalists, 1682-1932*, pp. 222-223. Frans A. Stafleu, Utrecht.

Forbes, C. 1984. *The Journals of Cochran Forbes. Missionary to Hawaii 1831-1864*. HMCS, Honolulu.

Ford, A. 1964. *John James Audubon*, pp. 339-340, 342, 346, 348-350, 352-353, 364, 408. [A pro Audubon slant on the acquisition of the Townsend specimens.] University of Oklahoma Press.

Fries, W.H. 1973. *The Double Elephant Folio. The Story of Audubon's "Birds of America"*. American Library Association, Chicago.

Gill, F.B. 1995. Philadelphia: 180 years of Ornithology at the Academy of Natural Sciences *in* Davis, W.E. and Jackson, J.A. (Eds.) *Contributions to the History of North American Ornithology. Memoirs of the Nuttall Ornithological Club, No. 12,* Cambridge, Massachusetts.

Gollop, J.B., Barry, T.W. and Iversen, E.H. 1986. *Eskimo Curlew. A vanishing species?* Special Publication No. 17 of the Saskatchewan Natural History Society, Regina, Saskatchewan.

Goode, G.B. 1893. The Genesis of the National Museum. *Annual Report of the U.S. National Museum, 1891*. pp. 273-380.

Goodman, G.J. 1843. The Story of Parthenium alpinum. *Madrano* **7**: 115-118.

Graustein, J.E. 1967. *Thomas Nuttall, Naturalist. Explorations in America 1808-1841*. Harvard University Press, Cambridge.

Grinnell, J. 1909. Bibliography of California Ornithology [part 1]. *Pacific Coast Avifauna* **5**: 1-166.

Hall, E.R. 1981. *The Mammals of North America*, 2nd edition, 2 vols. John Wiley and Sons, New York.

Hall, F.S. 1937. Studies in the History of Ornithology in the State of Washington (1792-1932), with special reference to the discovery of new species. Part III. David Douglas, pioneer naturalist on the Columbia River – 1825-1833. *Murrelet* **15**: 2-19.

Hall, F.S. 1939. Studies in the History of Ornithology in the State of Washington (1792-1932), with special reference to the discovery of new species. Part IV. The overland journey of the naturalists Thomas Nuttall and John Kirk Townsend. *Murrelet* **18**: 2-13; **19**: 2-7. [Two portraits: the one labelled Townsend is also of Nuttall.]

Handley, C.O. 1959. A revision of the American bats of the genera Euderma and Plecotus. *Proceedings of the United States National Museum* **110**. No. 3417.

Handley, C.O. 1994. Identification of the Carolina shrews of Bachman 1837. *Carnegie Museum of Natural History Special Publication* **18**: 393-406.

Hanley, W. 1977. *Natural History in America. From Mark Catesby to Rachel Carson*. Quadrangle, The New York Times Book Co., New York.

Harris, H. 1919. Historical notes on Harris's Sparrow (*Zonotrichia querula*). *Auk* **44**: 180-190.

Harris, H. 1941. The annals of *Gymnogyps* to 1900. *Condor* **43**: 2-55.

Harrison, C. and Walker, C. 1997. *in* Audubon, J.J. *The Birds of America*. Wordsworth Editions Ltd, Ware, Hertfordshire.

Harston, T. 1999. *Squirrels of the West*. Lone Pine Publishing, Edmonton.

Hatch, M.H. 1953. History of the study of the Coleoptera in the Pacific Northwest, pp. 6-10, *in* The Beetles of the Pacific Northwest. *University of Washington Publications in Biology* **16**: 1-340.

Herrick, F.H. 1938. *Audubon the Naturalist. A History of his Life and Time*. Second edition, two volumes in one. D. Appleton-Century Co., New York & London.

Houston, C.S. 1998. List of New Species Added to North American Natural History, 1825-1827, by Surgeon-Naturalist John Richardson, pp. 391-400. In appendix to Davis, R.C. (Ed,) *Sir John Franklin's Journals and Correspondence: the second Arctic land expedition 1825-1827*. The Champlain Society, Toronto.

Huber, W. 1930. A method of salting and preparing water bird skins [Townsend's Tundra Swan from Columbia River]. *Auk* **47**: 408-411.

Hulbert, A.B (Ed.). 1934. [Townsend's Original Journal. Extracts for 6th November 1834, and for 10th July-15th September 1834, as they appeared in *Waldie's Select Circulating Library* 1835.] In *The Call of the Columbia. Overland to the Pacific*, vol 4, pp. 184-226. Stewart Commission of Colorado College and the Denver Public Library.

Jackson, C.E and Davis, P. 2001. *Sir William Jardine. A Life in Natural History*. Leicester University Press, London and New York.

Jardine, W. 1874. *A Catalogue of the Birds contained in the Collection of Sir William Jardine*. Printed by Taylor & Francis, London.

[Jardine, W.] 1886. *A [sale] Catalogue of the valuable collection of Bird Skins formed by the late Sir William Jardine … of Jardine Hall, Lockerbie, Dumfriesshire, June 17th 1886*. Taylor and Francis, London.

Jepson, L.J. 1934. The Overland journey of Thomas Nuttall. *Madrano* **2**: 143-147.

Jobanek, G.A. 1986. John Kirk Townsend in the Northwest. *Oregon Birds* **12**: 253-276.

Jobanek, G.A. 1994. Dubious Records in the Early Oregon Bird Literature. *Oregon Birds* **20**: 3-23.

Jobanek, G. A. 1997. *An annotated bibliography or Oregon bird literature published before 1935*. Entries 330, 331. 1333, 1334. Oregon State University Press. [not seen].

Jobanek, G.A. and Marshall, D. B. 1992. John K. Townsend's 1836 report of the birds of the Columbia River region, Oregon and Washington. *Northwestern Naturalist* **73**: 1-14.

Jordan, J.W. 1911. *Colonial Families of Philadelphia*, vol 2: 1536-1539. The Lewis Publishing Company, Philadelphia.

Koopman, K.F. 1976. Catalog of type specimens of recent mammals in the Academy of Natural Sciences at Philadelphia. *Proceedings of the Academy of Natural Sciences of Philadelphia* **128**: 1-24.

Lea, I. 1839. New Freshwater and Land Shells. *Transactions of the American Philosophical Society*, pp. 73-108 [not seen].

Lee, D. and Frost, J.H. 1884. *Ten Years in Oregon*. Published for the authors by J. Collard, Printer, New York.

Lee, J. 1916. Diary of Rev. Jason Lee. *Oregon Historical Quarterly* 17: 116-266.

Low, S.M. 1988. *An Index and Guide to Audubon's Birds of America*. AMNH, Abbeville Press, New York.

McDaniel, W.B. 1963 – see miscellaneous unpublished sources.

Meany, E.S. 1923. *Origin of Washington Geographic Names*. Washington University Press, Seattle.

Mearns, B. and Mearns, R. 1988. *Biographies for Birdwatchers. The Lives of Those Commemorated in Western Palearctic Bird Names*. Academic Press, London.

Mearns, B. and Mearns, R. 1992. *Audubon to Xantus. The Lives of those Commemorated in North American Bird Names*. Academic Press, London.

Mearns, B. and Mearns, R. 1995. A wrongly identified Audubon bird painting: Brewer's Sparrow mistaken for Clay-colored Sparrow. *Archives of Natural History* 22: 153-158.

Mearns, B. and Mearns, R. 1998. *The Bird Collectors*. Academic Press, London.

Medway, D.G. 1981. The contribution of Cook's third voyage to the ornithology of the Hawaiian Islands. *Pacific Science* 35: 105-175.

Merriam, C.H. 1891. Mammals of Idaho. *North American Fauna* 5: 75-78.

Michener, E. 1863. Agricultural Ornithology. Insectivorous birds of Chester County, Pennsylvania. *U.S. Agricultural Report*, pp. 287-307.

Michener, E. 1893. *Autobiographical notes from the life and letters of Ezra Michener, M.D.* pp. 48-49. Friends' Book Association, Philadelphia.

Morton, S.G. 1839. Flat-head tribes of Columbia River, pp. 202-216. In *Crania Americana. Or, a Comparative View of the Skulls of Various Aboriginal Nations of North and South America*. J. Dobson, Philadelphia.

Morton, S.G. 1840. *Catalogue of Skulls of Man, and the Inferior Animals in the Collection of Samuel Gorge Morton, M. D.* Printed by Turner & Fisher, Philadelphia.

Morton, S.G. 1849. *Catalogue of Skulls of Man, and the Inferior Animals in the Collection of Samuel Gorge Morton, M. D., Penn. and Edinb.* 3rd edition. Merrihew & Thompson, Printers, Philadelphia.

Nolan, E.J. 1909. *A short history of the Academy of Natural Sciences of Philadelphia*. The Academy of Natural Sciences, Philadelphia.

Nuttall, T. 1840. *A Manual of the Ornithology of the United States and of Canada. Second Edition, With Additions. The Land Birds*. Hilliard, Gray and Company, Boston.

Olson, R.J.M. and Mazzitelli, A. 2003. Audubons' Bats: Like father, like son? *New-York Journal of American History* 65: 68-89 [formerly *New-York Historical Society Quarterly*].

Olson, S.L. 1989. David Douglas and the Original Description of the Hawaiian Goose. *Elepaio* 49: 49-51.

Olson, S.L. 1996. The contribution of the voyage of H.M.S. Blonde (1825) to Hawaiian ornithology. *Archives of Natural History* 23: 1-42.

Olson, S.L. and James, H. F. 1994. A chronology of ornithological exploration in the Hawaiian Islands, from Cook to Perkins. *Studies in Avian Biology* 15: 91-102.

Orton, J. 1871. Birds in the museum of Vassar College. *American Naturalist* 4: 716.

Overmeyer, P.H. 1933. Nathaniel Jarvis Wyeth. His First Expedition. His Second Expedition and Later Life. *Washington Historical Quarterly* 24: 28-48.

Palmer, T.S. 1928. Notes on persons whose names appear in the nomenclature of California birds. *Condor* 30: 261-307.

Parker, S. 1838. *Journal of an Exploring Tour Beyond the Rocky Mountains*. 1990 reprint with introduction by Larry R. Jones. University of Idaho Press. [pp. 215-217 includes a section on birds extracted from Townsend's 1836 MS report on the birds of the lower Columbia River region, supplied directly to Parker.]

Parkes, K.C. 1985. Audubon's Mystery Birds. *Natural History* 94: 88-92.

Peale, T.R. 1848. *United States Exploring Expedition during the years 1838, 1839, 1840, 1841, 1842. Vol VIII: Mammalia and Ornithology*. C. Sherman, Philadelphia.

Pearse, T. 1968. *Birds of the Early Explorers in the Northern Pacific*. T. Pearse, Comox, B.C.

Peck, R.M. 1991. Introduction, pp. I-3 to I-38. In Cassin, J. 1856. *Illustrations of the Birds of California, Texas, Oregon, British and Russian America*. J.B. Lippincott, Philadelphia (1991 facsimile edition by The Texas State Historical Association, Austin, Texas).

Peck, R.M. 1999. George Ord. In Garraty, J.A. and Carnes, M.C. (Eds.) *American National Biography*, vol. 16, pp. 755-756. OUP, New York and Oxford.

Peck, R.M. 2000. Audubon and Bachman; a Collaboration in Science, pp. 71-115. *In* Boehme, S.E. (Ed,) *John James Audubon in the West. The Last Expedition. Mammals of North America*. Harry N. Abrams, Inc., Publishers, in association with the Buffalo Bill Historical Center.

Philbrick, N. 2003. *Sea of Glory. America's voyage of discovery, the U.S. Exploring Expedition, 1838-1842*. Viking, Penguin Group (USA) Inc., New York.

Porter, C.M. 1979. 'Subsilentio': discouraged works of early nineteenth-century American Natural History. *Archives of Natural History* 9: 109-119.

Porter, M.R. and Davenport, O. 1963. *Scotsman in Buckskin. Sir William Drummond Stewart and the Rocky Mountain Fur Trade*. [especially pp. 75-98]. Hastings House, New York.

Randall, J.W. 1839. Catalogue of the Crustacea brought by Thomas Nuttall and J.K. Townsend, from the West Coast of North America and the Sandwich Islands. *Journal of the Academy of Natural Sciences of Philadelphia*, pp. 106-157.

Rhoads, S.N. 1897. A revision of the west American flying squirrels. *Proceedings of the Academy of Natural Sciences of Philadelphia* 49: 314-327.

Rhoads, S.N. 1903. Auduboniana [three Audubon letters to Edward Harris]. *Auk* 20: 377-383.

Rice, H. 1960. The World of John James Audubon: catalogue of an exhibition in Princeton University Library, 15 May-30 September 1959. *Princeton University Library Chronicle* 21 (Autumn 1959-Winter 1960): 8-88.

Rich, E.E. (Ed.). 1941. *The Letters of John McLoughlin from Fort Vancouver to the Governor and Committee, first series, 1825-38*. Volume IV, pp. cxxi, 126, 203, 344. Published by The Champlain Society for the Hudson's Bay Record Society.

Ross, A. 1855. *The Fur Hunters of the Far West; a Narrative of Adventures in the Oregon and Rocky Mountains*. Smith, Elder and Co., London.

Russell, O. 1955, *Journal of a Trapper. Or Nine Years Residence among the Rocky Mountains between the years of 1834 and 1843.* Edited from Osborne Russell's original MS by A.L. Haines. University of Nebraska Press, Lincoln (Bison Book edition 1986).

Salvin, O. 1882. *A catalogue of the collection of birds formed by the late Hugh Edwin Strickland.* University Press, Cambridge [UK].
Sass, H.R. 1908. An interesting Audubon specimen [Black-headed Grosbeak at Charleston Museum]. *Auk* **25**: 228-229.

Scheffer, T.H. 1946. Re-allocation of the Townsend Ground Squirrel. *Journal of Mammalogy* **27**: 395-396.

Sharpe, R.B. 1906. Birds, *in* Gunther, A. (Ed.) 1904-12. *The History of the Collections contained in the Natural History Departments of the British Museum*, vol. 2. BM publication, London.

Shepard, C. 1986. *Diary of Cyrus Shepard. March 4, 1834 – December 20, 1835.* Typescript, compiled by Gerry Gilman. Clark County Geneological Society, Vancouver, Washington.

Shuler, J. 1995. *Had I the Wings: the friendship of Bachman and Audubon.* University of Georgia Press, Athens, Georgia.

Simpson, M.B. 1999. *John Kirk Townsend.* In Garraty, J.A. and Carnes, M.C. (Eds.) *American National Biography*, vol. 21, pp. 787-788. OUP, New York and Oxford.

Slipp, J.W. 1947. Swainson's Hawk in western Washington with a note on the type locality. *Auk* **64**: 389-400.

Sokal, M.M. 1999. *Samuel George Morton.* In Garraty, J.A. and Carnes, M.C. (Eds.) *American National Biography*, vol. 15, pp. 959-961. OUP, New York and Oxford.

Sorensen, W.C. 1995. *Brethren of the Net. American Entomology, 1840-1880.* University of Alabama Press, Tuscaloosa and London.

Stone, W. 1892. Catalogue of the Corvidae, Paradisidae and Oriolidae in the collection of the Academy of Natural Sciences of Philadelphia. *Proceedings of the Academy of Natural Sciences of Philadelphia, 1891.* **43**: 441-450.

Stone, W. 1899a. A study of the type specimens of birds in the collection of the Academy of Natural Sciences of Philadelphia, with a brief history of the collection. *Proceedings of the Academy of Natural Sciences of Philadelphia* **51** : 5-62.

Stone, W. 1899b. Some Philadelphia Ornithological Collections and Collectors, 1784-1850. *Auk* **16**: 166-177.

Stone, W. 1903. John Kirk Townsend. *Cassinia* **7**: 1-5.

Stone, W. 1906. A bibliography and nomenclator of the ornithological works of John James Audubon. *Auk* **23**: 298-312.

Stone, W. 1916. Philadelphia to the coast in early days, and the development of western ornithology prior to 1850. *Condor* **18**: 3-14.

Stone, W. 1923. [Review of Bent's *Life Histories of North American Petrels and Pelicans and their allies.*] *Auk* **40**: 149-150.

Stone, W. 1930a. Townsend's Oregon Tubinares. *Auk* **47**: 414-415.

Stone, W. 1930b. Townsend's bird studies in Oregon, 1834-36. *Oregon Historical Quarterly* **31**: 51-54. [This is a reprint of Stone 1903, with slight alterations (e.g. the name of Townsend's wife is corrected).]

Stone, W. 1931. [Reply to F.L. Burns's "Townsend's Oregon Tubinares."] *Auk* **48**: 108-109.

Stone, W. 1934. [Reply to F.L. Burns's "Townsend's Sooty Albatross."] *Auk* **51**: 225-226.

Stone, W. 1936a. On the types of J.K. Townsend's Birds. *Auk* **53**: 242.

Stone, W. 1936b. John Kirk Townsend. In *Dictionary of American Biography*, vol XVIII, pp. 617-618.

Street, P.B. 1948. The Edward Harris collection of birds. *Wilson Bulletin* **60**: 167-184.
Stresemann, E. 1954. Ferdinand Deppe's travels in Mexico. *Condor* **56**: 86-92.

Thayer, J.E. 1916. Auduboniana [Townsend's Parkman's Wren secured by Thayer]. *Auk* **33**: 115-118.

Tolmie, W. F. 1963. *The Journals of William Fraser Tolmie. Physician and Fur Trader.* Mitchell Press, Ltd., Vancouver, Canada. [No journal entries between Dec 1835 and Oct 1836 (and few entries to end of 1836) accounts for lack of any reference to Townsend who was with Tolmie for part of that period.]

Townsend, H. 1852. *History of England in Verse.* Lindsay & Blakiston, Philadelphia. [Not seen].

Townsend, J.K. 1837. Description of Twelve New Species of Birds, chiefly from the vicinity of the Columbia river [plus list of species from the Columbia drawn up by the Ornithological Committee]. *Journal of the Academy of Natural Sciences of Philadelphia* **7**: 187-193.

Townsend, J.K. 1839a. *Narrative of a Journey across the Rocky Mountains, to the Columbia River, and a Visit to the Sandwich Islands, Chili, &c. with a Scientific Appendix.* Henry Perkins, Philadelphia.

Townsend, J.K. 1839b. *Catalogue of Quadrupeds found in the Territory of the Oregon* in *Narrative of a Journey across the Rocky Mountains ... &c*, pp. 311-331. Henry Perkins, Philadelphia.

Townsend, J.K. 1839c. *Catalogue of Birds found in the Territory of the Oregon* in *Narrative of a Journey across the Rocky Mountains ... &c*, pp. 331-352. Henry Perkins, Philadelphia.

Townsend, J.K. 1839d. Description of a New Species of CYPCELUS, from the Columbia River [Vaux's Swift *Chaetura vauxi*]. *Journal of the Academy of Natural Sciences of Philadelphia* **8**: 148.

Townsend, J.K. 1839e. Description of a New Species of Sylvia, from the Columbia River [MacGillivray's Warbler *Oporornis tolmiei*]. Note on Sylvia Tolmoei. *Journal of the Academy of Natural Sciences of Philadelphia* **8**: 149-150, 159.

Townsend, J.K. 1839f. List of the Birds Inhabiting the Region of the Rocky Mountains, The Territory of the Oregon, and the North West Coast of America. *Journal of the Academy of Natural Sciences of Philadelphia* **8**: 151-158.

Townsend, J.K. 1839g. *Ornithology of the United States of North America; or descriptions of the birds inhabiting the States and Territories of The Union.* Vol 1, pp. 1-12 [all published]. J.B. Chevalier, Philadelphia.

Townsend, J.K. 1840. *Sporting Excursions in the Rocky Mountains, including a Journey to the Columbia River, and a Visit to the Sandwich Islands, Chili, etc.* 2 vols. Henry Colburn, London.

Townsend, J.K. 1847a. Sketches of a voyage, and residence in the South Sea islands, *Literary Record and Journal of the Linnaean Association of Pennsylvania College*, **3**: 88-92, 113-120, 121-126, 160-166.

Townsend, J.K. 1847b. On a singing mouse. *Proceedings of the Academy of Natural Sciences of Philadelphia*, p. 262.

Townsend, J.K. 1848. Popular monograph of the Accipitrine birds of N. A. Parts 1 and II, *Literary Record and Journal of the Linnaean Association of Pennsylvania College*, **4**: 249-255, 265-272.

Townsend, J.K. 1849. Sketches of a visit to the Sandwich Islands. *Saturday Evening Post*, vol. XXVIII, 24th and 31st March 1849. Philadelphia. [A slightly re-worded version of Townsend 1847a.]

Townsend, J.K. 1850. On the Giant Wolf of North America. - Lupus gigas. *Journal of the Academy of Natural Sciences of Philadelphia*, 2nd series, 2: 75.

Townsend, J.K. 1905. *Narrative of a Journey across the Rocky Mountains … &c.* In Thwaites, R.G. *Early Western Travels*, Vol. XXI, Arthur H. Clark Co., Cleveland. [Omits JKT's visits to the Hawaiian Islands , his journey home and scientific appendix.]

Townsend, J. K. 1970. *Narrative of a Journey across the Rocky Mountains … &c.* In *Oregon, or a short History of a Long Journey* [Wyeth's first expedition] *by J.B. Wyeth and Narrative of a Journey by John K. Townsend*. Introduction by G. Thomas Edwards. Ye Galleon Press, Fairfield, Washington. [Text follows R.G. Thwaites' version.]

Townsend, J.K. 1978. *Across the Rockies to the Columbia, formerly titled Narrative of a Journey across the Rocky Mountains to the Columbia River*. Introduction by D. Jackson. University of Nebraska Press. [Omits JKT's visits to the Hawaiian Islands , his journey home and scientific appendix.]

Townsend, J.K. 1999. *Narrative of a Journey across the Rocky Mountains, to the Columbia River, and a Visit to the Sandwich Islands, Chili, &c.* Introduction and notes by G. A. Jobanek. Oregon State University Press. [Includes all original text and the scientific appendix.]

Townsend, M. 1844. *Life in the Insect World; or Conversations upon Insects, between an Aunt and her Nieces*. Lindsay & Blakiston, Philadelphia. [Not seen]

True, F.W. 1896. Revision of American Moles. *Proceedings of the [U.S.] National Museum* **19**: 60-67.

Tyler, R. 2000. The Publication of *The Viviparous Quadrupeds of North America*, pp. 119-182. *In* Boehme, S.E. (Ed,) *John James Audubon in the West. The Last Expedition. Mammals of North America*. Harry N. Abrams, Inc., Publishers, in association with the Buffalo Bill Historical Center.

Viola, H.J. and Margolis, C. (Eds.) 1985. *Magnificent Voyagers. The US Exploring Expedition 1838-1842*. SI Press, Washington, D.C.

Waldie's Select Circulating Library 1835, part II, pp. 424-432. A Townsend letter of 6th Oct 1834, quoted in Hall 1937, p. 10 [the text is similar to the entry in the *Narrative*]; 12th July 1836 [reprinted in Graustein 1967, pp. 444-445].

Weigley, R.F. (Ed.) 1982. *Philadelphia: a 300-year history*. W.W. Norton and Co., New York and London.

Wheeler, A. 1997. Zoological collections in the early British Museum: the Zoological Society's Museum. *Archives of Natural History* **24**: 89-126.

Whitman, N. 1891. [Narcissa Whitman's journal, 1st-17th Sept 1836.] *Transactions of the Oregon Pioneer Association, pp.* 57-65.

Wilson, S.B. and Evans, A.H. 1890-99. *Aves Hawaiiensis. The Birds of the Sandwich Islands*. R.H. Porter, London.

Wyeth, N.J. 1899. *The Journals of Captain Nathaniel J. Wyeth's Expedition to the Oregon Country 1831-1836*. University Press, Eugene, Oregon. (1984 reprint, edited by D. Johnson, Ye Galleon Press, Fairfield, Washington).

INDEX OF BIRDS AND MAMMALS

SELECT GENERAL INDEX